A PROBLEM-SOLVING APPROACH TO
INTRODUCTORY ALGEBRA

SECOND EDITION

MERVIN L. KEEDY
Purdue University

MARVIN L. BITTINGER
Indiana University—Purdue University at Indianapolis

▲ ADDISON-WESLEY PUBLISHING COMPANY
Reading, Massachusetts • Menlo Park, California • Don Mills, Ontario
Wokingham, England • Amsterdam • Sydney • Singapore • Tokyo
Mexico City • Bogotá • Santiago • San Juan

Sponsoring Editor • *Susan Zorn*
Production Supervisor • *Susanah H. Michener*
Production Editor • *Martha Morong, Quadrata, Inc.*
Text Designer • *Vanessa Piñeiro, Piñeiro Design Associates*
Illustrators • *VAP Group Limited, Belanger Associates*
Art Consultant • *Joseph Vetere*
Manufacturing Supervisor • *Ann DeLacey*
Cover Designer • *Marshall Henrichs*

Photo Credits

1, AP/Wide World Photos   **4**, Belanger Associates   **44**, AP/Wide World Photos   **48**, Belanger Associates   **52**, Boston Museum of Science   **59**, Michael Hayman/Stock Boston   **73**, Belanger Associates   **101**, Owen Franklin/Stock Boston   **126**, Lionel J-M Delevingne/Stock Boston   **135**, Peter Manzel/Stock Boston   **143**, Phyllis Graber Jensen/Stock Boston   **145**, Barbara Alper/Stock Boston   **175**, Belanger Associates   **204**, Bohdan Hynewych/Stock Boston   **205**, Joseph A. Kovacs/Stock Boston   **221**, Geoffery Clements/Collection of the Whitney Museum of American Art, New York   **257**, AP/Wide World Photos   **259**, W. B. Finch/Stock Boston   **261**, Belanger Associates   **268**, AP/Wide World Photos   **283**, Belanger Associates   **292**, Belanger Associates   **294**, Tim Carlson/Stock Boston   **307**, Ellis Herwig/Stock Boston   **308**, Belanger Associates   **309**, AP/Wide World Photos   **314**, Belanger Associates   **318**, Ira Kirschenbaum/Stock Boston   **319**, Belanger Associates   **321**, NASA   **324**, AP/Wide World Photos   **329**, Peter Manzel/Stock Boston   **367**, Belanger Associates   **370**, Pamela R. Schoylea/Stock Boston   **393**, Belanger Associates   **397**, Belanger Associates   **435**, Bohdan Hrynewych/Stock Boston   **438**, Bohdan Hrynewych/Stock Boston   **440**, AP/Wide World Photos   **460**, Ellis Herwig/Stock Boston

Library of Congress Cataloging-in-Publication Data

Keedy, Mervin Laverne.
  A problem solving approach to introductory algebra.

  1. Algebra.   2. Algebra—Problems, exercises, etc.   3. Problem solving.   I. Bittinger, Marvin L.   II. Title.
QA152.2.K44   1986   512   85-18592
ISBN 0-201-12968-X

Reprinted with corrections August, 1986

Copyright © 1986 by Addison-Wesley Publishing Company, Inc. All rights reserved. No part of this publication may be reproduced, stored in a retrieval system, or transmitted, in any form or by any means, electronic, mechanical, photocopying, recording, or otherwise, without the prior written permission of the publisher. Printed in the United States of America. Published simultaneously in Canada.

BCDEFGHIJ-HA-89876

PREFACE

Problem Solving

The primary distinguishing feature of this book is its treatment of and emphasis on problem solving. The approach to problem solving is different here, not only because of the increased space devoted to it, but also because of the way in which students are guided in formulating problems and translating the situation of a problem to mathematical language. Problem-solving activity occurs at the beginning of Chapter 1, and then recurs, to some extent or other, in *all* of the later chapters. Problem-solving practice is provided in nearly every exercise set, and the rationale for learning to factor, solve equations, and the like is that those skills invariably enable the student to solve problems better or to solve more sophisticated problems.

Mathematical Content and Level

The mathematical content of this text is introductory algebra. It is intended for use by students who have a firm background in arithmetic. Students whose background is weak are advised to use the companion volume, *Arithmetic, Fourth Edition*, by the same authors (Addison-Wesley, 1983).

There is also a succeeding text, *A Problem-Solving Approach to Intermediate Algebra, Second Edition,* written in the same style as this one. A placement test is available to help determine which book is appropriate for the student (see the list of supplements that follows this preface).

This book contains a brief chapter on functions. While the first two sections can be considered optional if time is short, the later two sections on variation should still be covered, especially by students planning further study in algebra.

For students whose mathematical study is not recent, it may be helpful to begin not with Section 1.1, but rather with a review of some arithmetic manipulation first, and then returning to the introductory section on problem solving.

Readability and Teachability

The style and format of the book are designed to make the text easy to read and comprehend. Examples to illustrate each skill and concept have been chosen carefully to make sure that they are appropriate and to the point. The exercises in the exercise sets are very much like the examples. Exercises are ample and they are paired and graded, with answers to the odd-numbered exercises provided at the back of the book. Answers to the other exercises can be found in the *Student's Guide to Exercises* (see the list of supplements).

The exercises that follow the symbolism ○ are generally a bit different from the examples in the text. Thus they offer some challenge. Some are challenging enough for the best students. Exercises that are designed to be done with a calculator, designated by 📷 , are also included.

End-of-Chapter Materials

At the end of each chapter is a summary and review. This feature has been designed for maximum student help. In boldface type the objectives of the chapter are stated. Following each list of objectives, some exercises on those objectives are given. Answers are at the back of the book, together with section references so that the student can restudy the material pertaining to a missed item. The summary and review is then followed by a practice test, whose answers are not in the book.

The text also contains cumulative reviews every three to four chapters. These reviews cover all material from Chapter 1 up to that point in the text and are intended for skill maintenance. All answers together with section references are at the back of the book.

ACKNOWLEDGMENTS

We wish to thank many people without whose committed efforts our work could not have been completed successfully. Gloria Schnippel and Karen Anderson did a tremendous job checking the manuscript. Randy Becker, Judy Beecher, Virginia Hamilton (Ball State), Judy Penna, and Julie Stephenson were superb in their work on the supplements.

Many instructors provided reviewing information. We thank them for their many suggestions for improvement. They are Ruth Afflack (California State University at Long Beach), Don Albers (Menlo College), Jean Berdon (Cañada College), Randy Davidson (University of New Orleans), Arthur Dull (Diablo Valley College), Robert Limburg (St. Louis Community College at Florissant Valley), John Samoylo (Delaware County Community College), Dick Spangler (Tacoma Community College), Louise Dyson (Clark College), Eleanor Kendrick (San Jose City College), Roger Judd (Chemeketa Community College), and Kathleen Thestia (Phoenix College).

M.L.K.
M.L.B.

SUPPLEMENTS

The following supplementary materials are available.

- *Student's Guide to Exercises*. Contains answers for all the exercises in the exercise sets. Solutions accompany all odd-numbered exercises.
- *Instructor's Manual with Tests and Computer Programs*. Contains six alternate forms of the chapter tests and six final exams. All answers to these tests, as well as the answers to the chapter tests found in the text, are also included in this manual. A black-line grid master for graphing is included. Another option is the inclusion of computer programs, about one per chapter, which when used by the student can allow an appreciation of the power of the computer in relation to the mathematics being considered at that point in the text. The programs are in BASIC.
- *Instructional Software for Algebra*. A computer software supplement to many parts of the book. This software is highly interactive and covers topics such as simplifying algebraic expressions, factoring, equations, inequalities, 2×2 linear systems, radicals, and quadratic equations. It is available for the Apple II series and the IBM-PC.
- *Computerized Testbank*. Contains test items that can be randomly selected and a test constructed by a computer using the testbank. The testbank is available for the Apple II series and IBM-PC computers.
- *Videotape Cassettes*. Supplements the regular lectures and can be used in a number of teaching situations, such as learning labs or as fill-in lectures when instructors are absent. John Jobe of Oklahoma State University speaks to students and works out examples with lucid explanations.
- *Placement Test*. Helps students to be placed properly in either this book or other books by the same authors. The test is handy because it is short, although its brevity may make it appear less useful than it actually is. It was thoroughly evaluated statistically and is known to give a good screening with respect to placement into one of four categories into which the books of the series fall.

    I *Arithmetic*, Fourth Edition
   II *Introductory Algebra*, Fourth Edition
       A Problem-Solving Approach to Introductory Algebra, Second Edition
  III *Intermediate Algebra*, Fourth Edition
      A Problem-Solving Approach to Intermediate Algebra, Second Edition
  IV *College Algebra*, Fourth Edition
      Fundamental College Algebra, Second Edition
      Algebra and Trigonometry, Fourth Edition
      Fundamental Algebra and Trigonometry, Second Edition
      Trigonometry: Triangles and Functions, Fourth Edition

CONTENTS

1 INTRODUCTION TO ALGEBRA AND PROBLEM SOLVING 1

- **1.1** Introduction to Problem Solving 2
- **1.2** Symbols and Expressions in Algebra 5
- **1.3** Factorizations and LCMs 9
- **1.4** Properties of Numbers and Algebraic Expressions 16
- **1.5** Exponential Notation 24
- **1.6** The Associative Laws 28
- **1.7** The Distributive Law 32
- **1.8** Equations and Translation 37
- **1.9** Problem Solving 42
- **1.10** Problem Solving Involving Percents 48

Summary and Review 55
Test 57

2 INTEGERS, RATIONAL NUMBERS, AND REAL NUMBERS 59

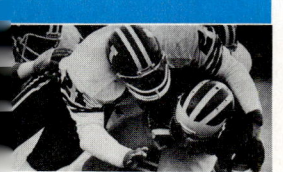

- **2.1** Integers and the Number Line 60
- **2.2** Rational Numbers and Real Numbers 64
- **2.3** Addition 67
- **2.4** Subtraction 74
- **2.5** Multiplication 79
- **2.6** Division 82
- **2.7** Using the Distributive Laws 88
- **2.8** Multiplying by -1 and Simplifying 91

Summary and Review 98
Test 99

3 SOLVING EQUATIONS AND PROBLEMS 101

3.1 The Addition Principle 102
3.2 The Multiplication Principle 104
3.3 Using the Principles Together 108
3.4 Equations Containing Parentheses 113
3.5 Problem Solving 115
3.6 Formulas 123
Summary and Review 130
Test 131

Cumulative Review: Chapters 1–3 132

4 POLYNOMIALS 135

4.1 Properties of Exponents 136
4.2 Polynomials 141
4.3 More on Polynomials 147
4.4 Addition of Polynomials 151
4.5 Subtraction of Polynomials 155
4.6 Multiplication of Polynomials 160
4.7 Special Products of Polynomials 164
4.8 More Special Products 167
Summary and Review 171
Test 173

5 POLYNOMIALS AND FACTORING 175

5.1 Factoring Polynomials 176
5.2 Differences of Squares 179
5.3 Trinomial Squares 182
5.4 Factoring Trinomials of the Type $x^2 + bx + c$ 185
5.5 Factoring Trinomials of the Type $ax^2 + bx + c, a \neq 1$ 189
5.6 Factoring: A General Strategy 195
5.7 Solving Equations by Factoring 198
5.8 Problem Solving 202
5.9 Polynomials in Several Variables 208
Summary and Review 217
Test 219

6 GRAPHS, SYSTEMS OF EQUATIONS, AND PROBLEM SOLVING 221

6.1 Graphs and Equations 222
6.2 Graphing Equations 225
6.3 Linear Equations 232
6.4 Translating Problems to Equations 236
6.5 Systems of Equations 239
6.6 Solving by Substitution: Problem Solving 243
6.7 The Addition Method: Problem Solving 248

6.8 More Problem Solving 254
6.9 Problems Involving Motion 264
6.10 Equations of Lines and Slope 269
Summary and Review 277
Test 280

Cumulative Review: Chapters 1–6 281

7 INEQUALITIES AND SETS 283

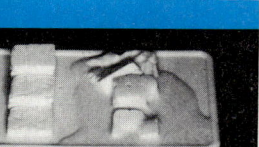

7.1 Using the Addition Principle 284
7.2 Using the Multiplication Principle 287
7.3 Problem Solving Using Inequalities 291
7.4 Graphs of Inequalities 295
7.5 Sets 300
Summary and Review 304
Test 305

8 FUNCTIONS AND VARIATION 307

8.1 Functions 308
8.2 Functions and Graphs 315
8.3 Direct Variation 318
8.4 Inverse Variation 322
Summary and Review 325
Test 327

9 FRACTIONAL EXPRESSIONS AND EQUATIONS 329

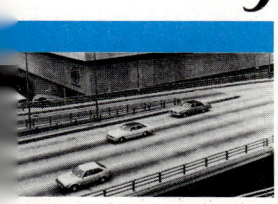

9.1 Multiplying and Simplifying 330
9.2 Division and Reciprocals 335
9.3 Addition and Subtraction 338
9.4 Least Common Multiples 343
9.5 Addition with Different Denominators 346
9.6 Subtraction with Different Denominators 350
9.7 Complex Fractional Expressions 352
9.8 Division of Polynomials 355
9.9 Solving Fractional Equations 359
9.10 Problem Solving 364
9.11 Formulas 374
9.12 Negative Exponents and Scientific Notation 377
Summary and Review 385
Test 387

Cumulative Review: Chapters 1–9 389

10 RADICAL EXPRESSIONS AND EQUATIONS   393

- **10.1** Square Roots and Real Numbers   394
- **10.2** Radical Expressions   399
- **10.3** Multiplying and Factoring   402
- **10.4** Simplifying Radical Expressions   405
- **10.5** Simplifying Square Roots of Quotients   408
- **10.6** Division   413
- **10.7** Addition and Subtraction   415
- **10.8** Right Triangles and Problem Solving   419
- **10.9** Equations with Radicals   426

Summary and Review   430

Test   432

11 QUADRATIC EQUATIONS   435

- **11.1** Introduction to Quadratic Equations   436
- **11.2** Solving by Factoring   441
- **11.3** Completing the Square   444
- **11.4** Solving by Completing the Square   448
- **11.5** The Quadratic Formula   451
- **11.6** Fractional and Radical Equations   456
- **11.7** Formulas   459
- **11.8** Problem Solving   462
- **11.9** Graphs of Quadratic Equations and Functions   468

Summary and Review   473

Test   474

Cumulative Review: Chapters 1–11   475

Table 1 Fractional and Decimal Equivalents   479
Table 2 Square Roots   480
Table 3 Geometric Formulas   481

Answers   A-1

Index   I-1

INTRODUCTION TO ALGEBRA AND PROBLEM SOLVING

1

The height of the Eiffel Tower is 295 m. It is 203 m higher than the Statue of Liberty. How can we use equations to find the height of the Statue of Liberty?

Problem solving is the main idea of this text. The first section of this first chapter is designed to introduce you to the process of solving problems, particularly in algebra. You will find additional and increasing emphasis on problem solving throughout the book.

The rest of this chapter is devoted to a review of certain aspects of arithmetic together with an introduction to algebraic symbolism and manipulations. As you will see, the manipulations of algebra, such as simplifying expressions and solving equations, are based on the properties of numbers.

If your previous study of arithmetic is not recent, you may feel more secure starting with Section 1.2. Then, after you have studied Sections 1.2–1.7, you can return to Section 1.1 and the more interesting aspect of algebra—problem solving.

1.1 INTRODUCTION TO PROBLEM SOLVING

There are many kinds of problems. Generally speaking, when you have a problem to solve, there is some kind of question. To solve the problem, in some fashion or other, you find an answer to the question. Here are some examples.

EXAMPLES OF PROBLEMS

1. How do I get a business degree at this university?
2. What is the best route to drive to Pittsburgh?
3. How can I make my apartment burglar proof?
4. If I get a 16% raise, will I be able to afford a motorcycle?
5. How many dimples are on a golf ball?
6. Is a golf ball with more dimples better than one with fewer dimples?
7. I have 80 ft of fencing. How can I fence a rectangular garden with the greatest area?
8. Is there a number which when multiplied by itself gives 10?

Although the preceding problems are quite different, there are also some similarities. We will look at some of the similarities and see if we can come up with the beginnings of a plan, or *process*, for attacking problems. As you will see later, not all these examples can be solved using algebra. Nonetheless, problem solving with algebra has its similarities with problem solving without algebra.

The First Step in Problem Solving: To Familiarize

To solve any problem, the first step is to become familiar with the situation. You should make sure that you know what information is available and that you know what the question or unknown is. Sometimes you must find information for yourself. The following are some things that you might do to help familiarize yourself with a problem situation. You might not need to use them all for a specific problem.

THE FIRST STEP IN PROBLEM SOLVING WITH ALGEBRA

Familiarize yourself with a problem situation.

1. If a problem is given in words, read it carefully.
2. List the information given and the question to be answered.
3. Find further information.
4. Make a table of the information given and the information you have collected.
5. Make a drawing and label it with known information. Also indicate unknown information.
6. Guess or estimate the answer.

1.1 INTRODUCTION TO PROBLEM SOLVING

Once you have become familiar with a problem situation, you are well on the way toward solving the problem. In this section, you will get some practice in this all-important first step. Later you will learn the complete problem-solving process.

EXAMPLE 1 How might you familiarize yourself with the situation of Problem 1: "How do I get a business degree at this university?"

Obviously further information is needed. You will have to find it. You might:

a) Talk to an upperclassman who is a business major.
b) Get a university catalog and study it.
c) Talk to a counselor in the business department.

When the information is known, it would be a good idea to list it in a table.

The problem of Example 1 is a bit unusual because once you have familiarized yourself with the situation, the problem is already solved.

EXAMPLE 2 How might you familiarize yourself with the situation of problem 7, "If I have 80 feet of fencing, how can I fence a rectangular garden with the greatest area?"

The problem is given in words. You should by all means read it carefully, perhaps several times. Probably the next thing to do is to list the information.

Fencing—80 ft available

Garden—to be rectangular

Area—to be the greatest possible

A drawing is highly important in this case. You should make a sketch and mark the information on it.

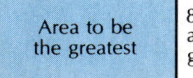

Area to be the greatest | 80 ft of fencing around the garden

Rectangular garden

How can you find the area? You probably already know that the area of a rectangle is found by multiplying the length by the width. If you ever need a formula that you do not know, you should look it up. (In doing so, you are finding further information.) It would be a *great* idea, at this point, to add this information to your drawing.

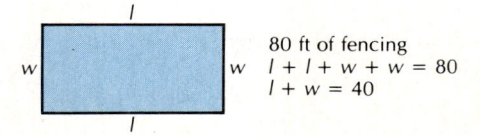

80 ft of fencing
$l + l + w + w = 80$
$l + w = 40$

From your drawing you see that $l + w$ will have to be 40 ft. At this point you might guess what the length and width might be to give the greatest area. You could try several guesses and compare them. You might complete a table as follows.

① Pick values of l and w whose sum is 40.
② Multiply to find area and compare.

l	w	Area
15	25	375
18	22	396
13.4	26.6	356.44
20	20	400

Go back and read the problem again. Now do you feel more familiar with the problem?

Remember that problems can be very, very different. There are no strict rules for solving problems; we can only give you guidelines. The first and most important one is to thoroughly familiarize yourself with the situation. You can even familiarize yourself with a problem in many different ways.

EXERCISE SET 1.1

Describe, in appropriate detail, how you would go about familiarizing yourself with each problem situation. If tables or drawings are appropriate, make them. The way you carry out the familiarization may not be the same as another person would do it. You need not find an *answer* to the problem.

1. What is the best way to get to Pittsburgh?

2. If I get a 16% raise, will I be able to afford a motorcycle?

3. How many dimples are on a golf-ball?

4. Find a number which when multiplied by itself gives 10. (Note that problems are not always phrased as questions.)

5. Using any combination of nickels, dimes, or quarters, how can you get $1 with seven coins?

6. About how many times does your heart beat in one month?

7. Fred has a triangular flower garden. Jill wants to plant a garden with the same area, but circular. How should she lay out the garden?

8. How long would it take to travel from Cincinnati to Louisville on the Ohio River?

9. Race car number 45 in the Indianapolis 500 qualified at a speed 10 mph faster than car 23. How much longer do you expect car 23 will take to finish the race?

10. John is twice as old as his sister Mary. How many years from now will he be $\frac{5}{4}$ as old?

11. How many bricks does it take to build a house?

12. How many ping pong balls can a bathtub hold?

13. Describe the sum of two odd numbers.

14. If 3 hens can lay 3 eggs in 3 days, how many eggs can 300 hens lay in 300 days?

1.2 SYMBOLS AND EXPRESSIONS IN ALGEBRA

The purpose of this section is to introduce you to the types of expressions encountered in algebra. We will study translating to expressions and evaluating expressions.

Algebraic Expressions

In arithmetic we use symbols such as 3, $\frac{1}{4}$, or 45.87, which are names for numbers. In algebra we use those same symbols, but we also use letters to stand for numbers. Sometimes a letter can stand for various numbers. In that case we call the letter a *variable*. When we replace a variable by a number, we say that we are *substituting* for the variable.

EXAMPLE 1 The length of a rectangle is 4 inches longer than the width. Suppose we use the letter w to stand for the width. Then $w + 4$ stands for the length. What is the length when the width is 6 in.? 10 in.? 13 in.?

The letter w is a variable because it can stand for various widths, which can change. To find the length, we substitute for w in $w + 4$.

a) When the width is 6, the length is $6 + 4$, or 10.   Substituting 6 for w
b) When the width is 10, the length is $10 + 4$, or 14.   Substituting 10 for w
c) When the width is 13, the length is $13 + 4$, or 17.   Substituting 13 for w

An *algebraic expression* is a symbol consisting of variables, numerals, and operation signs. When we substitute for the variables and calculate the results, we get a number. This process is called *evaluating the expression*, and the number we get is called the *value* of the expression. In Example 1, when we substituted 10 for w in the expression, we got the value 14.

EXAMPLE 2 The television set in the ad shown here costs $283. Suppose we use the letter n to stand for the number of sets purchased. Then $283 \times n$ stands for the total cost of n sets. Find the cost of 7 television sets. That is, evaluate $283 \times n$ when $n = 7$ (when n stands for 7).

> **ELECTRONIC TUNING**
> **13" diag. meas.**
> **COLOR**
> **SAVE $66.95**
> **13" Diagonal Color Portable TV**
> 1-knob electronic tune w/perma-set fine tuning.
> Everyday $349.95
> **$283**

We substitute 7 for n in the expression $283 \times n$:

$$283 \times 7 = 1981.$$

The cost of 7 television sets is $1981.

EXAMPLE 3 A baseball player's batting average is h/a, where h is the number of hits and a is the number of "at bats." Find the batting average of a batter who had 13 hits in 25 at bats.

The number of hits is 13, so we substitute 13 for h. The number of at bats is 25, so we substitute 25 for a:

$$\frac{h}{a} = \frac{13}{25} = 0.520.$$

The Second Step in Problem Solving: To Translate

After familiarizing yourself with a problem situation, you are ready for the next step in problem solving. In many problems, a mathematical expression corresponds to the situation of the problem. The second step in the problem-solving process is to *translate* the situation to mathematical language of some kind. In Example 1, for example, we wrote an algebraic expression. We translated the conditions of the problem to mathematical language.

When we say *translate* to mathematical language, we do not necessarily mean to write an algebraic expression, but in some cases a situation can be translated to an algebraic expression.

1.2 SYMBOLS AND EXPRESSIONS IN ALGEBRA 7

> **THE SECOND STEP IN PROBLEM SOLVING WITH ALGEBRA**
>
> Translate the situation of the problem to mathematical language. In some cases, translation can be done by writing an algebraic expression.

EXAMPLE 4 Write an expression that stands for twice (or two times) some number.

Think of some number, say 8. What number is twice 8? It is 16. How did you get 16? You multiplied by 2. Do the same thing using a variable. Let's use y to stand for some number. Multiply by 2. We get an expression

$$y \times 2, \quad 2 \times y, \quad \text{or} \quad 2 \cdot y.$$

In Example 4 we used x for a multiplication sign as in arithmetic. We also used a dot.

$$3 \cdot 7 \quad \text{also means} \quad 3 \times 7$$
$$3 \cdot a \quad \text{also means} \quad 3 \times a$$

When two letters or a *number* and a letter are written together, that also means that they are to be multiplied. Numbers to be multiplied are called *factors*. We usually write a numerical factor before any factor named by a letter.

$$5y \quad \text{means} \quad 5 \cdot y \quad \text{or} \quad 5 \times y.$$
$$ab \quad \text{means} \quad a \cdot b \quad \text{or} \quad a \times b.$$

EXAMPLE 5 Suppose that m stands for some number that is 5 more than some other number. Write an expression for the smaller number.

First, think of a number, say 12. It is 5 more than another number. That number must be 7. How did you get the other number? You subtracted 5 from 12, that is, $12 - 5$. Doing the same thing using m instead of 12, we get an expression for the smaller number:

$$m - 5.$$

Here are some further examples of translations to algebraic expressions. When you translate in this way, you may use different letters as variables. We generally use letters near the end of the alphabet, such as x, y, and z, for variables, although we sometimes use others also.

Five more than some number	$n + 5$ or $5 + n$
Half of a number	$\frac{1}{2}t$ or $\frac{t}{2}$
Five more than three times some number	$5 + 3p$ or $3p + 5$
The difference of two numbers	$x - y$
Six less than the product of two numbers	$mn - 6$
Seventy-six percent of some number	$76\%z$ or $0.76z$

EXERCISE SET 1.2

Substitute to find values of the expressions.

1. Theresa is 6 years younger than her husband Frank. Suppose the variable x stands for Frank's age. Then $x - 6$ stands for Theresa's age. How old is Theresa when Frank is 29? 34? 47?

2. Employee A took five times as long to do a job as employee B. Suppose t stands for the time it takes B to do the job. Then $5t$ stands for the time it takes A. How long did it take A if B took 30 seconds? 90 seconds? 2 minutes?

3. The area of a rectangle of length l and width w is lw. Find the area when $l = 14$ ft and $w = 8$ ft.

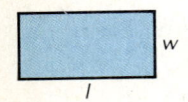

4. The area of a triangle with base b and height h is $\frac{1}{2}bh$. Find the area when $b = 45$ m (meters) and $h = 86$ m.

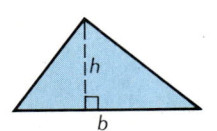

Evaluate.

5. $6x$ for $x = 7$

6. $7y$ for $y = 7$

7. $\dfrac{x}{y}$ for $x = 9$ and $y = 3$

8. $\dfrac{m}{n}$ for $m = 14$ and $n = 2$

9. $\dfrac{3 \cdot p}{q}$ for $p = 2$ and $q = 6$

10. $\dfrac{5 \cdot y}{z}$ for $y = 15$ and $z = 25$

11. $\dfrac{x + y}{5}$ for $x = 10$ and $y = 20$

12. $\dfrac{p + q}{2}$ for $p = 2$ and $q = 16$

13. $\dfrac{x - y}{8}$ for $x = 20$ and $y = 4$

14. $\dfrac{m - n}{5}$ for $m = 16$ and $n = 6$

15. $\dfrac{x}{y}$ for $x = 3$ and $y = 6$

16. $\dfrac{p}{q}$ for $p = 4$ and $q = 16$

17. $\dfrac{5z}{y}$ for $z = 8$ and $y = 2$

18. $\dfrac{9m}{q}$ for $m = 4$ and $q = 18$

Translate to an algebraic expression.

19. 6 more than b

20. 8 more than t

21. 9 less than c

22. 4 less than d

23. 6 greater than q

24. 11 greater than z

25. b more than a

26. c more than d

27. x less than y

28. c less than h

29. x added to w

30. s added to t

31. m subtracted from n

32. p subtracted from q

33. the sum of r and s

34. the sum of d and f

35. twice x

36. three times p

37. 5 multiplied by t

38. 9 multiplied by d

39. the product of 3 and b

40. A number y is 6 less than a larger number. Write an expression for the larger number.

42. A number x is 4 more than a smaller number. Write an expression for the smaller number.

41. A number m is 1 less than a larger number. Write an expression for the larger number.

43. A number h is 43 more than a smaller number. Write an expression for the smaller number.

Translate to an algebraic expression.

44. A number x plus three times y.

45. A number y plus two times x.

46. A number a plus 2 plus b.

47. A number that is 3 less than twice x.

48. Your age in 5 years, if you are a years old now.

49. Your age two years ago, if you are b years old now.

50. A number x increased by itself.

51. A number that is 98% of x.

52. The perimeter of a square with side s.

53. Evaluate $\frac{x+y}{4}$ when $y = 8$ and x is twice y.

54. Evaluate $\frac{x-y}{7}$ when $y = 35$ and x is twice y.

55. Evaluate $\frac{y-x}{3}$ when $x = 9$ and y is three times x.

56. Evaluate $\frac{256y}{32x}$ for $y = 1$ and $x = 4$.

57. Evaluate $\frac{y+x}{2} + \frac{3 \cdot y}{x}$ for $x = 2$ and $y = 4$.

Answer each question with an algebraic expression.

58. If $w + 3$ is a whole number, what is the next whole number after it?

59. If $d + 2$ is an odd whole number, what is the preceding odd number?

60. The difference between two numbers is 3. One number is t. What are two possible values for the other number?

61. Two numbers are $v + 2$ and $v - 2$. Write an expression for their sum.

62. Two numbers are $2 + w$ and $2 - w$. Write an expression for their sum.

63. You invest n dollars at 10% interest. Write an expression for the number of dollars in the bank a year from now.

1.3 FACTORIZATIONS AND LCMs

In algebra the notion of factoring is quite important. You will eventually learn to factor algebraic expressions. For now, we review factoring of numbers in order to review addition and subtraction using fractional notation.

Factoring

In this section we will be considering the set of *natural numbers*:

$$1, 2, 3, 4, 5, \text{ and so on.}$$

> To *factor* a number means to name it as a product.

EXAMPLE 1 Factor the number 8.

The number 8 can be named as a product in several ways:

$$2 \cdot 4, \quad 1 \cdot 8, \quad 2 \cdot 2 \cdot 2.$$

A symbol that names a number as a product is called a *factorization* of the number.

EXAMPLE 2 Write several factorizations of the number 12.

$$1 \cdot 12, \quad 2 \cdot 6, \quad 3 \cdot 4, \quad 2 \cdot 2 \cdot 3$$

Prime Numbers

Some numbers have exactly two factors, themselves and 1. Such numbers are called *prime*.

> A *prime* number is a natural number that has exactly two different factors.

EXAMPLE 3 Which of these numbers are prime? 7, 4, 11, 16, 1

7 is prime. It has exactly two different factors, 7 and 1.

4 is not prime. It has three different factors, 1, 2, and 4.

11 is prime. It has exactly two different factors, 11 and 1.

16 is not prime. It has factors 1, 2, 4, 8, and 16.

1 is not prime. It has only itself as a factor.

The following is a table of the prime numbers from 2 to 101. There are more extensive tables, but these prime numbers will be the most helpful to you in this book.

> **A TABLE OF PRIMES**
>
> 2, 3, 5, 7, 11, 13, 17, 19, 23, 29, 31, 37, 41,
> 43, 47, 53, 59, 61, 67, 71, 73, 79, 83, 89, 97, 101

Prime Factorizations

If a natural number, other than 1, is not prime, we call it *composite*. Every composite number can be factored into prime numbers.

EXAMPLE 4 Find the prime factorization of 36.

We begin by factoring 36 any way we can. One way is like this:

$$36 = 4 \cdot 9.$$

The factors 4 and 9 are not prime, so we factor them:

$$36 = 4 \cdot 9 = \underbrace{2 \cdot 2}_{4} \cdot \underbrace{3 \cdot 3}_{9}$$

The factors in the last factorization are all prime, so we have the *prime factorization* of 36.

EXAMPLE 5 Find the prime factorization of 60.

This time, we use our list of primes from the table.

Divide 60 by 2: $2 \overline{)60}^{\,30}$

$$60 = 2 \cdot 30$$

Divide 30 by 2: $2 \overline{)30}^{\,15}$

$$60 = 2 \cdot \underbrace{2 \cdot 15}_{30}$$

We cannot divide 15 by 2 without remainder. So we try dividing by the next prime, 3: $3 \overline{)15}^{\,5}$

$$60 = 2 \cdot 2 \cdot \underbrace{3 \cdot 5}_{15} \quad \text{This is the prime factorization.}$$

Multiples

The *multiples* of a number all have that number as a factor. For example, the multiples of 2 are

$$2, 4, 6, 8, 10, 12, 14, 16, \ldots .$$

We could name them in such a way as to show 2 as a factor, as follows:

$$2 \cdot 1, 2 \cdot 2, 2 \cdot 3, 2 \cdot 4, 2 \cdot 5, 2 \cdot 6, \ldots .$$

The multiples of 3 all have 3 as a factor. They are

$$3, 6, 9, 12, 15, 18, \ldots .$$

We could also name them as follows:

$$3 \cdot 1, \ 3 \cdot 2, \ 3 \cdot 3, \ 3 \cdot 4, \ 3 \cdot 5, \ldots.$$

Common Multiples. Two or more numbers always have a great many multiples in common. From lists of multiples, we can find common multiples.

EXAMPLE 6 Find some of the multiples that 2 and 3 have in common.

We make lists of their multiples and circle the multiples that appear in both lists.

2, 4, ⑥, 8, 10, ⑫, 14, 16, ⑱, 20, 22, ㉔, 26, 28, ㉚, 32, 34, ㊱, . . .

3, ⑥, 9, ⑫, 15, ⑱, 21, ㉔, 27, ㉚, 33, ㊱, . . .

The common multiples of 2 and 3 are

$$6, \ 12, \ 18, \ 24, \ 30, \ 36, \ldots.$$

Least Common Multiples

The numbers 2 and 3 have many common multiples. The *least*, or smallest of those common multiples is 6. We abbreviate *least common multiple* as LCM.

Here is a method of finding LCMs of two or more numbers.

> **To find the LCM of two or more numbers:**
>
> a) Determine whether the largest of the numbers is a multiple of the others. If it is, then it is the LCM.
>
> b) If not, make a list of multiples of the largest number, until you get one that is a multiple of the other numbers. That number is the LCM.

EXAMPLE 7 Find the LCM of 12 and 15.

a) 15 is not a multiple of 12.

b) We list multiples of 15, the larger number.

$15 \cdot 2 = 30$, but 30 is not a multiple of 12.

$15 \cdot 3 = 45$, but 45 is not a multiple of 12.

$15 \cdot 4 = 60$, and 60 is a multiple of 12, because $60 = 5 \cdot 12$.

The LCM of 12 and 15 is 60.

EXAMPLE 8 Find the LCM of 4 and 20.

a) 20 is a multiple of 4, because $20 = 4 \cdot 5$.

The LCM of 4 and 20 is 20.

EXAMPLE 9 Find the LCM of 6, 10, and 12.

    a) The largest number, 12, is not a multiple of 6 and 10.

    b) We list multiples of 12

        $12 \cdot 2 = 24$, and 24 is a multiple of 6 but not a multiple of 10.

        $12 \cdot 3 = 36$, and 36 is a multiple of 6 but not a multiple of 10.

        $12 \cdot 4 = 48$, and 48 is a multiple of 6 but not a multiple of 10.

        $12 \cdot 5 = 60$, and 60 is a multiple of both 6 and 10.

The LCM of 6, 10, and 12 is 60.

Finding LCMs by Factoring

The method we have just used works for natural numbers, but it does not work in algebra. We now learn another method that will work in arithmetic *and also in algebra*. To show how the other method works, we make lists, but we write factorizations. Let's look for the LCM of 9 and 15.

    The multiples of 9 are

$$3 \cdot 3 \cdot 2,\ 3 \cdot 3 \cdot 3,\ 3 \cdot 3 \cdot 4,\ 3 \cdot 3 \cdot 5,\ 3 \cdot 3 \cdot 6, \ldots .$$

    The multiples of 15 are

$$3 \cdot 5 \cdot 2,\ 3 \cdot 5 \cdot 3,\ 3 \cdot 5 \cdot 4,\ 3 \cdot 5 \cdot 5, \ldots .$$

The LCM of 9 and 15 is $3 \cdot 3 \cdot 5$. The LCM must have all of the factors of 9, and it must have all of the factors of 15.

To find the LCM of several numbers:

1. Write the prime factorization of each number.
2. Write each factor the greatest number of times it appears in any one factorization.

EXAMPLE 10 Find the LCM of 24 and 36.

    a) We find the prime factorizations.

$$24 = 2 \cdot 2 \cdot 2 \cdot 3$$

$$36 = 2 \cdot 2 \cdot 3 \cdot 3$$

    b) We write 2 as a factor 3 times (the greatest number of times it occurs).

    c) We write 3 as a factor 2 times (the greatest number of times it occurs).

The LCM is $2 \cdot 2 \cdot 2 \cdot 3 \cdot 3$, or 72.

EXAMPLE 11 Find the LCM of 27, 90, and 84.

a) We factor.
$$27 = 3 \cdot 3 \cdot 3$$
$$90 = 2 \cdot 3 \cdot 3 \cdot 5$$
$$84 = 2 \cdot 2 \cdot 3 \cdot 7$$

b) We write 2 as a factor 2 times, 3 as a factor 3 times, 5 one time, and 7 one time. The LCM is $2 \cdot 2 \cdot 3 \cdot 3 \cdot 3 \cdot 5 \cdot 7$, or 3780.

EXAMPLE 12 Find the LCM of 7 and 21.

Since 7 is prime, it has no prime factorization. We still need it as a factor.
$$7 = \phantom{3 \cdot{}} 7$$
$$21 = 3 \cdot 7$$

The LCM is $3 \cdot 7$, or 21.

• •

If one number is a factor of another, the LCM is the larger of the numbers.

EXAMPLE 13 Find the LCM of 8 and 9.
$$8 = 2 \cdot 2 \cdot 2$$
$$9 = 3 \cdot 3$$

The LCM is $2 \cdot 2 \cdot 2 \cdot 3 \cdot 3$, or 72.

• •

If two or more numbers have no common prime factor, the LCM is the product of the numbers.

EXERCISE SET 1.3

Write at least one factorization of each number.

1. 21
2. 30
3. 45
4. 35
5. 49
6. 52
7. 28
8. 39
9. 76
10. 42
11. 56
12. 102
13. 93
14. 88
15. 144
16. 121

1.3 FACTORIZATIONS AND LCMs

Find the prime factorization of each number.

17. 14	18. 15	19. 33	20. 55
21. 9	22. 25	23. 49	24. 121
25. 18	26. 24	27. 40	28. 56
29. 90	30. 120	31. 210	32. 330
33. 91	34. 143	35. 119	36. 221

Make lists of multiples, and from them find the first three common multiples.

| 37. 3, 7 | 38. 5, 7 | 39. 8, 12 | 40. 12, 16 |
| 41. 10, 12 | 42. 18, 27 | 43. 20, 30 | 44. 24, 36 |

Find the LCM using a list of multiples.

| 45. 3, 15 | 46. 20, 40 | 47. 30, 40 | 48. 50, 60 |
| 49. 13, 23 | 50. 18, 24 | 51. 60, 70 | 52. 35, 45 |

Find the prime factorization of each number. Then find the LCM.

53. 12, 18	54. 18, 30	55. 45, 72	56. 30, 36
57. 30, 50	58. 24, 36	59. 30, 40	60. 13, 23
61. 18, 24	62. 12, 28	63. 35, 45	64. 60, 70
65. 2, 3, 5	66. 3, 5, 7	67. 24, 36, 12	68. 8, 16, 22
69. 5, 12, 15	70. 12, 18, 40	71. 6, 12, 18	72. 18, 24, 30

73. Consider 8 and 12. Determine whether each of the following is the LCM of 8 and 12. Tell why or why not.
a) $2 \cdot 2 \cdot 3 \cdot 3$ b) $2 \cdot 2 \cdot 3$ c) $2 \cdot 3 \cdot 3$ d) $2 \cdot 2 \cdot 2 \cdot 3$

Use your calculator and the list-of-multiples method to find the LCM of each pair of numbers.
74. 288, 324   75. 2700, 7800

76. A cigar company uses two sizes of boxes, 6 in. and 8 in. long. These are packed in bigger cartons to be shipped. What is the shortest length carton that will accommodate boxes of either size without any room left over? (Each carton can contain only boxes of one size; no mixing is allowed.)

Planet Orbits and LCMs. The Earth, Jupiter, Saturn, and Uranus all revolve around the sun. The Earth takes 1 year, Jupiter takes 12 years, Saturn takes 30 years, and Uranus takes 84 years. On a certain night you look at all the planets, and you wonder how many years it will be before they have the same position again. To find out, you find the LCM of 12, 30, and 84. It will be that number of years.

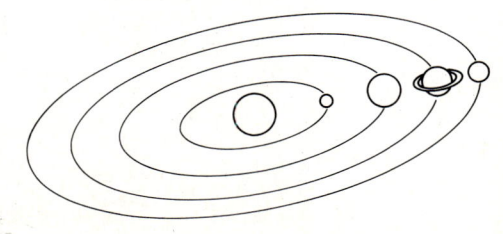

77. How often will Jupiter and Saturn appear in the same direction?
78. How often will Saturn and Uranus appear in the same direction?
79. How often will Jupiter, Saturn, and Uranus appear in the same direction?
80. In the table, the top number has been factored in such a way that the sum of the factors is the bottom number. For example, in the first column 56 has been factored as 7 · 8, and 7 + 8 = 15, the bottom number. Find the missing numbers in the table.

Product	56	63	36	72	140	96		168	110			
Factor	7									9	24	3
Factor	8					8	8			10	18	
Sum	15	16	20	38	24	20	14		21			24

1.4 PROPERTIES OF NUMBERS AND ALGEBRAIC EXPRESSIONS

When you have translated a problem situation to mathematical language, you may have an algebraic expression. Then what do you do? One thing you may wish to do in the course of solving a problem is to find a simpler expression that has the same values as the one to which you translated. Algebraic expressions can be manipulated in other ways to help give answers to problems. To handle expressions we study the properties of numbers.

We use several different kinds of numbers in arithmetic and algebra. Here are some that you already know.

Natural numbers are the numbers used for counting, 1, 2, 3, 4, 5, and so on.
Whole numbers are the natural numbers and 0. The whole numbers are 0, 1, 2, 3, and so on.
Numbers of arithmetic are the whole numbers and the fractions, such as $\frac{2}{3}$, $\frac{4}{1}$, or $\frac{6}{5}$. All of these numbers can be named with fractional notation $\frac{a}{b}$.

Note that all whole numbers are also numbers of arithmetic. In this chapter we use the word *number* to mean a number of arithmetic.

The Commutative Laws

Let us examine the expressions $x + y$ and $y + x$.

1.4 PROPERTIES OF NUMBERS AND ALGEBRAIC EXPRESSIONS

EXAMPLE 1 Evaluate $x + y$ and $y + x$ for $y = 7$ and $x = 4$.

We substitute 7 for y and 4 for x in both expressions:

$$x + y = 4 + 7 = 11; \qquad y + x = 7 + 4 = 11.$$

EXAMPLE 2 Evaluate $x + y$ and $y + x$ for $y = 12$ and $x = 23$.

We substitute 12 for y and 23 for x in both expressions:

$$x + y = 23 + 12 = 35; \qquad y + x = 12 + 23 = 35.$$

Note that the expressions

$$x + y \quad \text{and} \quad y + x$$

have the same values, no matter what the variables stand for. We know that because when we add two numbers, it does not matter which comes first. Similarly, xy and yx have the same values, no matter what the variables stand for.

These are examples of general patterns or laws.

COMMUTATIVE LAWS

Addition For any numbers a and b,

$$a + b = b + a.$$

(We can change the order when adding without affecting the answer.)

Multiplication For any numbers a and b,

$$ab = ba.$$

(We can change the order when multiplying without affecting the answer.)

Expressions that have the same value for all replacements are called *equivalent expressions*.

Using a commutative law, we know that $x + 2$ and $2 + x$ are equivalent. Similarly $3 \cdot x$ and $x \cdot 3$ are equivalent.

EXAMPLE 3 Use a commutative law to write an expression equivalent to $y + 5$, xy, and $7 + ab$.

An expression equivalent to $y + 5$ is $5 + y$ by the commutative law of addition.
An expression equivalent to xy is yx by the commutative law of multiplication.
An expression equivalent to $7 + ab$ is $ab + 7$ by the commutative law of addition.
Another expression equivalent to $7 + ab$ is $7 + ba$ by the commutative law of multiplication.
An expression equivalent to $7 + ab$ is $ba + 7$ by both commutative laws.

Properties of 0 and 1

Some simple but powerful properties of numbers are the properties of 0 and 1. We shall see some ways to use both of these properties in algebra and arithmetic.

THE ADDITIVE PROPERTY OF 0

For any number a,
$$a + 0 = a.$$
(Adding 0 to any number gives that same number.)

THE PROPERTY OF 1

For any number a,
$$1 \cdot a = a.$$
(Multiplying a number by 1 gives that same number.)

Here are some of the many ways to name the number 1:
$$\tfrac{5}{5} \quad \tfrac{3}{3} \quad \tfrac{26}{26}.$$

The following algebraic expressions have the value 1 for all replacements, except that we do not allow a replacement that results in division by 0. (Later, in Chapter 2, we shall see why such division is not allowed.)

$$\frac{n}{n} \qquad \frac{x-2}{x-2} \qquad \frac{5y+4}{5y+4}$$

For any number a, except 0,
$$\frac{a}{a} = 1.$$

We can use the property of 1 to write equivalent expressions.

EXAMPLE 4 Write an expression equivalent to $\tfrac{2}{3}$ by multiplying by 1. Use $\tfrac{5}{5}$ for 1.

Recall from arithmetic that to multiply with fractional notation we multiply numerators and denominators.

$$\tfrac{2}{3} = \tfrac{2}{3} \cdot 1 \qquad \text{Using the property of 1}$$
$$= \tfrac{2}{3} \cdot \tfrac{5}{5} \qquad \text{Using } \tfrac{5}{5} \text{ for 1}$$
$$= \tfrac{10}{15} \qquad \text{Multiplying numerators and denominators}$$

1.4 PROPERTIES OF NUMBERS AND ALGEBRAIC EXPRESSIONS

EXAMPLE 5 Write an expression equivalent to $\frac{x}{2}$ by multiplying by $\frac{y}{y}$.

$$\frac{x}{2} = \frac{x}{2} \cdot 1 \qquad \text{Using the property of 1}$$

$$= \frac{x}{2} \cdot \frac{y}{y} \qquad \text{Using } \frac{y}{y} \text{ for 1}$$

$$= \frac{xy}{2y}$$

The expressions $x/2$ and $xy/2y$ will have the same value for all replacements of x and y. We do not divide by 0, so we will never replace y by 0.

Simplifying Expressions

We know that $\frac{1}{2}, \frac{2}{4}, \frac{4}{8}$, etc., all name the same number. Any number of arithmetic can be named in many ways. The *simplest fractional notation* is the notation that has the smallest numerator and denominator. We call the process of finding the simplest fractional notation *simplifying*. We begin by factoring the numerator and denominator. Then we factor the fractional expression and use the property of 1.

EXAMPLE 6 Simplify: $\frac{10}{15}$.

$$\frac{10}{15} = \frac{2 \cdot 5}{3 \cdot 5} \qquad \text{Factoring numerator and denominator}$$

$$= \frac{2}{3} \cdot \frac{5}{5} \qquad \text{Factoring the fractional expression}$$

$$= \frac{2}{3} \qquad \text{Using the property of 1 (removing a factor of 1)}$$

EXAMPLE 7 Simplify: $\frac{36}{24}$.

$$\frac{36}{24} = \frac{6 \cdot 6}{4 \cdot 6} \qquad \text{Factoring numerator and denominator}$$

$$= \frac{3 \cdot 2 \cdot 6}{2 \cdot 2 \cdot 6} \qquad \text{Further factoring}$$

$$= \frac{3}{2} \cdot \frac{2 \cdot 6}{2 \cdot 6} \qquad \text{Factoring the fractional expression}$$

$$= \frac{3}{2} \qquad \text{Using the property of 1}$$

The number of factors in the numerator and denominator may not always be the same. If not, we can always insert the number 1 as a factor. The property of 1 allows us to do that.

EXAMPLE 8 Simplify: $\frac{18}{72}$.

$$\frac{18}{72} = \frac{2 \cdot 9}{4 \cdot 2 \cdot 9} = \frac{1 \cdot 2 \cdot 9}{4 \cdot 2 \cdot 9} \quad \text{Using the property of 1 to insert the factor 1}$$

$$= \frac{1}{4} \cdot \frac{2 \cdot 9}{2 \cdot 9} \quad \text{Factoring the fractional expression}$$

$$= \frac{1}{4}$$

EXAMPLE 9 Simplify: $\frac{72}{9}$.

$$\frac{72}{9} = \frac{8 \cdot 9}{1 \cdot 9} \quad \text{Factoring and inserting a factor of 1 in the denominator}$$

$$= \frac{8}{1} \cdot \frac{9}{9} = \frac{8}{1}$$

$$= 8$$

Now let us simplify some algebraic expressions. We will use the property of 1 as in Examples 6–9. We factor the numerator and denominator. Then we factor the fractional expression and "remove" factors of 1.

EXAMPLE 10 Simplify: $\frac{xy}{3y}$.

$$\frac{xy}{3y} = \frac{x \cdot y}{3 \cdot y} \quad \text{Factoring numerator and denominator}$$

$$= \frac{x}{3} \cdot \frac{y}{y} \quad \text{Factoring the fractional expression}$$

$$= \frac{x}{3} \quad \text{Using the property of 1 (removing a factor of 1)}$$

EXAMPLE 11 Simplify: $\frac{x}{5xy}$.

$$\frac{x}{5xy} = \frac{1 \cdot x}{5 \cdot x \cdot y} \quad \text{Inserting a factor of 1 in the numerator}$$

$$= \frac{1}{5y} \cdot \frac{x}{x} \quad \text{Using the commutative property and factoring the fractional expression}$$

$$= \frac{1}{5y}$$

Multiplication, Addition, Subtraction, and Division

After we have performed an operation of multiplication, addition, subtraction, or division, the answer may not be simplified. In such cases, we should simplify. We continue as we did in Examples 6–9.

EXAMPLE 12 Multiply and simplify: $\frac{5}{6} \cdot \frac{9}{25}$.

$$\frac{5}{6} \cdot \frac{9}{25} = \frac{5 \cdot 9}{6 \cdot 25} \qquad \text{Multiplying numerators and denominators}$$

$$= \frac{1 \cdot 5 \cdot 3 \cdot 3}{2 \cdot 3 \cdot 5 \cdot 5} \qquad \text{Factoring numerator and denominator}$$

$$= \frac{3 \cdot 5 \cdot 1 \cdot 3}{3 \cdot 5 \cdot 2 \cdot 5} = \frac{3 \cdot 5}{3 \cdot 5} \cdot \frac{1 \cdot 3}{2 \cdot 5} \qquad \text{Factoring the fractional expression}$$

$$= \frac{3}{10} \qquad \text{"Removing" a factor of 1}$$

When denominators are the same, we can add by adding numerators and keeping the same denominator. That is,

$$\frac{a}{c} + \frac{b}{c} = \frac{a+b}{c}.$$

When denominators are different, we use the property of 1 and multiply to find a common denominator. The smallest such denominator is called the *least common denominator*. That number is the least common multiple of the original denominators.

EXAMPLE 13 Add and simplify: $\frac{3}{8} + \frac{5}{12}$.

The least common multiple of the denominators is 24. We multiply by 1 to obtain the least common multiple for each denominator:

$$\frac{3}{8} + \frac{5}{12} = \frac{3}{8} \cdot \frac{3}{3} + \frac{5}{12} \cdot \frac{2}{2} \qquad \text{Multiplying by 1. Since } 3 \cdot 8 = 24, \text{ we multiply the}$$
$$= \frac{9}{24} + \frac{10}{24} \qquad \qquad \text{first number by } \frac{3}{3}. \text{ Since } 2 \cdot 12 = 24, \text{ we multiply}$$
$$= \frac{19}{24}. \qquad \qquad \text{the second number by } \frac{2}{2}.$$

EXAMPLE 14 Subtract and simplify: $\frac{9}{8} - \frac{4}{5}$.

$$\frac{9}{8} - \frac{4}{5} = \frac{9}{8} \cdot \frac{5}{5} - \frac{4}{5} \cdot \frac{8}{8} \qquad \text{The LCM of 8 and 5 is 40.}$$
$$= \frac{45}{40} - \frac{32}{40}$$
$$= \frac{13}{40}$$

In arithmetic you usually write $1\frac{1}{8}$ rather than $\frac{9}{8}$. In algebra you will find that the so-called *improper* symbols such as $\frac{9}{8}$ are more useful and are quite *proper*.

EXAMPLE 15 Subtract and simplify: $\frac{7}{10} - \frac{1}{5}$.

$$\frac{7}{10} - \frac{1}{5} = \frac{7}{10} - \frac{1}{5} \cdot \frac{2}{2} \qquad \text{The LCM of 10 and 5 is 10.}$$
$$= \frac{7}{10} - \frac{2}{10}$$
$$= \frac{5}{10} = \frac{1}{2} \cdot \frac{5}{5} = \frac{1}{2}$$

Reciprocals

Two numbers whose product is 1 are called *reciprocals* of each other. All the numbers of arithmetic, except zero, have reciprocals.

EXAMPLES
16. The reciprocal of $\frac{2}{3}$ is $\frac{3}{2}$ because $\frac{2}{3} \cdot \frac{3}{2} = \frac{6}{6} = 1$.
17. The reciprocal of 9 is $\frac{1}{9}$ because $9 \cdot \frac{1}{9} = \frac{9}{9} = 1$.
18. The reciprocal of $\frac{1}{4}$ is 4 because $\frac{1}{4} \cdot 4 = 1$.

Reciprocals and Division

The number 1 and reciprocals can be used to explain division of numbers of arithmetic. To divide, we can multiply by 1, choosing carefully the symbol for 1.

EXAMPLE 19 Divide $\frac{2}{3}$ by $\frac{7}{5}$.

$$\frac{\frac{2}{3}}{\frac{7}{5}} = \frac{\frac{2}{3}}{\frac{7}{5}} \times \frac{\frac{5}{7}}{\frac{5}{7}} \qquad \text{Multiplying by } \frac{\frac{5}{7}}{\frac{5}{7}}. \text{ We use } \frac{5}{7} \text{ because it is the reciprocal of } \frac{7}{5}.$$
$$= \frac{\frac{2}{3} \times \frac{5}{7}}{\frac{7}{5} \times \frac{5}{7}} \qquad \text{Multiplying numerators and denominators}$$
$$= \frac{\frac{10}{21}}{1} = \frac{10}{21} \qquad \text{Simplifying}$$

After multiplying we got 1 for a denominator. That was because we used the reciprocal of the divisor, $\frac{7}{5}$, for both the numerator and denominator of the symbol for 1.

When multiplying by 1 to divide, we get a denominator of 1. What do we get in the numerator? In Example 19, we got $\frac{2}{3} \times \frac{5}{7}$. This is the product of $\frac{2}{3}$, the dividend, and $\frac{5}{7}$, the reciprocal of the divisor.

• • • • • • • • • • • • • • • • • • • •

To divide, multiply by the reciprocal of the divisor:

$$\frac{a}{b} \div \frac{c}{d} = \frac{a}{b} \cdot \frac{d}{c}.$$

1.4 PROPERTIES OF NUMBERS AND ALGEBRAIC EXPRESSIONS

EXAMPLE 20 Divide by multiplying by the reciprocal of the divisor: $\frac{1}{2} \div \frac{3}{5}$.

$$\frac{1}{2} \div \frac{3}{5} = \frac{1}{2} \cdot \frac{5}{3} = \frac{5}{6} \qquad \frac{5}{3} \text{ is the reciprocal of } \frac{3}{5}$$

After dividing, simplification is often possible and should be done.

EXAMPLE 21 Divide: $\frac{2}{3} \div \frac{4}{9}$.

$$\frac{2}{3} \div \frac{4}{9} = \frac{2}{3} \cdot \frac{9}{4} = \underbrace{\frac{2 \cdot 3 \cdot 3}{3 \cdot 2 \cdot 2} = \frac{2 \cdot 3}{2 \cdot 3} \cdot \frac{3}{2}}_{\text{Simplifying}} = \frac{3}{2}$$

EXERCISE SET 1.4

Write an expression equivalent to each of the following. Use a commutative law.

1. $y + 8$
2. $x + 3$
3. mn
4. ab
5. $9 + xy$
6. $11 + ab$
7. $ab + c$
8. $rs + t$

Write an expression equivalent to each of the following. Use the indicated name for 1.

9. $\frac{5}{6}$ (Use $\frac{8}{8}$ for 1.)
10. $\frac{9}{10}$ (Use $\frac{11}{11}$ for 1.)
11. $\frac{6}{7}$ (Use $\frac{100}{100}$ for 1.)
12. $\frac{y}{10}$ (Use $\frac{z}{z}$ for 1.)
13. $\frac{s}{20}$ (Use $\frac{t}{t}$ for 1.)
14. $\frac{m}{3n}$ (Use $\frac{p}{p}$ for 1.)

Simplify.

15. $\frac{18}{45}$
16. $\frac{16}{56}$
17. $\frac{49}{14}$
18. $\frac{72}{27}$
19. $\frac{6}{42}$
20. $\frac{13}{104}$
21. $\frac{56}{7}$
22. $\frac{132}{11}$
23. $\frac{5y}{5}$
24. $\frac{ab}{9b}$
25. $\frac{x}{9xy}$
26. $\frac{q}{8pq}$
27. $\frac{8a}{3ab}$
28. $\frac{9p}{17pq}$
29. $\frac{3pq}{6q}$
30. $\frac{51d}{17sd}$
31. $\frac{9nz}{19tn}$
32. $\frac{13rv}{3vh}$

Compute and simplify.

33. $\frac{1}{4} \cdot \frac{1}{2}$
34. $\frac{11}{10} \cdot \frac{8}{5}$
35. $\frac{17}{2} \cdot \frac{3}{4}$
36. $\frac{11}{12} \cdot \frac{12}{11}$
37. $\frac{1}{2} + \frac{1}{2}$
38. $\frac{1}{2} + \frac{1}{4}$
39. $\frac{4}{9} + \frac{13}{18}$
40. $\frac{4}{5} + \frac{8}{15}$
41. $\frac{3}{10} + \frac{8}{15}$
42. $\frac{9}{8} + \frac{7}{12}$
43. $\frac{5}{4} - \frac{3}{4}$
44. $\frac{12}{5} - \frac{2}{5}$
45. $\frac{13}{18} - \frac{4}{9}$
46. $\frac{13}{15} - \frac{8}{45}$
47. $\frac{11}{12} - \frac{2}{5}$
48. $\frac{15}{16} - \frac{2}{3}$
49. $\frac{7}{6} \div \frac{3}{5}$
50. $\frac{7}{5} \div \frac{3}{4}$
51. $\frac{8}{9} \div \frac{4}{15}$
52. $\frac{3}{4} \div \frac{3}{7}$
53. $\frac{1}{4} \div \frac{1}{2}$
54. $\frac{1}{10} \div \frac{1}{5}$
55. $\frac{\frac{13}{12}}{\frac{39}{5}}$
56. $\frac{\frac{17}{6}}{\frac{3}{8}}$
57. $100 \div \frac{1}{5}$
58. $78 \div \frac{1}{6}$
59. $\frac{3}{4} \div 10$
60. $\frac{5}{6} \div 15$

Tell which of the following expressions are equivalent. Also, tell why.

61. $3t + 5$ and $3 \cdot 5 + t$
62. $4x$ and $x + 4$
63. $5m + 6$ and $6 + 5m$
64. $(x + y) + z$ and $z + (x + y)$
65. $bxy + bx$ and $yxb + bx$
66. $ab + bc$ and $ac + db$
67. $a + c + e + g$ and $ae + cg$
68. $abc \cdot de$ and $a \cdot b \cdot c \cdot ed$

Simplify.

69. $\dfrac{128}{192}$

70. $\dfrac{pqrs}{qrst}$

71. $\dfrac{33sba}{2(11a)}$

72. $\dfrac{4 \cdot 9 \cdot 16}{2 \cdot 8 \cdot 15}$

73. $\dfrac{36 \cdot (2rh)}{8 \cdot (9hg)}$

74. $\dfrac{3 \cdot (4xy) \cdot (5)}{2 \cdot (3x) \cdot (4y)}$

75. Is there a commutative law for division of whole numbers? If not, give an example to show that there is no such law.

1.5 EXPONENTIAL NOTATION

Algebraic expressions can be written using exponential notation. In this section we learn how to write exponential notation and to evaluate algebraic expressions involving exponential notation.

Exponential Notation

Shorthand notation for $10 \cdot 10 \cdot 10$ is called *exponential notation*.

For $\underbrace{10 \cdot 10 \cdot 10}_{3 \text{ factors}}$ we write 10^3.

This is read "ten cubed" or "ten to the third power." We call the number 3 an *exponent* and we say that 10 is the *base*. For $10 \cdot 10$ we write 10^2, read "ten squared," or "ten to the second power."

EXAMPLE 1 Write exponential notation for $10 \cdot 10 \cdot 10 \cdot 10 \cdot 10$.

$$10 \cdot 10 \cdot 10 \cdot 10 \cdot 10 = 10^5$$

An exponent of 2 or greater tells us how many times the base is used as a factor.

1.5 EXPONENTIAL NOTATION

EXAMPLE 2 What is the meaning of 3^5? n^4? $(2n)^3$? $(50x)^2$?

3^5 means $3 \cdot 3 \cdot 3 \cdot 3 \cdot 3$
n^4 means $n \cdot n \cdot n \cdot n$
$(2n)^3$ means $2n \cdot 2n \cdot 2n$
$(50x)^2$ means $50x \cdot 50x$

1 and 0 as Exponents

Look for a pattern.

$$10 \cdot 10 \cdot 10 \cdot 10 = 10^4$$
$$10 \cdot 10 \cdot 10 = 10^3$$
$$10 \cdot 10 = 10^2$$
$$10 = 10^?$$
$$1 = 10^?$$

We are dividing by 10 each time. The exponents decrease by 1 each time. To continue the pattern we would say that

$$10 = 10^1 \quad \text{and} \quad 1 = 10^0.$$

We read exponential notation as follows:

b^n is read the *nth power of b*, or simply *b to the nth*, or *b to the n*.
We may also read b^2 *b-squared*.
b^3 may also be read *b-cubed*.

We now summarize our definition of exponential notation.

---

b^0 means 1, for any number b, except 0.
b^1 means b, for any number b.
If n is a whole number greater than 1,

$$b^n \text{ means } \underbrace{b \cdot b \cdot b \cdots b}_{n \text{ factors}}.$$

EXAMPLE 3 Write exponential notation for each of the following.

$7 \cdot 7 \cdot 7 \cdot 7$, $7 \cdot 7 \cdot 7 \cdot 7$ means 7^4,
$2y \cdot 2y \cdot 2y$, $2y \cdot 2y \cdot 2y$ means $(2y)^3$,
$n \cdot n \cdot n \cdot n \cdot n \cdot n$, $n \cdot n \cdot n \cdot n \cdot n \cdot n$ means n^6,
$10x \cdot 10x$, $10x \cdot 10x$ means $(10x)^2$.

Evaluating Expressions

We now consider evaluating expressions containing exponential notation.

EXAMPLE 4   Evaluate x^4 for $x = 2$.

$$x^4 = 2^4 \qquad \text{Substituting}$$
$$= 2 \cdot 2 \cdot 2 \cdot 2$$
$$= 16$$

EXAMPLE 5   The area of a circle is given by $A = \pi r^2$, where r is the radius. Find the area of a circle with a radius of 10 cm. Use 3.14 for π.

$$A = \pi r^2 \approx 3.14 \times (10 \text{ cm})^2$$
$$= 3.14 \times 100 \text{ cm}^2$$
$$= 314 \text{ cm}^2$$

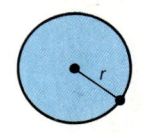

Here "cm^2" means "square centimeters," and "$\approx$" means "approximately equal."

EXAMPLE 6   Evaluate $m^3 + 5$ for $m = 4$.

We agree to evaluate $m^3 + 5$ by evaluating m^3 first and then adding 5:

$$m^3 + 5 = 4^3 + 5 \qquad \text{Substituting}$$
$$= (4 \cdot 4 \cdot 4) + 5$$
$$= 64 + 5$$
$$= 69$$

EXERCISE SET 1.5

What is the meaning of each expression?

1. 2^4
2. 5^3
3. $(1.4)^5$
4. $(2.5)^4$
5. n^5
6. m^6
7. $(7p)^2$
8. $(11c)^3$
9. $(19k)^4$
10. $(104d)^5$
11. $(10pq)^3$
12. $(24ct)^3$

Write exponential notation for each of the following.

13. $10 \times 10 \times 10 \times 10 \times 10 \times 10$
14. $6 \times 6 \times 6 \times 6$
15. $x \cdot x \cdot x \cdot x \cdot x \cdot x \cdot x$
16. $y \cdot y \cdot y$
17. $3y \cdot 3y \cdot 3y \cdot 3y$
18. $5m \cdot 5m \cdot 5m \cdot 5m \cdot 5m$

1.5 EXPONENTIAL NOTATION

Evaluate each expression.

19. m^3 for $m = 3$
20. x^6 for $x = 2$
21. p^1 for $p = 19$
22. x^{19} for $x = 0$
23. x^4 for $x = 4$
24. y^{15} for $y = 1$
25. n^0 for $n = 5$
26. $y^2 - 7$ for $y = 10$
27. $z^5 + 5$ for $z = 2$
28. Find the area of a circle when $r = 34$ ft. Use 3.14 for π.
29. The area of a square with sides of length s is given by $A = s^2$. Find the area of a square with sides of length 24 m (meters).

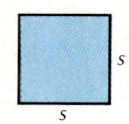

Write each of the following with a single exponent. For example,
$$\frac{3^5}{3^3} = \frac{3 \cdot 3 \cdot 3 \cdot 3 \cdot 3}{3 \cdot 3 \cdot 3} = 3 \cdot 3 = 3^2.$$

30. $\dfrac{10^5}{10^3}$
31. $\dfrac{10^7}{10^2}$
32. $\dfrac{5^4}{5^2}$
33. $\dfrac{2^6}{8^2}$

34. Evaluate $x^3 y^2 + zx$ for $x = 2$, $y = 1$, and $z = 3$.
35. Evaluate $c^2 a^3 + ba$ for $a = 3$, $b = 1$, and $c = 2$.
36. Evaluate $x^2 + 2xy + y^2$ for $x = 7$ and $y = 8$.

$(3n)^3$ means $3n \cdot 3n \cdot 3n$; $3n^3$ means $3 \cdot n \cdot n \cdot n$. Evaluate $(3n)^3$ and $3 \cdot n^3$ for each of the following.

37. When $n = 2$
38. When $n = 3$
39. When $n = 4$
40. When $n = 5$

Evaluate $(5p)^2$ and $5 \cdot p^2$ for each of the following.

41. When $p = 4$
42. When $p = 7$
43. When $p = 11$
44. When $p = 26$

Write exponential notation for each of the following.

45. $x \cdot x \cdot x \cdot y \cdot y \cdot y$
46. $3a \cdot 3a \cdot 3a \cdot 2b \cdot 2b$
47. Find $x^{149} y$ for $x = 13$ and $y = 0$.
48. Find $x^{410} y^2$ for $x = 1$ and $y = 3$.
49. 10^{127} is one followed by how many zeros?
50. Find $(x^2)^2$ if $x = 3$.
51. What powers of 6 are too large for your calculator readout?

Which of the following pairs of expressions are equivalent?

52. $\dfrac{5n}{n}$, $5n^0$ ($n \neq 0$)
53. n^3, $\dfrac{n \cdot n \cdot n \cdot n}{n}$ ($n \neq 0$)
54. $\dfrac{4^5}{2}$, 2^5
55. 2^2, $2 \cdot 2 \cdot 2^0$
56. xy^2, $x \cdot x \cdot y \cdot y$
57. $4x$, $x \cdot x \cdot x \cdot x$

1.6 THE ASSOCIATIVE LAWS

In this section we study two more number properties: the associative laws. We continue using number properties to manipulate algebraic expressions.

Parentheses

What does $5 \times 2 + 4$ mean? If we multiply 5 by 2 and add 4, we get 14. If we add 2 and 4 and multiply by 5, we get 30. To tell which operation to do first, we use parentheses. For example,

$$(3 \times 5) + 6 \quad \text{means} \quad 15 + 6, \text{ or } 21;$$

and

$$3 \times (5 + 6) \quad \text{means} \quad 3 \times 11, \text{ or } 33.$$

Parentheses tell us what to do first. If there are no parentheses, we multiply or divide first working from left to right, and then add or subtract working from left to right.

EXAMPLE 1 Calculate: $15 - 2 \times 5 + 3$.

$$15 - 2 \times 5 + 3$$
$$15 - 10 + 3 \quad \text{Multiplying}$$
$$5 + 3 \quad \text{Subtracting and adding, from left to right}$$
$$8$$

Always calculate within parentheses first. When there are exponents and no parentheses, simplify powers before multiplying or dividing.

EXAMPLE 2 Calculate: $(3 \times 4)^2$.

$$(3 \times 4)^2 = 12^2 \quad \text{Working within parentheses first}$$
$$= 144$$

EXAMPLE 3 Calculate: $3 \times 4^2 - 29$.

$$3 \times 4^2 - 29 = 3 \times 16 - 29 \quad \text{There are no parentheses, so find } 4^2 \text{ first}$$
$$= 48 - 29 \quad \text{Multiplying second}$$
$$= 19 \quad \text{Subtracting}$$

1.6 THE ASSOCIATIVE LAWS

EXAMPLE 4 Evaluate $(3x)^3 - 2$ for $x = 2$.

$$(3x)^3 - 2 = (3 \cdot 2)^3 - 2 \quad \text{Substituting}$$
$$= 6^3 - 2 \quad \text{Multiplying within parentheses first}$$
$$= 216 - 2$$
$$= 214$$

EXAMPLE 5 Calculate $(2 + x) \cdot (y - 1)$ for $x = 3$ and $y = 5$.

$$(2 + x) \cdot (y - 1) = (2 + 3) \cdot (5 - 1)$$
$$= 5 \cdot 4 \quad \text{Working within parentheses first}$$
$$= 20$$

Using the Associative Laws

EXAMPLE 6 Calculate and compare: $3 + (8 + 5)$ and $(3 + 8) + 5$.

$$3 + (8 + 5) = 3 + 13 \quad \text{Calculating within parentheses first}$$
$$= 16$$

$$(3 + 8) + 5 = 11 + 5 \quad \text{Calculating within parentheses first}$$
$$= 16$$

The two expressions are equivalent. Moving the parentheses did not affect the expression.

EXAMPLE 7 Calculate and compare: $3 \cdot (4 \cdot 2)$ and $(3 \cdot 4) \cdot 2$.

$$3 \cdot (4 \cdot 2) = 3 \cdot 8 \qquad (3 \cdot 4) \cdot 2 = 12 \cdot 2$$
$$= 24 \qquad\qquad\qquad = 24$$

When only addition is involved, parentheses can be placed any way we please without affecting the answer. When only multiplication is involved, parentheses can be placed any way we please without affecting the answer.

- - -

ASSOCIATIVE LAWS

Addition For any numbers a, b, and c,
$$a + (b + c) = (a + b) + c.$$
(Numbers can be grouped in any manner for addition.)

Multiplication For any numbers a, b, and c,
$$a \cdot (b \cdot c) = (a \cdot b) \cdot c.$$
(Numbers can be grouped in any manner for multiplication.)

EXAMPLE 8 Use an associative law to write an expression equivalent to $y + (z + 3)$.

An equivalent expression is
$$(y + z) + 3$$
by the associative law of addition.

EXAMPLE 9 Use an associative law to write an expression equivalent to $8 \cdot (x \cdot y)$.

An equivalent expression is
$$(8 \cdot x) \cdot y$$
by the associative law of multiplication.

When only additions or only multiplications are involved, parentheses may be placed any way we please. So we often omit them. For example,

$$x + (y + 7) \quad \text{means} \quad x + y + 7, \quad \text{and} \quad l(wh) \quad \text{means} \quad lwh.$$

Using the Laws Together

If addition or multiplication is the only operation in an expression, then the associative and commutative laws allow us to group and change order as we please. For instance, in a calculation like $(5 + 2) + (3 + 5) + 8$, addition is the only operation. So we can change grouping and order to make easy combinations: $5 + 5 + 2 + 8 + 3 = 10 + 10 + 3 = 23$.

EXAMPLE 10 Use the commutative and associative laws to write at least three expressions equivalent to $(x + 5) + y$.

a) $(x + 5) + y = x + (5 + y)$ Using the associative law first and then using the
    $ = x + (y + 5)$ commutative law

b) $(x + 5) + y = y + (x + 5)$ Using the commutative law and then the
    $ = y + (5 + x)$ commutative law again

c) $(x + 5) + y = 5 + (x + y)$ Using the commutative law first and then the
    associative law

EXAMPLE 11 Use the commutative and associative laws to write at least three expressions equivalent to $(3 \cdot x) \cdot y$.

a) $(3 \cdot x) \cdot y = 3 \cdot (x \cdot y)$ Using the associative law first and then the
    $ = 3 \cdot (y \cdot x)$ commutative law

b) $(3 \cdot x) \cdot y = y \cdot (x \cdot 3)$ Using the commutative law twice

c) $(3 \cdot x) \cdot y = x \cdot (y \cdot 3)$ Using the commutative law, then the associative,
    and then the commutative law again

EXERCISE SET 1.6

Calculate.

1. $7 + 2 \times 6$
2. $11 + 4 \times 4$
3. $8 \times 7 + 6 \times 5$
4. $10 \times 5 + 1 \times 1$
5. $19 - 5 \times 3 + 3$
6. $14 - 2 \times 6 + 7$
7. $9 \div 3 + 16 \div 8$
8. $32 - 8 \div 4 - 2$
9. $7 + 10 - 10 \div 2$
10. $(5 \cdot 4)^2$
11. $(6 \cdot 3)^2$
12. $3 \cdot 2^3$
13. $4 \cdot 5^2$
14. $(8 + 2)^2$
15. $(5 + 3)^3$
16. $7 + 2^2$
17. $6 + 4^2$
18. $(5 - 2)^2$
19. $(3 - 2)^2$
20. $10 - 3^2$
21. $12 - 2^3$
22. $20 + 4^3 \div 8$
23. $2 \times 10^3 - 500$
24. $7 \times 3^4 + 18$
25. $80 - 6^2 \div 9$

Evaluate each expression.

26. $3 \cdot (a + 10)$ for $a = 12$
27. $b \cdot (7 + b)$ for $b = 5$
28. $(t + 3)^3$ for $t = 4$
29. $(12 - w)^3$ for $w = 7$
30. $(x + 5) \cdot (12 - x)$ for $x = 7$
31. $(y - 4) \cdot (y + 6)$ for $y = 10$
32. $(5y)^3 - 75$ for $y = 2$
33. $(7x)^2 + 59$ for $x = 3$
34. $\dfrac{y + 3}{2y}$ for $y = 5$
35. $\dfrac{(4x) + 2}{2x}$ for $x = 5$
36. $\dfrac{w^2 + 4}{5w}$ for $w = 4$
37. $\dfrac{b^2 + b}{2b}$ for $b = 5$
38. $(x - 4) \cdot (8 + y)$ for $x = 12$ and $y = 2$
39. $(y + 6) \cdot (9 - x)$ for $x = 7$ and $y = 10$

Use the associative laws to write an expression equivalent to each of the following.

40. $(a + b) + 3$
41. $(5 + x) + y$
42. $3 \cdot (a \cdot b)$
43. $(6 \cdot x) \cdot y$

Use the commutative and associative laws to write three equivalent expressions.

44. $(a + b) + 2$
45. $(3 + x) + y$
46. $5 + (v + w)$
47. $6 + (x + y)$
48. $(x \cdot y) \cdot 3$
49. $(a \cdot b) \cdot 5$
50. $7 \cdot (a \cdot b)$
51. $5 \cdot (x \cdot y)$
52. $2 \cdot c \cdot d$

○ ─────────────────────

Find a value of the variable that shows that the two expressions are *not* equivalent.

53. $3x^2$; $(3x)^2$
54. $(a + 2)^3$; $a^3 + 2^3$
55. $\dfrac{x + 2}{2}$; x
56. $\dfrac{y^6}{y^3}$; y^2

Write an algebraic expression for each of the following.

57. A number squared plus 7
58. A number plus the square of 7
59. The square of a sum of 7 and some number
60. A number squared plus 7 squared

61. The numerator is 3 more than some number and the denominator is the square of the numerator.

62. Two numbers are multiplied. One of them is 5 more than the other.

63. Carole is twice as old as Victor was a year ago. Victor's age is now x. Write an expression for Carole's age.

64. Is there an associative law for subtraction of whole numbers? If not, give counterexamples.

1.7 THE DISTRIBUTIVE LAW

The distributive law is the basis of many procedures in both arithmetic and algebra. It is probably the most important law that we use to manipulate algebraic expressions. It involves two operations: addition and multiplication.

Multiplying a Sum by a Factor

If we wish to multiply a sum of several numbers by a factor, we can either add and then multiply, or multiply and then add.

EXAMPLE 1 Compute two ways: $5 \cdot (4 + 8)$.

a) $5 \cdot (4 + 8)$ Adding within parentheses first, and then multiplying
  $5 \cdot 12$
  60

b) $(5 \cdot 4) + (5 \cdot 8)$ Distributing the multiplication to terms within parentheses first and then adding
  $20 + 40$
  60

THE DISTRIBUTIVE LAW

For any numbers a, b, and c,

$$a(b + c) = ab + ac.$$

In the statement of the distributive law we know that in an expression such as $ab + ac$, the multiplications are to be done first. We can also omit the multiplication dot before or between parentheses. So, instead of writing $(4 \cdot 5) + (4 \cdot 7)$ we can write $4 \cdot 5 + 4 \cdot 7$. Instead of $3 \cdot (x + 2)$ we can write $3(x + 2)$. However, in $a(b + c)$ we cannot omit the parentheses. If we did we would have $ab + c$ which means $(ab) + c$. For example, $3(4 + 2) = 18$, but $3 \cdot 4 + 2 = 14$.

1.7 THE DISTRIBUTIVE LAW

Note that the distributive law can be extended to more than two numbers inside the parentheses:
$$a(b + c + d) = ab + ac + ad.$$

The distributive law would apply to the following situation. Someone decides to invest $1000 in one bank at 8% and $2000 in another bank at 8%. At the end of one year the total interest from the two investments would be
$$(8\% \cdot 1000) + (8\% \cdot 2000).$$

The same interest would also have been made by investing the entire $3000 in just one bank. The interest is
$$8\% \cdot (1000 + 2000), \quad \text{or} \quad 8\% \cdot 3000.$$

In the expression $x + y + z$, the parts separated by plus signs are called *terms*. Thus x, y, and z are terms in $x + y + z$. The distributive law is the basis of an algebraic procedure called "multiplying." For example, consider
$$8(a + b).$$

Using the distributive law we multiply each term of $(a + b)$ by 8:
$$8 \cdot (a + b) = 8 \cdot a + 8 \cdot b.$$

EXAMPLE 2 Multiply: $3(x + 2)$.

$$3(x + 2) = 3x + 3 \cdot 2 \quad \text{Using the distributive law}$$
$$= 3x + 6$$

EXAMPLE 3 Multiply: $6(s + 2t + 5w)$.

$$6(s + 2t + 5w) = 6s + 6 \cdot 2t + 6 \cdot 5w \quad \text{Using the distributive law}$$
$$= 6s + 12t + 30w$$

Factoring

If we reverse the statement of the distributive law, we have the basis of a process called *factoring*: $ab + ac = a(b + c)$. To *factor* an expression means to write an equivalent expression which is a product.

EXAMPLE 4 Factor: $3x + 3y$.

By the distributive law,
$$3x + 3y = 3(x + y).$$

When we write $3(x + y)$, we say we have *factored* $3x + 3y$. That is, we have written an equivalent expression which is a product. It is important to realize that, when we factor an expression like $3x + 3y$, the factored expression is equivalent to the original one.

EXAMPLE 5 Factor: $5x + 5y + 5z$.

$$5x + 5y + 5z = 5(x + y + z)$$

In the expression $6x + 3y + 9z$ the terms are $6x$, $3y$, and $9z$. In this case, the terms are products. To factor, look for a factor common to all the terms. Then "remove" it, so to speak, using the distributive law.

EXAMPLE 6 Factor: $6x + 3y + 9z$.

$$\begin{aligned}6x + 3y + 9z &= 3 \cdot 2x + 3 \cdot y + 3 \cdot 3z \quad &\text{The common factor is 3.}\\ &= 3(2x + y + 3z) \quad &\text{Using the distributive law}\end{aligned}$$

EXAMPLE 7 Factor: $7y + 21z + 7$.

$$\begin{aligned}7y + 21z + 7 &= 7 \cdot y + 7 \cdot 3z + 7 \cdot 1 \quad &\text{The common factor is 7.}\\ &= 7(y + 3z + 1)\end{aligned}$$

Be sure not to omit the 1 or the common factor 7.

Factoring can be checked by multiplying. We multiply the factored expression to see if we get the original expression.

EXAMPLE 8 Factor $5x + 10$ and check by multiplying.

$$5x + 10 = 5(x + 2)$$

Check: $\quad 5(x + 2) = 5x + 5 \cdot 2$
$\quad\quad\quad\quad\quad\quad = 5x + 10$

Collecting Like Terms

Terms such as $5x$ and $4x$, whose variable factors are exactly the same, are called *like terms*. Similarly $3y^2$ and $9y^2$ are like terms because the variables are the same and they are raised to the same power. Terms such as $4y$ and $5y^2$ are not like terms. We often simplify expressions by using the distributive law to *collect* or *combine like terms*.

EXAMPLE 9 Collect like terms: $x + x$.

$$\begin{aligned}x + x &= 1 \cdot x + 1 \cdot x \quad &\text{Using the property of 1}\\ &= (1 + 1)x \quad &\text{Using the distributive law}\\ &= 2x\end{aligned}$$

Note that we have applied the distributive law "on the right." This works because of the commutative law of multiplication:

$$ba + ca = ab + ac = a(b + c) = (b + c)a.$$

1.7 THE DISTRIBUTIVE LAW

EXAMPLE 10 Collect like terms: $3x + 4x$.

$$3x + 4x = (3 + 4)x \qquad \text{Using the distributive law}$$
$$= 7x$$

EXAMPLE 11 Collect like terms: $2x + 3y + 5x + 8y$.

$$2x + 3y + 5x + 8y = 2x + 5x + 3y + 8y \qquad \text{Regrouping and reordering using the associative and commutative laws}$$
$$= (2 + 5)x + (3 + 8)y \qquad \text{Factoring}$$
$$= 7x + 11y$$

EXAMPLE 12 Collect like terms: $5x^2 + x^2$.

$$5x^2 + x^2 = 5x^2 + 1 \cdot x^2 \qquad \text{Using the property of 1}$$
$$= (5 + 1)x^2$$
$$= 6x^2$$

EXAMPLE 13 Collect like terms: $x + 0.08x$.

$$x + 0.08x = 1 \cdot x + 0.08x = (1 + 0.08)x = 1.08x$$

With practice we can leave out some steps, collecting like terms mentally. Note that the numbers, *constants*, like 4 and 7 in $4 + x + 7$ are considered like terms.

EXAMPLES Collect like terms.

14. $5y + 2y + 4y = 11y$
15. $3x + 7x + 2y = 10x + 2y$
16. $3x^2 + 7x^2 + 2y = 10x^2 + 2y$
17. $8p + q + p + 0.3q = 9p + 1.3q$
18. $4 + x + 7 = x + 11$
19. $3x + 25 + 7y + 8x + 11 = 11x + 7y + 36$

EXERCISE SET 1.7

Multiply.

1. $2(b + 5)$
2. $4(x + 3)$
3. $7(1 + t)$
4. $6(v + 4)$
5. $3(x + 1)$
6. $7(x + 8)$
7. $4(1 + y)$
8. $9(s + 1)$
9. $6(5x + 2)$
10. $9(6m + 7)$
11. $7(x + 4 + 6y)$
12. $4(5x + 8 + 3p)$

Factor.

13. $2x + 4$
14. $5y + 20$
15. $30 + 5y$
16. $7x + 28$
17. $14x + 21y$
18. $18a + 24b$
19. $5x + 10 + 15y$
20. $9a + 27b + 81$

Factor and check by multiplying.

21. $9x + 27$
22. $6x + 24$
23. $9x + 3y$
24. $15x + 5y$
25. $8a + 16b + 64$
26. $5 + 20x + 35y$
27. $11x + 44y + 121$
28. $7 + 14b + 56w$

Collect like terms.

29. $9a + 10a$
30. $12x + 2x$
31. $10a + a$
32. $16x + x$
33. $2x + 9z + 6x$
34. $3a + 5b + 7a$
35. $7x + 6y^2 + 9y^2$
36. $12m^2 + 6q + 9m^2$
37. $41a + 90 + 60a + 2$
38. $42x + 6 + 4x + 2$
39. $8a + 8b + 3a + 3b$
40. $100y + 200z + 190y + 400z$
41. $8u^2 + 3t + 10t + 6u^2 + 2$
42. $5 + 6h + t + 8 + 9h$
43. $23 + 5t + 7y + t + y + 27$
44. $45 + 90d + 87 + 9d + 3 + 7d$
45. $\frac{1}{2}b + \frac{1}{2}b$
46. $\frac{2}{3}x + \frac{1}{3}x$
47. $2y + \frac{1}{4}y + y$
48. $\frac{1}{2}a + a + 5a$

---

49. When you put money in the bank and draw simple interest, the amount in your account later on is given by the expression $P + Prt$, where P is the principal, r is the rate of interest, and t is the time. Factor the expression.

50. Solve.
  a) Factor $17x + 34$. Then evaluate both expressions when $x = 10$ (the value of x is 10).
  b) Will you get the same answer for both expressions no matter what the value of x? Why?

51. Find a simpler expression that always has the same value as

$$\frac{3a + 6}{2a + 4}.$$

(*Hint:* Factor numerator and denominator and factor the fractional expression.)

52. Find a simpler expression that always has the same value as

$$\frac{4x + 12y}{3x + 9y}.$$

Collect like terms if possible and factor the result.

53. $1x + 2x^2 + 3x^3 + 4x^2 + 5x$
54. $q + qr + qrs + qrst$
55. $21x + 44xy + 15y - 16x - 8y - 38xy + 2x + xy$
56. Expand:

$$a\{1 + b[1 + c(1 + d)]\}.$$

(*Hint:* Begin with $c(1 + d)$ and work outward.)

1.8 EQUATIONS AND TRANSLATION

You have had practice in translating to algebraic expressions. In this section, you will extend that skill. You will learn to translate a problem situation to a mathematical sentence called an *equation*. The skills you have already learned will be most useful, and when you learn this extension of those skills, you will be in a stronger position to attack problem solving.

First, we define what we mean by an equation. An *equation* is a number sentence with an equals sign, =, for its verb. Here are some examples:

$$3 + 2 = 5, \quad 7 - 3 = 4, \quad x + 6 = 13, \quad 3x - 2 = 7 - x.$$

It often happens that an equation will have some algebraic expression with variables on one or both sides.

Translating to Equations

We shall look at some examples of problem situations and how they can be translated to an equation.

EXAMPLE 1 Translate the following problem to an equation.

What number plus 478 is 1019?

We shall use a variable to represent "what number." In this example, the translation comes almost directly from the English sentence.

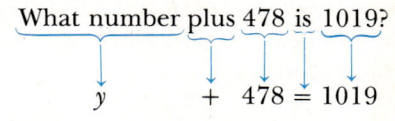

$$\underbrace{\text{What number}}_{y} \underbrace{\text{plus}}_{+} \underbrace{478}_{478} \underbrace{\text{is}}_{=} \underbrace{1019?}_{1019}$$

Note that "is" translates to "=" and "plus" translates to "+".

Sometimes it helps to reword a problem before translating.

EXAMPLE 2 When 54 is multiplied by a number, the result is 7896. Find the number.

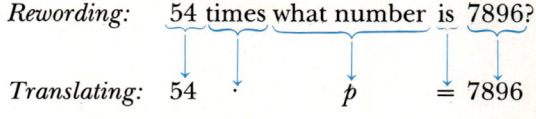

Rewording: 54 times what number is 7896?

Translating: $54 \cdot p = 7896$

Note that "times" translates to "·".

Solutions of Equations

An equation says that the symbols on either side of the equals sign, =, stand for the same number. Equations may be true. They may be false. They may be neither true nor false.

EXAMPLES Determine whether these equations are true, false, or neither.

3. $3 + 2 = 5$ The equation is *true*.
4. $7 - 2 = 4$ The equation is *false*.
5. $x + 6 = 13$ The equation is *neither* true nor false, because we do not know what number x represents.

> Any replacement for the variable that makes an equation true is called a *solution* of the equation. To solve an equation means to find *all* of its solutions.

One way to solve equations is by trial.

EXAMPLE 6 Solve $x + 6 = 13$ by trial.

If we replace x by 2 we get $2 + 6 = 13$, a false equation.
If we replace x by 8 we get $8 + 6 = 13$, a false equation.
If we replace x by 7 we get $7 + 6 = 13$, a true equation.

No other number makes the equation true, so the only solution is the number 7.

The Third Step in Problem Solving: To Carry Out

The first step in problem solving is to *familiarize* yourself with the situation. The second step is to *translate* to mathematical language. The third step is to *carry out* some mathematical manipulation. If you have translated to an equation, that step will consist of solving the equation.

> **THE THIRD STEP IN PROBLEM SOLVING WITH ALGEBRA**
>
> Carry out some mathematical manipulation. If you have translated to an equation, this means to solve the equation.

Solving equations is an important part of problem solving. Equations can sometimes be complex and not too easy to solve. For now, let's see how to solve some of the simpler ones.

Solving Equations of the Type $x + a = b$

Think of a very simple equation such as $x + 2 = 5$. We have added 2 to x to get 5. Since addition and subtraction are opposite operations, we can subtract 2 to "undo" the addition. The equation says that $x + 2$ and 5 represent the same number. To

1.8 EQUATIONS AND TRANSLATION 39

"undo" the addition we subtract on both sides:
$$x + 2 - 2 = 5 - 2$$
$$x = 3.$$

The solution of the equation $x = 3$ is obvious. It is 3. It is also the solution of the original equation $x + 2 = 5$. We can check this by substituting 3 for x in the original equation. Now let us use this method to solve some equations that are not as simple.

• •

To solve an equation of the type $x + a = b$, subtract a on both sides of the equation.

EXAMPLE 7 Solve: $x + 17 = 50$.

We subtract 17 on both sides of the equation:
$$x + 17 = 50$$
$$x + 17 - 17 = 50 - 17 \qquad \text{Subtracting 17 on both sides}$$
$$x = 33.$$

We now check to see if 33 is really a solution. We substitute 33 into the original equation.

Check:
$x + 17 = 50$	Writing the original equation
$33 + 17$ | 50	Substituting 33 for x
50	Calculating

Since the left-hand and right-hand sides match, 33 checks. It is the solution.

EXAMPLE 8 Solve: $x + 1.6 = 9.8$.

$$x + 1.6 = 9.8$$
$$x = 9.8 - 1.6 \qquad \text{Subtracting 1.6 on both sides}$$
$$x = 8.2$$

Check:
$x + 1.6 = 9.8$
$8.2 + 1.6$ | 9.8
9.8

The solution is 8.2.

EXAMPLE 9 Solve: $x + \frac{3}{4} = \frac{5}{6}$.

$$x = \frac{5}{6} - \frac{3}{4} \qquad \text{Subtracting } \tfrac{3}{4}$$
$$x = \frac{5}{6} \cdot \frac{2}{2} - \frac{3}{4} \cdot \frac{3}{3} \qquad \text{Multiplying by 1 to get the least common denominator, 12}$$
$$x = \frac{10}{12} - \frac{9}{12}$$
$$x = \frac{1}{12}$$

Check:

$$\begin{array}{c|c} x + \frac{3}{4} = \frac{5}{6} \\ \hline \frac{1}{12} + \frac{3}{4} & \frac{5}{6} \\ \frac{1}{12} + \frac{9}{12} & \\ \frac{10}{12} & \\ \frac{5}{6} & \end{array}$$

The solution is $\frac{1}{12}$.

Solving Equations of the Type $ax = b$

Think of a very simple equation such as $2x = 8$. We have multiplied x by 2 to get 8. Since multiplication and division are opposite operations, we can divide by 2 to "undo" the multiplication. The equation says that $2x$ and 8 represent the same number. To "undo" the multiplication, we divide on both sides:

$$\frac{2x}{2} = \frac{8}{2}$$

$$\frac{2}{2}x = \frac{8}{2}$$

$$x = 4.$$

The solution of the equation $x = 4$ is obvious. It is 4. It is also the solution of the original equation $2x = 8$.

• •

> To solve an equation of the type $ax = b$, where a is not 0, divide by a on both sides of the equation.

EXAMPLE 10 Solve: $5x = 15$.

$$5x = 15$$

$$\frac{5x}{5} = \frac{15}{5} \qquad \text{Dividing by 5 on both sides}$$

$$x = \frac{15}{5}$$

$$x = 3$$

Check:

$$\begin{array}{c|c} 5x = 15 \\ \hline 5 \cdot 3 & 15 \\ 15 & \end{array} \qquad \text{Substituting 3 for } x$$

The solution is 3.

1.8 EQUATIONS AND TRANSLATION

EXAMPLE 11 Solve: $4.7y = 40.42$.

$$4.7y = 40.42$$
$$y = \frac{40.42}{4.7} \quad \text{Dividing by 4.7 on both sides}$$
$$y = 8.6$$

Check:
$$\begin{array}{c|c} 4.7y = 40.42 \\ \hline 4.7(8.6) & 40.42 \\ 40.42 & \end{array}$$

The solution is 8.6.

EXERCISE SET 1.8

Translate each problem to an equation. Do not solve.

1. What number added to 60 is 112?

2. Seven times what number is 2233?

3. When 42 is multiplied by a number, the result is 2352. Find the number.

4. When 345 is added to a number, the result is 987. Find the number.

5. A game board has 64 squares. If you win 35 squares, how many does your opponent get?

6. A consultant charges $80 an hour. How many hours did the consultant work to make $53,400?

Solve by trial.

7. $x + 17 = 32$ **8.** $y + 28 = 92$ **9.** $x - 7 = 12$ **10.** $y - 8 = 19$

11. $6x = 54$ **12.** $8y = 72$ **13.** $\frac{x}{6} = 5$ **14.** $\frac{y}{8} = 6$

15. $5x + 7 = 107$ **16.** $9x + 5 = 86$ **17.** $7x - 1 = 48$ **18.** $4y - 2 = 10$

Solve by subtracting on both sides. Be sure to check.

19. $x + 7 = 24$ **20.** $y + 26 = 43$ **21.** $x + 19 = 105$ **22.** $y + 37 = 212$

23. $x + 99 = 476$ **24.** $x + 112 = 1001$ **25.** $x + 5064 = 7882$ **26.** $x + 4112 = 8007$

27. $x + 2.78 = 8.44$ **28.** $x + 3.04 = 4.69$ **29.** $x + \frac{1}{7} = \frac{6}{7}$ **30.** $x + \frac{3}{13} = \frac{11}{13}$

Solve by dividing on both sides. Be sure to check.

31. $15x = 90$ **32.** $26z = 182$ **33.** $4x = 5$ **34.** $6y = 27$

35. $10x = 2.4$ **36.** $9y = 3.6$ **37.** $2.9y = 8.99$ **38.** $5.5y = 34.1$

39. $6.2y = 52.7$ **40.** $9.4x = 23.5$ **41.** $117t = 2106$ **42.** $193c = 4053$

○ ─────────────

Solve.

43. $5x + 3x = 10$ **44.** $9y + 4y = 26$

45. $225a = 27$ **46.** $1000x = 300$

47. $0.0592y = 0.4736$

48. $1.009x = 14.126$

49. $0.125n = 1$

50. $0.004t = 1$

51. Write an equation that has *no* whole number solution.

52. Write an equation for which *every* whole number is a solution.

Solve.

53. ▦ $x + 506{,}233 = 976{,}421$

54. ▦ $0.1265x = 1065.636$

55. $2(x + 1) = 12$

56. $3(y + 2) + 3 = 15$

1.9 PROBLEM SOLVING

The Five-step Process for Problem Solving

We have discussed three steps in problem solving with algebra, the most recent being to carry out some mathematical manipulation. Often that means to solve an equation. There are two more steps in the problem-solving process. The next one is to *check* your possible answer by going to the conditions of the original problem. That way you will know whether you have an answer to the problem itself. The last step is to *state* clearly the answer to the problem. We list the five steps below. You should learn them well, and remember to apply them as you work problems in algebra.

> **THE FIVE STEPS IN PROBLEM SOLVING IN ALGEBRA**
>
> 1. *Familiarize* yourself with the problem situation.
> 2. *Translate* to mathematical language.
> 3. *Carry out* some mathematical manipulation.
> 4. *Check* your possible answer in the original problem.
> 5. *State* the answer clearly.

Problem Solving

In this chapter we have reviewed some arithmetic skills, introduced some algebraic tools, and presented a five-step process for problem solving. Now we solve some problems using these skills. The problems in this section are simple ones. You may be able to solve some of them without using equations, but it is better if you do not. You are learning how algebra works. As we proceed through the book, you will continue to learn methods of algebra that will allow you to solve harder problems.

1.9 PROBLEM SOLVING 43

EXAMPLE 1 What number plus 478.6 is 1019.2?

1. **Familiarize** yourself with the situation. We identify all the pertinent information. There are two numbers in the problem, 478.6 and 1019.2. We want to find what number added to 478.6 gives 1019.2. We can try some possibilities:

$$478.6 + 500 = 978.6, \quad 478.6 + 600 = 1078.6, \quad 478.6 + 580.2 = 1058.8.$$

You might find the answer this way, but let us continue with the process.

2. **Translate** the problem to mathematical language. We translate as follows:

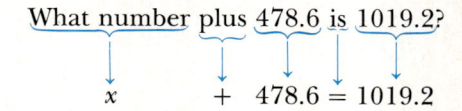

What number plus 478.6 is 1019.2?
$$x + 478.6 = 1019.2$$

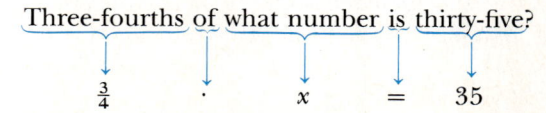

3. **Carry out** some mathematical manipulation. The translation gives us the equation

$$x + 478.6 = 1019.2.$$

To carry out some mathematical manipulation, we solve the equation.

$$x = 1019.2 - 478.6 \quad \text{Subtracting 478.6}$$
$$x = 540.6$$

4. **Check** the answer in the original problem. To do this we add 540.6 to 478.6:

$$540.6 + 478.6 = 1019.2.$$

We see that 540.6 checks in the original problem.

5. **State** the answer clearly. The answer is 540.6.

EXAMPLE 2 Three-fourths of what number is thirty-five?

1. **Familiarize** yourself with the situation. We identify all the pertinent information. We sketch and compare the "three-fourths" and "thirty-five." We try to draw a picture of the situation:

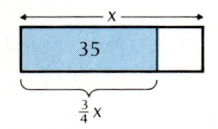

[diagram: a rectangle labeled with total length x, containing 35 in the left portion, with $\frac{3}{4}x$ indicated below]

We have let x represent the unknown number.

2. **Translate** the problem to mathematical language.

Three-fourths of what number is thirty-five?
$$\tfrac{3}{4} \cdot x = 35$$

3. **Carry out** some mathematical manipulation. The translation gives us the equation

$$\tfrac{3}{4}x = 35.$$

We solve the equation:

$$x = \frac{35}{\tfrac{3}{4}} \qquad \text{Dividing by } \tfrac{3}{4}$$

$$x = 35 \cdot \tfrac{4}{3} = \tfrac{140}{3}.$$

4. **Check** the answer in the original problem. To check, we find out if $\tfrac{3}{4}$ of this number is 35:

$$\frac{3}{4} \cdot \frac{140}{3} = \frac{3 \cdot 140}{4 \cdot 3}$$

$$= \frac{3 \cdot 35 \cdot 4}{4 \cdot 3}$$

$$= 35.$$

5. **State** the answer clearly. The number $\tfrac{140}{3}$ is the answer.

> Recall that in translating, *is* translates to =. The word *of* translates to ·, and the unknown number translates to a variable.

EXAMPLE 3 Hank Aaron hit 755 home runs in his career. Babe Ruth hit 714 in his career. How many more home runs did Aaron hit than Ruth?

Aaron hit 755 home runs

Ruth hit 714 home runs

1. **Familiarize** The pertinent information is that Aaron hit 755 home runs and Ruth 714. We can draw a picture to show this. What we are trying to find out is how many more home runs Aaron hit than Ruth. We are asking what we should add to 714 to get 755.

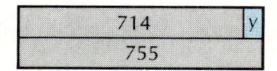

714	y
755	

2. **Translate** Sometimes it helps to reword a problem before translating.

   Rewording: Ruth's home runs plus how many is Aaron's home runs

   Translating: 714 + y = 755

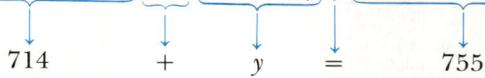

3. **Carry out** We solve the equation as follows:

   $$y = 755 - 714 \quad \text{Subtracting 714}$$
   $$y = 41.$$

4. **Check** We check by adding 41 to 714: $714 + 41 = 755$.
5. **State** Aaron hit 41 more home runs than Ruth.

EXAMPLE 4 Which of the positions of these ladders appears to be safest?

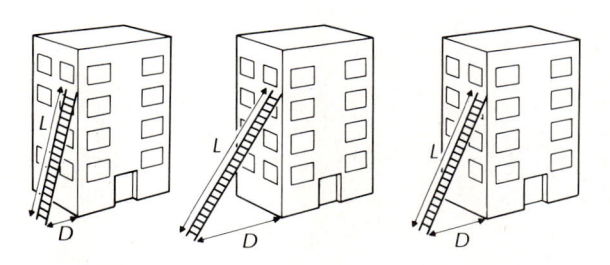

Actually it has been determined that the safest position of a ladder is when it satisfies the condition

$$L = 4D,$$

where L is the distance along the ladder to where it rests against a wall, or other object, and D is the distance of the bottom of the ladder out from the wall.

Suppose the distance L is 30 ft. How far should the bottom of the ladder be from the wall to be safest?

1. **Familiarize** We can familiarize ourselves with the problem situation by drawing a picture.

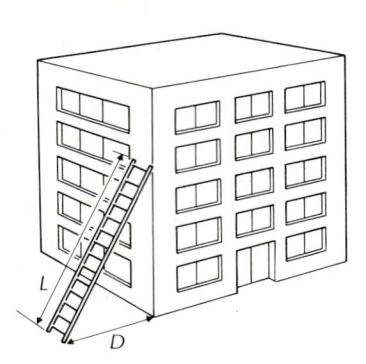

We can also make some computations for various values of D. For $D = 5$ ft, we get $L = 4(5) = 20$ ft. For $D = 10$ ft, $L = 4(10) = 40$ ft. Thus it seems reasonable that the answer to our problem is between 5 ft and 10 ft.

2. **Translate** We know that $L = 4D$. We substitute 30 for L and obtain the equation

$$30 = 4D.$$

3. **Carry out** We solve the equation:

$$30 = 4D$$

$$\frac{30}{4} = \frac{4D}{4} \quad \text{Dividing by 4 on both sides}$$

$$7.5 = D$$

4. **Check** We check by substituting 7.5 for D in the equation $L = 4D$:

$$L = 4(7.5) = 30.$$

5. **State** If the distance L is 30 ft, the distance D should be 7.5 ft.

EXERCISE SET 1.9

Solve these problems. Even though you might find the answer quickly some other way, practice using the five-step problem-solving process.

1. What number added to 60 gives 112?

2. What number added to 45.3 gives 53.1?

3. The result of adding 29 to a number is 171. Find the number.

4. The result of adding 123 to a number is 987. Find the number.

5. Seven times what number is 2233?

6. Four times what number is 8944?

7. When 42 is multiplied by a number the result is 2352. Find the number.

8. When 48 is multiplied by a number the result is 624. Find the number.

9. Two-thirds of what number is forty-eight?

10. One-eighth of what number is fifty-six?

11. A student missed a perfect quiz paper by 5 problems. There were 8 problems on the quiz. How many did the student get right?

12. A football player caught 3 passes for a total of 55 yards. The first two were for 23 and 8 yards. Find the third.

13. The New York Yankees won 37 more games than the Minnesota Twins. The Yankees won 101. How many did the Twins win?

14. A game board has 64 squares. If you win 35 squares and your opponent wins the rest, how many does your opponent get?

15. Cash register A contains $48 less than cash register B. If B has $115, how much does A have?

16. There are 352,198 people in a city. 187,804 are at least 28 years old. How many have not reached age 28?

17. A dozen bagels cost $3.12. How much is each bagel?

18. A movie theater took in $438.75 from 117 customers. All the tickets were the same price. What was the price of a ticket?

19. A consultant charges $80 an hour. How many hours did the consultant work to make $53,400?

20. The area of Lake Superior is about four times the area of Lake Ontario. The area of Lake Superior is 78,114 km^2. What is the area of Lake Ontario?

21. It takes a 60-watt bulb about 16.6 hours to use one kilowatt-hour of electricity. That is about 2.5 times as long as it takes a 150-watt bulb to use one kilowatt-hour. How long does it take a 150-watt bulb to use one kilowatt-hour?

22. The area of Alaska is about 483 times the area of Rhode Island. The area of Alaska is 1,519,202 km^2. What is the area of Rhode Island?

23. The boiling point of ethyl alcohol is 78.3°C. That is 13.5°C higher than the boiling point of methyl alcohol. What is the boiling point of methyl alcohol?

24. The height of the Eiffel Tower is 295 m. It is about 203 m higher than the Statue of Liberty. What is the height of the Statue of Liberty?

25. The distance from the earth to the sun is about 150,000,000 km. That is about 391 times the distance from the earth to the moon. What is the distance from the earth to the moon?

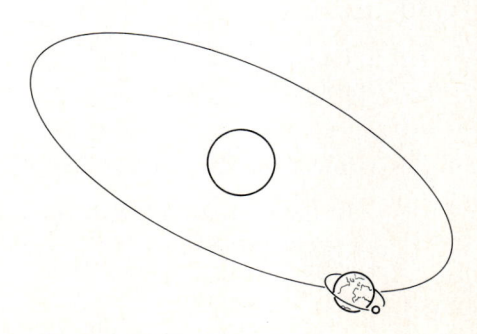

Solve.

26. In baseball, "batting average" times "at bats" equals hits. Reggie has 125 at bats and 36 hits. Find his batting average.

27. The equation for converting Celsius temperature to Fahrenheit is $F = 1.8C + 32$. Find F if the temperature is 15°C.

28. In three-way light bulbs, the highest wattage is a sum of the two lower wattages. If the lowest is 30 watts and the highest is 150 watts, what is the middle wattage?

29. A roll of film cost $3.14 and development cost $6.13. What was the cost for each of the 36 prints?

30. Franklin Laundry dryers cost a dime for 7 minutes. How many dimes will it take to dry your clothes in 45 minutes?

31. One inch = 2.54 cm. A meter is 100 cm. Find the number of inches in a meter.

32. If sound travels at 1087 feet per second, how long does it take the sound an airplane makes to travel to your ear when it is 10,000 feet overhead?

These problems are impossible to solve because some piece of information is missing. Tell what you would need to know to solve the problem.

33. A person makes three times the salary of ten years ago. What was the salary ten years ago?

34. Records were on sale for 75¢ off the marked price. After buying four records, a person has $8.72 left. How much was there to begin with?

1.10 PROBLEM SOLVING INVOLVING PERCENTS

Many problems involve percents. We can use our knowledge of equations and our problem-solving process to solve such problems.

Converting to Decimal Notation

On the average, a family spends 26% of its income for food. What does this mean? It means that out of every $100 earned, $26 is spent for food. Thus 26% is a ratio of 26 to 100.

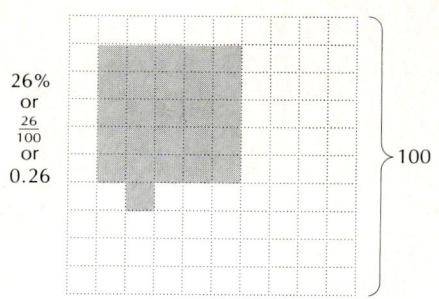

1.10 PROBLEM SOLVING INVOLVING PERCENTS

26%
or
$\frac{26}{100}$
or
0.26 $\Big\}$ 100

The percent symbol % means "per hundred." We can regard the percent symbol as part of a name for a number. For example,

$$26\% \quad \text{is defined to mean} \quad 26 \times 0.01 \quad \text{or} \quad 26 \times \tfrac{1}{100}.$$

In general,

> $n\%$ means $n \times 0.01$, or $n \times \tfrac{1}{100}$.

EXAMPLE 1 Find decimal notation for 78.5%.

$$78.5\% = 78.5 \times 0.01 \qquad \text{Replacing \% by } \times 0.01$$
$$= 0.785$$

Converting to Fractional Notation

EXAMPLE 2 Find fractional notation for 88%.

$$88\% = 88 \times \tfrac{1}{100} \qquad \text{Replacing \% by } \times \tfrac{1}{100}$$
$$= \tfrac{88}{100} \qquad \text{You need not simplify.}$$

EXAMPLE 3 Find fractional notation for 34.7%.

$$34.7\% = 34.7 \times \tfrac{1}{100} \qquad \text{Replacing \% by } \times \tfrac{1}{100}$$
$$= \tfrac{34.7}{100}$$
$$= \tfrac{34.7}{100} \cdot \tfrac{10}{10} \qquad \text{Multiplying by 1 to get a whole number in the numerator}$$
$$= \tfrac{347}{1000}$$

There is a table of decimal and percent equivalents at the back of the book. If you do not already know these facts, you should memorize them.

Converting from Decimal to Percent Notation

By applying the definition of % in reverse, we can convert from decimal notation to percent notation. We multiply by 1, naming it 100×0.01.

EXAMPLE 4 Find percent notation for 0.93.

$$\begin{aligned}
0.93 &= 0.93 \times 1 \\
&= 0.93 \times (100 \times 0.01) && \text{Replacing 1 by } 100 \times 0.01 \\
&= (0.93 \times 100) \times 0.01 && \text{Using associativity} \\
&= 93 \times 0.01 \\
&= 93\% && \text{Replacing } \times 0.01 \text{ by } \%
\end{aligned}$$

EXAMPLE 5 Find percent notation for 0.002.

$$\begin{aligned}
0.002 &= 0.002 \times (100 \times 0.01) \\
&= (0.002 \times 100) \times 0.01 \\
&= 0.2 \times 0.01 \\
&= 0.2\% && \text{Replacing } \times 0.01 \text{ by } \%
\end{aligned}$$

Converting from Fractional to Percent Notation

We can also convert from fractional to percent notation. Again we multiply by 1, but this time we use $100 \times \frac{1}{100}$.

EXAMPLE 6 Find percent notation for $\frac{5}{8}$.

$$\begin{aligned}
\tfrac{5}{8} &= \tfrac{5}{8} \times (100 \times \tfrac{1}{100}) \\
&= (\tfrac{5}{8} \times 100) \times \tfrac{1}{100} \\
&= \tfrac{500}{8} \times \tfrac{1}{100} \\
&= \tfrac{500}{8}\%, \quad \text{or} \quad 62.5\%
\end{aligned}$$

The result of this example says that the ratio of 5 to 8 is the same as the ratio of 62.5 to 100.

Problem Solving Involving Percents

Let's solve some problems involving percents. Our five-step problem-solving process will again be helpful.

EXAMPLE 7 What percent of 45 is 15?

1. **Familiarize** This type of problem is stated so explicitly that we can go right to the translation.

1.10 PROBLEM SOLVING INVOLVING PERCENTS

2. **Translate** We translate as follows.

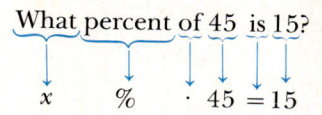

$$\underbrace{\text{What}}_{x} \underbrace{\text{percent}}_{\%} \underbrace{\text{of}}_{\cdot} \underbrace{45}_{45} \underbrace{\text{is}}_{=} \underbrace{15?}_{15}$$

3. **Carry out** We solve the equation:

$$x\% \cdot 45 = 15$$
$$(x \times 0.01) \times 45 = 15$$
$$x(0.45) = 15$$
$$x = \tfrac{15}{0.45} \qquad \text{Dividing by 0.45}$$
$$x = \tfrac{15}{0.45} \times \tfrac{100}{100} = \tfrac{1500}{45}$$
$$x = 33\tfrac{1}{3}.$$

4. **Check** We check by finding $33\tfrac{1}{3}\%$ of 45:

$$33\tfrac{1}{3}\% \cdot 45 = \tfrac{1}{3} \cdot 45 = 15. \qquad \text{See Table 1.}$$

5. **State** The answer is $33\tfrac{1}{3}\%$.

EXAMPLE 8 3 is 16 percent of what?

1. **Familiarize** This problem is stated so explicitly that we can go right to the translation.

2. **Translate**

$$\underbrace{3}_{3} \underbrace{\text{is}}_{=} \underbrace{16}_{16} \underbrace{\text{percent}}_{\%} \underbrace{\text{of}}_{\cdot} \underbrace{\text{what?}}_{y}$$

3. **Carry out** We solve the equation:

$$3 = 16\% \cdot y$$
$$3 = 16 \times 0.01 \times y$$
$$3 = 0.16y$$

$$0.16y = 3$$
$$y = \tfrac{3}{0.16} \qquad \text{Dividing by 0.16}$$
$$y = 18.75.$$

4. **Check** We check by finding 16% of 18.75:

$$16\% \times 18.75 = 0.16 \times 18.75 = 3.$$

5. **State** The answer is 18.75.

52   INTRODUCTION TO ALGEBRA AND PROBLEM SOLVING

Perhaps you have noticed in Examples 7 and 8 that to handle percents in such problems, you can convert to decimal notation and then go ahead.

EXAMPLE 9   Blood is 90% water. The average adult has 5 quarts of blood. How much water is in the average adult's blood?

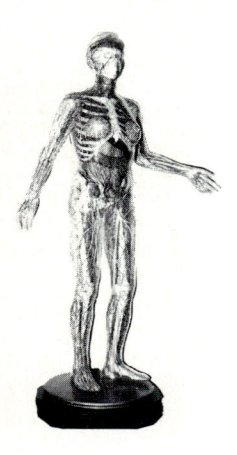

1. **Familiarize**   The translation step is quite easy, and we can go to it without familiarization.
2. **Translate**   Sometimes it is helpful to reword the problem before translating.

$$\text{Rewording:} \quad 90\% \text{ of } 5 \text{ is what?}$$

$$\text{Translating:} \quad 90\% \cdot 5 = x$$

3. **Carry out**   We solve the equation:

$$90\% \cdot 5 = x$$
$$90 \times 0.01 \times 5 = x$$
$$0.90 \times 5 = x \quad \text{Converting 90\% to decimal notation}$$
$$4.5 = x.$$

4. **Check**   The check is actually the computation we use to solve the equation:

$$90\% \cdot 5 = 0.90 \times 5 = 4.5.$$

5. **State**   The answer is that there are 4.5 quarts of water in a human being who has 5 quarts of blood.

EXAMPLE 10   An investment is made at 8% simple interest for 1 year. It grows to $783. How much was originally invested (the principal)?

1. **Familiarize** Suppose that $100 was invested. Recalling the formula for simple interest, $I = Prt$, we know that the interest for 1 year on $100 at 8% simple interest is given by $I = \$100 \cdot 8\% \cdot 1 = \$8$. Then, at the end of the year, the *amount* in the account is found by adding principal and interest:

$$(\text{Principal}) + (\text{Interest}) = \text{Amount}$$
$$\$100 \quad + \quad \$8 \quad = \quad \$108.$$

In this problem we are working backwards. We are trying to find the principal.

2. **Translate** We reword the problem and then translate:

Rewording: (Principal) + (Interest) = Amount

Translating: $\quad x \quad + \quad 8\%x \quad = \quad 783 \quad$ Interest is 8% of the principal

3. **Carry out** We solve the equation:

$$x + 8\%x = 783$$
$$x + 0.08x = 783 \quad \text{Converting}$$
$$1.08x = 783 \quad \text{Collecting like terms}$$
$$x = \tfrac{783}{1.08} \quad \text{Dividing by 1.08}$$
$$x = 725.$$

4. **Check** We check by taking 8% of $725 and adding it to $725:

$$8\% \times \$725 = 0.08 \times 725 = \$58.$$

Then $725 + $58 = $783, so $725 checks.

5. **State** The original investment was $725.

EXERCISE SET 1.10

Find decimal notation.

1. 76%
2. 54%
3. 54.7%
4. 96.2%
5. 100%
6. 1%
7. 0.61%
8. 125%

Find fractional notation.

9. 20%
10. 80%
11. 78.6%
12. 13.5%
13. $12\tfrac{1}{2}\%$
14. 120%
15. 0.042%
16. 0.68%

Find percent notation.

17. 4.54
18. 1
19. 0.998
20. 0.73
21. 2
22. 0.0057
23. 0.072
24. 1.34

Find percent notation.

25. $\frac{1}{8}$
26. $\frac{1}{3}$
27. $\frac{17}{25}$
28. $\frac{11}{20}$
29. $\frac{17}{100}$
30. $\frac{119}{100}$
31. $\frac{7}{10}$
32. $\frac{8}{10}$
33. $\frac{3}{5}$
34. $\frac{17}{50}$
35. $\frac{2}{3}$
36. $\frac{3}{8}$
37. $\frac{7}{4}$
38. $\frac{7}{8}$
39. $\frac{3}{4}$
40. $\frac{99.4}{100}$

Problem solving

41. What percent of 68 is 17?
42. What percent of 75 is 36?
43. What percent of 125 is 30?
44. What percent of 300 is 57?
45. 45 is 30% of what number?
46. 20.4 is 24% of what number?
47. 0.3 is 12% of what number?
48. 7 is 175% of what number?
49. What number is 65% of 840?
50. What number is 1% of a million?
51. What percent of 80 is 100?
52. What percent of 10 is 205?
53. What is 2% of 40?
54. What is 40% of 2?
55. 2 is what percent of 40?
56. 40 is 2% of what number?
57. On a test of 88 items, a student got 76 correct. What percent were correct?
58. A baseball player had 13 hits in 25 times at bat. What percent were hits?
59. A family spent $208 one month for food. This was 26% of its income. What was their monthly income?
60. The sales tax rate in New York City is 8%. How much would be charged on a purchase of $428.86? How much will the total cost of the purchase be?
61. Water volume increases 9% when it freezes. If 400 cubic centimeters of water is frozen, how much will its volume increase? What will be the volume of the ice?
62. An investment is made at 9% simple interest for 1 year. It grows to $8502. How much was originally invested?
63. An investment is made at 8% simple interest for 1 year. It grows to $7776. How much was originally invested?
64. Due to inflation the price of an item rose 8%, which was 12¢. What was the old price? the new price?

Simplify.

65. 12% + 14%
66. 84% − 16%
67. 1 − 10%
68. 81% − 10%
69. 12 × 100%
70. 42% − (1 − 58%)
71. 3(1 + 15%)
72. 7(1% + 13%)
73. $\frac{100\%}{40}$

74. Twenty-seven people make a certain amount of money at a sale. What percentage does each receive if they share the profit equally?
75. A meal came to $16.41 without tax. Calculate 6% sales tax and then calculate a 15% tip based on the sum of the meal price and the tax. What is the total paid?
76. Rollie's Records charges $7.99 for an album. Warped Records charges $9.95 but you have a coupon for $2 off. 7% sales tax is charged on the *regular* prices. How much does the record cost at each store?
77. The weather report is "a 60% chance of showers during the day, 30% tonight, and 5% tomorrow morning." What are the chances it won't rain during the day? tonight? tomorrow morning?

78. If x is 160% of y, y is what % of x?

79. The new price of a car is 25% higher than the old price of $8800. The old price is what percent lower than the new price?

80. A distributing company gives successive discounts to dealers and computes prices as follows:

List price (price printed in the catalog) less successive discounts of 10%, 20%, and 10%. To find the actual price, take 10% off. Then take 20% off what is left and then 10% off that amount.

A list price is $140. What is the actual price?

SUMMARY AND REVIEW: CHAPTER 1

The following contains a summary of what you should be able to do after completing this chapter. The review exercises are for practice. Answers are at the back of the book. If you miss an exercise, restudy the section indicated alongside the answer.

You should be able to:

Evaluate simple algebraic expressions.

Evaluate.

1. $3a$ for $a = 5$
2. $\dfrac{x}{y}$ for $x = 12$ and $y = 2$
3. $\dfrac{2 \cdot p}{q}$ for $p = 20$ and $q = 8$
4. $\dfrac{x - y}{3}$ for $x = 17$ and $y = 5$
5. $n^3 + 1$ for $n = 2$
6. x^0 for $x = 6$
7. $(x + 1)^2$ for $x = 4$
8. $(y - 1)(y + 3)$ for $y = 5$

Calculate with and without parentheses.

Calculate and compare answers to 9–11.

9. $120 - 6^2 \div 4 + 8$
10. $(120 - 6^2) \div 4 + 8$
11. $(120 - 6^2) \div (4 + 8)$

Translate phrases to algebraic expressions.

Translate to an algebraic expression.

12. 8 less than z
13. Three times x
14. Nineteen percent of some number

15. A number x is 1 more than a smaller number. Write an expression for the smaller number.

Tell the meaning of exponential notation such as 4^6 and write expressions such as $y \cdot y \cdot y \cdot y$ in exponential notation.

16. What is the meaning of the expression $(2m)^3$?
17. Write $6z \cdot 6z$ in exponential notation.

Write equivalent expressions using the commutative laws, the associative laws, and the property of 1.

Use a commutative law to write an expression equivalent to each of the following.

18. $4 + y$
19. ab
20. $pq + 2$

Use an associative law to write an expression equivalent to each of the following.

21. $(3 + x) + 1$　　　　　　　　　　　　　　　　**22.** $m \cdot (4 \cdot n)$

Use the commutative and associative laws to write three expressions equivalent to each of the following.

23. $(1 + m) + n$　　　　　　　　　　　　　　　　**24.** $4 \cdot (x \cdot y)$

Write an equivalent expression. Use the indicated name for 1.

25. $\dfrac{2}{5} \left(\text{Use } \dfrac{6}{6} \text{ for 1.} \right)$　　　　　　　　　　**26.** $\dfrac{2x}{y} \left(\text{Use } \dfrac{z}{z} \text{ for 1.} \right)$

Find prime factorization of a number and find the least common multiple of two or more numbers using both the list-of-multiples method and the prime-factorization method.

Find the prime factorization of each number.

27. 92　　　　　　　　　　　　　　　　　　　　　**28.** 1400

Find the LCM.

29. 12, 32　　　　　　　　　　　　　　　　　　　**30.** 5, 18, 45

Simplify fractional notation and multiply, add, subtract, and divide with fractional notation.

Simplify.

31. $\dfrac{20}{48}$　　　　　　　　　　　　　　　　　　**32.** $\dfrac{10ac}{18bc}$

Compute and simplify.

33. $\dfrac{4}{9} + \dfrac{5}{12}$　　　**34.** $\dfrac{3}{4} \div 3$　　　**35.** $\dfrac{2}{3} - \dfrac{1}{15}$　　　**36.** $\dfrac{9}{10} \cdot \dfrac{16}{5}$

Use the distributive law to multiply and factor algebraic expressions and to simplify expressions by collecting like terms.

Multiply.

37. $6(3x + 5y)$　　　　　　　　　　　　　　　　**38.** $8(5x + 3y + 2)$

Factor.

39. $18x + 6y$　　　　　　　　　　　　　　　　　**40.** $36x + 16 + 4y$

Collect like terms.

41. $36y + 8a + 4y + 2a$　　　　　　　　　　　　**42.** $54x + 3b + 9b + 6x$

Convert percent notation to decimal and fractional notation and convert decimal and fractional notation to percent notation.

43. Find decimal notation: 4.7%.　　　　　　　　**44.** Find fractional notation: 60%.

Find percent notation.

45. 0.886　　　　　　　**46.** $\dfrac{5}{8}$　　　　　　　　**47.** $\dfrac{29}{25}$

Solve equations of the types $x + a = b$ and $ax = b$.

Solve.

48. $x + 11 = 50$　　　**49.** $x + \dfrac{3}{8} = \dfrac{4}{8}$　　　**50.** $7n = 126$　　　**51.** $1.3z = 33.8$

State and apply the five-step problem-solving process.

52. State the five-step problem-solving process.　　**53.** What number added to 35 is 102?

54. Three-fifths of what number is 30?

55. An artist charges $75 for each drawing. How many drawings did the artist complete to earn $2400?

56. 25 is 10% of what number?

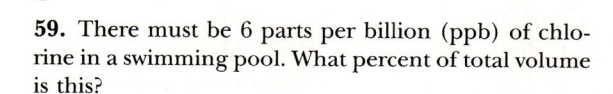

57. 40% of 75 is what number?

58. A government employee received an 8% raise. The new salary is $15,336. What was the original salary?

59. There must be 6 parts per billion (ppb) of chlorine in a swimming pool. What percent of total volume is this?

60. Which is better, a discount of 40% or successive discounts of 20% and 20%?

61. Which is better, successive discounts of 10%, 10%, and 20% or of 20%, 10%, and 10%?

62. Evaluate $a^{50} - 20a^{25}b^4 + 100b^8$ when $a = 1$ and $b = 2$.

63. Solve: $4(y + 5) + 4 = 40$.

64. Solve using this information.
  a) "Everyone will get 0.4% of the profits," says Big Louie. How many people must there be for all the profits to be shared equally?
  b) It turned out that Louie was taking 10% of the profits. Everyone else received 0.4% of the 90% that remained. Including Louie, how many people were there?

TEST: CHAPTER 1

Evaluate.

1. $\dfrac{3x}{y}$ for $x = 10$ and $y = 5$

2. $x^2 - 5$ for $x = 8$

3. $(y + 3)(y - 4)$ for $y = 6$

4. x^1 for $x = 8$

5. Calculate: $200 - 2^3 + 5 + 10$.

6. Write an algebraic expression for 8 less than n.

Use a commutative law to write an expression equivalent to each of the following.

7. pq

8. $z + 3$

Use an associative law to write an expression equivalent to each of the following.

9. $x \cdot (4 \cdot y)$

10. $a + (b + 1)$

11. Write an expression equivalent to $\frac{3}{7}$ using $\frac{7}{7}$ as a name for 1.

12. Find the prime factorization of 300.

13. Find the LCM: 15, 24, 60.

Simplify.

14. $\dfrac{16}{24}$

15. $\dfrac{9xy}{15yz}$

Compute and simplify.

16. $\frac{10}{27} \div \frac{8}{3}$

17. $\frac{9}{10} - \frac{5}{8}$

Multiply.

18. $10(9x + 3y)$

19. $7(9m + 2x + 1)$

Factor.

20. $15y + 5$

21. $24x + 16 + 8y$

Collect like terms.

22. $21x + 96 + 5x + 6$

23. $18y + 30a + 9a + 4y$

24. Find decimal notation: 0.7%.

25. Find fractional notation: 91%.

26. Find percent notation: $\frac{11}{25}$.

Solve.

27. $x + 9 = 11$

28. $5t = 95$

29. $y + \frac{3}{10} = \frac{9}{10}$

30. $0.3z = 1.5$

31. State the five-step problem-solving process.

32. What number added to 43 is 60?

33. Four times what number is 56?

34. 16 is 25% of what number?

35. 62% of 125 is what number?

36. Tom's salary is seven-eighths of Jeff's. Tom's salary is $14,700. What is Jeff's salary?

37. On a test of 100 points, student A missed 18 more than student B. If B's score is 83, what is A's score?

38. An investment is made at 12% simple interest for 1 year. It grows to $28,000. How much was originally invested?

39. Evaluate $\frac{5y - x}{4}$ when $x = 20$ and y is 4 less than x.

40. Simplify: $\frac{13,860}{42,000}$.

41. Solve using this information.

a) Your boss says, "I'll raise your $1000 salary 50% this month but lower that salary 50% next month." Would you receive more money by taking the offer or by keeping your old salary for two months?

b) Would you receive more money by continuing to alternate 50% increases and decreases indefinitely or by keeping your old salary? (*Hint:* Calculate each for six months.)

INTEGERS, RATIONAL NUMBERS, AND REAL NUMBERS

2

A football team makes many gains and losses of yards during a game. Integers can be used to solve the problem of finding the total gain or loss.

The problems we solved in Chapter 1 and the manipulations we performed used the numbers of arithmetic. These numbers consisted of the whole numbers 0, 1, 2, 3, 4, and so on, together with fractions such as $\frac{1}{2}, \frac{2}{3}, \frac{4}{5}, \frac{9}{7}$, and so on.

In this chapter we expand to larger sets of numbers. This will allow us in Chapter 3 to solve equations that have no solution in the set of numbers of arithmetic.

In the latter part of Chapter 2 we learn to add, subtract, multiply, and divide and to carry out other algebraic manipulations with these new numbers.

2.1 INTEGERS AND THE NUMBER LINE

The numbers of arithmetic can be represented on a number line.

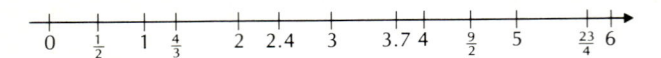

$$0 \quad \tfrac{1}{2} \quad 1 \quad \tfrac{4}{3} \quad 2 \quad 2.4 \quad 3 \quad 3.7 \quad 4 \quad \tfrac{9}{2} \quad 5 \quad \tfrac{23}{4} \quad 6$$

In this chapter we use the number line to learn about other sets of numbers. The numbers of arithmetic can be used to solve many kinds of problems. However, we can increase our problem-solving ability by using numbers called *integers*.

The Set of Integers

To create the set of integers, we begin with the set of *whole* numbers 0, 1, 2, 3, and so on. For each natural number 1, 2, 3, and so on, we invent a new number.

For the number 1 there will be a new number named -1 (negative 1).

For the number 2 there will be a new number named -2 (negative 2), and so on.

The *integers* consist of the whole numbers and these new numbers. We picture them on a number line as follows.

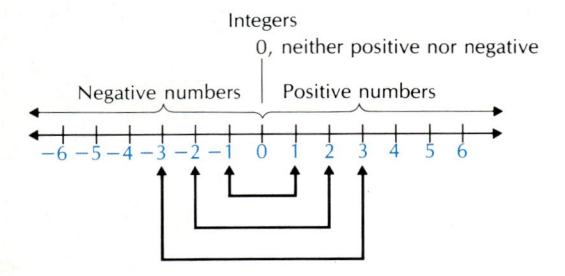

Integers
0, neither positive nor negative

Negative numbers | Positive numbers

$$-6 \; -5 \; -4 \; -3 \; -2 \; -1 \quad 0 \quad 1 \quad 2 \quad 3 \quad 4 \quad 5 \quad 6$$

We call the newly invented numbers *negative* integers. The natural numbers are called *positive* integers. Zero is neither positive nor negative.

Problem Solving and Integers

Integers can be associated with many real-world situations and used in problem solving. The following examples will help you get ready to translate problem situations to mathematical language.

EXAMPLE 1 Tell which integer corresponds to this situation: The temperature is 3 degrees below zero.

$$3° \text{ below zero is } -3°$$

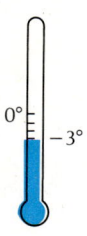

EXAMPLE 2 Tell which integer corresponds to this situation: Getting set 21 points in a card game.

Getting set 21 points in a card game gives you -21 points.

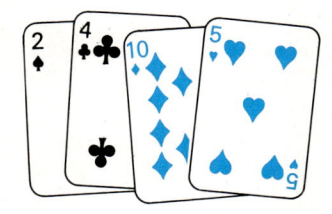

EXAMPLE 3 Tell which integer corresponds to this situation: Death Valley is 280 feet below sea level.

The integer -280 corresponds to the situation. The elevation is -280 ft.

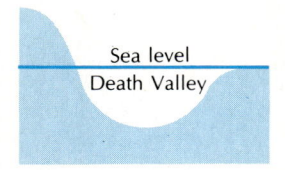

EXAMPLE 4 Tell which integers correspond to this situation: A salesperson made $78 on Monday, but lost $57 on Tuesday.

The integers 78 and -57 correspond to the situation. The integer 78 corresponds to the profit on Monday and -57 corresponds to the loss on Tuesday.

Order on the Number Line

Numbers are named in order on the number line, with larger numbers named further to the right. For any two numbers on the line, the one to the left is less than the one to the right.

We use the symbol $<$ to mean "is less than." The sentence $-6 < 8$ means "-6 is less than 8." The symbol $>$ means "is greater than." The sentence $-3 > -7$ means "-3 is greater than -7."

EXAMPLES Use either $<$ or $>$ to write true sentences.

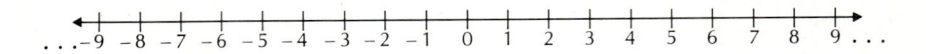

$\dots -9\ -8\ -7\ -6\ -5\ -4\ -3\ -2\ -1\ 0\ 1\ 2\ 3\ 4\ 5\ 6\ 7\ 8\ 9 \dots$

5.   2   9   Since 9 is the right of 2, 9 is greater than 2 or 2 is less than 9. The answer is $2 < 9$.
6. -7 3   Since -7 is to the left of 3, we have $-7 < 3$.
7.   6 -12 Since 6 is to the right of -12, then $6 > -12$.
8. -18 -5 Since -18 is to the left of -5, we have $-18 < -5$.

In the following examples we see some ways in which integers are used in translating to mathematical language. Recall that this is a step in our five-step problem-solving process.

EXAMPLE 9 Translate to mathematical language: A number is less than -3.

> Translate   The translation is $x < -3$.

EXAMPLE 10 Translate to mathematical language: A temperature of $-20°$ is higher than a temperature of $-36°$.

> Translate   The translation is $-20 > -36$.

EXAMPLE 11 Translate to mathematical language: A debt of $150 is worse than a debt of $100.

> Translate   The translation is $-150 < -100$.

Note that all positive integers are greater than zero and all negative integers are less than zero.

Absolute Value

From the number line we see that numbers like 4 and -4 are the same distance from zero. We call the distance from zero the *absolute value* of the number.

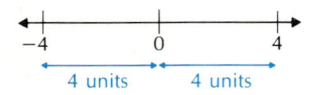

$-4 \quad 0 \quad 4$
4 units   4 units

2.1 INTEGERS AND THE NUMBER LINE

The *absolute value* of a number is its distance from 0 on a number line. We use the symbol $|n|$ to represent the absolute value of a number n.

To find absolute value:
1. If a number is negative, make it positive.
2. If a number is positive or zero, leave it alone.

EXAMPLES Find the absolute value.

12. $|-7|$ The distance of -7 from 0 is 7, so $|-7|$ is 7.
13. $|12|$ The distance of 12 from 0 is 12, so $|12|$ is 12.
14. $|0|$ The distance of 0 from 0 is 0, so $|0|$ is 0.

EXERCISE SET 2.1

Tell which integers correspond to each situation.

1. In a game a player won 5 points. In the next game a player lost 12 points.

2. The temperature on Wednesday was 18° above zero. On Thursday it was 2° below zero.

3. A family owes $17. The same family has $12 in its bank account.

4. A business earned $1200 one week and lost $560 the next.

5. The Dead Sea, between Jordan and Israel, is 1286 feet below sea level, whereas Mt. Everest is 29,028 feet above sea level.

6. In bowling, team A is 34 pins behind team B after the first game. Describe this situation in two ways.

7. A student deposited $750 in a savings account. Two weeks later the student withdrew $125.

8. During a certain time period, the United States had a deficit of $3 million in foreign trade.

9. During a video game a player intercepted a missile worth 20 points, lost a starship worth 150 points, and captured a base worth 300 points.

10. 3 seconds before liftoff of a rocket occurs. 128 seconds after the liftoff of a rocket.

Write a true sentence using $<$ or $>$.

11. 5   0
12. 9   0
13. -9 5
14. 8 -8
15. -6 6
16. 0 -7
17. -8 -5
18. -4 -3
19. -5 -11
20. -3 -4
21. -6 -5
22. -10 -14

Find the absolute value.

23. $|-3|$
24. $|-7|$
25. $|10|$
26. $|11|$
27. $|0|$
28. $|-4|$
29. $|-24|$
30. $|325|$

31. $|x|$ when $x = 5$

32. $|b|$ when $b = -3$

Translate to mathematical language.

33. -5 is greater than some number.

34. Some number is less than -1.

35. In pinochle a score of 120 is better than a score of -20.

36. A deposit of $20 in a savings account is better than a withdrawal of $25.

37. In trade, a deficit of $500,000 is worse than an excess of $1,000,000.

38. In bowling, it is better to be 60 pins ahead than to be 20 pins ahead.

39. On a test, it is better to have 2 points taken off than 10 points taken off.

List each set of integers from least to greatest.

40. $13, -12, 5, -17$

41. $-23, 4, 0, -17$

42. Explain why in golf or dieting it is better to be -2 or -3 with respect to par or a certain weight.

Solve.

43. $|x| = 7$

44. $|x| < 2$ (consider only integer replacements)

Write a true sentence using $<$, $>$, or $=$.

45. $|-3|$ 5

46. 2 $|-4|$

47. 0 $|0|$

48. $|-5|$ $|-2|$

49. $|4|$ $|-7|$

50. $|-8|$ $|8|$

List in order from least to greatest.

51. $7^1, -5, |-6|, 4, |3|, -100, 0, 1^7, \frac{14}{4}$

2.2 RATIONAL NUMBERS AND REAL NUMBERS

We created the set of integers by inventing a negative number for each natural number. To create a bigger number system, called the set of *rational numbers,* we invent negative numbers that correspond with the nonzero numbers of arithmetic.

For $\frac{2}{3}$ there will be a new number named $-\frac{2}{3}$ (negative $\frac{2}{3}$).

For $\frac{13}{5}$ there will be a new number named $-\frac{13}{5}$ (negative $\frac{13}{5}$).

For 1.5 there will be a new number named -1.5 (negative 1.5).

For 73 there will be a new number named -73 (negative 73).

And so on.

Note that this new set of numbers, the rational numbers, contains the whole numbers, the integers, and the numbers of arithmetic.

2.2 RATIONAL NUMBERS AND REAL NUMBERS

Graphing Rational Numbers

We picture the rational numbers on a number line as follows. There is a point on the line for every rational number.

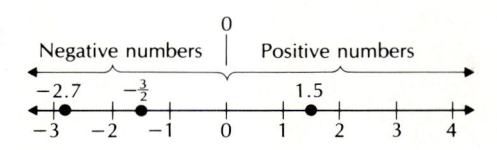

To *graph* a number means to find and mark its point on the line. Some numbers are graphed in the preceding figure.

EXAMPLE 1 Graph: $\frac{5}{2}$.

The number $\frac{5}{2}$ can be named $2\frac{1}{2}$ or 2.5. Its graph is halfway between 2 and 3.

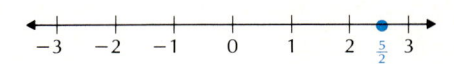

EXAMPLE 2 Graph: -3.2.

The graph of -3.2 is $\frac{2}{10}$ of the way from -3 to -4.

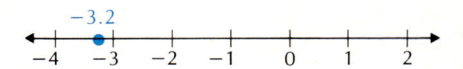

EXAMPLE 3 Graph: $\frac{13}{8}$.

The number $\frac{13}{8}$ can be named $1\frac{5}{8}$ or 1.625. The graph is about $\frac{6}{10}$ of the way from 1 to 2.

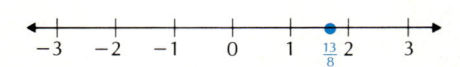

Order of the Rational Numbers

The relations < (is less than) and > (is greater than) are the same for rational numbers as they are for integers. Recall that numbers on the line increase from left to right.

EXAMPLES Use either < or > to write a true sentence.

    4.   1.38   1.83   The answer is $1.38 < 1.83$.

5. $-3.45 \quad 1.32$ The answer is $-3.45 < 1.32$ because -3.45 is to the left of 1.32.

6. $-3.33 \quad -4.44$ The answer is $-3.33 > -4.44$.

7. $\frac{5}{8} \quad \frac{7}{11}$ We convert to decimal notation. $\frac{5}{8} = 0.625$, $\frac{7}{11} = 0.6363,\ldots$ Thus $\frac{5}{8} < \frac{7}{11}$.

In Example 7 we can abbreviate $0.6363\ldots$ by putting a bar over the repeating part.

$$\frac{7}{11} = 0.6\overline{36}3$$

Absolute Value

The absolute value of a rational number is its distance from zero, just as for integers.

EXAMPLES Find the absolute value.

8. $\left|\frac{3}{2}\right| = \frac{3}{2}$
9. $|-2.73| = 2.73$

Real Numbers

The number line has a point for every rational number. Is there a rational number for every point of the line? The answer is *no*. There are some points of the line for which there is no rational number.

We can create a number system in which all points have numbers. That new number system is called the system of *real numbers*. Later in this book you will work with some real numbers that are not rational, but for most of this book we shall work only with rational numbers.

The following figure shows the relationship between various kinds of numbers.

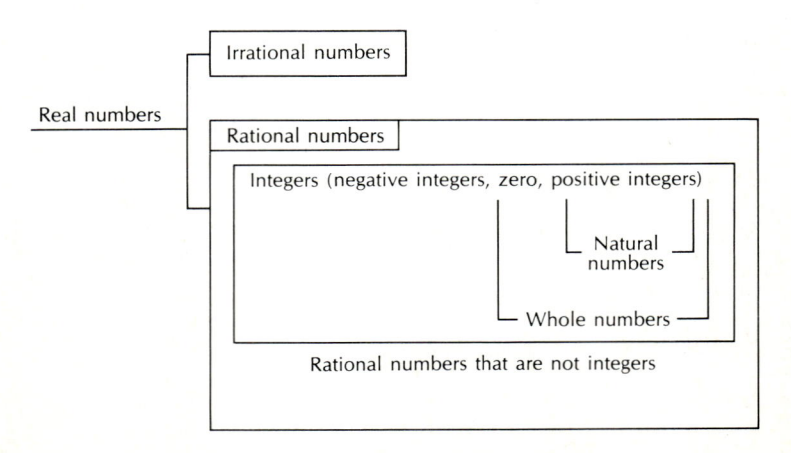

EXERCISE SET 2.2

1. List ten examples of rational numbers.

2. List ten examples of rational numbers that are *not* integers.

Graph each number on a number line.

3. $\frac{10}{3}$ **4.** $-\frac{17}{5}$ **5.** -4.3 **6.** 6.45

Write a true sentence using $<$ or $>$.

7. $2.14 \quad 1.24$ **8.** $-3.3 \quad -2.2$ **9.** $-14.5 \quad 0.011$ **10.** $17.2 \quad -1.67$

11. $-12.88 \quad -6.45$ **12.** $-14.34 \quad -17.88$ **13.** $\frac{5}{12} \quad \frac{11}{25}$ **14.** $-\frac{14}{17} \quad -\frac{27}{35}$

Find the absolute value.

15. $\left|-\frac{2}{3}\right|$ **16.** $\left|-\frac{10}{7}\right|$ **17.** $\left|\frac{0}{4}\right|$ **18.** $|14.8|$

○ ──────────────────────────

Evaluate these expressions for $x = 5$, $y = -3$, and $z = 4$.

19. $\dfrac{x + |y|}{2z}$ **20.** $2|x| + |y| - |z|$

Solve.

21. $|n| = \frac{13}{5}$ **22.** $|x| = -4$

23. What number corresponds to point A in the figure?

24. What number corresponds to point B in the figure?

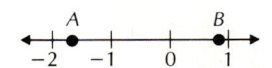

```
    A       B
◄───●───┼───●───►
   -2  -1   0   1
```

25. We know that $0.3333\ldots$ is $\frac{1}{3}$ and $0.6666\ldots$ is $\frac{2}{3}$. What rational number is named by:

a) $0.9999\ldots$?

b) $0.1111\ldots$?

2.3 ADDITION

In this section we consider addition of rational numbers, but what we say also applies to real numbers. First, to gain an understanding, we add using a number line. Then we consider rules for addition.

Addition on a Number Line

Addition of numbers can be illustrated on a number line. To do the addition $a + b$, we start at a, and then move according to b.

a) If b is positive, we move to the right.
b) If b is negative, we move to the left.
c) If b is 0, we stay at a.

EXAMPLE 1 Add: $3 + (-5)$.

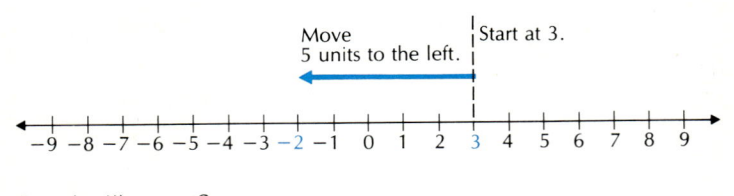

$3 + (-5) = -2$

EXAMPLE 2 Add: $-4 + (-3)$.

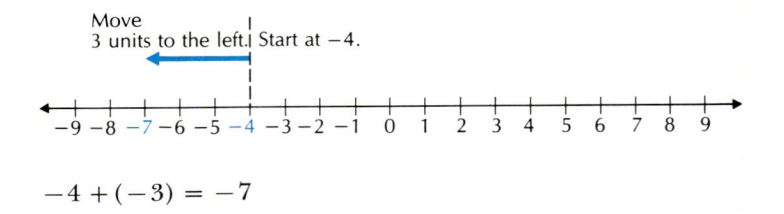

$-4 + (-3) = -7$

EXAMPLE 3 Add: $-4 + 9$.

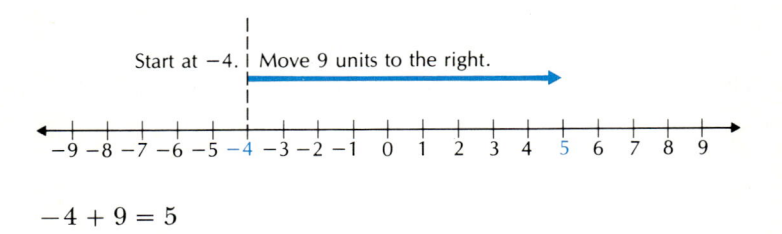

$-4 + 9 = 5$

EXAMPLE 4 Add: $-5.2 + 0$.

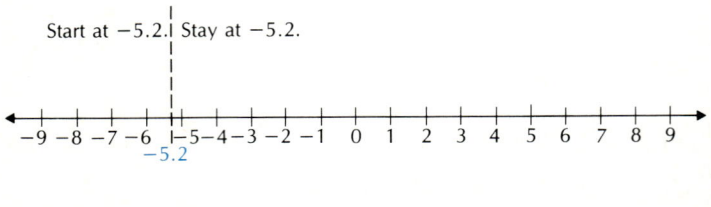

$-5.2 + 0 = -5.2$

2.3 ADDITION

Adding Without a Number Line

You may have noticed some patterns in the previous examples. These lead us to rules for adding without using a number line.

> **RULES FOR ADDITION**
>
> 1. *Positive numbers:* Add the same as numbers of arithmetic.
> 2. *Negative numbers:* Add absolute values. Make the answer negative.
> 3. *A positive and a negative number:* Subtract absolute values. Then:
>   a) If the positive number has greater absolute value, make the answer positive.
>   b) If the negative number has the greater absolute value, make the answer negative.
>   c) If the numbers have the same absolute value, make the answer 0.
> 4. *One number is zero:* The sum is the other number.

Rule 4 is known as the *Additive Property of Zero*. It says that for any rational number a, $a + 0 = a$.

EXAMPLES Add without using a number line.

5. $-12 + (-7)$ Two negatives. Think: Add the absolute values, getting 19. Make the answer *negative*, -19.

6. $-1.4 + 8.5$ The absolute values are 1.4 and 8.5. The difference is 7.1. The positive addend has the larger absolute value, so the answer is *positive*, 7.1.

7. $-36 + 21$ The absolute values are 36 and 21. The difference is 15. The negative addend has the larger absolute value, so the answer is *negative*, -15.

8. $1.5 + (-1.5)$ The sum is 0.

9. $-\frac{7}{8} + 0$ The sum is $-\frac{7}{8}$.

EXAMPLES Add.

10. $-9.2 + 3.1 = -6.1$
11. $-\frac{3}{2} + \frac{9}{2} = \frac{6}{2} = 3$
12. $-\frac{2}{3} + \frac{5}{8} = -\frac{16}{24} + \frac{15}{24} = -\frac{1}{24}$

Suppose we wish to add several numbers, some positive and some negative, as follows. How can we proceed?

$$13 + (-4) + 17 + (-5) + (-2)$$

The commutative and associative laws of addition hold for rational numbers as they do for other kinds of numbers. That means that we can change grouping and order as we please. For instance, we can group the positive numbers together and the negative numbers together and add them separately. Then we add the two results.

EXAMPLE 13 Add: $15 + (-2) + 7 + 14 + (-5) + (-12)$.

a) $15 + 7 + 14 = 36$ Adding the positive numbers
b) $-2 + (-5) + (-12) = -19$ Adding the negative numbers
c) $36 + (-19) = 17$ Adding the results

Additive Inverses

When two numbers such as 6 and -6 are added, the result is 0. Such numbers are called *additive inverses* of each other. Every rational number has an additive inverse.

> Two numbers whose sum is 0 are called *additive inverses* (or simply, *inverses*) of each other.

EXAMPLES Find the additive inverse of each number.

14. 34   The inverse of 34 is -34 because $34 + (-34) = 0$.
15. -8 The inverse of -8 is 8 because $-8 + 8 = 0$.
16. 0   The inverse of 0 is 0 because $0 + 0 = 0$.
17. $-\frac{7}{8}$ The inverse of $-\frac{7}{8}$ is $\frac{7}{8}$ because $-\frac{7}{8} + \frac{7}{8} = 0$.

To name the additive inverse we use the symbol $-$, as follows.

> The additive inverse of a number a can be named $-a$ (read "the inverse of a").

A symbol such as -8 is usually read "negative 8." It could be read "the inverse of 8," because the inverse of 8 is negative 8. Thus a symbol like -8 can be read in two ways. A symbol like $-x$, however, should be read "the inverse of x" and *not* "negative x," because we do not know whether it represents a positive number, a negative number, or 0.

Note that if we take a number, say, 8, and find its inverse, -8, and then find the inverse of the result, we will have the original number, 8, again.

> The inverse of the inverse of a number is the number itself. That is, for any number a,
> $$-(-a) = a.$$

2.3 ADDITION

EXAMPLE 18 Find $-x$ and $-(-x)$ when $x = 16$.

a) If $x = 16$, then $-x = -16$.   The inverse of 16 is -16.
b) If $x = 16$, then $-(-x) = -(-16) = 16$.   The inverse of the inverse of 16 is 16.

EXAMPLE 19 Find $-x$ and $-(-x)$ when $x = -3$.

a) If $x = -3$, then $-x = -(-3) = 3$.
b) If $x = -3$, then $-(-x) = -(-(-3)) = -3$.

We can use the symbolism $-a$ for the additive inverse of a to restate the definition of additive inverse.

> For any rational number a,
> $$a + (-a) = (-a) + a = 0.$$

Signs of Numbers. A negative number is sometimes said to have a "negative sign." A positive number is said to have a "positive sign." When we replace a number by its additive inverse, we can say that we have "changed its sign."

EXAMPLES Change the sign. (Find the additive inverse.)

20. -3 $-(-3) = 3$
21. -10 $-(-10) = 10$
22. 0 $-0 = 0$
23. 14 $-(14) = -14$

Problem Solving

Let us now see how we can use addition of rational numbers to solve problems.

EXAMPLE 24 A business made a profit of $18 on Monday. There was a loss of $7 on Tuesday. On Wednesday there was a loss of $5, and on Thursday there was a profit of $11. Find the total profit or loss.

1. **Familiarize** We can represent a loss with a negative number and a profit with a positive number. Thus we have the following losses and profits:

$$18, \quad -7, \quad -5, \quad 11.$$

2. **Translate** To translate to mathematical language we write the following addition sentence, which represents the total profit or loss:

$$18 + (-7) + (-5) + 11 = x.$$

3. **Carry out** We carry out some mathematical manipulation by adding the numbers on the left side of the equation. This gives us a solution:

$$18 + (-7) + (-5) + 11 = x$$
$$29 + (-12) = x$$
$$17 = x.$$

4. **Check** In this case the check amounts to no more than going back over the addition to be sure it is correct.

5. **State** The total profit was $17.

EXERCISE SET 2.3

Add using a number line.

1. $-9 + 2$
2. $2 + (-5)$
3. $-10 + 6$
4. $8 + (-3)$
5. $-8 + 8$
6. $6 + (-6)$
7. $-3 + (-5)$
8. $-4 + (-6)$

Add without using a number line.

9. $-7 + 0$
10. $-13 + 0$
11. $0 + (-27)$
12. $0 + (-35)$
13. $17 + (-17)$
14. $-15 + 15$
15. $-17 + (-25)$
16. $-24 + (-17)$
17. $18 + (-18)$
18. $-13 + 13$
19. $-18 + 18$
20. $11 + (-11)$
21. $8 + (-5)$
22. $-7 + 8$
23. $-4 + (-5)$
24. $10 + (-12)$
25. $13 + (-6)$
26. $-3 + 14$
27. $11 + (-9)$
28. $-14 + (-19)$
29. $-20 + (-6)$
30. $19 + (-19)$
31. $-15 + (-7)$
32. $23 + (-5)$
33. $40 + (-8)$
34. $-23 + (-9)$
35. $-25 + 25$
36. $40 + (-40)$
37. $63 + (-18)$
38. $85 + (-65)$
39. $-6.5 + 4.7$
40. $-3.6 + 1.9$
41. $-2.8 + (-5.3)$
42. $-7.9 + (-6.5)$
43. $-\frac{3}{5} + \frac{2}{5}$
44. $-\frac{4}{3} + \frac{2}{3}$
45. $-\frac{3}{7} + (-\frac{5}{7})$
46. $-\frac{4}{9} + (-\frac{6}{9})$
47. $-\frac{5}{8} + \frac{1}{4}$
48. $-\frac{5}{6} + \frac{2}{3}$
49. $-\frac{3}{7} + (-\frac{2}{5})$
50. $-\frac{5}{8} + (-\frac{1}{3})$
51. $-\frac{3}{5} + (-\frac{2}{15})$
52. $-\frac{5}{9} + (-\frac{1}{18})$
53. $-5.7 + (-7.2) + 6.6$
54. $-10.3 + (-7.5) + 3.1$
55. $-8.5 + 7.9 + (-3.7)$
56. $-9.6 + 8.4 + (-11.8)$
57. $-\frac{7}{16} + \frac{7}{8}$
58. $-\frac{3}{28} + \frac{5}{42}$
59. $75 + (-14) + (-17) + (-5)$
60. $28 + (-44) + 17 + 31 + (-94)$
61. $-44 + (-\frac{3}{8}) + 95 + (-\frac{5}{8})$
62. $24 + 3.1 + (-44) + (-8.2) + 63$
63. $98 + (-54) + 113 + (-998) + 44 + (-612) + (-18) + 334$
64. $-455 + (-123) + 1026 + (-919) + 213 + 111 + (-874)$

Find the additive inverse of each number.

65. 24
66. -64
67. -9
68. $\frac{7}{2}$
69. -26.9
70. 48.2

Find $-x$ when x is:

71. 9 **72.** -26 **73.** $-\frac{14}{3}$ **74.** $\frac{1}{328}$ **75.** 0.101 **76.** 0

Find $-(-x)$ when x is:

77. -65 **78.** 29 **79.** $\frac{5}{3}$ **80.** -9.1

Change the sign. (Find the additive inverse.)

81. -1 **82.** -7 **83.** 7 **84.** 10
85. -14 **86.** -22.4 **87.** 0 **88.** $-\frac{7}{8}$

Problem solving

89. In a football game, the quarterback attempted passes with the following results.

First try	13-yd gain
Second try	incomplete
Third try	12-yd loss (tackled behind the line)
Fourth try	21-yd gain
Fifth try	14-yd loss

Find the total gain (or loss).

90. The following table shows the profits and losses of a company over a five-year period. Find the profit or loss after this period of time.

Year	Profit or loss
1983	$+32,056$
1984	$-2,925$
1985	$+81,429$
1986	$-19,365$
1987	$-13,875$

91. The barometric pressure at Omaha was 1012 millibars (mb). The pressure dropped 6 mb, then it rose 3 mb. After that it dropped 14 mb and then rose 4 mb. What was the pressure then?

92. One day the value of a share of IBM stock was $253\frac{1}{4}$. That day it rose in value $\$\frac{5}{8}$. The next day it lost $\$\frac{3}{8}$. What was the value of the stock at the end of the two days?

93. For what numbers x is $-x$ negative?

94. For what numbers x is $-x$ positive?

Tell whether the sum is positive, negative, or zero.

95. If n and m are positive, $n + m$ is _____.

96. If n is positive and m is negative, $n + (-m)$ is _____.

97. If n is positive and m is negative, $-n + m$ is _____.

98. If $n = m$ and n and m are negative, $-n + (-m)$ is _____.

99. If $n = m$ and n and m are negative, $n + (-m)$ is _____.

2.4 SUBTRACTION

We now consider subtraction of rational numbers, but what we say also applies to real numbers. First, to gain an understanding, we subtract using a number line. Then we learn a general rule for subtracting.

Subtraction Using a Number Line

Subtraction is defined as follows.

> The difference $a - b$ is the number which when added to b gives a.

To see that the definition is right, think of $5 - 2$. When you subtract, you get 3. To check, you add 3 to 2, to see if 3 is the number which when added to 2 gives 5.

EXAMPLE 1 Subtract: $10 - 4$.

We know that $10 - 4$ is the number which when added to 4 gives 10. So we start at 4 and go to 10. That takes a move of 6 units to the right, so the answer is 6.

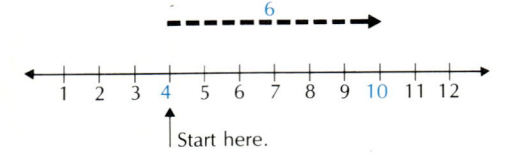

$10 - 4 = 6$

EXAMPLE 2 Subtract: $6 - 8$.

Since $6 - 8$ is the number which when added to 8 gives 6, we start at 8 and go to 6. That takes a move of 2 units to the left, so the answer is -2.

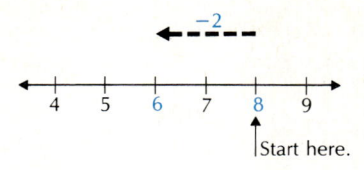

2.4 SUBTRACTION

$6 - 8 = -2$

EXAMPLE 3 Subtract: $-1 - 5$. (We read this as "negative one minus five.")

Since $-1 - 5$ is the number which when added to 5 gives -1, we start at 5 and go to -1. That takes a move of 6 units to the left, so the answer is -6.

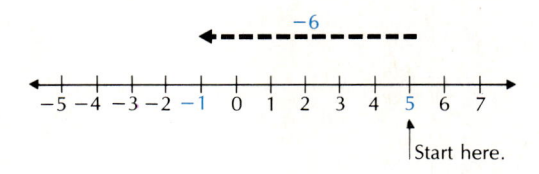

$-1 - 5 = -6$

EXAMPLE 4 Subtract: $-3 - (-7)$. (We read this as "negative three minus negative seven.")

Since $-3 - (-7)$ is the number which when added to -7 gives -3, we start at -7 and go to -3. That takes a move of 4 units to the right, so the answer is 4.

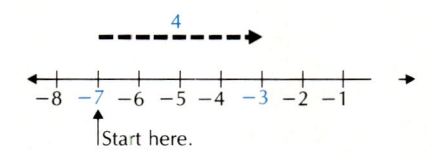

$-3 - (-7) = 4$

Subtracting Without a Number Line

To see how to subtract without using a number line, look for a pattern in the following examples.

EXAMPLES
5. a) $3 - 8 = -5$ b) $3 + (-8) = -5$
6. a) $-5 - 4 = -9$ b) $-5 + (-4) = -9$
7. a) $-7 - (-10) = 3$ b) $-7 + 10 = 3$
8. a) $-7 - (-2.4) = -4.6$ b) $-7 + 2.4 = -4.6$

Perhaps you noted in the preceding examples that we can subtract by adding an additive inverse. This can always be done.

> For any integers a and b,
> $$a - b = a + (-b).$$
> **(To subtract, we can add the inverse of the subtrahend.)**

EXAMPLES Subtract. Check by addition.

9. $2 - 6 = 2 + (-6) = -4$ Changing the sign of 6 and changing the subtraction to addition

    Check: $6 + (-4) = 2$ $2 - 6$, or -4, is that number which when added to 6 gives 2.

10. $-4 - (-9) = -4 + 9 =$ Changing the sign of -9 and changing the subtraction to addition

    Check: $-9 + 5 = -4$ $-4 - (-9)$, or 5, is that number which when added to -9 gives -4.

11. $-4.2 - (-3.6) = -4.2 + 3.6 = -0.6$
    Check: $-3.6 + (-0.6) = -4.2$

12. $-\frac{1}{2} - (-\frac{3}{4}) = -\frac{1}{2} + \frac{3}{4} = \frac{1}{4}$
    Check: $(-\frac{3}{4}) + \frac{1}{4} = -\frac{1}{2}$

EXAMPLES Subtract.

13. $3 - 5 = 3 + (-5) = -2$
14. $\frac{1}{8} - \frac{7}{8} = \frac{1}{8} + (-\frac{7}{8}) = -\frac{6}{8}$, or $-\frac{3}{4}$
15. $-4.6 - (-9.8) = -4.6 + 9.8 = 5.2$
16. $-\frac{3}{4} - \frac{7}{5} = -\frac{15}{20} + (-\frac{28}{20}) = -\frac{43}{20}$

When several additions and subtractions occur together, we can make them all additions.

EXAMPLES Simplify.

17. $8 - (-4) - 2 - (-4) + 2$
    $8 + 4 + (-2) + 4 + 2$
    16

18. $8.2 - (-6.1) + 2.3 - (-4)$
    $8.2 + 6.1 + 2.3 + 4$
    20.6

2.4 SUBTRACTION

19. $-4 - (-2x) + x - (-5)$
$-4 + 2x + x + 5$
$3x + 1$

Problem Solving

Let us now see how we can use subtraction of rational numbers to solve problems.

EXAMPLE 20 The lowest point in Asia is the Dead Sea, which is 400 meters (m) below sea level. The lowest point in the United States is Death Valley, which is 86 meters below sea level. How much higher is Death Valley than the Dead Sea?

1. **Familiarize** In this case it is helpful to draw a picture of the situation.

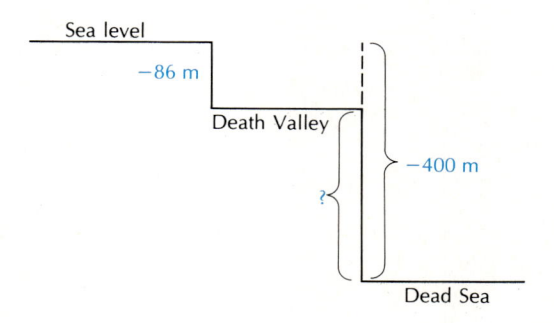

2. **Translate** We see that -86 is the higher altitude at Death Valley and -400 is the lower altitude at the Dead Sea. To find how much higher Death Valley is we subtract. This gives the subtraction equation

$$-86 - (-400) = x.$$

3. **Carry out** To solve the equation, we carry out the subtraction on the left:

$$-86 - (-400) = x$$
$$-86 + 400 = x$$
$$314 = x.$$

4. **Check** We can check in the original problem by adding 314, the difference, to the lower altitude, -400:

$$314 + (-400) = -86.$$

The result is -86, the altitude of Death Valley.

5. **State** Death Valley is 314 m higher than the Dead Sea.

EXERCISE SET 2.4

Subtract using a number line.

1. $3 - 7$
2. $4 - 9$
3. $0 - 7$
4. $0 - 10$
5. $-8 - (-2)$
6. $-6 - (-8)$
7. $-10 - (-10)$
8. $-8 - (-8)$

Subtract by adding the inverse of the subtrahend.

9. $3 - 7$
10. $4 - 9$
11. $0 - 7$
12. $0 - 10$
13. $-8 - (-2)$
14. $-6 - (-8)$
15. $-10 - (-10)$
16. $-8 - (-8)$
17. $7 - 7$
18. $9 - 9$
19. $7 - (-7)$
20. $4 - (-4)$
21. $8 - (-3)$
22. $-7 - 4$
23. $-6 - 8$
24. $6 - (-10)$
25. $-4 - (-9)$
26. $-14 - 2$
27. $2 - 9$
28. $2 - 8$
29. $-6 - (-5)$
30. $-4 - (-3)$
31. $8 - (-10)$
32. $5 - (-6)$
33. $0 - 5$
34. $0 - 6$
35. $-5 - (-2)$
36. $-3 - (-1)$
37. $-7 - 14$
38. $-9 - 16$
39. $0 - (-5)$
40. $0 - (-1)$
41. $-8 - 0$
42. $-9 - 0$
43. $7 - (-5)$
44. $8 - (-3)$
45. $2 - 25$
46. $18 - 63$
47. $-42 - 26$
48. $-18 - 63$
49. $-71 - 2$
50. $-49 - 3$
51. $24 - (-92)$
52. $48 - (-73)$
53. $-50 - (-50)$
54. $-70 - (-70)$
55. $\frac{3}{8} - \frac{5}{8}$
56. $\frac{3}{9} - \frac{9}{9}$
57. $\frac{3}{4} - \frac{2}{3}$
58. $\frac{5}{8} - \frac{3}{4}$
59. $-\frac{3}{4} - \frac{2}{3}$
60. $-\frac{5}{8} - \frac{3}{4}$
61. $-\frac{5}{8} - (-\frac{3}{4})$
62. $-\frac{3}{4} - (-\frac{2}{3})$
63. $6.1 - (-13.8)$
64. $1.5 - (-3.5)$
65. $-3.2 - 5.8$
66. $-2.7 - 5.9$
67. $0.99 - 1$
68. $0.87 - 1$
69. $-79 - 114$
70. $-197 - 216$
71. $0 - (-500)$
72. $500 - (-1000)$
73. $-2.8 - 0$
74. $6.04 - 1.1$
75. $7 - 10.53$
76. $8 - (-9.3)$
77. $\frac{1}{6} - \frac{2}{3}$
78. $-\frac{3}{8} - (-\frac{1}{2})$
79. $-\frac{4}{7} - (-\frac{10}{7})$
80. $\frac{12}{5} - \frac{12}{5}$
81. $-\frac{7}{10} - \frac{10}{15}$
82. $-\frac{4}{18} - (-\frac{2}{9})$
83. $\frac{1}{13} - \frac{1}{12}$
84. $-\frac{1}{7} - (-\frac{1}{6})$

Simplify.

85. $18 - (-15) - 3 - (-5) + 2$
86. $22 - (-18) + 7 + (-42) - 27$
87. $-31 + (-28) - (-14) - 17$
88. $-43 - (-19) - (-21) + 25$
89. $-34 - 28 + (-33) - 44$
90. $39 + (-88) - 29 - (-83)$
91. $-93 - (-84) - 41 - (-56)$
92. $84 + (-99) + 44 - (-18) - 43$
93. $-5 - (-3x) + 3x + 4x - (-12)$
94. $14 - (-5x) + 2x - (-32)$
95. $13x - (-2x) + 45 - (-21)$
96. $8x - (-2x) - 14 - (-5x) + 53$

Problem solving

97. Your total assets are $619.46. You borrow $950 for the purchase of a stereo system. What are your total assets now?

98. You owe a friend $420. The friend decides to cancel $156 of the debt. How much do you owe now?

99. You are in debt $215.50. How much money will you need to make your total assets y dollars?

100. On a winter night the temperature dropped from $-5°C$ to $-12°C$. How many degrees did it drop?

101. The lowest point in Africa is Lake Assal, which is 156 m below sea level. The lowest point in South America is the Valdes Peninsula, which is 40 m below sea level. How much lower is Lake Assal than the Valdes Peninsula?

102. The deepest point in the Pacific Ocean is the Marianas Trench with a depth of 10,415 m. The deepest point in the Atlantic Ocean is the Puerto Rico Trench with a depth of 8,648 m. How much higher is the Puerto Rico Trench than the Marianas Trench?

Subtract.

103. 🧮 $123{,}907 - 433{,}789$

104. 🧮 $23{,}011 - (-60{,}432)$

Simplify.

105. $-(-3 + 2)$

106. $|3 - 7| - 4$

107. $-5|2| + 3|4|$

108. $-|39 - (-12) - 60|$

Tell whether each of the following statements is true or false for all integers m and n. If false, give a counterexample.

109. $n - 0 = 0 - n$.

110. $0 - n = n$.

111. If $m \neq n$, then $m - n \neq 0$.

112. If $m = -n$, then $m + n = 0$.

113. If $m + n = 0$, then m and n are additive inverses.

114. If $m - n = 0$, then $m = -n$.

2.5 MULTIPLICATION

We consider multiplication of rational numbers, but what we say also applies to real numbers. Multiplication of rational numbers is very much like multiplication of numbers of arithmetic. The only difference is that we must determine whether the answer is positive or negative.

Multiplication

To see how to multiply a positive number and a negative number, consider the pattern in the following.

This number decreases by 1 each time.　　　This number decreases by 5 each time.

$$4 \cdot 5 = 20$$
$$3 \cdot 5 = 15$$
$$2 \cdot 5 = 10$$
$$1 \cdot 5 = 5$$
$$0 \cdot 5 = 0$$
$$-1 \cdot 5 = -5$$
$$-2 \cdot 5 = -10$$
$$-3 \cdot 5 = -15$$

According to this pattern, it looks as though the product of a negative and a positive number is negative. That is the case, and we have the first part of the rule for multiplying rational numbers.

> **To multiply a positive and a negative number, multiply their absolute values. The answer is negative.**

EXAMPLES Multiply.

1. $8(-5) = -40$
2. $-\frac{1}{3} \cdot \frac{5}{7} = -\frac{5}{21}$
3. $(-7.2)5 = -36$

How do we multiply two negative numbers? We again look for a pattern.

This number decreases by 1 each time.   This number increases by 5 each time.

$$4 \cdot (-5) = -20$$
$$3 \cdot (-5) = -15$$
$$2 \cdot (-5) = -10$$
$$1 \cdot (-5) = -5$$
$$0 \cdot (-5) = 0$$
$$-1 \cdot (-5) = 5$$
$$-2 \cdot (-5) = 10$$
$$-3 \cdot (-5) = 15$$

According to the pattern it looks as if the product of two negative numbers should be positive. That is actually so, and we have the second part of the rule for multiplying rational numbers.

> **To multiply two negative numbers, multiply their absolute values. The answer is positive.**

We already know how to multiply two positive numbers, of course. The only case we have not considered is multiplying by 0. As with other numbers, the product of any rational number and 0 is 0.

> **THE MULTIPLICATIVE PROPERTY OF ZERO**
>
> For any rational number n,
>
> $$n \cdot 0 = 0.$$
>
> (The product of 0 and any rational number is 0.)

2.5 MULTIPLICATION

Just as for numbers of arithmetic, the commutative and associative laws of multiplication hold for rational numbers. Thus we can choose order and grouping as we please when multiplying.

EXAMPLES Multiply.

4. $-8 \cdot 2(-3) = -16(-3)$ Multiplying the first two numbers
   $= 48$ Multiplying the results

5. $-8 \cdot 2(-3) = 24 \cdot 2$ Multiplying the negatives
   $= 48$

6. $-3(-2)(-5)(4) = 6(-5)(4)$ Multiplying the first two numbers
   $= (-30)4$
   $= -120$

7. $(-\frac{1}{2})(8)(-\frac{2}{3})(-6) = (-4)4$
   $= -16$

8. $-5 \cdot (-2) \cdot (-3) \cdot (-6) = 10 \cdot 18$ Multiplying the first two numbers and the last two numbers
   $= 180$

9. $(-3)(-5)(-2)(-3)(-6) = (-30)(18) = -540$

We can see the following pattern in the results of Examples 8 and 9.

• •

The product of an even number of negative numbers is positive.
The product of an odd number of negative numbers is negative.

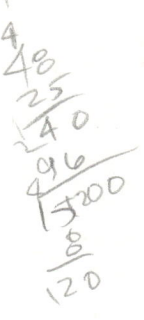

EXERCISE SET 2.5

Multiply.

1. $-8 \cdot 2$
2. $-2 \cdot 5$
3. $-7 \cdot 6$
4. $-9 \cdot 2$
5. $8 \cdot (-3)$
6. $9 \cdot (-5)$
7. $-9 \cdot 8$
8. $-10 \cdot 3$
9. $-8 \cdot (-2)$
10. $-2 \cdot (-5)$
11. $-7 \cdot (-6)$
12. $-9 \cdot (-2)$
13. $15 \cdot (-8)$
14. $-12 \cdot (-10)$
15. $-14 \cdot 17$
16. $-13 \cdot (-15)$
17. $-25 \cdot (-48)$
18. $39 \cdot (-43)$
19. $-3.5 \cdot (-28)$
20. $97 \cdot (-2.1)$
21. $9 \cdot (-8)$
22. $7 \cdot (-9)$
23. $4 \cdot (-3.1)$
24. $3 \cdot (-2.2)$
25. $-6 \cdot (-4)$
26. $-5 \cdot (-6)$
27. $-7 \cdot (-3.1)$
28. $-4 \cdot (-3.2)$
29. $\frac{2}{3} \cdot (-\frac{3}{5})$
30. $\frac{5}{7} \cdot (-\frac{2}{3})$
31. $-\frac{3}{8} \cdot (-\frac{2}{9})$
32. $-\frac{5}{8} \cdot (-\frac{2}{5})$
33. -6.3×2.7
34. -4.1×9.5
35. $-\frac{5}{9} \cdot \frac{3}{4}$
36. $-\frac{8}{3} \cdot \frac{9}{4}$
37. $7 \cdot (-4) \cdot (-3) \cdot 5$
38. $9 \cdot (-2) \cdot (-6) \cdot 7$
39. $-\frac{2}{3} \cdot \frac{1}{2} \cdot (-\frac{6}{7})$
40. $-\frac{1}{8} \cdot (-\frac{1}{4}) \cdot (-\frac{3}{5})$
41. $-3 \cdot (-4) \cdot (-5)$
42. $-2 \cdot (-5) \cdot (-7)$

43. $-2 \cdot (-5) \cdot (-3) \cdot (-5)$
44. $-3 \cdot (-5) \cdot (-2) \cdot (-1)$
45. $\frac{1}{5}(-\frac{2}{9})$
46. $-\frac{3}{5}(-\frac{2}{7})$
47. $-7 \cdot (-21) \cdot 13$
48. $-14 \cdot (34) \cdot 12$
49. $-4 \cdot (-1.8) \cdot 7$
50. $-8 \cdot (-1.3) \cdot (-5)$
51. $-\frac{1}{9}(-\frac{2}{3})(\frac{5}{7})$
52. $-\frac{7}{2}(-\frac{5}{7})(-\frac{2}{5})$
53. $4 \cdot (-4) \cdot (-5) \cdot (-12)$
54. $-2 \cdot (-3) \cdot (-4) \cdot (-5)$
55. $0.07 \cdot (-7) \cdot 6 \cdot (-6)$
56. $80 \cdot (-0.8) \cdot (-90) \cdot (-0.09)$
57. $(-\frac{5}{6})(\frac{1}{8})(-\frac{3}{7})(-\frac{1}{7})$
58. $(\frac{4}{5})(-\frac{2}{3})(-\frac{15}{7})(\frac{1}{2})$
59. $(-14) \cdot (-27) \cdot 0$
60. $7 \cdot (-6) \cdot 5 \cdot (-4) \cdot 3 \cdot (-2) \cdot 1 \cdot 0$
61. $(-8)(-9)(-10)$
62. $(-7)(-8)(-9)(-10)$
63. $(-6)(-7)(-8)(-9)(-10)$
64. $(-5)(-6)(-7)(-8)(-9)(-10)$

Simplify.

65. $-6[(-5) + (-7)]$
66. $7[(-16) + 9]$
67. $-3[(-8) + (-6)](-\frac{1}{7})$
68. $8[17 - (-3)](-\frac{1}{4})$
69. $-(3^5) \cdot [-(2^3)]$
70. $4(2^4) \cdot [-(3^3)] \cdot 6$
71. $(-2)^4$
72. -2^4
73. $(-1)^{23}$
74. -1^{23}

In Exercises 75–78, evaluate when $x = -2$, $y = -4$, and $z = 5$.

75. $xy + z$
76. $-4y + 3x + z$
77. $-6(3x - 5y) + z$
78. $(-9z)(-5x)(-7y)$

79. What must be true of m and n if $-mn$ is to be (a) positive? (b) zero? (c) negative?

2.6 DIVISION

We consider division of rational numbers, but what we say also applies to the real numbers. Division is defined in terms of multiplication. We shall see that this results in rules for division very much like those for multiplication.

Division

Division is defined as follows.

> The quotient $\frac{a}{b}$ (or $a \div b$) is the number, if there is one, which when multiplied by b gives a.

2.6 DIVISION 83

EXAMPLES Divide, if possible. Check your answer.

1. $14 \div (-7)$ — We look for a number which when multiplied by -7 gives 14. That number is -2. Thus $14 \div (-7) = -2$. *Check:* $(-2)(-7) = 14$.

2. $\dfrac{-32}{-4}$ — We look for a number which when multiplied by -4 gives -32. That number is 8. Thus $\dfrac{-32}{-4} = 8$. *Check:* $8(-4) = -32$.

3. $-10 \div 7$ — We look for a number which when multiplied by 7 gives -10. That number is $-\frac{10}{7}$. *Check:* $-\frac{10}{7} \cdot 7 = -10$.

4. $-17 \div 0$ — We look for a number which when multiplied by 0 gives -17. There is no such number because the product of 0 and *any* number is 0.

• •

When we divide a positive number by a negative or a negative number by a positive, the answer is negative. When we divide two negative numbers, the answer is positive.

Example 4 shows why we cannot divide -17 by 0. We can use the same argument to show why we cannot divide any nonzero number b by 0. For $b \div 0$ we look for a number which when multiplied by 0 gives b. There is no such nonzero number because the product of 0 and any number is 0.

On the other hand, if we divide 0 by 0, we look for a number r, such that $r \cdot 0 = 0$. But, $0 \cdot r = 0$ for any number r. Thus it appears that $0 \div 0$ could be any number we choose. This would be very confusing, getting any answer we want when we divide 0 by 0. Thus we agree to exclude division by zero.

• •

We never divide any number by 0.

Reciprocals

When two numbers such as $\frac{1}{2}$ and 2 are multiplied, the result is 1. Such numbers are called *reciprocals* of each other. Every nonzero rational number has a reciprocal, also called a *multiplicative inverse*.

• •

Two numbers whose product is 1 are called *reciprocals* of each other.

EXAMPLES Find the reciprocal of each number.

5. $\frac{7}{8}$ — The reciprocal of $\frac{7}{8}$ is $\frac{8}{7}$ because $\frac{7}{8} \cdot \frac{8}{7} = 1$.

6. -5 — The reciprocal of -5 is $-\frac{1}{5}$ because $-5(-\frac{1}{5}) = 1$.

7. 3.9   The reciprocal of 3.9 is $\frac{1}{3.9}$ because $3.9(\frac{1}{3.9}) = 1$.
8. $-\frac{1}{2}$ The reciprocal of $-\frac{1}{2}$ is -2 because $(-\frac{1}{2})(-2) = 1$.
9. $-\frac{2}{3}$ The reciprocal of $-\frac{2}{3}$ is $-\frac{3}{2}$ because $(-\frac{2}{3})(-\frac{3}{2}) = 1$.
10. $\frac{1}{3/4}$ The reciprocal of $\frac{1}{3/4}$ is $\frac{3}{4}$ because $(\frac{1}{3/4})(\frac{3}{4}) = 1$.

> The reciprocal of a nonzero number a can be named $\frac{1}{a}$.
>
> The reciprocal of a nonzero number $\frac{a}{b}$ can be named $\frac{b}{a}$.

The reciprocal of a positive number is also a positive number, because their product must be a positive number, 1. The reciprocal of a negative number is also a negative number, because their product must be the positive number 1.

It is important *not* to confuse *additive inverse* with *reciprocal*. Keep in mind that the additive inverse of a number is what we add to it to get 0, whereas a reciprocal is what we multiply the number by to get 1. Compare the following.

Number	Additive inverse	Reciprocal
$-\frac{3}{8}$	$\frac{3}{8}$	$-\frac{8}{3}$
19	-19	$\frac{1}{19}$
$\frac{18}{7}$	$-\frac{18}{7}$	$\frac{7}{18}$
-7.9	7.9	$-\frac{1}{7.9}$ or $-\frac{10}{79}$
0	0	Does not exist

Division and Reciprocals

We know we can subtract by adding an inverse. Similarly we can divide by multiplying by a reciprocal.

> For any rational number a and any rational number b,
>
> $$\frac{a}{b} = a \cdot \frac{1}{b}.$$
>
> (To divide, we can multiply by a reciprocal.)

EXAMPLES   Rewrite each division as multiplication.

11. $-4 \div 3$ $-4 \div 3$ is the same as $-4 \cdot \frac{1}{3}$

2.6 DIVISION

12. $\dfrac{6}{-7}$ $\dfrac{6}{-7} = 6\left(-\dfrac{1}{7}\right)$

13. $\dfrac{x+2}{5}$ $\dfrac{x+2}{5} = (x+2)\dfrac{1}{5}$ Parentheses are necessary here.

14. $\dfrac{-17}{\frac{1}{b}}$ $\dfrac{-17}{\frac{1}{b}} = -17 \cdot b$

15. $\dfrac{3}{5} \div \left(-\dfrac{9}{7}\right)$ $\dfrac{3}{5} \div \left(-\dfrac{9}{7}\right) = \dfrac{3}{5}\left(-\dfrac{7}{9}\right)$

When actually doing division calculations, we sometimes multiply by a reciprocal and we sometimes divide directly. With fractional notation, it is usually better to multiply by a reciprocal. With decimal notation, it is usually better to divide directly.

EXAMPLES Divide.

16. $\frac{2}{3} \div \left(-\frac{5}{4}\right) = \frac{2}{3} \cdot \left(-\frac{4}{5}\right) = -\frac{8}{15}$

17. $-\frac{5}{6} \div \left(-\frac{3}{4}\right) = -\frac{5}{6} \cdot \left(-\frac{4}{3}\right) = \frac{20}{18} = \dfrac{10 \cdot 2}{9 \cdot 2} = \frac{10}{9} \cdot \frac{2}{2} = \frac{10}{9}$

Be careful not to change the sign when taking a reciprocal.

18. $-\frac{3}{4} \div \frac{3}{10} = -\frac{3}{4} \cdot \left(\frac{10}{3}\right) = -\frac{30}{12} = -\frac{5}{2} \cdot \frac{6}{6} = -\frac{5}{2}$

19. $-27.9 \div (-3) = \dfrac{-27.9}{-3} = 9.3$ Do the long division $3\overline{\smash{)}27.9}$ giving 9.3.

20. $-6.3 \div 2.1 = -3$ Do the long division $2.1\overline{\smash{)}6.3}$ giving 3. Make the answer negative.

Rational Numbers as Quotients of Integers

We can also describe the rational numbers as quotients of integers.

> The set of rational numbers consists of all numbers that can be named with fractional notation $\dfrac{a}{b}$, where a and b are integers and b is not 0.

The following are examples of rational numbers:

$\frac{5}{4}$, $\frac{-3}{8}$, $\frac{10}{1}$,   8, -467,   0, $\frac{13}{-5}$, $-\frac{2}{3}$, -1.25, 0.3333.

When we multiply numbers of arithmetic using fractional notation, we multiply numerators and multiply denominators. The same is true for rational numbers, even if the numerators or denominators are negative.

EXAMPLE 21   Multiply: $\dfrac{-2}{5} \cdot \dfrac{-8}{-3}$.

$$\dfrac{-2}{5} \cdot \dfrac{-8}{-3} = \dfrac{(-2)(-8)}{5(-3)} = \dfrac{16}{-15}$$ Multiplying numerators and multiplying denominators

EXAMPLE 22   Multiply: $\dfrac{4}{-7} \cdot \dfrac{-1}{-1}$.

$$\dfrac{4}{-7} \cdot \dfrac{-1}{-1} = \dfrac{4(-1)}{(-7)(-1)} = \dfrac{-4}{7}$$

In Example 22, we have verified that $4/-7 = -4/7$ because $-1/-1 = 1$. Similarly, $-2/-3 = 2/3$ because

$$\dfrac{-2}{-3} = 1 \cdot \dfrac{-2}{-3} = \dfrac{-1}{-1} \cdot \dfrac{-2}{-3} = \dfrac{(-1)(-2)}{(-1)(-3)} = \dfrac{2}{3}.$$

We can generalize this to a method for making sign changes in fractional notation.

For any numbers a and b, $b \neq 0$,

$$-\dfrac{a}{b} = \dfrac{-a}{b} = \dfrac{a}{-b}, \quad \dfrac{-a}{-b} = \dfrac{a}{b}, \quad \text{and} \quad -\dfrac{-a}{-b} = -\dfrac{a}{b}.$$

EXERCISE SET 2.6

Divide, if possible. Check each answer.

1. $36 \div (-6)$
2. $\dfrac{28}{-7}$
3. $\dfrac{26}{-2}$
4. $26 \div (-13)$

5. $\dfrac{-16}{8}$
6. $-22 \div (-2)$
7. $\dfrac{-48}{-12}$
8. $-63 \div (-9)$

9. $\dfrac{-72}{9}$
10. $\dfrac{-50}{25}$
11. $-100 \div (-50)$
12. $\dfrac{-200}{8}$

13. $-108 \div 9$
14. $\dfrac{-64}{-7}$
15. $\dfrac{200}{-25}$
16. $-300 \div (-13)$

17. $\dfrac{75}{0}$
18. $\dfrac{0}{-5}$
19. $\dfrac{88}{-9}$
20. $\dfrac{-145}{-5}$

2.6 DIVISION

Find the reciprocal.

21. $\dfrac{15}{7}$
22. $\dfrac{3}{8}$
23. $-\dfrac{47}{13}$
24. $-\dfrac{31}{12}$
25. 13
26. -10
27. 4.3
28. -8.5
29. $-\dfrac{1}{7.1}$
30. $\dfrac{1}{-4.9}$
31. $\dfrac{p}{q}$
32. $\dfrac{s}{t}$
33. $\dfrac{1}{4y}$
34. $\dfrac{-1}{8a}$
35. $\dfrac{2a}{3b}$
36. $\dfrac{-4y}{3x}$

Rewrite each division as multiplication.

37. $3 \div 19$
38. $4 \div (-9)$
39. $\dfrac{6}{-13}$
40. $-\dfrac{12}{41}$
41. $\dfrac{13.9}{-1.5}$
42. $-\dfrac{47.3}{21.4}$
43. $\dfrac{x}{\frac{1}{y}}$
44. $\dfrac{13}{x}$
45. $\dfrac{3x+4}{5}$
46. $\dfrac{4y-8}{-7}$
47. $\dfrac{5a-b}{5a+b}$
48. $\dfrac{2x+x^2}{x-5}$

Divide.

49. $\dfrac{3}{4} \div \left(-\dfrac{2}{3}\right)$
50. $\dfrac{7}{8} \div \left(-\dfrac{1}{2}\right)$
51. $-\dfrac{5}{4} \div \left(-\dfrac{3}{4}\right)$
52. $-\dfrac{5}{9} \div \left(-\dfrac{5}{6}\right)$
53. $-\dfrac{2}{7} \div \left(-\dfrac{4}{9}\right)$
54. $-\dfrac{3}{5} \div \left(-\dfrac{5}{8}\right)$
55. $-\dfrac{3}{8} \div \left(-\dfrac{8}{3}\right)$
56. $-\dfrac{5}{8} \div \left(-\dfrac{6}{5}\right)$
57. $-6.6 \div 3.3$
58. $-44.1 \div (-6.3)$
59. $\dfrac{-11}{-13}$
60. $\dfrac{-1.9}{20}$
61. $\dfrac{48.6}{-3}$
62. $\dfrac{-17.8}{3.2}$

Simplify.

63. $\dfrac{(-9)(-8)+(-3)}{25}$
64. $\dfrac{-3(-9)+7}{-4}$
65. $\dfrac{(-2)^7}{(-4)^2}$
66. $\dfrac{(-3)^4}{-9}$
67. $\left(-5\dfrac{3}{7}\right) \div 4\dfrac{2}{5}$
68. $\dfrac{10}{7} \div (-0.25)$
69. $\left(-2\dfrac{2}{3}\right) \div \left(-\dfrac{40}{15}\right)$

70. ▦ Find the reciprocal of -10.5.

71. Determine whether division of rational numbers is commutative. That is, is it true that $a \div b = b \div a$ for all rational numbers a and b?

72. Determine whether division of rational numbers is associative. That is, is it true that $a \div (b \div c) = (a \div b) \div c$?

73. Determine those rational numbers that are their own reciprocals.

74. Determine those rational numbers a for which the additive inverse of a is the same as the reciprocal of a.

75. ▦ What should happen if you enter a number on a calculator and press the reciprocal key twice? Why?

Tell whether each expression represents a positive number or a negative number when m and n are positive.

76. $\dfrac{-n}{m}$ 77. $\dfrac{-n}{-m}$ 78. $-\left(\dfrac{-n}{m}\right)$ 79. $-\left(\dfrac{n}{-m}\right)$ 80. $-\left(\dfrac{-n}{-m}\right)$

2.7 USING THE DISTRIBUTIVE LAWS

We studied a distributive law in Chapter 1. There are two distributive laws that hold for the rational numbers and real numbers. Here we learn how to use them to multiply, factor, and collect like terms.

Multiplying

For numbers of arithmetic we know that multiplication is distributive over addition. For rational numbers the distributive law also holds.

> **THE DISTRIBUTIVE LAW OF MULTIPLICATION OVER ADDITION**
>
> For any rational numbers a, b, and c,
>
> $$a(b + c) = ab + ac.$$

There is another distributive law that relates multiplication and subtraction. This law says that to multiply by a difference we can either subtract and then multiply, or multiply and then subtract.

> **THE DISTRIBUTIVE LAW OF MULTIPLICATION OVER SUBTRACTION**
>
> For any rational numbers a, b, and c,
>
> $$a(b - c) = ab - ac.$$

What do we mean by the *terms* of an expression? When they are all separated by plus signs, it is easy to tell. If there are subtraction signs, we can find an equivalent expression that uses addition signs.

EXAMPLE 1 What are the terms of $3x - 4y + 2z$?

$$3x - 4y + 2z = 3x + (-4y) + 2z \quad \text{Separating parts with + signs}$$

The terms are $3x$, $-4y$, and $2z$.

The distributive laws are a basis for a procedure called *multiplying*. In an expression such as $8(a + 2b - 7)$, we multiply each term inside the parentheses by 8:

$$8 \cdot (a + 2b - 7) = 8 \cdot a + 8 \cdot 2b - 8 \cdot 7 = 8a + 16b - 56.$$

2.7 USING THE DISTRIBUTIVE LAWS

EXAMPLES Multiply.

2. $9(x - 5) = 9x - 9(5)$ Using the distributive law of multiplication
   $ = 9x - 45$ over subtraction

3. $\frac{4}{3}(s - t + w) = \frac{4}{3}s - \frac{4}{3}t + \frac{4}{3}w$ Using both distributive laws

4. $-4(x - 2y + 3z) = -4 \cdot x - (-4)(2y) + (-4)(3z)$
   $ = -4x - (-8y) + (-12z)$
   $ = -4x + 8y - 12z$

Factoring

Factoring is the reverse of multiplying. To factor, we can use the distributive laws in reverse:

$$ab + ac = a(b + c),$$
$$ab - ac = a(b - c).$$

Look at Example 2. To *factor* $9x - 45$, we find an equivalent expression that is a product, $9(x - 5)$. When all the terms of an expression have a factor in common, we can "factor it out" using the distributive laws. Note that:

$9x$ has the factors $9, x, 3, -3, 1, -1, -9$;
-45 has the factors $45, -1, 9, -5, -9, 5, 3, -3, 15, -15, -45, 1$.

We usually remove the largest common factor. In the case of $9x - 45$, that factor is 9.

Remember that an expression is factored when we find an equivalent expression that is a product.

EXAMPLES Factor.

5. $5x - 10 = 5 \cdot x - 5 \cdot 2$
   $ = 5(x - 2)$

6. $ax - ay + az = a(x - y + z)$

7. $9x + 27y - 9 = 9 \cdot x + 9 \cdot 3y - 9 \cdot 1$
   $ = 9(x + 3y - 1)$

Note that $3(3x + 9y - 3)$ is also equivalent to $9x + 27y - 9$, but it is *not* the desired form. Factor out the largest common factor.

EXAMPLES Factor.

8. $5x - 5y = 5(x - y)$
9. $-3x + 6y - 9z = -3(x - 2y + 3z)$
10. $18z - 12x - 24 = 6(3z - 2x - 4)$
11. $\frac{1}{2}x + \frac{3}{2}y - \frac{1}{2} = \frac{1}{2}(x + 3y - 1)$

Collecting Like Terms

The process of *collecting like terms* is also based on the distributive laws. In particular, the distributive law of multiplication over subtraction allows us to collect like terms without rewriting subtractions as additions. We can again apply the distributive law "on the right" because of the commutative law of multiplication.

EXAMPLES Collect like terms.

12. $4x - 2x = (4 - 2)x = 2x$ Factoring out the x
13. $2x + 3y - 5x - 2y = 2x - 5x + 3y - 2y$
    $= (2 - 5)x + (3 - 2)y = -3x + y$
14. $3x - x = (3 - 1)x = 2x$
15. $x - 0.24x = 1 \cdot x - 0.24x = (1 - 0.24)x = 0.76x$
16. $x - 6x = 1 \cdot x - 6 \cdot x = (1 - 6)x = -5x$
17. $4x - 7y + 9x - 5 + 3y - 8 = 13x - 4y - 13$

EXERCISE SET 2.7

Multiply.

1. $7(4 - 3)$
2. $15(8 - 6)$
3. $-3(3 - 7)$
4. $1.2(5 - 2.1)$
5. $4.1(6.3 - 9.4)$
6. $-\frac{8}{9}(\frac{2}{3} - \frac{5}{3})$
7. $7(x - 2)$
8. $5(x - 8)$
9. $-7(y - 2)$
10. $-9(y - 7)$
11. $-9(-5x - 6y + 8)$
12. $-7(-2x - 5y + 9)$
13. $-4(x - 3y - 2z)$
14. $8(2x - 5y - 8z)$
15. $3.1(-1.2x + 3.2y - 1.1)$
16. $-2.1(-4.2x - 4.3y - 2.2)$

Factor.

17. $8x - 24$
18. $10x - 50$
19. $32 - 4y$
20. $24 - 6m$
21. $8x + 10y - 22$
22. $9a + 6b - 15$
23. $ax - 7a$
24. $by - 9b$
25. $ax - ay - az$
26. $cx + cy - cz$

Give the terms of each expression.

27. $4x + 3z$
28. $8x - 1.4y$
29. $7x + 8y - 9z$
30. $8a + 10b - 18c$
31. $12x - 13.2y + \frac{5}{8}z - 4.5$
32. $-7.8a - 3.4y - 8.7z - 12.4$

Collect like terms.

33. $11x - 3x$
34. $9t - 17t$
35. $6n - n$
36. $y - 17y$

37. $9x + 2y - 5x$
38. $8y - 3z + 4y$
39. $11x + 2y - 4x - y$
40. $13a + 9b - 2a - 4b$
41. $2.7x + 2.3y - 1.9x - 1.8y$
42. $6.7a + 4.3b - 4.1a - 2.9b$
43. $\frac{1}{5}x + \frac{4}{5}y + \frac{2}{5}x - \frac{1}{5}y$
44. $\frac{7}{8}x + \frac{5}{8}y + \frac{1}{8}x - \frac{3}{8}y$

Factor.

45. $2\pi r + \pi rs$
46. $\frac{1}{2}ah + \frac{1}{2}bh$

Simplify.

47. $8x - 9 - 2(7 - 5x)$
48. $-9(y + 7) - 6(y + 3)$
49. $\dfrac{5x - 15}{5} + \dfrac{2x + 6}{2}$
50. $\dfrac{x^2 - 2x}{x} + \dfrac{3 - 9x}{3}$

Write an algebraic expression. Simplify each expression, if possible.

51. A principal of P dollars was invested in a savings account at 8% simple interest. How much was in the account after 1 year?

52. The population of a town is P. After a 6% increase, what was the new population?

53. Eight times the difference of x and y.

54. Nine times the difference of y and z, increased by $3z$.

55. Three times the sum of a and b, decreased by $7a$.

56. The total cost if you buy x cassette tapes at $2.95 on Monday and y cassettes at the same price on Wednesday.

57. The total intake of a store when branch A sells x microcomputers at $2500 and branch B sells y microcomputers at the same price.

58. An investor has 5420 shares of a stock bought at $41\frac{1}{8}$. The stock is now worth $37\frac{3}{4}$. Show two ways of determining how much has been lost. Solve.

2.8 MULTIPLYING BY −1 AND SIMPLIFYING

We expand our ability to manipulate expressions in this section. We first consider additive inverses of sums and differences. Then we simplify expressions involving parentheses.

Inverses of Sums

What happens when we multiply a rational number by -1?

EXAMPLES Multiply.

1. $-1 \cdot 7 \qquad -1 \cdot 7 = -7$
2. $-1(-5) \qquad -1(-5) = 5$
3. $-1 \cdot 0 \qquad -1 \cdot 0 = 0$

From these examples it appears that when we multiply a number by -1, we get the additive inverse of that number.

THE PROPERTY OF -1

For any rational number a,
$$-1 \cdot a = -a.$$
(Negative one times a is the additive inverse of a.)

The property of -1 enables us to find certain expressions equivalent to additive inverses of sums.

EXAMPLES Rename each additive inverse without parentheses.

4. $-(3 + x) = -1(3 + x)$ Using the property of -1
   $= -1 \cdot 3 + (-1)x$ Using a distributive law
   $= -3 + (-x)$ Using the property of -1
   $= -3 - x$

5. $-(3x + 2y + 4) = -1(3x + 2y + 4)$ Using the property of -1
   $= -1(3x) + (-1)(2y) + (-1)4$ Using a distributive law
   $= -3x - 2y - 4$ Using the property of -1

Examples 4 and 5 illustrate an important property of rational numbers.

THE INVERSE OF A SUM

For any rational numbers a and b,
$$-(a + b) = -a + (-b).$$
(The inverse of a sum is the sum of the inverses.)

If we want to remove parentheses in an expression like
$$-(x - 5),$$
we can do so by replacing each term in the parentheses by its additive inverse ("change the sign of every term"). Doing so for $-(x - 5)$ we obtain $-x + 5$ as an equivalent expression. The same is true for a sum or difference of more than two terms.

EXAMPLES   Rename each additive inverse without parentheses.

6. $-(5 - y) = -5 + y$ Changing the sign of each term
7. $-(2a - 7b - 6) = -2a + 7b + 6$

Removing Parentheses and Simplifying

When a sum is added as in $5x + (2x + 3)$ we can simply remove, or drop, the parentheses and collect like terms. On the other hand, when a sum is subtracted, as in $3x - (4x + 2)$, we can subtract by adding an inverse, as usual. We then remove parentheses by changing the sign of each term inside the parentheses and collect like terms.

EXAMPLE 8   Remove parentheses and simplify.

$$\begin{aligned}
3x - (4x + 2) &= 3x + (-(4x + 2)) &&\text{Adding the inverse of } (4x + 2)\\
&= 3x + (-4x - 2) &&\text{Changing the sign of each term inside}\\
& &&\text{the parentheses}\\
&= 3x - 4x - 2\\
&= -x - 2 &&\text{Collecting like terms}
\end{aligned}$$

EXAMPLES   Remove parentheses and simplify.

9. $5y - (3y + 4) = 5y - 3y - 4$ Removing parentheses by changing the sign of every term
$= 2y - 4$

10. $3y - 2 - (2y - 4) = 3y - 2 - 2y + 4$
$= y + 2$

Next, consider subtracting an expression consisting of several terms preceded by a number other than 1 or -1.

EXAMPLES   Remove parentheses and simplify.

11. $\begin{aligned}[t]x - 3(x + y) &= x + (-3(x + y)) &&\text{Adding the inverse of } 3(x + y)\\
&= x + (-3x - 3y) &&\text{Multiplying } x + y \text{ by } -3\\
&= x - 3x - 3y &&\text{Removing parentheses}\\
&= -2x - 3y &&\text{Collecting like terms}\end{aligned}$

12. $\begin{aligned}[t]3y - 2(4y - 5) &= 3y + (-2(4y - 5)) &&\text{Adding the inverse of } 2(4y - 5)\\
&= 3y + (-8y + 10)\\
&= 3y - 8y + 10\\
&= -5y + 10\end{aligned}$

Parentheses Within Parentheses

Sometimes parentheses occur within parentheses. When that happens we can use parentheses of different shapes, such as [ ], called "brackets," or { }, called "braces."

> When parentheses occur within parentheses, do the computations in the innermost ones first. Then work from the inside out.

EXAMPLES Simplify.

13. $[3 - (7 + 3)] = [3 - 10]$ Working with the innermost parentheses first. Computing $7 + 3$.
    $= -7$

14. $\{8 - [9 - (12 + 5)]\} = \{8 - [9 - 17]\}$ Computing $12 + 5$
    $= \{8 - [-8]\}$ Computing $9 - 17$
    $= 16$

15. $[(-4) \div (-\frac{1}{4})] \div \frac{1}{4} = [(-4) \cdot (-4)] \div \frac{1}{4}$ Working with the innermost parentheses first
    $= 16 \div \frac{1}{4}$ Computing $(-4) \div (-\frac{1}{4})$
    $= 16 \cdot 4$
    $= 64$

16. $4(2 + 3) - \{7 - [4 - (8 + 5)]\}$
    $= 4 \cdot 5 - \{7 - [4 - 13]\}$ Working with the innermost parentheses first
    $= 20 - \{7 - [-9]\}$ Computing $4 \cdot 5$ and $4 - 13$
    $= 20 - 16$ Computing $7 - [-9]$
    $= 4$

17. $[5(x + 2) - 3x] - [3(y + 2) - 7(y - 3)]$
    $= [5x + 10 - 3x] - [3y + 6 - 7y + 21]$ Working with the innermost parentheses first
    $= [2x + 10] - [-4y + 27]$ Collecting like terms within parentheses
    $= 2x + 10 + 4y - 27$ Removing parentheses
    $= 2x + 4y - 17$ Collecting like terms

Computer Language and Order of Operations

When several operations are to be done in a calculation or a problem, in what order should they be done? We have agreements about such calculations, and we also use parentheses. In handwritten algebra, little, if any, difficulty arises in knowing in which order operations are to be performed. Consider, for example,

$$8 + 5^3.$$

It is rather natural to first find the power 5^3, and then add it to 8:

$$8 + 5^3 = 8 + 125 = 133.$$

2.8 MULTIPLYING BY −1 AND SIMPLIFYING

With computers, however, there may be some problem in knowing in which order operations are to be performed because the symbolism is different. The order in which a computer performs operations is the same as it is in ordinary algebra, but because of the different symbolism it is important to give some thought to order of operations when using a computer language. The computer symbols we will use to illustrate are from the computer language known as BASIC.

BASIC notation	Algebra notation
A + B	$a + b$
A − B	$a - b$
A * B	ab, or $a \cdot b$
A/B	$\dfrac{a}{b}$, or $a \div b$
A^B	a^b
( )	( )

We now state the rules for order of operations.

RULES FOR ORDER OF OPERATIONS

1. **Do all calculations within parentheses before operations outside.**
2. **Evaluate all exponential expressions.**
3. **Do all multiplications and divisions.**
4. **Do all additions and subtractions.**

If you are computing by hand, it can be helpful to think of each division as a multiplication by a reciprocal. Then, by the commutative and associative laws, you know that you can do the results in any order you prefer. We will do these operations in order from left to right as a computer would do them. A similar statement holds with respect to the order of addition and subtraction.

Let us apply these rules to sentences we see in arithmetic or algebra and then with sentences in BASIC language.

EXAMPLE 18 Simplify: $-34 \cdot 56 - 17$.

There are no parentheses or powers so we start with the third step.

$-34 \cdot 56 - 17 = -1904 - 17$ Carrying out all multiplications and divisions in order from left to right

$\qquad\qquad\quad = -1921$ Carrying out all additions and subtractions in order from left to right

EXAMPLE 19 Simplify: $2^4 + 51 \cdot 4 - (37 + 23 \cdot 2)$.

$$2^4 + 51 \cdot 4 - (37 + 23 \cdot 2) = 2^4 + 51 \cdot 4 - (37 + 46)$$ Carry out all operations inside parentheses first. Multiply 23 by 2.

$$= 2^4 + 51 \cdot 4 - 83$$ Complete the addition inside parentheses.

$$= 16 + 51 \cdot 4 - 83$$ Evaluate exponential expressions.

$$= 16 + 204 - 83$$ Do all multiplications.

$$= 220 - 83$$ Do all additions and subtractions in order from left to right.

$$= 137$$

Now let us apply the rules to BASIC language.

EXAMPLE 20 Simplify: $3 + 4 * 7 - (7 + 5)\char`\^2$.

$$3 + 4 * 7 - (7 + 5)\char`\^2 = 3 + 4 * 7 - 12\char`\^2$$ Carry out all operations inside parentheses first.

$$= 3 + 4 * 7 - 144$$ Evaluate exponential expressions.

$$= 3 + 28 - 144$$ Do the multiplication.

$$= 31 - 144$$ Do all additions and subtractions in order from left to right.

$$= -113$$

(*Optional*) To program a computer to perform a task, it is important to be able to write BASIC notation for algebraic symbolism.

EXAMPLES Write BASIC notation for each of the following.

21. $\left(\dfrac{54 \cdot 37}{2}\right)^3$ ((54*37)/2)^3

22. $b^2 - 4ac$ B^2 − 4 * A * C

23. $c + d - \dfrac{a^2}{b}$ C + D − A^2/B

EXERCISE SET 2.8

Rename each additive inverse without parentheses.

1. $-(2x + 7)$
2. $-(3x + 5)$
3. $-(5x - 8)$
4. $-(6x - 7)$
5. $-(4a - 3b + 7c)$
6. $-(5x - 2y - 3z)$
7. $-(6x - 8y + 5)$
8. $-(8x + 3y + 9)$
9. $-(3x - 5y - 6)$
10. $-(6a - 4b - 7)$
11. $-(-8x - 6y - 43)$
12. $-(-2a + 9b - 5c)$

Remove parentheses and simplify.

13. $9x - (4x + 3)$
14. $7y - (2y + 9)$
15. $2a - (5a - 9)$
16. $11n - (3n - 7)$
17. $2x + 7x - (4x + 6)$
18. $3a + 2a - (4a + 7)$
19. $2x - 4y - 3(7x - 2y)$
20. $3a - 7b - 1(4a - 3b)$
21. $15x - y - 5(3x - 2y + 5z)$
22. $4a - b - 4(5a - 7b + 8c)$

Simplify.

23. $[9 - 2(5 - 4)]$
24. $[6 - 5(8 - 4)]$
25. $8[7 - 6(4 - 2)]$
26. $10[7 - 4(7 - 5)]$
27. $[4(9 - 6) + 11] - [14 - (6 + 4)]$
28. $[7(8 - 4) + 16] - [15 - (7 + 3)]$
29. $[10(x + 3) - 4] + [2(x - 1) + 6]$
30. $[9(x + 5) - 7] + [4(x - 12) + 9]$
31. $[7(x + 5) - 19] - [4(x - 6) + 10]$
32. $[6(x + 4) - 12] - [5(x - 8) + 11]$
33. $3\{[7(x - 2) + 4] - [2(2x - 5) + 6]\}$
34. $4\{[8(x - 3) + 9] - [4(3x - 7) + 2]\}$
35. $[(-24) \div (-3)] \div (-\frac{1}{2})$
36. $[32 \div (-2)] \div (-2)$
37. $16 \cdot (-24) + 50$
38. $10 \cdot 20 - 15 \cdot 24$
39. $2^4 + 2^3 - 10$
40. $40 - 3^2 - 2^3$
41. $5^3 + 26 \cdot 71 - (16 + 25 \cdot 3)$
42. $4^3 + 10 \cdot 20 + 8^2 - 23$
43. ▦ $3000 \cdot (1 + 0.16)^3$
44. ▦ $2000 \cdot (3 + 1.14)^2$

Simplify.

45. $4 * 5 - 2 * 6 + 4$
46. $8 * (7 - 3)/4$
47. $4 * (6 + 8)/(4 + 3)$
48. $4\textasciicircum 3/8$
49. $(2 * (5 - 3))\textasciicircum 2$
50. $5\textasciicircum 3 - 7\textasciicircum 2$

Write BASIC notation for each of the following.

51. $a^2 + 2ab + b^2$
52. $a^3 - b^3$
53. $\dfrac{2(3 - b)}{c}$
54. $\dfrac{a + b}{c - d}$
55. $\dfrac{a}{b} - \dfrac{c}{d}$
56. $\left(\dfrac{a^2}{b^3}\right)^3$

Find an equivalent expression by enclosing the last three terms in parentheses preceded by a minus sign.

57. $6y + 2x - 3a + c$
58. $x - y - a - b$
59. $6m + 3n - 5m + 4b$

60. If $-(a + b)$ is $-a + (-b)$ what should be the sum of $(a + b)$ and $-a + (-b)$? Show that your answer is correct.

Simplify.

61. $z - \{2z - [3z - (4z - 5z) - 6z] - 7z\} - 8z$
62. $\{x - [f - (f - x)] + [x - f]\} - 3x$
63. $x - \{x - 1 - [x - 2 - (x - 3 - \{x - 4 - [x - 5 - (x - 6)]\})]\}$

If $n > 0$, $m > 0$, and $n \neq m$, which of the following are true?

64. $-n + m = n - m$
65. $-n + m = -(n + m)$
66. $-n - m = -(n + m)$
67. $-n - m = -(n - m)$

68. $n(-n - m) = -n^2 + nm$
69. $-m(n - m) = -(mn + m^2)$
70. $-m(-n + m) = m(n - m)$
71. $-n(-n - m) = n(n + m)$
72. Determine whether it is true that, for any rational number x, $(-x)^2 = x^2$. Explain why or why not.
73. Determine whether it is true that, for any rational numbers a and b, $ab = (-a)(-b)$. Explain why or why not.
74. Determine whether it is true that, for any rational numbers a and b, $-(ab) = (-a)b = a(-b)$. Explain why or why not.

SUMMARY AND REVIEW: CHAPTER 2

The following contains a summary of what you should be able to do after completing this chapter. The review exercises are for practice. Answers are at the back of the book. If you miss an exercise, restudy the section indicated alongside the answer.

You should be able to:

Tell which integers correspond to real-world situations.

1. Tell which integers correspond to this situation: Mike has a debt of $45 and Joe has $72 in his savings account.

Find the absolute value of any rational number.

Find the absolute value.

2. $|-38|$
3. $|q|$ when $q = 7$
4. $|\frac{5}{2}|$

Graph rational numbers on a number line and tell which of two numbers is greater writing a true sentence using < or >.

Graph each number on a number line.

5. -2.5
6. $\frac{8}{9}$

Write a true sentence using < or >.

7. -3 \quad 10
8. -1 \quad -6
9. 0.126 \quad -12.6
10. $-\frac{2}{3}$ \quad $-\frac{1}{10}$

Find the additive inverse and the reciprocal of a rational number.

Find the additive inverse of each.

11. 3.8
12. $-\frac{3}{4}$
13. Find $-x$ when x is -34.
14. Find $-(-x)$ when x is 5.

Find the reciprocal.

15. $\frac{3}{8}$
16. -7
17. $\frac{x}{2y}$

Add, subtract, multiply, and divide rational numbers.

Compute and simplify.

18. $4 + (-7)$
19. $-\frac{2}{3} + \frac{1}{12}$
20. $6 + (-9) + (-8) + 7$
21. $-3.8 + 5.1 + (-12) + (-4.3) + 10$
22. $-3 - (-7)$
23. $-\frac{9}{10} - \frac{1}{2}$

TEST: CHAPTER 2

24. $-3.8 - 4.1$
25. $-9 \cdot (-6)$
26. $-2.7(3.4)$
27. $\frac{2}{3} \cdot (-\frac{3}{7})$
28. $3 \cdot (-7) \cdot (-2) \cdot (-5)$
29. $35 \div (-5)$
30. $-5.1 \div 1.7$
31. $-\frac{3}{5} \div (-\frac{4}{5})$

Solve problems using the five-step problem-solving process.

32. On first, second, and third down a football team had these gains and losses: 5-yd gain, 12-yd loss, and 15-yd gain. Find the total gain (or loss).

33. Your total assets are $170. You borrow $300. What are your total assets now?

Use the distributive laws to multiply and factor algebraic expressions and to simplify expressions by collecting like terms.

Multiply.
34. $5(3x - 7)$
35. $-2(4x - 5)$
36. $10(0.4x + 1.5)$
37. $-8(3 - 6x)$

Factor.
38. $2x - 14$
39. $6x - 6$
40. $5x + 10$
41. $12 - 3x$

Collect like terms.
42. $11a + 2b - 4a - 5b$
43. $7x - 3y - 9x + 8y$
44. $6x + 3y - x - 4y$
45. $-3a + 9b + 2a - b$

Simplify expressions by removing parentheses and collecting like terms.

Remove parentheses and simplify.
46. $13 - 4 + 8 - (-2)$
47. $20 - 17 - 12 + 13 - (-4)$
48. $-5 - 3x + 8 - (-9x)$
49. $4y - 19 - (-7y) + 3$
50. $2a - (5a - 9)$
51. $3(b + 7) - 5b$
52. $3[11 - 3(4 - 1)]$
53. $2[6(y - 4) + 7]$
54. $[8(x + 4) - 10] - [3(x - 2) + 4]$
55. $5\{[6(x - 1) + 7] - [3(3x - 4) + 8]\}$

56. Simplify: $-|\frac{7}{8} - (-\frac{1}{2}) - \frac{3}{4}|$.

57. If $0.090909\ldots = \frac{1}{11}$ and $0.181818\ldots = \frac{2}{11}$, what rational number is named by:
  a) $0.272727\ldots$?
  b) $0.909090\ldots$?

58. Solve: $|y| = -1.237$.

59. Simplify: $(|2.7 - 3| + 3^2 - |-3|) \div (-3)$.

TEST: CHAPTER 2

Write a true sentence using $<$ or $>$.
1. $-4 \quad 0$
2. $-3 \quad -8$
3. $-0.78 \quad -0.87$
4. $-\frac{1}{8} \quad \frac{1}{2}$

Find the absolute value.
5. $|-7|$
6. $|\frac{9}{4}|$

Find the additive inverse of each.

7. $\frac{2}{3}$

8. -1.4

9. Find $-x$ when x is -8.

Find the reciprocal.

10. -2

11. $\frac{4}{7}$

Compute and simplify.

12. $3.1 + (-4.7)$
13. $-8 + 4 + (-7) + 3$
14. $-\frac{1}{5} + \frac{3}{8}$
15. $2 - (-8)$
16. $3.2 - 5.7$
17. $\frac{1}{8} - (-\frac{3}{4})$
18. $4 \cdot (-12)$
19. $-\frac{1}{2} \cdot (-\frac{3}{8})$
20. $-45 \div 5$
21. $-\frac{3}{5} \div (-\frac{4}{5})$
22. $4.864 \div (-0.5)$

23. Wendy had $43 in her savings account. She withdrew $25. Then she made a deposit of $30. How much was in her savings account?

Multiply.

24. $3(6 - x)$
25. $-5(y - 1)$

Factor.

26. $12 - 22x$
27. $7x + 21 + 14y$

Simplify.

28. $6 + 7 - 4 - (-3)$
29. $5x - (3x - 7)$
30. $4(2a - 3b) + a - 7$
31. $4\{3[5(y - 3) + 9] + 2(y + 8)\}$

SOLVING EQUATIONS AND PROBLEMS

3

The price of an automobile was decreased to a sale price of $13,559. This was a 9% reduction. What was the original price of the automobile? An equation can be used to solve such a problem.

In this chapter we reformulate the equation-solving procedures we introduced in Chapter 1. Then we consider other manipulations for equation solving. All this greatly expands our equation-solving ability.

In the latter part of the chapter we introduce formulas. Formulas arise in many kinds of applications. The skill of solving a formula for a certain letter is particularly important and is the main objective. Finally, we use our new equation-solving ability to expand our problem-solving ability.

3.1 THE ADDITION PRINCIPLE

In this section we reformulate an equation-solving principle called the *addition principle*. We actually used this principle in Chapter 1, but now we can add negative numbers on both sides of an equation.

Using the Addition Principle

Recall the following.

> The replacements that make an equation true are called *solutions*. To *solve* an equation means to find all of its solutions.

There are various ways to solve equations. We develop one now, using an idea called the *addition principle*. An equation $a = b$ says that a and b stand for the same number. Suppose this is true, and we add a number c to the number a. We get the same answer if we add c to b, because a and b are the same number.

> **THE ADDITION PRINCIPLE**
>
> If an equation $a = b$ is true, then
> $$a + c = b + c$$
> is true for any number c.

The idea in using this principle is to obtain an equation for which you can *see* what the solution is.

EXAMPLE 1 Solve: $x + 5 = -7$.

$$x + 5 = -7$$
$$x + 5 + (-5) = -7 + (-5) \quad \text{Using the addition principle;}$$
$$ \quad \text{adding } -5 \text{ on both sides}$$
$$x + 0 = -7 + (-5) \quad \text{Simplifying}$$
$$x = -12$$

We can see that the solution of $x = -12$ is the number -12. To check the answer we substitute -12 in the original equation.

Check:
$$\begin{array}{c|c} x + 5 = -7 \\ \hline -12 + 5 & -7 \\ -7 & \end{array}$$

The solution of the original equation is -12.

3.1 THE ADDITION PRINCIPLE

In Example 1, to get x alone, we added the inverse of 5. This "got rid of" the 5 on the left. When using the addition principle, we sometimes say we "add the same number on both sides of an equation." We started with $x + 5 = -7$, and using the addition principle we derived a simple equation $x = -12$ for which it was easy to "*see*" the solution. Equations with the same solutions, such as $x + 5 = -7$ and $x = -12$, are called *equivalent equations*.

The addition principle also allows us to subtract on both sides of an equation. This is because subtracting is the same as adding an inverse. In Example 1 we could have subtracted 5 on both sides of the equation.

Now we "undo" a subtraction using the addition principle.

EXAMPLE 2 Solve: $y - 8.4 = -6.5$.

$$y - 8.4 = -6.5$$
$$y - 8.4 + 8.4 = -6.5 + 8.4 \qquad \text{Adding 8.4 to get rid of } -8.4 \text{ on the left}$$
$$y = 1.9$$

Check:
$$\begin{array}{c|c} y - 8.4 = -6.5 \\ \hline 1.9 - 8.4 & -6.5 \\ -6.5 & \end{array}$$

The solution is 1.9.

EXAMPLE 3 Solve: $x - \frac{2}{3} = 2\frac{1}{2}$.

$$x - \tfrac{2}{3} = 2\tfrac{1}{2}$$
$$x - \tfrac{2}{3} + \tfrac{2}{3} = 2\tfrac{1}{2} + \tfrac{2}{3} \qquad \text{Adding } \tfrac{2}{3}$$
$$x = 2\tfrac{1}{2} + \tfrac{2}{3} \qquad \text{Simplifying}$$
$$x = \tfrac{5}{2} + \tfrac{2}{3} \qquad \text{Converting to fractional notation to carry out the addition}$$
$$x = \tfrac{5}{2} \cdot \tfrac{3}{3} + \tfrac{2}{3} \cdot \tfrac{2}{2} \qquad \text{Multiplying by 1 to obtain the least common denominator}$$
$$x = \tfrac{15}{6} + \tfrac{4}{6}$$
$$x = \tfrac{19}{6}$$

The check is left to the student. The solution is $\frac{19}{6}$.

EXERCISE SET 3.1

Solve using the addition principle. Don't forget to check!

1. $x + 2 = 6$
2. $x + 5 = 8$
3. $x + 15 = -5$
4. $y + 9 = 43$
5. $x + 6 = -8$
6. $t + 9 = -12$
7. $x + 16 = -2$
8. $y + 25 = -6$
9. $x - 9 = 6$
10. $x - 8 = 5$
11. $x - 7 = -21$
12. $x - 3 = -14$
13. $5 + t = 7$
14. $8 + y = 12$
15. $-7 + y = 13$

16. $-9 + z = 15$
17. $-3 + t = -9$
18. $-6 + y = -21$
19. $r + \frac{1}{3} = \frac{8}{3}$
20. $t + \frac{3}{8} = \frac{5}{8}$
21. $m + \frac{5}{6} = -\frac{11}{12}$
22. $x + \frac{2}{3} = -\frac{5}{6}$
23. $x - \frac{5}{6} = \frac{7}{8}$
24. $y - \frac{3}{4} = \frac{5}{6}$
25. $-\frac{1}{5} + z = -\frac{1}{4}$
26. $-\frac{1}{8} + y = -\frac{3}{4}$
27. $x + 2.3 = 7.4$
28. $y + 4.6 = 9.3$
29. $x - 4.8 = 7.6$
30. $y - 8.3 = 9.5$
31. $-9.7 = -4.7 + y$
32. $-7.8 = 2.8 + x$
33. $5\frac{1}{6} + x = 7$
34. $5\frac{1}{4} = 4\frac{2}{3} + x$
35. $q + \frac{1}{3} = -\frac{1}{7}$
36. $47\frac{1}{8} = -76 + z$

Problem-solving practice

37. In Churchill, Manitoba, the average daily low temperature in January is $-31°C$. This is 50° less than the average daily low temperature in Key West, Florida. What is the average daily low temperature in Key West in January?

38. After depositing paychecks of $232.58 and $486.79, a family had a balance of $1279.88 in its checking account. How much was in the account before the deposits?

39. Solve: $-356.788 = -699.034 + t$.

Solve.

40. $8 - 25 = 8 + x - 21$
41. $16 + x - 22 = -16$
42. $x + x = x$
43. $x + 3 = 3 + x$
44. $x + 4 = 5 + x$

Solve for x.

45. $x + 7 = b + 10$
46. $1 - c = a + x$
47. $x - 4 = a$
48. $|x| + 6 = 19$
49. Solve: $-\frac{3}{2} + x = -\frac{5}{17} - \frac{3}{2}$.
50. If $x - 4720 = 1634$, find $x + 4720$.
51. Explain why it is not necessary to prove a subtraction principle: if $a = b$ is true, then
52. Solve: $|x| + 17 = 4$.

$$a - c = b - c \text{ is true,}$$

for any number c.

3.2 THE MULTIPLICATION PRINCIPLE

Another principle for solving equations, similar to the addition principle, uses multiplication.

Using the Multiplication Principle

Suppose $a = b$ is true, and we multiply a by some number c. We get the same answer if we multiply b by c, because a and b are the same number.

3.2 THE MULTIPLICATION PRINCIPLE

THE MULTIPLICATION PRINCIPLE

If an equation $a = b$ is true, then
$$a \cdot c = b \cdot c$$
is true for any number c.

When using the multiplication principle, we sometimes say that we "multiply on both sides by the same number."

EXAMPLE 1 Solve: $3x = 9$.

$$3x = 9$$
$$\tfrac{1}{3} \cdot 3x = \tfrac{1}{3} \cdot 9 \quad \text{Using the multiplication principle; multiplying by } \tfrac{1}{3} \text{ on both sides}$$
$$1 \cdot x = 3 \quad \text{Simplifying}$$
$$x = 3$$

It is easy to see that the solution of $x = 3$ is 3.

Check: $\begin{array}{c|c} 3x = 9 \\ \hline 3 \cdot 3 & 9 \\ 9 & \end{array}$

The solution of the original equation is 3.

In Example 1, to get x alone, we multiplied by the reciprocal of 3. When we multiplied we got $1 \cdot x$, which simplified to x. This enabled us to "get rid of" the 3 on the left. The multiplication principle also allows us to divide on both sides of an equation by a nonzero number. This is because division by a number c is the same as multiplying by a reciprocal.

EXAMPLE 2 Solve: $\tfrac{3}{8} = -\tfrac{5}{4}x$.

$$\tfrac{3}{8} = -\tfrac{5}{4}x$$
$$-\tfrac{4}{5} \cdot \tfrac{3}{8} = -\tfrac{4}{5} \cdot (-\tfrac{5}{4}x) \quad \begin{array}{l}\text{Multiplying by } -\tfrac{4}{5} \text{ to get rid of } -\tfrac{5}{4} \text{ on the right} \\ \text{(this is the same as dividing by } -\tfrac{5}{4}\text{)}\end{array}$$
$$-\tfrac{3}{10} = x \quad \text{Simplifying}$$

Check: $\begin{array}{c|c} \tfrac{3}{8} = -\tfrac{5}{4}x \\ \hline \tfrac{3}{8} & -\tfrac{5}{4}(-\tfrac{3}{10}) \\ & \tfrac{3}{8} \end{array}$

The solution is $-\tfrac{3}{10}$.

EXAMPLE 3 Solve: $-x = 9$.

$$-x = 9$$
$$-1 \cdot x = 9 \qquad \text{Using the property of } -1$$
$$-1 \cdot (-1 \cdot x) = -1 \cdot 9 \qquad \text{Multiplying on both sides by } -1, \text{ the reciprocal of itself, or dividing by } -1$$
$$1 \cdot x = -9$$
$$x = -9$$

Check:
$$\begin{array}{c|c} -x = 9 \\ \hline -(-9) & 9 \\ 9 & \end{array}$$

The solution is -9.

Now we "undo" a division using the multiplication principle.

EXAMPLE 4 Solve: $\dfrac{-y}{9} = 14$.

$$\frac{-y}{9} = 14$$
$$9 \cdot \frac{-y}{9} = 9 \cdot 14 \qquad \text{Think of } \frac{-y}{9} \text{ as } \frac{1}{9} \cdot -y; \text{ multiply by 9 on both sides.}$$
$$-y = 126$$
$$y = -126$$

Check:
$$\begin{array}{c|c} \dfrac{-y}{9} = 14 \\ \hline \dfrac{-(-126)}{9} & 14 \\ \dfrac{126}{9} & \\ 14 & \end{array}$$

The solution is -126.

EXAMPLE 5 Solve: $1.16y = 9744$.

$$1.16y = 9744$$
$$\tfrac{1}{1.16} \cdot (1.16y) = \tfrac{1}{1.16} \cdot (9744) \qquad \text{Multiplying by } \tfrac{1}{1.16} \text{ (or dividing by 1.16)}$$
$$y = \tfrac{9744}{1.16}$$
$$y = 8400$$

Check: $\quad\dfrac{1.16y = 9744}{1.16(8400) \mid 9744}$
$\qquad\qquad\qquad 9744$

The solution is 8400.

Note that equations are reversible. That is, if $a = b$ is true, then $b = a$ is true. Thus, when we solve $15 = 3x$, we can reverse it and solve $3x = 15$ if we wish.

EXERCISE SET 3.2

Solve using the multiplication principle. Don't forget to check!

1. $6x = 36$
2. $3x = 39$
3. $5x = 45$
4. $9x = 72$
5. $84 = 7x$
6. $56 = 8x$
7. $-x = 40$
8. $100 = -x$
9. $-x = -1$
10. $-68 = -r$
11. $7x = -49$
12. $9x = -36$
13. $-12x = 72$
14. $-15x = 105$
15. $-21x = -126$
16. $-13x = -104$
17. $\dfrac{t}{7} = -9$
18. $\dfrac{y}{-8} = 11$
19. $\dfrac{3}{4}x = 27$
20. $\dfrac{4}{5}x = 16$
21. $\dfrac{-t}{3} = 7$
22. $\dfrac{-x}{6} = 9$
23. $-\dfrac{m}{3} = \dfrac{1}{5}$
24. $\dfrac{1}{9} = -\dfrac{z}{7}$
25. $-\dfrac{3}{5}r = -\dfrac{9}{10}$
26. $-\dfrac{2}{5}y = -\dfrac{4}{15}$
27. $-\dfrac{3}{2}r = -\dfrac{27}{4}$
28. $\dfrac{5}{7}x = -\dfrac{10}{14}$
29. $6.3x = 44.1$
30. $2.7y = 54$
31. $-3.1y = 21.7$
32. $-3.3y = 6.6$
33. $38.7m = 309.6$
34. $29.4x = 235.2$
35. $-\dfrac{2}{3}y = -10.6$
36. $-\dfrac{9}{7}y = 12.06$

Problem-solving practice

37. Apollo 10 reached a speed of 24,790 miles per hour. That was 37 times the speed of the first supersonic flight in 1947. What was the speed of the first supersonic flight?

38. Roger Staubach completed 1685 passes in his pro football career. This is about 57% of the number he attempted. How many did he attempt?

39. Solve: $-0.2344m = 2028.732$.

Solve.

40. $0 \cdot x = 0$
41. $0 \cdot x = 9$

Solve for x.

42. $ax = 5a$
43. $3x = \dfrac{b}{a}$

44. $cx = a^2 + 1$
45. $\dfrac{a}{b}x = 4$

46. $4|x| = 48$
47. $2|x| = -12$

48. Determine whether you can square both sides of an equation and get an equivalent equation.

49. A student makes a calculation and gets an answer of 22.5. On the last step the student multiplies by 0.3 when a division by 0.3 should have been done. What should the correct answer be?

3.3 USING THE PRINCIPLES TOGETHER

In this section we consider equation solving where we may need to use both the addition and multiplication principles. We also consider equations where collecting like terms is useful.

Applying Both Principles

Let's consider an equation where we apply both principles. We usually apply the addition principle first. Then we apply the multiplication principle.

EXAMPLE 1 Solve: $3x + 4 = 13$.

$$3x + 4 = 13$$
$$3x + 4 + (-4) = 13 + (-4) \quad \text{Using the addition principle;}$$
$$\qquad\qquad\qquad\qquad\qquad\text{adding } -4 \text{ on both sides}$$
$$3x = 9 \quad \text{Simplifying}$$
$$\tfrac{1}{3} \cdot 3x = \tfrac{1}{3} \cdot 9 \quad \text{Using the multiplication principle;}$$
$$\qquad\qquad\qquad\text{multiplying by } \tfrac{1}{3} \text{ on both sides}$$
$$x = 3 \quad \text{Simplifying}$$

Check:
$$\begin{array}{c|c} 3x + 4 = 13 \\ \hline 3 \cdot 3 + 4 & 13 \\ 9 + 4 & \\ 13 & \end{array}$$

The solution is 3.

EXAMPLE 2 Solve: $-5x - 6 = 16$.

$$-5x - 6 = 16$$
$$-5x - 6 + 6 = 16 + 6 \quad \text{Adding 6 on both sides}$$
$$-5x = 22$$
$$-\tfrac{1}{5} \cdot (-5x) = -\tfrac{1}{5} \cdot 22 \quad \text{Multiplying by } -\tfrac{1}{5} \text{ on both sides}$$
$$x = -\tfrac{22}{5} \quad \text{or} \quad -4\tfrac{2}{5}$$

3.3 USING THE PRINCIPLES TOGETHER

Check:
$$\begin{array}{r|l} -5x - 6 = 16 \\ \hline -5(-\tfrac{22}{5}) - 6 & 16 \\ 22 - 6 \\ 16 \end{array}$$

The solution is $-\tfrac{22}{5}$.

EXAMPLE 3 Solve: $45 - x = 13$.

$$45 - x = 13$$
$$-45 + 45 - x = -45 + 13 \qquad \text{Adding } -45 \text{ on both sides}$$
$$-x = -32$$
$$-1 \cdot x = -32 \qquad \text{Using the property of } -1: -x = -1 \cdot x$$
$$x = \tfrac{-32}{-1} \qquad \text{Dividing on both sides by } -1 \text{ (You could have multiplied on both sides by } -1 \text{ instead. That would change the sign on both sides.)}$$
$$x = 32$$

Check:
$$\begin{array}{r|l} 45 - x = 13 \\ \hline 45 - 32 & 13 \\ 13 \end{array}$$

The solution is 32.

EXAMPLE 4 Solve: $16.3 - 7.2y = -8.18$.

$$16.3 - 7.2y = -8.18$$
$$-7.2y = -16.3 + (-8.18) \qquad \text{Adding } -16.3 \text{ on both sides}$$
$$-7.2y = -24.48$$
$$y = \tfrac{-24.48}{-7.2} \qquad \text{Dividing by } -7.2 \text{ on both sides}$$
$$y = 3.4$$

Check:
$$\begin{array}{r|l} 16.3 - 7.2y = -8.18 \\ \hline 16.3 - 7.2(3.4) & -8.18 \\ 16.3 - 24.48 \\ -8.18 \end{array}$$

The solution is 3.4.

Collecting Like Terms

If there are like terms on one side of the equation, we collect them before using the principles.

EXAMPLE 5 Solve: $3x + 4x = -14$.

$$3x + 4x = -14$$
$$7x = -14 \quad \text{Collecting like terms}$$
$$\tfrac{1}{7} \cdot 7x = \tfrac{1}{7} \cdot (-14)$$
$$x = -2$$

The number -2 checks, so the solution is -2.

If there are like terms on opposite sides of the equation, we get them on the same side by using the addition principle. Then we collect them.

EXAMPLE 6 Solve: $2x - 2 = -3x + 3$.

$$2x - 2 = -3x + 3$$
$$2x - 2 + 2 = -3x + 3 + 2 \quad \text{Adding 2}$$
$$2x = -3x + 5 \quad \text{Simplifying}$$
$$2x + 3x = -3x + 3x + 5 \quad \text{Adding } 3x$$
$$5x = 5 \quad \text{Collecting like terms and simplifying}$$
$$\tfrac{1}{5} \cdot 5x = \tfrac{1}{5} \cdot 5 \quad \text{Multiplying by } \tfrac{1}{5}$$
$$x = 1 \quad \text{Simplifying}$$

Check:

$$\begin{array}{c|c} \multicolumn{2}{c}{2x - 2 = -3x + 3} \\ \hline 2 \cdot 1 - 2 & -3 \cdot 1 + 3 \\ 2 - 2 & -3 + 3 \\ 0 & 0 \end{array}$$

The solution is 1.

In Example 6, we used the addition principle to get all terms with a variable on one side and all numbers on the other side. Then we collected like terms and proceeded as before. If there are like terms on one side at the outset, they should be collected first.

EXAMPLE 7 Solve: $6x + 5 - 7x = 10 - 4x + 3$.

$$6x + 5 - 7x = 10 - 4x + 3$$
$$-x + 5 = 13 - 4x \quad \text{Collecting like terms}$$
$$-x + 4x = 13 - 5 \quad \text{Adding } 4x \text{ and subtracting 5 to get all terms with variables on one side and all other terms on the other}$$
$$3x = 8 \quad \text{Collecting like terms}$$
$$\tfrac{1}{3} \cdot 3x = \tfrac{1}{3} \cdot 8 \quad \text{Multiplying by } \tfrac{1}{3}$$
$$x = \tfrac{8}{3} \quad \text{Simplifying}$$

The number $\tfrac{8}{3}$ checks, so it is the solution.

Using the Multiplication Principle First

We have stated that we generally use the addition principle first. In some situations it is to our advantage to use the multiplication principle first. If we multiply by 4 on both sides of $\frac{1}{2}x = \frac{3}{4}$, we get $2x = 3$, which has no fractions. We have "cleared the fractions." If we multiply by 10 on both sides of $2.3x = 5$, we get $23x = 50$, which has no decimal points. We have "cleared the decimals." The equations are then easier to solve. You do not have to clear the fractions or decimals, but doing so eases computations.

In what follows we use the multiplication principle first to "clear," or "get rid of," fractions or decimals. For fractions, the number we multiply by is either the product of all the denominators or the least common multiple of all the denominators.

EXAMPLE 8 Solve: $\frac{2}{3}x - \frac{1}{6} + \frac{1}{2}x = \frac{7}{6} + 2x$.

The number 6 is the least common multiple of all the denominators. We multiply on both sides by 6.

$6(\frac{2}{3}x - \frac{1}{6} + \frac{1}{2}x) = 6(\frac{7}{6} + 2x)$ Multiplying by 6 on both sides

$6 \cdot \frac{2}{3}x - 6 \cdot \frac{1}{6} + 6 \cdot \frac{1}{2}x = 6 \cdot \frac{7}{6} + 6 \cdot 2x$ Using the distributive laws (*Caution!* Be sure to multiply all the terms by 6.)

$4x - 1 + 3x = 7 + 12x$ Simplifying. Note that the fractions are cleared.

$7x - 1 = 7 + 12x$ Collecting like terms

$7x - 12x = 7 + 1$ Subtracting $12x$ and adding 1 to get all the terms with variables on one side and all the other terms on the other

$-5x = 8$ Collecting like terms

$-\frac{1}{5} \cdot (-5x) = -\frac{1}{5} \cdot 8$ Multiplying by $-\frac{1}{5}$ or dividing by -5

$x = -\frac{8}{5}$

The number $-\frac{8}{5}$ checks and is the solution.

Here is a procedure for solving the equations in this section.

> 1. Multiply on both sides to clear the equation of fractions or decimals. (This is optional, but it can ease computations.)
> 2. Collect like terms on each side, if necessary.
> 3. Get all terms with variables on one side and all the other terms on the other side.
> 4. Collect like terms again, if necessary.
> 5. Multiply or divide to solve for the variable.

We illustrate this by repeating Example 4, but we clear the equation of decimals first.

EXAMPLE 9 Solve: $16.3 - 7.2y = -8.18$.

The greatest number of decimal places in any one number is two. Multiplying by 100, which has two 0s, will clear the decimals.

$$100(16.3 - 7.2y) = 100(-8.18) \quad \text{Multiplying by 100 on both sides}$$
$$100(16.3) - 100(7.2y) = 100(-8.18) \quad \text{Using a distributive law}$$
$$1630 - 720y = -818 \quad \text{Simplifying}$$
$$-720y = -818 - 1630 \quad \text{Subtracting 1630 on both sides}$$
$$-720y = -2448 \quad \text{Collecting like terms}$$
$$y = \frac{-2448}{-720} = 3.4 \quad \text{Dividing by } -720 \text{ on both sides}$$

The number 3.4 checks and is the solution.

EXERCISE SET 3.3

Solve and check.

1. $5x + 6 = 31$
2. $3x + 6 = 30$
3. $8x + 4 = 68$
4. $7z + 9 = 72$
5. $4x - 6 = 34$
6. $6x - 3 = 15$
7. $3x - 9 = 33$
8. $5x - 7 = 48$
9. $7x + 2 = -54$
10. $5x + 4 = -41$
11. $6y + 3 = -45$
12. $9t + 8 = -91$
13. $-4x + 7 = 35$
14. $-5x - 7 = 108$
15. $-7x - 24 = -129$
16. $-6z - 18 = -132$
17. $-4x + 71 = -1$
18. $-8y + 83 = -85$

Solve and check.

19. $5x + 7x = 72$
20. $4x + 5x = 45$
21. $8x + 7x = 60$
22. $3x + 9x = 96$
23. $4x + 3x = 42$
24. $6x + 19x = 100$
25. $4y - 2y = 10$
26. $8y - 5y = 48$
27. $-6y - 3y = 27$
28. $-4y - 8y = 48$
29. $-7y - 8y = -15$
30. $-10y - 3y = -39$
31. $10.2y - 7.3y = -58$
32. $6.8y - 2.4y = -88$
33. $x + \frac{1}{3}x = 8$
34. $x + \frac{1}{4}x = 10$
35. $8y - 35 = 3y$
36. $4x - 6 = 6x$
37. $4x - 7 = 3x$
38. $9x - 6 = 3x$
39. $8x - 1 = 23 - 4x$
40. $5y - 2 = 28 - y$
41. $2x - 1 = 4 + x$
42. $5x - 2 = 6 + x$
43. $6x + 3 = 2x + 11$
44. $5y + 3 = 2y + 15$
45. $5 - 2x = 3x - 7x + 25$
46. $10 - 3x = 2x - 8x + 40$

47. $4 + 3x - 6 = 3x + 2 - x$

48. $5 + 4x - 7 = 4x + 3 - x$

49. $4y - 4 + y = 6y + 20 - 4y$

50. $5y - 7 + y = 7y + 21 - 5y$

Solve and check. Clear fractions or decimals first.

51. $\frac{7}{2}x + \frac{1}{2}x = 3x + \frac{3}{2} + \frac{5}{2}x$

52. $\frac{7}{8}x - \frac{1}{4} + \frac{3}{4}x = \frac{1}{16} + x$

53. $\frac{2}{3} + \frac{1}{4}t = \frac{1}{3}$

54. $-\frac{3}{2} + x = -\frac{5}{6} - \frac{4}{3}$

55. $\frac{2}{3} + 3y = 5y - \frac{2}{15}$

56. $\frac{1}{2} + 4m = 3m - \frac{5}{2}$

57. $\frac{5}{3} + \frac{2}{3}x = \frac{25}{12} + \frac{5}{4}x + \frac{3}{4}$

58. $1 - \frac{2}{3}y = \frac{9}{5} - \frac{y}{5} + \frac{3}{5}$

59. $2.1x + 45.2 = 3.2 - 8.4x$

60. $0.96y - 0.79 = 0.21y + 0.46$

61. $1.03 - 0.62x = 0.71 - 0.22x$

62. $1.7t + 8 - 1.62t = 0.4t - 0.32 + 8$

63. $0.42 - 0.03y = 3.33 - y$

64. $0.7n - 15 + n = 2n - 8 - 0.4n$

65. $\frac{2}{7}x + \frac{1}{2}x = \frac{3}{4}x + 1$

66. $\frac{5}{16}y + \frac{3}{8}y = 2 + \frac{1}{4}y$

67. $\frac{4}{5}x - \frac{3}{4}x = \frac{3}{10}x - 1$

68. $\frac{8}{5}y - \frac{2}{3}y = 23 - \frac{1}{15}y$

Problem-solving practice

69. The population of the world in 1980 was 4.4 billion. This was a 23% increase over the population in 1970. What was the population in 1970, to the nearest tenth of a billion?

70. The population of the United States in 1980 was 224 million. This was a 48% increase over the population in 1950. What was the population in 1950, to the nearest million?

71. Solve: $0.008 + 9.62x - 42.8 = 0.944x + 0.0083 - x$.

Solve the first equation for x. Then substitute this number into the second equation and solve it for y.

72. $9x - 5 = 22$,
$4x + 2y = 2$

73. $9x + 2 = -1$,
$4x - y = \frac{11}{3}$

74. $0.2x + 0.12 = 0.146$,
$0.17x + 0.03y = 0.01238$

Solve for y.

75. $\dfrac{y-2}{3} = \dfrac{2-y}{5}$

76. $\dfrac{y}{a} - 3y = 1$

77. $0 = y - (-14) - (-3y)$

78. $\dfrac{5+2y}{3} = \dfrac{25}{12} + \dfrac{5y+3}{4}$

79. $0.05y - 1.82 = 0.708y - 0.504$

80. Solve the equation $4x - 8 = 32$ by first using the addition principle. Then solve it by first using the multiplication principle.

3.4 EQUATIONS CONTAINING PARENTHESES

Here we consider certain kinds of equations that contain parentheses. To solve such equations we use the distributive laws to first remove the parentheses. Then we proceed as before.

EXAMPLE 1 Solve: $4x = 2(12 - 2x)$.

$$4x = 2(12 - 2x)$$
$$4x = 24 - 4x \quad \text{Using a distributive law to multiply and remove parentheses}$$
$$4x + 4x = 24 \quad \text{Adding } 4x \text{ to get all } x\text{-terms on one side}$$
$$8x = 24 \quad \text{Collecting like terms}$$
$$x = 3 \quad \text{Multiplying by } \tfrac{1}{8}$$

Check:
$$\begin{array}{c|c} 4x = 2(12 - 2x) \\ \hline 4 \cdot 3 & 2(12 - 2 \cdot 3) \\ 12 & 2(12 - 6) \\ & 2 \cdot 6 \\ & 12 \end{array}$$

The solution is 3.

EXAMPLE 2 Solve: $3(x - 2) - 1 = 2 - 5(x + 5)$.

$$3(x - 2) - 1 = 2 - 5(x + 5)$$
$$3x - 6 - 1 = 2 - 5x - 25 \quad \text{Using the distributive laws to multiply and remove parentheses}$$
$$3x - 7 = -5x - 23 \quad \text{Simplifying}$$
$$3x + 5x = -23 + 7 \quad \text{Adding } 5x \text{ and also } 7, \text{ to get all } x\text{-terms on one side and all other terms on the other side}$$
$$8x = -16 \quad \text{Simplifying}$$
$$x = -2 \quad \text{Multiplying by } \tfrac{1}{8}$$

Check:
$$\begin{array}{c|c} 3(x - 2) - 1 = 2 - 5(x + 5) \\ \hline 3(-2 - 2) - 1 & 2 - 5(-2 + 5) \\ 3 \cdot (-4) - 1 & 2 - 5(3) \\ -12 - 1 & 2 - 15 \\ -13 & -13 \end{array}$$

The solution is -2.

EXERCISE SET 3.4

Solve the following equations. Check.

1. $3(2y - 3) = 27$
2. $4(2y - 3) = 28$
3. $40 = 5(3x + 2)$
4. $9 = 3(5x - 2)$
5. $2(3 + 4m) - 9 = 45$
6. $3(5 + 3m) - 8 = 88$

7. $5r - (2r + 8) = 16$
8. $6b - (3b + 8) = 16$
9. $3g - 3 = 3(7 - g)$
10. $3d - 10 = 5(d - 4)$
11. $6 - 2(3x - 1) = 2$
12. $10 - 3(2x - 1) = 1$
13. $5(d + 4) = 7(d - 2)$
14. $3(t - 2) = 9(t + 2)$
15. $3(x - 2) = 5(x + 2)$
16. $5(y + 4) = 3(y - 2)$
17. $8(2t + 1) = 4(7t + 7)$
18. $7(5x - 2) = 6(6x - 1)$
19. $3(r - 6) + 2 = 4(r + 2) - 21$
20. $5(t + 3) + 9 = 3(t - 2) + 6$
21. $19 - (2x + 3) = 2(x + 3) + x$
22. $13 - (2c + 2) = 2(c + 2) + 3c$
23. $\frac{1}{4}(8y + 4) - 17 = -\frac{1}{2}(4y - 8)$
24. $\frac{1}{3}(6x + 24) - 20 = -\frac{1}{4}(12x - 72)$

Problem-solving practice

25. A bottle factory has 59 breaks on the assembly line during a business day. The expected breakage rate is 1.3%. About how many bottles were produced?

26. Bowling at Chan's Bowling Lanes cost Steve and Trina $9.00. This included shoe rental at 75¢ a pair for each of them. Steve bowled 3 games and Trina bowled 2 games. How much was each game?

Solve for x.

27. $475(54x + 7856) + 9762 = 402(83x + 975)$
28. $-2[3(x - 2) + 4] = 4(1 - x) + 8$
29. $x(x - 4) = 3x(x + 1) - 2(x^2 + x - 5)$
30. $3(x + 4) = 3(4 + x)$
31. $4(x - a) = 16$
32. ▦ $30{,}000 + 20{,}000x = 55{,}000(1 + 12{,}500x)$

3.5 PROBLEM SOLVING

We have studied many new equation-solving tools in this chapter. We now apply them to problem solving.

The Five-step Process for Problem Solving in Algebra

Let us review the process for solving problems in algebra.

> **THE FIVE STEPS IN PROBLEM SOLVING IN ALGEBRA**
>
> 1. *Familiarize* yourself with the problem situation.
> 2. *Translate* to mathematical language.
> 3. *Carry out* some mathematical manipulation.
> 4. *Check* your possible answer in the original problem.
> 5. *State* the answer clearly.

EXAMPLE 1 A 6-ft board is cut into two pieces, one twice as long as the other. How long are the pieces?

1. **Familiarize** We first draw a picture. Note that we have let x represent the length of one piece, and $2x$, the length of the other.

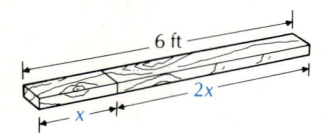

2. **Translate** From the figure, we can see that the lengths of the two pieces add up to 6 ft. That gives us our translation.

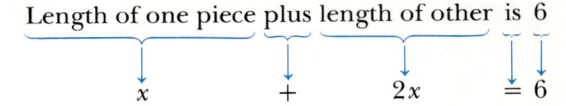

$$\underbrace{\text{Length of one piece}}_{x} \underbrace{\text{plus}}_{+} \underbrace{\text{length of other}}_{2x} \underbrace{\text{is}}_{=} \underbrace{6}_{6}$$

3. **Carry out** We solve the equation:

$$x + 2x = 6$$
$$3x = 6 \quad \text{Collecting like terms}$$
$$x = 2. \quad \text{Multiplying by } \tfrac{1}{3} \text{ on both sides}$$

4. **Check** Do we have an answer to the *problem*? If one piece is 2 ft long, then the other, to be twice as long, must be 4 ft long. The lengths of the pieces add up to 6 ft. This checks.

5. **State** One piece is 2 ft long and the other is 4 ft long.

EXAMPLE 2 Five plus three more than a number is nineteen. What is the number?

1. **Familiarize** This problem is stated explicitly enough that we can go right to the translation.

2. **Translate**

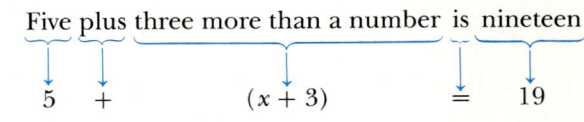

$$\underbrace{\text{Five}}_{5} \underbrace{\text{plus}}_{+} \underbrace{\text{three more than a number}}_{(x + 3)} \underbrace{\text{is}}_{=} \underbrace{\text{nineteen}}_{19}$$

3. **Carry out** We solve the equation:

$$5 + (x + 3) = 19$$
$$x + 8 = 19 \quad \text{Collecting like terms}$$
$$x = 11. \quad \text{Adding } -8$$

4. **Check** Three more than 11 is 14. Adding 5 to 14, we get 19. This checks.

5. **State** The number is 11.

3.5 PROBLEM SOLVING   117

EXAMPLE 3 The price of an automobile was decreased to a sale price of $13,559. This was a 9% reduction. What was the former price?

BRAND-NEW 1986 CAVIAT
REDUCED FROM X?
9% OFF TO ONLY $13,559

1. **Familiarize** Suppose the former price was $16,000. A 9% reduction would be found by taking 9% of $16,000, that is,

$$9\% \text{ of } \$16{,}000 = 0.09(\$16{,}000) = \$1440.$$

Then the sale price is found by subtracting the amount of reduction:

(Former price) − (Reduction) = (Sale price)
$16,000 − $1440 = $14,560

Our guess of $16,000 was too high; we are getting familiar with the problem.

2. **Translate** We reword and then translate.

(Former price) − (Reduction) = Sale price   *Rewording*

x − 9%x = $13,559   *Translating*

3. **Carry out** We solve the equation:

$$x - 9\%x = 13{,}559$$
$$x - 0.09x = 13{,}559 \quad \text{Converting to decimal notation}$$
$$0.91x = 13{,}559 \quad \text{Collecting like terms}$$
$$x = \frac{13{,}559}{0.91} \quad \text{Dividing by 0.91}$$
$$x = \$14{,}900.$$

4. **Check** To check we find 9% of $14,900 and subtract:

$$9\% \times \$14{,}900 = 0.09 \times \$14{,}900 = \$1341$$
$$\$14{,}900 - \$1341 = \$13{,}559.$$

Since we get the sale price, $13,559, the $14,900 checks.

5. **State** The former price was $14,900.

118   SOLVING EQUATIONS AND PROBLEMS

This problem is easy with algebra. Without algebra it is not. A common error in a problem like this is to take 9% of the sale price and subtract.

EXAMPLE 4   The sum of two consecutive integers is 29. What are the integers?

1. **Familiarize**   *Consecutive* integers are next to each other, such as 3 and 4, or -6 and -5. The larger is 1 plus the smaller. Thus, if x represents the smaller number, then $x + 1$ represents the larger number. Another way such numbers could be represented is to let y represent the larger number, and $y - 1$, the smaller.

   To get more familiar with the problem we can make a table. How do we get the entries in the table? First, we just guess a value for x. Then we find $x + 1$. Finally, we add the two numbers and see what happens. You might actually solve the problem this way, even though we want you to practice using algebra.

x	$x + 1$	Sum of x and $x + 1$
3	4	7
-6	-5	-11
19	20	39
13	14	27
-1	0	-1

2. **Translate**   We reword the problem and translate as follows.

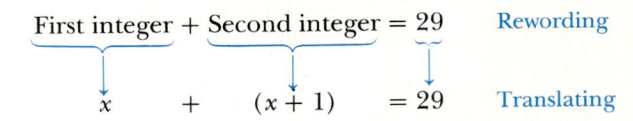

   First integer + Second integer = 29   Rewording

   $$x + (x + 1) = 29 \quad \text{Translating}$$

   We have let x represent the smaller integer. Then $x + 1$ represents the larger. Note that it is a good idea to write down what your letters represent in a problem.

3. **Carry out**   We solve the problem:

   $$x + (x + 1) = 29$$
   $$2x + 1 = 29 \quad \text{Collecting like terms}$$
   $$2x = 28 \quad \text{Adding } -1$$
   $$x = 14. \quad \text{Multiplying by } \tfrac{1}{2}$$

   Now if x is 14, then $x + 1$ is 15.

4. **Check**   Our possible answers are 14 and 15. These are consecutive integers. Their sum is 29, so the answers check in the *original problem*.

5. **State**   The consecutive integers are 14 and 15.

3.5 PROBLEM SOLVING

EXAMPLE 5   Acme Rent-A-Car rents an intermediate-size car (such as a Chevrolet, Ford, or Plymouth) at a daily rate of $44.95 plus 29¢ per mile. A businessperson is not to exceed a daily budget of $100. What mileage will allow the businessperson to stay within budget?

ACME Rent-a-Car $44.95 Plus 29¢ Per Mile

1. **Familiarize**   Suppose the businessperson drives 75 miles. Then the cost is

   Daily charge plus mileage charge

   or

   ($44.95) plus (Cost per mile) times (Number of miles driven)
   $44.95 + $0.29   ·   (75)

   which is $44.95 + $21.75, or $66.70. This familiarizes us with the way in which a calculation is made.

2. **Translate**   We reword the problem and translate as follows.

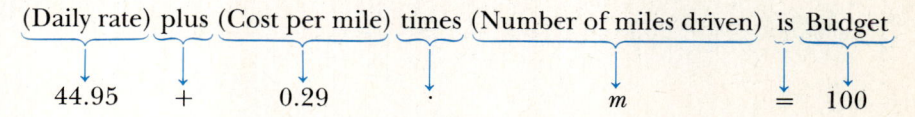

   (Daily rate) plus (Cost per mile) times (Number of miles driven) is Budget
   44.95   +   0.29   · m =   100

   We have let m represent the number of miles driven.

3. **Carry out**   We solve the equation:

   $$44.95 + 0.29m = 100$$
   $$100(44.95 + 0.29m) = 100(100) \quad \text{Multiplying by 100 on both sides to clear the decimals}$$
   $$100(44.95) + 100(0.29m) = 10{,}000 \quad \text{Using a distributive law}$$
   $$4495 + 29m = 10{,}000$$
   $$29m = 5505 \quad \text{Adding } -4495$$
   $$m = \tfrac{5505}{29} \quad \text{Dividing by 29}$$
   $$m \approx 189.8. \quad \text{Rounding to the nearest tenth}$$

4. **Check**   We check in the original problem. We multiply 189.8 by $0.29, obtaining $55.042. Then we add $55.042 to $44.95 and get $99.992, which is just under $100, the budget. At least the businessperson now knows to stay under this mileage.

5. **State** The businessperson should stay under 189.8 miles in order not to exceed the budget.

EXAMPLE 6 The perimeter of a rectangle is 150 cm (centimeters). The length is 15 cm greater than the width. Find the dimensions.

1. **Familiarize** We first draw a picture. We have let x represent the width, and $x + 15$, the length.

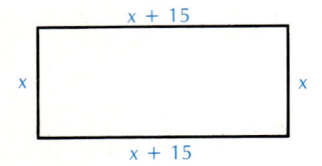

```
         x + 15
     ┌───────────┐
   x │           │ x
     └───────────┘
         x + 15
```

The perimeter of a polygon is the distance around it. (You may need to look up this word in a geometry book.) We can find the perimeter by adding the lengths of the sides.

2. **Translate** The definition of perimeter leads us to a rewording of the problem and a translation.

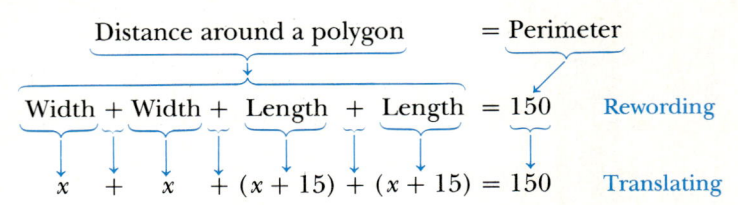

$$\underbrace{\text{Distance around a polygon}}_{} = \underbrace{\text{Perimeter}}_{}$$

$$\underbrace{\text{Width}}_{x} + \underbrace{\text{Width}}_{x} + \underbrace{\text{Length}}_{(x+15)} + \underbrace{\text{Length}}_{(x+15)} = 150 \quad \text{Rewording}$$

$$x + x + (x + 15) + (x + 15) = 150 \quad \text{Translating}$$

3. **Carry out** We solve the equation:

$$x + x + (x + 15) + (x + 15) = 150$$
$$4x + 30 = 150$$
$$4x = 120$$
$$x = 30.$$

Possible dimensions are $x = 30$ and $x + 15 = 45$.

4. **Check** If the width is 30 and the length is $30 + 15$, or 45, then the perimeter is $30 + 30 + 45 + 45$, or 150. This checks.

5. **State** The width is 30 cm and the length is 45 cm.

EXAMPLE 7 The second angle of a triangle is twice as large as the first. The measure of the third angle is 20° greater than that of the first angle. How large are the angles?

1. **Familiarize** We draw a picture. We use x for the measure of the first angle. The second is twice as large, so its measure will be $2x$. The third angle is 20° greater than the first angle, so its measure will be $x + 20$.

3.5 PROBLEM SOLVING

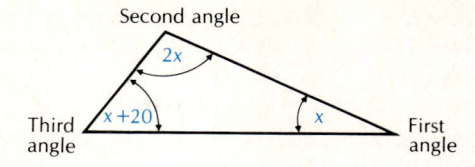

```
            Second angle
                2x
     Third  /x+20\      x    First
     angle                   angle
```

2. **Translate** To translate we need to recall a geometric fact (you might, as part of step (1), look it up in a geometry book or in the list of formulas at the back of this book). The measures of any triangle add up to 180°.

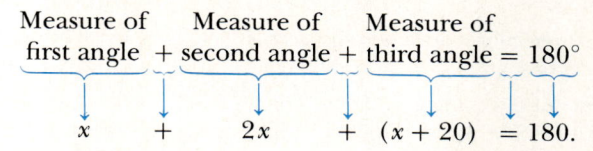

$$\underbrace{\text{Measure of}}_{x} + \underbrace{\text{Measure of}}_{2x} + \underbrace{\text{Measure of}}_{(x+20)} = 180°$$
$$\text{first angle} \quad \text{second angle} \quad \text{third angle}$$

3. **Carry out** We solve:

$$x + 2x + (x + 20) = 180$$
$$4x + 20 = 180$$
$$4x = 160$$
$$x = 40.$$

Possible answers for the angle measures are as follows:

First angle: $x = 40°$
Second angle: $2x = 80°$
Third angle: $x + 20 = 60°$.

4. **Check** Consider 40°, 80°, and 60°. The second is twice the first, and the third is 20° greater than the first. The sum is 180°. These numbers check.

5. **State** The measures are 40°, 80°, and 60°.

EXERCISE SET 3.5

1. When 18 is subtracted from six times a certain number, the result is 96. What is the number?

2. When 28 is subtracted from five times a certain number, the result is 232. What is the number?

3. If you double a number and then add 16, you get $\frac{2}{5}$ of the original number. What is the original number?

4. If you double a number and then add 85, you get $\frac{3}{4}$ of the original number. What is the original number?

5. If you add two-fifths of a number to the number itself, you get 56. What is the number?

6. If you add one-third of a number to the number itself, you get 48. What is the number?

7. A 180-m rope is cut into three pieces. The second piece is twice as long as the first. The third piece is three times as long as the second. How long is each piece of rope?

8. A 480-m wire is cut into three pieces. The second piece is three times as long as the first. The third piece is four times as long as the second. How long is each piece?

9. Consecutive odd integers are next to each other, such as 5 and 7. The larger is 2 plus the smaller. The sum of two consecutive odd integers is 76. What are the integers?

10. The sum of two consecutive odd integers is 84. What are the integers?

11. Consecutive even integers are next to each other, like 6 and 8. The larger is 2 plus the smaller. The sum of two consecutive even integers is 114. What are the integers?

12. The sum of two consecutive even integers is 106. What are the integers?

13. The sum of three consecutive integers is 108. What are the integers?

14. The sum of three consecutive integers is 126. What are the integers?

15. The sum of three consecutive odd integers is 189. What are the integers?

16. The sum of three consecutive odd integers is 255. What are the integers?

17. The perimeter of a rectangle is 310 m. The length is 25 m greater than the width. Find the width and length of the rectangle.

18. The perimeter of a rectangle is 304 cm. The length is 40 cm greater than the width. Find the width and length of the rectangle.

19. The perimeter of a rectangle is 152 m. The width is 22 m less than the length. Find the width and the length.

20. The perimeter of a rectangle is 280 m. The width is 26 m less than the length. Find the width and the length.

21. The second angle of a triangle is four times as large as the first. The third angle is 45° less than the sum of the other two angles. Find the measure of the first angle.

22. The second angle of a triangle is three times as large as the first. The third angle is 25° less than the sum of the other two angles. Find the measure of the first angle.

23. Money is invested in a savings account at 12% simple interest. After 1 year there is $4928 in the account. How much was originally invested?

24. Money is borrowed at 10% simple interest. After 1 year $7194 pays off the loan. How much was originally borrowed?

25. After a 40% reduction, a shirt is on sale at $9.60. What was the marked price (that is, the price before reduction)?

26. After a 34% reduction, a blouse is on sale at $9.24. What was the marked price?

27. Badger Rent-A-Car rents an intermediate-size car at a daily rate of $34.95 plus 10¢ per mile. A businessperson is not to exceed a daily car rental budget of $80. What mileage will allow the businessperson to stay within budget?

28. Badger also rents compact cars at $43.95 plus 10¢ per mile. What mileage will allow the businessperson to stay within the budget of $90?

29. The second angle of a triangle is three times as large as the first. The measure of the third angle is 40° greater than that of the first angle. How large are the angles?

30. One angle of a triangle is 32 times as large as another. The measure of the third angle is 10° greater than that of the smallest angle. How large are the angles?

31. The equation

$$R = -0.028t + 20.8$$

can be used to predict the world record in the 200-meter dash. R stands for the record in seconds, and t stands for the number of years since 1920. In what year will the record be 19.0 seconds?

32. The equation

$$F = \tfrac{1}{4}N + 40$$

can be used to determine temperatures given how many times a cricket chirps per minute, where F represents temperature in degrees and N is the number of chirps per minute. Determine the chirps per minute necessary in order for the temperature to be 80°.

33. Abraham Lincoln's 1863 Gettysburg Address refers to the year 1776 as "Four *score* and seven years ago." Write an equation and find what a *score* is.

35. If the daily rental for a car is $18.90 plus a certain price per mile and a person must drive 190 miles and stay within a $55.00 budget, what is the highest price per mile the person can afford?

37. The width of a rectangle is $\frac{3}{4}$ the length. The perimeter of the rectangle becomes 50 cm when the length and width are each increased by 2 cm. Find the length and width.

39. In a basketball league, the Falcons won 15 out of their first 20 games. How many more games will they have to play where they win only half the time in order to win 60% of the total games?

41. The buyer of a piano priced at $2000 is given the choice of paying cash at the time of purchase or $2150 at the end of one year. What rate of interest is the buyer being charged if payment is made at the end of one year?

43. A storekeeper goes to the bank to get $10 worth of change. The storekeeper requests twice as many quarters as half dollars, twice as many dimes as quarters, three times as many nickels as dimes, and no pennies or dollars. How many of each coin did the storekeeper get?

34. One number is 25% of another. The larger number is 12 more than the smaller. What are the numbers?

36. A student scored 78 on a test that had 4 seven-point fill-ins and 24 three-point multiple choice questions. The student had one fill-in wrong. How many multiple choice questions did the student get right?

38. Apples are collected in a basket for six people. One third, one fourth, one eighth, and one fifth are given to four people, respectively. The fifth person gets ten apples with one apple remaining for the sixth person. Find the original number of apples in the basket.

40. In one city, a sales tax of 9% was added to the price of gasoline as registered on the pump. Suppose a driver asked for $10 worth of regular. The attendant filled the tank until the pump read $9.10 and charged the driver $10. Something was wrong. Use algebra to correct the error.

42. A student has an average score of 82 on three tests. The student's average score on the first two tests is 85. What was the score on the third test?

44. The area of this triangle is 2.9047 in^2. Find x.

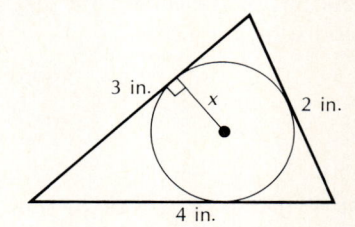

3.6 FORMULAS

A formula is a kind of "recipe" for doing a certain kind of calculation. In this section we learn how to calculate with formulas and how to solve a formula for a certain letter.

We have already used some formulas in this book. For example, the formula

$$L = 4D$$

is used to determine the safest position of a ladder, where L is the distance along the ladder to where it rests against a wall or other object, and D is the distance that the bottom of the ladder is out from the wall.

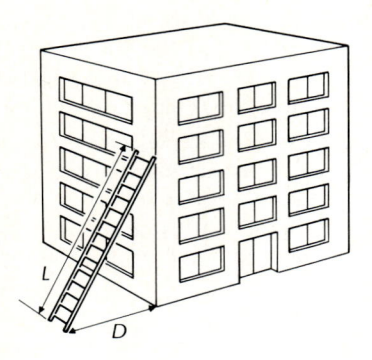

Suppose we know that $D = 3$ ft. Then we can find L by substituting 3 for D in the formula:

$$L = 4(3) = 12 \text{ ft.}$$

Thus, if a ladder is to be 3 ft out from a wall, the safest position of the ladder occurs when length L is 12 ft.

Suppose we know the length L and want to know how far out from the wall, that is, D, we should set the ladder. For example, say we have a 40-ft ladder that is to touch a wall at the top of the ladder. How far out from the wall should the bottom of the ladder be placed to be safest? To find out, we substitute 40 for L and solve for D:

$$40 = 4D$$
$$\tfrac{1}{4} \cdot 40 = \tfrac{1}{4} \cdot 4D \qquad \text{Multiplying by } \tfrac{1}{4}$$
$$10 = D.$$

If we had to do this many times, it might be faster to find a formula with D by itself on one side. Then computations could be done directly. Consider the equation

$$L = 4D.$$

We multiply on both sides by $\tfrac{1}{4}$ and get a formula for D in terms of L:

$$L = 4D \qquad \text{We want to get this letter alone.}$$
$$\tfrac{1}{4} \cdot L = \tfrac{1}{4}(4D) \qquad \text{Multiplying by } \tfrac{1}{4}$$
$$\tfrac{1}{4}L = D.$$

Now to find D when $L = 40$, we substitute 40 for L in the formula $D = \tfrac{1}{4}L$:

$$D = \tfrac{1}{4}(40) = 10.$$

3.6 FORMULAS

EXAMPLE 1 A formula for the circumference C of a circle of radius r is
$$C = 2\pi r.$$

a) Find the circumference when the radius is 5 ft.
b) Solve the formula for the radius r.
c) Use the formula in part (b) to find the radius when the circumference is 300 cm.

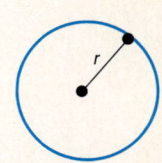

a) To find the circumference when the radius is 5 ft, we substitute 5 for r in the formula $C = 2\pi r$:
$$C = 2\pi r$$
$$= 2\pi(5)$$
$$= 10\pi.$$

To get an approximation for the answer we could use 3.14 for π. Then
$$C = 10\pi \approx 10(3.14) = 31.4 \text{ ft.}$$

b) $C = 2\pi r$ We want this letter alone.

$$\frac{1}{2\pi} \cdot C = \frac{1}{2\pi} \cdot 2\pi r \quad \text{Multiplying by } \frac{1}{2\pi}$$

$$\frac{C}{2\pi} = \frac{2\pi}{2\pi} \cdot r = r$$

We can write the answer as
$$r = \frac{C}{2\pi}.$$

c) To find the radius when the circumference is 300 cm, we substitute 300 for C:
$$r = \frac{C}{2\pi} = \frac{300}{2\pi} = \frac{150}{\pi}.$$

Remember, formulas are equations. We can use the same principles in solving a formula that you use for any other equations. To see how the principles apply to formulas, compare the following.

A. Solve.
$$5x + 2 = 12$$
$$5x = 12 - 2$$
$$5x = 10$$
$$x = \tfrac{10}{5}$$
$$x = 2$$

B. Solve.
$$5x + 2 = 12$$
$$5x = 12 - 2$$
$$x = \frac{12 - 2}{5}$$

C. Solve for x.
$$ax + b = c$$
$$ax = c - b$$
$$x = \frac{c - b}{a}$$

In (A) we solved as we did before. In (B) we did not carry out the calculations. In (C) we could not carry out the calculations because we had unknown numbers.

With the formulas in this section we can use a procedure like that described in Section 3.3.

> **To solve a formula for a given letter, identify the letter, and:**
> 1. **Multiply on both sides to clear fractions or decimals, if that is needed.**
> 2. **Collect like terms on each side, if necessary.**
> 3. **Get all terms with the letter to be solved for on one side of the equation and all other terms on the other side.**
> 4. **Collect like terms again, if necessary.**
> 5. **Solve for the letter in question.**

EXAMPLE 2 Solve for a: $A = \dfrac{a+b+c}{3}$.

This is a formula for the average A of three numbers a, b, and c.

$$A = \dfrac{a+b+c}{3} \qquad \text{We want this letter alone.}$$

$$3A = a + b + c \qquad \text{Multiplying by 3 to clear fractions}$$

$$3A - b - c = a$$

EXAMPLE 3 Solve for C: $Q = \dfrac{100M}{C}$.

This is a formula used in psychology for finding the intelligence quotient Q, where M is mental age and C is chronological, or actual, age.

We solve as follows:

$$Q = \frac{100M}{C} \quad \text{We want this letter alone.}$$

$$CQ = 100M \quad \text{Multiplying by } C \text{ to clear fractions}$$

$$C = \frac{100M}{Q} \quad \text{Multiplying by } \frac{1}{Q}$$

EXERCISE SET 3.6

1. The area A of a rectangle of length l and width w is given by

$$A = lw.$$

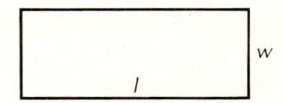

a) Find the area when the length is 17 ft and the width is 4 ft.
b) Solve the formula for w.
c) Using the formula found in part (b), find the width w when the area is 48 cm² (square centimeters) and the length is 6 cm.

3. The area A of a triangle with a base of length b and a height of length h is given by

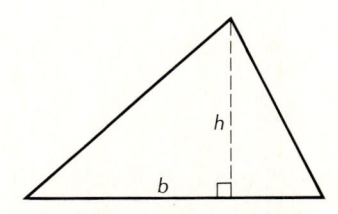

$$A = \tfrac{1}{2}bh.$$

a) Find the area of a triangle with a base of length $5\tfrac{1}{2}$ m (meters) and a height of $6\tfrac{1}{4}$ m.
b) Solve the formula for h.
c) Using the formula found in part (b), find the height h of a triangle that has an area of 504 yd² and a base of length 12 yd.

2. The simple interest I on an investment of principal P at interest rate r for time t, in years, is given by

$$I = Prt.$$

a) Find the interest I when principal P is $2000, interest rate r is 12%, and the time is 3 years.
b) Solve the formula for t.
c) Using the formula found in part (b), find the time t that it takes for a principal of $5000, at simple interest rate 11%, to earn $2200 in interest.

4. The perimeter P of a rectangle of length l and width w is given by

$$P = 2l + 2w.$$

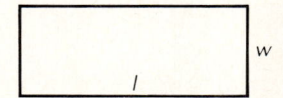

a) Find the perimeter of a rectangle of length 44 cm and width 25 cm.
b) Solve the formula for l.
c) Using the formula found in part (b), find the length of a rectangle whose perimeter is 250 mi and whose width is 14 mi.

Solve.

5. $A = bh$, for b. (This is the formula for the area of a parallelogram of base b and height h.)

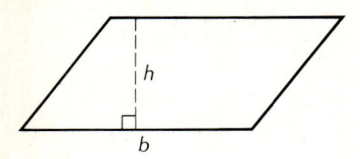

6. $A = bh$, for h.

7. $d = rt$, for r. (This is a formula for distance in terms of speed r and time t.)

8. $d = rt$, for t.

9. $I = Prt$, for P.

10. $I = Prt$, for r.

11. $F = ma$, for a. (This is a physics formula for force F in terms of mass m and acceleration a.)

12. $F = ma$, for m.

13. $P = 2l + 2w$, for w.

14. $A = \frac{1}{2}bh$, for b.

15. $A = \pi r^2$, for r^2. (This is the formula for the area A of a circle of radius r.)

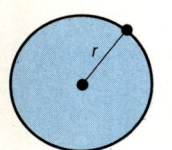

16. $A = \pi r^2$, for π.

17. $E = mc^2$, for m. (This is a relativity formula from physics.)

18. $E = mc^2$, for c^2.

19. $A = \dfrac{a+b+c}{3}$, for b.

20. $A = \dfrac{a+b+c}{3}$, for c.

21. $v = \dfrac{3k}{t}$, for t.

22. $P = \dfrac{ab}{c}$, for c.

23. $A = \frac{1}{2}ah + \frac{1}{2}bh$, for b.

24. $A = \frac{1}{2}ah + \frac{1}{2}bh$, for a.

25. The formula
$$H = \frac{D^2 N}{2.5}$$
is used to find the horsepower H of an N-cylinder engine. Solve for D^2.

26. Solve for N:
$$H = \frac{D^2 N}{2.5}.$$

27. The area of a sector of a circle is given by
$$A = \frac{\pi r^2 S}{360},$$
where r is the radius and S is the angle measure of the sector. Solve for S.

28. Solve for r^2:
$$A = \frac{\pi r^2 S}{360}.$$

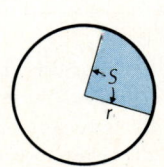

29. The formula

$$R = -0.0075t + 3.85$$

can be used to estimate the world record in the 1500-meter run t years after 1930. Solve for t.

30. The formula

$$F = \tfrac{9}{5}C + 32$$

can be used to convert from Celsius, or Centigrade, temperature C to Fahrenheit temperature F. Solve for C.

---

31. In $P = 2a + 2b$, P doubles. Do a and b both double?

32. In $A = lw$, l and w both double. What happens to A?

33. In $A = \tfrac{1}{2}bh$, b increases by 4 units, and h does not change. What happens to A?

34. In $T = 1.2a + 1.09b$, does an increase in a or an increase in b have more effect on T?

Solve.

35. $ax + b = c$, for b

36. $ax + b = c$, for a

37. $ax + b = 0$, for x

38. $A = \dfrac{1}{R}$, for R

39. $\dfrac{s}{t} = \dfrac{t}{v}$, for s

40. $\dfrac{a}{b} = \dfrac{c}{d}$, for $\dfrac{a}{c}$

41. $g = 40n + 20k$, for k

42. $r = 2h - \tfrac{1}{4}f$, for f

43. $y = a - ab$, for a

44. $x = a + b - 2ab$, for a

45. $d = \dfrac{1}{e + f}$, for f

46. $x = \dfrac{\left(\dfrac{y}{z}\right)}{\left(\dfrac{z}{t}\right)}$, for y

47. $m = ax^2 + bx + c$, for b

48. If $a^2 = b^2$, does $a = b$?

The sum A of the measures of the interior angles of a polygon of n sides is given by the formula

$$A = 180°(n - 2).$$

49. A polygon has 6 sides. What is the sum of the interior angle measures?

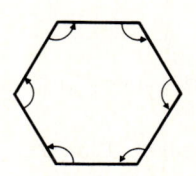

50. The sum of the measures of the angles of a polygon is 1440°. How many sides does it have?

51. Solve $A = 180°(n - 2)$ for n.

SUMMARY AND REVIEW: CHAPTER 3

The following contains a summary of what you should be able to do after completing this chapter. The review exercises are for practice. Answers are at the back of the book. If you miss an exercise, restudy the section indicated alongside the answer.

You should be able to:

Solve equations using the addition principle, the multiplication principle, the addition and multiplication principles together, and the distributive laws to first remove parentheses.

Solve.

1. $x + 5 = -17$
2. $-8x = -56$
3. $-\dfrac{x}{4} = 48$
4. $n - 7 = -6$
5. $15x = -35$
6. $x - 11 = 14$
7. $-\frac{2}{3} + x = -\frac{1}{6}$
8. $\frac{4}{5}y = -\frac{3}{16}$
9. $y - 0.9 = 9.09$
10. $5 - x = 13$
11. $5t + 9 = 3t - 1$
12. $7x - 6 = 25x$
13. $\frac{1}{4}x - \frac{5}{8} = \frac{3}{8}$
14. $14y = 23y - 17 - 10$
15. $0.22y - 0.6 = 0.12y + 3 - 0.8y$
16. $\frac{1}{4}x - \frac{1}{8}x = 3 - \frac{1}{16}x$
17. $4(x + 3) = 36$
18. $3(5x - 7) = -66$
19. $8(x - 2) = 5(x + 4)$
20. $-5x + 3(x + 8) = 16$

Solve a formula for a certain letter.

Solve.

21. $C = \pi d$, for d.
22. $V = \frac{1}{3}Bh$, for B.
23. $A = \dfrac{a + b}{2}$, for a.

Solve problems using the five-step problem-solving process.

Solve.

24. A color TV sold for $629 in May. This was $38 more than the January cost. Find the January cost.

25. Selma gets a $4 commission for each appliance that she sells. One week she got $108 in commissions. How many appliances did she sell?

26. An 8-m board is cut into two pieces. One piece is 2 m longer than the other. How long are the pieces?

27. If 14 is added to three times a certain number, the result is 41. Find the number.

28. The sum of two consecutive odd integers is 116. Find the integers.

29. The perimeter of a rectangle is 56 cm. The width is 6 cm less than the length. Find the width and the length.

30. After a 30% reduction, an item is on sale for $154. What was the marked price (the price before reducing)?

31. A businessperson's salary is $30,000. That is a 15% increase over the previous year's salary. What was the previous salary (to the nearest dollar)?

32. The measure of the second angle of a triangle is 50° more than that of the first. The measure of the third angle is 10° less than twice the first. Find the measures of the angles.

33. The total length of the Nile and Amazon Rivers is 13,108 km. If the Amazon were 234 km longer, it would be as long as the Nile. Find the length of each river.

34. One cashier works at a rate of 3 minutes per customer and a second express cashier works at a rate of 2 customers per minute. How many customers are served in an hour?

35. Consumer experts advise us never to pay the sticker price for a car. A rule of thumb is to pay the sticker price minus 20% of the sticker price, plus $200. A car is purchased for $11,520 using the rule. What was the sticker price?

Solve.

36. $2|n| + 4 = 50$

37. $|3n| = 60$

38. $\left|\dfrac{n}{2}\right| = 12$

39. Solve $y = 2a - ab + 3$, for a.

TEST: CHAPTER 3

Solve.

1. $x + 7 = 15$
2. $t - 9 = 17$
3. $3x = -18$
4. $-\frac{4}{7}x = -28$
5. $3t + 7 = 2t - 5$
6. $\frac{1}{2}x - \frac{3}{5} = \frac{2}{5}$
7. $8 - y = 16$
8. $-\frac{2}{5} + x = -\frac{3}{4}$
9. $3(x + 2) = 27$
10. $-3x + 6(x + 4) = 9$
11. $0.4p + 0.2 = 4.2p - 7.8 - 0.6p$

Problem solving

12. The perimeter of a rectangle is 36 cm. The length is 4 cm greater than the width. Find the width and length.

13. If you triple a number and then subtract 14, you get $\frac{2}{3}$ of the original number. What is the original number?

14. The sum of three consecutive odd integers is 249. Find the integers.

15. Money is invested in a savings account at 12% simple interest. After one year there is $840 in the account. How much was originally invested?

Solve the formulas for the given letter.

16. Solve $A = 2\pi rh$, for r.

17. Solve $w = \dfrac{P - 2l}{2}$, for l.

18. Solve $c = \dfrac{1}{a - d}$, for d.

19. Solve: $3|w| - 8 = 37$.

20. A movie theater had a certain number of tickets to give away. Five people got the tickets. The first got $\frac{1}{3}$ of the tickets, the second got $\frac{1}{4}$ of the tickets, and the third got $\frac{1}{5}$ of the tickets. The fourth person got eight tickets, and there were five tickets left for the fifth person. Find the total number of tickets given away.

CUMULATIVE REVIEW: CHAPTERS 1–3

Evaluate.

1. $\dfrac{y-x}{4}$, for $y = 12$ and $x = 6$
2. $\dfrac{3x}{y}$, for $x = 5$ and $y = 4$
3. $x^3 - 3$, for $x = 3$
4. $(x - 2)(x + 1)$, for $x = 6$

Calculate.

5. $200 - 5^2 \div 5 + 20$
6. $(200 - 5^2) \div 5 + 20$

Translate these phrases to algebraic expressions.

7. Four less than twice w
8. Three times the sum of x and y

9. Use the commutative and associative laws to write three expressions equivalent to
$$q + (r + 6).$$

10. Find the prime factorization: 648.
11. Find the LCM: 8, 15, 24.

Simplify.

12. $\dfrac{80}{144}$
13. $\dfrac{108x}{72xy}$

14. Find decimal notation: 2.6%.
15. Find fractional notation: 80%.

Find percent notation.

16. 1.9
17. $\dfrac{7}{8}$

Write a true sentence using < or >.

18. -4 -6
19. $\dfrac{1}{2}$ $-\dfrac{9}{8}$
20. $|-3|$ 0

21. Find the additive inverse and the reciprocal of $\dfrac{2}{5}$.

Compute and simplify.

22. $-6.7 + 2.3 - 4.8$
23. $-\dfrac{1}{6} + \dfrac{4}{3} - \dfrac{7}{6}$
24. $-\dfrac{5}{8}\left(-\dfrac{4}{3}\right)$
25. $(-7)(5)(-6)(-0.5)$
26. $81 \div (-9)$
27. $-10.8 \div 36$
28. $-\dfrac{4}{5} \div -\dfrac{25}{8}$

Multiply.

29. $5(3x + 5y + 2z)$
30. $4(-3x - 2)$
31. $-6(2y - 4x)$
32. $-5(-x - 1)$

Factor.

33. $64 + 18x + 24y$
34. $16y - 56$
35. $-2x - 8$
36. $5a - 15b + 25$

Collect like terms.

37. $9b + 18y + 6b + 4y$
38. $3y + 4 + 6z + 6y$
39. $-4d - 6a + 3a - 5d + 1$
40. $3.2x + 2.9y - 5.8x - 8.1y$

Simplify.

41. $7 - 2x - (-5x) - 8$

42. $-3x - (-x + y)$

43. $-3(x - 2) - 4x$

44. $10 - 2(5 - 4x)$

45. $[3(x + 6) - 10] - [5 - 2(x - 8)]$

Solve.

46. $x + 1.75 = 6.25$

47. $\frac{5}{2}y = \frac{2}{5}$

48. $-2.6 + x = 8.3$

49. $4\frac{1}{2} + y = 8\frac{1}{3}$

50. $-\frac{3}{4}x = 36$

51. $-2.2y = -26.4$

52. $5.8x = -35.96$

53. $-4x + 3 = 15$

54. $-3x + 5 = -8x - 7$

55. $4y - 4 + y = 6y + 20 - 4y$

56. $-3(x - 2) = -15$

57. $\frac{1}{3}x - \frac{5}{6} = \frac{1}{2} + 2x$

58. $-3.7x + 6.2 = -7.3x - 5.8$

59. Solve $A = \frac{1}{2}h(b + c)$, for h.

Problem solving

60. What percent of 60 is 18?

61. Two is 4 percent of what number?

62. If 25 is subtracted from seven times a certain number, the result is 129. Find the number.

63. Jane and Becky purchased identical dresses. Jane paid $17 more than Becky. Jane paid $62. What did Becky pay?

64. Money is invested in a savings account at 12% simple interest. After one year there is $1680 in the account. How much was originally invested?

65. A 143-m wire is cut into three pieces. The second is 3 m longer than the first. The third is $\frac{4}{5}$ as long as the first. How long is each piece?

○ ─────────────────────────────────

66. Evaluate $\frac{1}{2}a^2 - \frac{1}{4}ab + \frac{1}{16}b^2$ when $a = -\frac{1}{4}$ and $b = -\frac{1}{2}$.

67. Arrange in order from least to greatest.

$$-0.2^2, \quad |0|, \quad (-0.2)^2, \quad -\left|-\tfrac{5}{2}\right|, \quad \left(-\tfrac{2}{5}\right)^2, \quad -0.2^3, \quad \left|-\tfrac{1}{5}\right|$$

Solve.

68. $\frac{2}{3}|x| - \frac{1}{3} = 3$

69. $4(x + 2) = 4(x - 2) + 16$

70. $0(x + 3) + 4 = 0$

71. $\dfrac{2 + 5x}{4} = \dfrac{11}{28} + \dfrac{8x + 3}{7}$

POLYNOMIALS

4

Polynomial equations can be used to solve problems concerning interest on investments.

An algebraic expression like $3x^2 - 7x + 5$ is called a *polynomial*. One of the most important parts of introductory algebra is the study of polynomials. In this chapter we learn to add, subtract, and multiply polynomials.

Of particular importance in this chapter is the study of fast ways to find special products of polynomials, which will be helpful not only in this book but also in more advanced mathematics.

4.1 PROPERTIES OF EXPONENTS

Algebraic expressions like the following are called *monomials*:

$$5x^3, \quad 7y^4, \quad \tfrac{1}{4}t^2, \quad x^1, \quad x^0, \quad 8, \quad 34, \quad 0.$$

Each expression is a number or a number times a variable to some power. More formally, a monomial is an expression of the type ax^n, where n is a whole number.

We will learn to multiply and divide monomials and to raise a power to a power.

Multiplying Using Exponents

We know that an exponential expression, or monomial, like a^3 means $a \cdot a \cdot a$. We also know that $a^1 = a$ and that $a^0 = 1$ when $a \neq 0$. Now consider an expression like $a^3 \cdot a^2$:

$$a^3 \cdot a^2 \quad \text{means} \quad (a \cdot a \cdot a)(a \cdot a).$$

Since an exponent greater than 1 tells how many times we use a base as a factor, then $(a \cdot a \cdot a)(a \cdot a) = a^5$. Note that the exponent in a^5 is the sum of those in $a^3 \cdot a^2$. Likewise, $b^4 \cdot b^3 = (b \cdot b \cdot b \cdot b)(b \cdot b \cdot b) = b^7$. Again, the exponent in b^7 is the sum of those in $b^4 \cdot b^3$. Adding the exponents gives the correct result.

> For any number a and any whole numbers m and n,
> $$a^m \cdot a^n = a^{m+n}.$$
> (When multiplying with exponential notation, we can add the exponents if the bases are the same.)

EXAMPLES Multiply and simplify.

1. $8^4 \cdot 8^3 = 8^{4+3}$ Adding exponents
   $= 8^7$
2. $x^2 \cdot x^9 = x^{2+9}$
   $= x^{11}$
3. $(2x)^5 \cdot (2x)^{10} \cdot (2x)^3 = (2x)^{5+10+3}$
   $= (2x)^{18}$
4. $x \cdot x^8 = x^1 \cdot x^8 = x^{1+8}$
   $= x^9$
5. $y^{13} \cdot y^0 \cdot y = y^{13+0+1}$
   $= y^{14}$

Dividing Using Exponents

What happens when we divide using exponential notation? Look at these examples.

EXAMPLES

6. $\dfrac{4^5}{4^2} = \dfrac{4 \cdot 4 \cdot 4 \cdot 4 \cdot 4}{4 \cdot 4}$

$= 4 \cdot 4 \cdot 4 \cdot \dfrac{4 \cdot 4}{4 \cdot 4}$

$= 4 \cdot 4 \cdot 4 \quad$ Removing a factor of 1
$= 4^3$

Note that the exponent in 4^3 is the difference of those in $4^5 \div 4^2$. If we subtract exponents, we get $5 - 2$, which is 3.

7. $\dfrac{9^3}{9^6} = \dfrac{9 \cdot 9 \cdot 9}{9 \cdot 9 \cdot 9 \cdot 9 \cdot 9 \cdot 9}$

$= \dfrac{9 \cdot 9 \cdot 9}{9 \cdot 9 \cdot 9} \cdot \dfrac{1}{9 \cdot 9 \cdot 9}$

$= \dfrac{1}{9^3}$

Once again, the exponent in the denominator of the answer is the difference, namely $6 - 3$.

8. $\dfrac{5^3}{5^3} = \dfrac{5 \cdot 5 \cdot 5}{5 \cdot 5 \cdot 5} = 1 = 5^0 = 5^{3-3}$

The answer is really 1, but we can think of it as 5^0, or 5^{3-3}, where the exponent is $3 - 3$, or 0.

These examples suggest the following rule in which we subtract exponents.

For any nonzero number a and any whole numbers m and n, we can divide using exponential notation by subtraction as follows:

a) $\dfrac{a^m}{a^n} = a^{m-n} \quad$ if $m > n$ or $m = n$

If the exponent m of the numerator is greater than the exponent n of the denominator, or equal to it, then the answer is the base raised to the power $m - n$.

b) $\dfrac{a^m}{a^n} = \dfrac{1}{a^{n-m}} \quad$ if $m < n$

If the exponent m of the numerator is smaller than the exponent n of the denominator, then the answer is 1 over the base raised to the power $n - m$.

EXAMPLES Divide and simplify.

9. $\dfrac{6^5}{6^3} = 6^{5-3}$ $\quad$ 10. $\dfrac{19^5}{19^8} = \dfrac{1}{19^{8-5}}$ $\quad$ 11. $\dfrac{t^3}{t^{12}} = \dfrac{1}{t^{12-3}}$ $\quad$ 12. $\dfrac{m^9}{m^9} = m^{9-9}$

$= 6^2 \phantom{\dfrac{6^5}{6^3}}$ $= \dfrac{1}{19^3}$ $= \dfrac{1}{t^9}$ $= m^0$

$\phantom{12.\ \dfrac{m^9}{m^9}}= 1$

Raising Powers to Powers

Next, consider an expression like $(3^2)^4$. In this case, we are raising 3^2 to the fourth power:

$$(3^2)^4 = (3^2)(3^2)(3^2)(3^2)$$
$$= (3 \cdot 3)(3 \cdot 3)(3 \cdot 3)(3 \cdot 3)$$
$$= 3^8.$$

Note that in this case we could have multiplied the exponents:

$$(3^2)^4 = 3^{2 \cdot 4} = 3^8.$$

Likewise, $(y^8)^3 = (y^8)(y^8)(y^8) = y^{24}$. Once again we could have multiplied the exponents:

$$(y^8)^3 = y^{8 \cdot 3} = y^{24}.$$

For any number a and any whole numbers m and n,

$$(a^m)^n = a^{mn}.$$

(To raise a power to a power, we can multiply the exponents.)

EXAMPLES Simplify.

13. $(3^4)^5 = 3^{4 \cdot 5} = 3^{20}$ $\qquad$ Multiplying exponents
14. $(y^6)^7 = y^{6 \cdot 7} = y^{42}$ $\qquad$ Multiplying exponents

Sometimes there may be a product inside the parentheses.

For any numbers a and b and any whole number n,

$$(ab)^n = a^n b^n.$$

(To raise a product to the nth power, raise each factor to the nth power.)

EXAMPLES Simplify.

15. $(4x^2)^3 = 4^3 \cdot (x^2)^3$ Raising each factor to the third power
    $= 64x^6$

16. $(5x^3y^5z^2)^4 = 5^4(x^3)^4(y^5)^4(z^2)^4$ Raising each factor to the fourth power
    $= 625x^{12}y^{20}z^8$

17. $(-5x^4y^3)^3 = (-5)^3(x^4)^3(y^3)^3$
    $= -125x^{12}y^9$

18. $[(-x)^{25}]^2 = (-x)^{50}$
    $= (-1 \cdot x)^{50}$ Using the property of -1
    $= (-1)^{50}x^{50}$
    $= 1 \cdot x^{50}$ The product of an even number of negative factors is positive.
    $= x^{50}$

Problem Solving: Compound Interest

EXAMPLE 19 Suppose $1000 is invested at 8%, compounded annually. How much is in the account at the end of 3 years?

1. **Familiarize** Suppose we invest P dollars at an interest rate of 8%, compounded annually. The amount to which this grows at the end of one year is given by

$$P + 8\%P = P + 0.08P \quad \text{By definition of percent}$$
$$= (1 + 0.08)P \quad \text{Factoring}$$
$$= 1.08P. \quad \text{Simplifying}$$

Going into the second year, the new principal is $(1.08)P$ dollars since interest has been added to the account. By the end of the second year, the following amount will be in the account:

$$(1.08)[(1.08)P], \quad \text{or} \quad (1.08)^2P. \quad \text{New principal}$$

Going into the third year, the principal will be $(1.08)^2P$ dollars. At the end of the third year, the following amount will be in the account:

$$(1.08)[(1.08)^2P], \quad \text{or} \quad (1.08)^3P. \quad \text{New principal}$$

Note the pattern: At the end of years 1, 2, and 3, the amount is

$$(1.08)P, \quad (1.08)^2P, \quad (1.08)^3P, \quad \text{and so on.}$$

· ·

If principal P is invested at interest rate r, compounded annually, in t years it will grow to the amount A given by

$$A = P(1 + r)^t.$$

2. **Translate** The translation is the preceding formula:
$$A = P(1 + r)^t.$$

3. **Carry out** To carry out some mathematical manipulation in the case of this problem, we substitute 1000 for P, 0.08 for r, and 3 for t. We get
$$A = P(1 + r)^t$$
$$= 1000(1 + 0.08)^3 = 1000(1.08)^3 = 1000(1.259712)$$
$$\approx \$1259.71.$$

4. **Check** We can check by repeating the calculations. Another way to check is to think about the problem. The answer we get should be larger than the $1000 we started with. This is the case, so we have another partial check. An answer of $3000 would have been too large since that would have implied about a 100% rate of interest.

5. **State** The amount in the account is $1259.71.

EXERCISE SET 4.1

Multiply and simplify.

1. $2^4 \cdot 2^3$
2. $3^5 \cdot 3^2$
3. $8^5 \cdot 8^9$
4. $n^3 \cdot n^{20}$
5. $x^4 \cdot x^3$
6. $y^7 \cdot y^9$
7. $9^{17} \cdot 9^{21}$
8. $t^0 \cdot t^{16}$
9. $(3y)^4(3y)^8$
10. $(2t)^8(2t)^{17}$
11. $(7y)^1(7y)^{16}$
12. $(8x)^0(8x)^1$

Divide and simplify.

13. $\dfrac{7^5}{7^2}$
14. $\dfrac{4^7}{4^3}$
15. $\dfrac{8^{12}}{8^6}$
16. $\dfrac{9^{14}}{9^2}$
17. $\dfrac{y^9}{y^5}$
18. $\dfrac{x^{12}}{x^{11}}$
19. $\dfrac{16^2}{16^8}$
20. $\dfrac{5^4}{5^{10}}$
21. $\dfrac{m^6}{m^{12}}$
22. $\dfrac{p^4}{p^5}$
23. $\dfrac{(8x)^6}{(8x)^{10}}$
24. $\dfrac{(9t)^4}{(9t)^{11}}$
25. $\dfrac{18^9}{18^9}$
26. $\dfrac{(6y)^7}{(6y)^7}$
27. $\dfrac{(7x)^0}{(7x)^0}$

Simplify.

28. $(2^5)^2$
29. $(3^4)^3$
30. $(5^2)^3$
31. $(6^8)^9$
32. $(y^5)^9$
33. $(x^3)^{20}$
34. $(m^{18})^4$
35. $(n^5)^{21}$
36. $(3y^4)^3$
37. $(5a^5)^2$
38. $(2x^8y^3z^7)^5$
39. $(3x^5y^9z^3)^4$
40. $(-2x^8y^6)^3$
41. $[(-x)^{40}]^5$
42. $[(-y)^{18}]^3$

Problem solving

43. Suppose $2000 is invested at 12%, compounded annually. How much is in the account at the end of 2 years?

44. Suppose $2000 is invested at 15%, compounded annually. How much is in the account at the end of 3 years?

45. Suppose $10,400 is invested at 16.5%, compounded annually. How much is in the account at the end of 5 years?

46. Suppose $20,800 is invested at 20.5%, compounded annually. How much is in the account at the end of 6 years?

47. Write $4^3 \cdot 8 \cdot 16$ as a power of 2.

48. Write $2^8 \cdot 16^3 \cdot 64$ as a power of 4.

49. Determine whether $(5y)^0$ and $5y^0$ are equivalent expressions.

50. Simplify: $(y^{2x})(y^{3x})$.

Simplify.

51. $\dfrac{(5^{12})^2}{5^{25}}$

52. $\dfrac{a^{20+20}}{(a^{20})^2}$

53. $\dfrac{(3^5)^4}{3^5 \cdot 3^4}$

54. $\dfrac{(7^5)^{14}}{(7^{14})^5}$

55. $\dfrac{a^{22}}{(a^2)^{11}}$

56. $\dfrac{49^{18}}{7^{35}}$ (*Hint:* Study Exercise 55.)

57. Simplify: $\dfrac{(\frac{1}{2})^4}{(\frac{1}{2})^5}$.

58. Simplify: $\dfrac{(0.4)^5}{((0.4)^3)^2}$.

59. How might you define 3^{-2} so that your definition is consistent with our rules for operating with exponents?

60. Determine whether $(a + b)^n = a^n + b^n$ is true for all numbers a, b, and n. (*Hint:* Substitute and evaluate.)

61. Determine whether $(a^b)^c = a^{bc}$ for all numbers a, b, and c. (*Hint:* Try different values. Remember to work inside parentheses first.)

62. Solve for x: $\dfrac{w^{50}}{w^x} = w^x$.

63. Solve for a: $\dfrac{(9x)^{12}}{(9x)^{14}} = \dfrac{1}{ax^2}$.

Insert $>$, $<$, or $=$ between each pair of numbers to make a true sentence.

64. $3^5 \quad 3^4$

65. $4^2 \quad 4^3$

66. $4^3 \quad 5^3$

67. $4^3 \quad 3^4$

Determine whether each of the following is true for any pair of whole numbers m and n and any positive numbers x and y.

68. $x^m \cdot y^n = (xy)^{mn}$

69. $x^m \cdot y^m = (xy)^m$

70. $x^m \cdot y^m = (xy)^{2m}$

71. $x^m \cdot x^n = x^{mn}$

72. $\left(\dfrac{x}{y}\right)^n = \dfrac{x^n}{y^n}$

73. $(x - y)^m = x^m - y^m$

4.2 POLYNOMIALS

We have already learned how to evaluate certain kinds of algebraic expressions. We have also learned to collect like terms for certain kinds of algebraic expressions. Now we learn to evaluate polynomials and to collect like terms for polynomials.

Algebraic expressions like the following are *polynomials*.

$$\tfrac{3}{4}y^5, \quad -2, \quad 5y + 3, \quad 3x^2 + 2x - 5, \quad -7a^3 + \tfrac{1}{2}a, \quad 6x, \quad 37p^4, \quad x, \quad 0.$$

A *polynomial* is a monomial or a sum or difference of monomials or a combination of sums and differences of monomials.

The following algebraic expressions are *not* polynomials:

$$\frac{x+3}{x-4}, \qquad 5x^3 - 2x^2 + \frac{1}{x}, \quad \text{and} \quad \frac{1}{x^3 - 2}.$$

Polynomials are used often in algebra, especially in equation solving. We will learn how to add, subtract, and multiply polynomials.

Evaluating Polynomials and Polynomials in Problem Solving

When we replace the variable in a polynomial by a number, the polynomial then represents a number. Finding that number is called *evaluating the polynomial*.

EXAMPLES Evaluate each polynomial for $x = 2$.

1. $3x + 5$: $\quad 3 \cdot 2 + 5 = 6 + 5$
   $ = 11$
2. $2x^2 - 7x + 3$: $\quad 2 \cdot 2^2 - 7 \cdot 2 + 3 = 2 \cdot 4 - 14 + 3$
   $ = 8 - 14 + 3$
   $ = -3$

Polynomials arise in many real-world situations and are used in problem solving. The following examples are two such applications. Although the examples are problem solving in nature, they involve only the evaluation of a polynomial. For that reason we do not apply the entire problem-solving strategy, although you can if you wish.

EXAMPLE 3 The volume of a cube with side of length x is given by the polynomial

$$x^3.$$

Find the volume of a cube with side of length 5 cm.

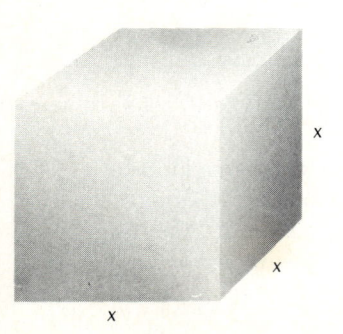

We evaluate the polynomial for $x = 5$:
$$x^3 = 5^3 = 125.$$
The volume is 125 cm³ (cubic centimeters).

EXAMPLE 4 In a sports league of n teams in which each team plays every other team twice, the total number of games to be played is given by the polynomial
$$n^2 - n.$$

A slow-pitch softball league has 10 teams. What is the total number of games to be played?

We evaluate the polynomial for $n = 10$:
$$n^2 - n = 10^2 - 10 = 100 - 10 = 90.$$
The league plays 90 games.

Identifying Terms

Subtractions can be rewritten as additions. We showed this in Section 2.4. For any polynomial we can find an equivalent polynomial using only additions.

EXAMPLES Find an equivalent polynomial using only additions.

5. $-5x^2 - x = -5x^2 + (-x)$
6. $4x^5 - 2x^6 - 4x = 4x^5 + (-2x^6) + (-4x)$

When a polynomial has only additions, the parts being added are called *terms*.

EXAMPLE 7 Identify the terms of the polynomial
$$4x^3 + 3x + 12 + 8x^3 + 5x.$$
Terms: $4x^3$, $3x$, 12, $8x^3$, $5x$

If there are subtractions you can *think* of them as additions without rewriting.

EXAMPLE 8 Identify the terms of the polynomial
$$3t^4 - 5t^6 - 4t + 2.$$
Terms: $3t^4$, $-5t^6$, $-4t$, 2

Note in an expression like $3x + 5y$ that the terms are the things added: $3x$ and $5y$. Note also that each term is a product.

Like Terms

Terms that have the same variable and the same exponent are called *like terms*, or *similar terms*.

EXAMPLES Identify the like terms in each polynomial.

9. $4x^3 + 5x - 4x^2 + 2x^3 + x^2$

   Like terms: $4x^3$ and $2x^3$ Same exponent and variable
   Like terms: $-4x^2$ and x^2 Same exponent and variable

10. $6 - 3a^2 + 8 - a - 5a$

    Like terms: 6 and 8 No variable at all
    Like terms: $-a$ and $-5a$

Collecting Like Terms

We can often simplify polynomials by *collecting like terms*, or *combining similar terms*. To do this we use the distributive laws.

EXAMPLES Collect like terms.

11. $2x^3 - 6x^3 = (2 - 6)x^3$ Using a distributive law
    $= -4x^3$

12. $5x^2 + 7 + 4x^4 + 2x^2 - 11 - 2x^4 = (5 + 2)x^2 + (4 - 2)x^4 + (7 - 11)$
    $= 7x^2 + 2x^4 - 4$

Note that using the distributive laws in this manner allows us to collect like terms by adding or subtracting the coefficients.

In collecting like terms we may get zero.

EXAMPLES Collect like terms.

13. $5x^3 - 5x^3 = (5-5)x^3 = 0x^3 = 0$
14. $3x^4 + 2x^2 - 3x^4 = (3-3)x^4 + 2x^2 = 0x^4 + 2x^2 = 2x^2$

Multiplying a term of a polynomial by 1 does not change the polynomial, but it may make it easier to factor.

EXAMPLES Collect like terms.

15. $5x^2 + x^2 = 5x^2 + 1x^2$ Replacing x^2 by $1x^2$
    $= (5+1)x^2$ Using a distributive law
    $= 6x^2$
16. $5x^4 - 6x^3 - x^4 = 5x^4 - 6x^3 - 1x^4$ $x^4 = 1x^4$
    $= (5-1)x^4 - 6x^3$
    $= 4x^4 - 6x^3$
17. $\frac{2}{3}x^4 - x^3 - \frac{1}{6}x^4 + \frac{2}{5}x^3 - \frac{3}{10}x^3 = \frac{3}{6}x^4 - \frac{9}{10}x^3$
    $= \frac{1}{2}x^4 - \frac{9}{10}x^3$

EXERCISE SET 4.2

Evaluate each polynomial for $x = 4$.

1. $-5x + 2$
2. $-3x + 1$
3. $2x^2 - 5x + 7$
4. $3x^2 + x + 7$
5. $x^3 - 5x^2 + x$
6. $7 - x + 3x^2$

The daily number of accidents (average number of accidents per day) involving drivers of age a is approximated by the polynomial $0.4a^2 - 40a + 1039$.

7. Evaluate the polynomial for $a = 18$ to find the number of daily accidents involving an 18-year-old driver.

8. Evaluate the polynomial for $a = 20$ to find the number of daily accidents involving a 20-year-old driver.

Evaluate each polynomial for $x = -1$.

9. $3x + 5$
10. $6 - 2x$
11. $x^2 - 2x + 1$
12. $5x - 6 + x^2$
13. $-3x^3 + 7x^2 - 3x - 2$
14. $-2x^3 - 5x^2 + 4x + 3$

The perimeter of a square of side x is given by the polynomial

$$4x.$$

15. Find the perimeter of a square of side 17 ft.
16. Find the perimeter of a square of side 28.5 m.

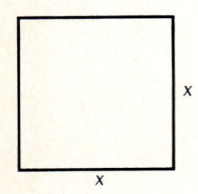

The distance s, in feet, traveled by a body falling freely from rest in t seconds is approximated by the polynomial

$$16t^2.$$

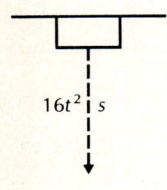

17. A stone is dropped from a cliff and takes 8 sec to hit the ground. How high is the cliff?
18. A brick is dropped from a building and takes 3 sec to hit the ground. How high is the building?

The amount of water, in gallons, in a tub after it has drained is given by the polynomial

$$400 - 200t + 25t^2,$$

where t is the time, in minutes, that the water has drained, and where t is greater than or equal to 0 and less than or equal to 4.

19. How much water was in the tub before it began draining?
20. How much water was in the tub after 1 min? 2 min? 3 min?

Identify the terms of each polynomial.

21. $2 - 3x + x^2$
22. $2x^2 + 3x - 4$

Identify the like terms in each polynomial.

23. $5x^3 + 6x^2 - 3x^2$
24. $3x^2 + 4x^3 - 2x^2$
25. $2x^4 + 5x - 7x - 3x^4$
26. $-3t + t^3 - 2t - 5t^3$

Collect like terms.

27. $2x - 5x$
28. $2x^2 + 8x^2$
29. $x - 9x$
30. $x - 5x$
31. $5x^3 + 6x^3 + 4$
32. $6x^4 - 2x^4 + 5$
33. $5x^3 + 6x - 4x^3 - 7x$
34. $3a^4 - 2a + 2a + a^4$
35. $6b^5 + 3b^2 - 2b^5 - 3b^2$
36. $2x^2 - 6x + 3x + 4x^2$
37. $\frac{1}{4}x^5 - 5 + \frac{1}{2}x^5 - 2x - 37$
38. $\frac{1}{3}x^3 + 2x - \frac{1}{6}x^3 + 4 - 16$
39. $6x^2 + 2x^4 - 2x^2 - x^4 - 4x^2$
40. $8x^2 + 2x^3 - 3x^3 - 4x^2 - 4x^2$
41. $\frac{1}{4}x^3 - x^2 - \frac{1}{6}x^2 + \frac{3}{8}x^3 + \frac{5}{16}x^3$
42. $\frac{1}{5}x^4 + \frac{1}{5} - 2x^2 + \frac{1}{10} - \frac{3}{15}x^4 + 2x^2 - \frac{3}{10}$

Combine like terms.

43. $3x^2 + 2x - 2 + 3x^0$
44. $\frac{9}{2}x^8 + \frac{1}{9}x^2 + \frac{1}{2}x^9 + \frac{9}{2}x^1 + \frac{9}{2}x^9 + \frac{8}{9}x^2 + \frac{1}{2}x - \frac{1}{2}x^8$
45. $(3x^2)^3 + 4x^2 \cdot 4x^4 - x^4(2x)^2 + ((2x)^2)^3 - 100x^2(x^2)^2$

Solve.

46. ▣ Evaluate $s^2 - 50s + 675$ and $-s^2 + 50s - 675$ for $s = 18$, $s = 25$, and $s = 32$.

47. ▣ The daily number of accidents involving drivers of age a is approximated by the polynomial $0.4a^2 - 40a + 1039$. For what age is the number of daily accidents smallest?

4.3 MORE ON POLYNOMIALS

We now consider the basic concepts regarding polynomials. Such concepts are important not only in future work in this chapter but also in other areas of mathematics.

Descending Order

Note in the following polynomial that the exponents decrease. We say that the polynomial is arranged in *descending order*:

$$8x^4 - 2x^3 + 5x^2 - x + 3.$$

The term with the largest exponent is first. The term with the next largest exponent is second, and so on. The associative and commutative laws allow us to arrange the terms of a polynomial in descending order.

EXAMPLES Arrange each polynomial in descending order.

1. $6x^5 + 4x^7 + x^2 + 2x^3 = 4x^7 + 6x^5 + 2x^3 + x^2$
2. $\frac{2}{3} + 4x^5 - 8x^2 + 5x - 3x^3 = 4x^5 - 3x^3 - 8x^2 + 5x + \frac{2}{3}$

We usually arrange polynomials in descending order. The opposite order is called *ascending*.

Collecting Like Terms and Descending Order

EXAMPLE 3 Collect like terms and then arrange in descending order.

$$2x^2 - 4x^3 + 3 - x^2 - 2x^3 = x^2 - 6x^3 + 3 \quad \text{Collecting like terms}$$
$$= -6x^3 + x^2 + 3 \quad \text{Arranging in descending order}$$

Degrees

The degree of the term $5x^3$ is 3. The *degree* of a term is its exponent.

EXAMPLE 4 Identify the degree of each term of $8x^4 + 3x + 7$.

The degree of $8x^4$ is 4.
The degree of $3x$ is 1.   Recall that $x = x^1$.
The degree of 7 is 0.   Think of 7 as $7x^0$. Recall that $x^0 = 1$.

The *degree of a polynomial* is its largest exponent, unless it is the polynomial 0. The polynomial 0 is a special case. Mathematicians agree that it has *no* degree either as a term or as a polynomial.

EXAMPLE 5 Identify the degree of $5x^3 - 6x^4 + 7$.

$$5x^3 - 6x^4 + 7 \quad \text{The largest exponent is 4.}$$

The degree of the polynomial is 4.

Coefficients

The coefficient of the term $5x^3$ is 5. In the following polynomial the color numbers are the *coefficients*:

$$3x^5 - 2x^3 + 5x + 4.$$

EXAMPLE 6 Identify the coefficient of each term in the polynomial

$$3x^4 - 4x^3 + 7x^2 + x - 8.$$

The coefficient of the first term is 3.
The coefficient of the second term is -4.
The coefficient of the third term is 7.

The coefficient of the fourth term is 1.

The coefficient of the fifth term is -8.

Missing Terms

If a coefficient is 0, we usually do not write the term. We say that we have a *missing term*.

EXAMPLE 7 In
$$8x^5 - 2x^3 + 5x^2 + 7x + 8,$$
there is no term with x^4. We say that the x^4 term (or the *fourth-degree term*) is missing.

We could write missing terms with zero coefficients or leave space. For example, we could write the polynomial $3x^2 + 9$ as
$$3x^2 + 0x + 9 \quad \text{or} \quad 3x^2 + 9,$$
but ordinarily we do not.

Monomials, Binomials, and Trinomials

Polynomials with just one term are called *monomials*. Polynomials with just two terms are called *binomials*. Those with just three terms are called *trinomials*.

EXAMPLE 8

Monomials	Binomials	Trinomials
$4x^2$	$2x + 4$	$3x^3 + 4x + 7$
9	$3x^5 + 6x$	$6x^7 - 7x^2 + 4$
$-23x^{19}$	$-9x^7 - 6$	$4x^2 - 6x - \frac{1}{2}$

EXERCISE SET 4.3

Arrange each polynomial in descending order.

1. $x^5 + x + 6x^3 + 1 + 2x^2$
2. $3 + 2x^2 - 5x^6 - 2x^3 + 3x$
3. $5x^3 + 15x^9 + x - x^2 + 7x^8$
4. $9x - 5 + 6x^3 - 5x^4 + x^5$
5. $8y^3 - 7y^2 + 9y^6 - 5y^8 + y^7$
6. $p^8 - 4 + p + p^2 - 7p^4$

Collect like terms and then arrange in descending order.

7. $3x^4 - 5x^6 - 2x^4 + 6x^6$
8. $-1 + 5x^3 - 3 - 7x^3 + x^4 + 5$
9. $-2x + 4x^3 - 7x + 9x^3 + 8$
10. $-6x^2 + x - 5x + 7x^2 + 1$

11. $3x + 3x + 3x - x^2 - 4x^2$
12. $-2x - 2x - 2x + x^3 - 5x^3$
13. $-x + \frac{3}{4} + 15x^4 - x - \frac{1}{2} - 3x^4$
14. $2x - \frac{5}{6} + 4x^3 + x + \frac{1}{3} - 2x$

Identify the degree of each term of the polynomial and the degree of the polynomial.

15. $2x - 4$
16. $6 - 3x$
17. $3x^2 - 5x + 2$
18. $5x^3 - 2x^2 + 3$
19. $-7x^3 + 6x^2 + 3x + 7$
20. $5x^4 + x^2 - x + 2$
21. $x^2 - 3x + x^6 - 9x^4$
22. $8x - 3x^2 + 9 - 8x^3$

Identify the coefficient of each term of the polynomial.

23. $-3x + 6$
24. $2x - 4$
25. $5x^2 + 3x + 3$
26. $3x^2 - 5x + 2$
27. $-7x^3 + 6x^2 + 3x + 7$
28. $5x^4 + x^2 - x + 2$
29. $-5x^4 + 6x^3 - 3x^2 + 8x - 2$
30. $7x^3 - 4x^2 - 4x + 5$

Identify the missing terms in the polynomial.

31. $x^3 - 27$
32. $x^5 + x$
33. $x^4 - x$
34. $5x^4 - 7x + 2$
35. $2x^3 - 5x^2 + x - 3$
36. $-6x^3$

Tell whether the polynomial is a monomial, binomial, trinomial, or none of these.

37. $x^2 - 10x + 25$
38. $-6x^4$
39. $x^3 - 7x^2 + 2x - 4$
40. $x^2 - 9$
41. $4x^2 - 25$
42. $2x^4 - 7x^3 + x^2 + x - 6$
43. $40x$
44. $4x^2 + 12x + 9$

Problem-solving practice

45. A family spent $2011 to drive a car one year, during which the car was driven 7400 miles. The family spent $972 for insurance and $114 for a license registration fee. The only other cost was for gasoline. How much did gasoline cost per mile?

46. Three tired campers stopped for the night. All they had to eat was a bag of apples. During the night one awoke and ate one third of the apples. Later, a second camper awoke and ate one third of the apples that remained. Much later, the third camper awoke and ate one third of those apples yet remaining after the other two had eaten. When they got up the next morning, 8 apples were left. How many did they have to begin with?

47. Construct a polynomial in x (meaning that x is the variable) of degree 5 with four terms and coefficients that are integers.

48. Construct a trinomial in y of degree 4 with coefficients that are rational numbers.

49. What is the degree of $(5m^5)^2$?

50. Construct three like terms of degree 4.

51. A polynomial in x has degree 3. The coefficient of x^2 is 3 less than the coefficient of x^3. The coefficient of x is 3 times the coefficient of x^2. The remaining coefficient is 2 more than the coefficient of x^3. The sum of the coefficients is -4. Find the polynomial.

4.4 ADDITION OF POLYNOMIALS

We now consider addition of polynomials. This addition is based on collecting like terms. We then use addition of polynomials in problem solving.

Addition

To add two polynomials we could think of writing a plus sign between them and then collecting like terms. Depending on the situation, you may see polynomials written in descending order, ascending order, or neither. Generally, if an exercise is written in one kind of order, we write the answer in that same order.

EXAMPLE 1 Add: $-3x^3 + 2x - 4$ and $4x^3 + 3x^2 + 2$.

$(-3x^3 + 2x - 4) + (4x^3 + 3x^2 + 2)$
$= (-3 + 4)x^3 + 3x^2 + 2x + (-4 + 2)$ Collecting like terms
(*No* signs are changed.)
$= x^3 + 3x^2 + 2x - 2$

EXAMPLE 2 Add: $\frac{2}{3}x^4 + 3x^2 - 2x + \frac{1}{2}$ and $-\frac{1}{3}x^4 + 5x^3 - 3x^2 + 3x - \frac{1}{2}$.

$(\frac{2}{3}x^4 + 3x^2 - 2x + \frac{1}{2}) + (-\frac{1}{3}x^4 + 5x^3 - 3x^2 + 3x - \frac{1}{2})$
$= (\frac{2}{3} - \frac{1}{3})x^4 + 5x^3 + (3 - 3)x^2 + (-2 + 3)x + (\frac{1}{2} - \frac{1}{2})$ Collecting like terms
$= \frac{1}{3}x^4 + 5x^3 + x$

We can add polynomials as we do because they represent numbers. After some practice you will be able to add mentally.

EXAMPLE 3 Add: $3x^2 - 2x + 2$ and $5x^3 - 2x^2 + 3x - 4$.

$(3x^2 - 2x + 2) + (5x^3 - 2x^2 + 3x - 4)$
$= 5x^3 + (3 - 2)x^2 + (-2 + 3)x + (2 - 4)$ You might do this step mentally.
$= 5x^3 + x^2 + x - 2$ Then you would write only this.

We can also add polynomials by writing like terms in columns.

EXAMPLE 4 Add: $9x^5 - 2x^3 + 6x^2 + 3$ and $5x^4 - 7x^2 + 6$ and $3x^6 - 5x^5 + x^2 + 5$.

Arrange the polynomials with like terms in columns.

$$\begin{array}{r} 9x^5 \qquad\quad -2x^3 + 6x^2 + 3 \\ 5x^4 \qquad\quad -7x^2 + 6 \\ 3x^6 - 5x^5 \qquad\qquad\quad + x^2 + 5 \\ \hline 3x^6 + 4x^5 + 5x^4 - 2x^3 \qquad\quad + 14 \end{array}$$

We leave spaces for missing terms.

We write the answer as $3x^6 + 4x^5 + 5x^4 - 2x^3 + 14$ without the missing space.

Problem Solving

The first two steps in our problem-solving process are **Familiarize** and **Translate**. In Example 5 we consider only these steps. You will see how a problem can be translated to a polynomial.

EXAMPLE 5 Find a polynomial for the sum of the areas of these rectangles.

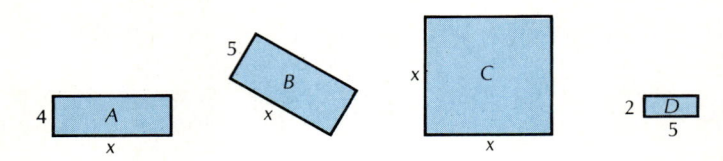

1. **Familiarize** Recall that the area of a rectangle is the product of the length and the width. (You may need to look up such a formula.)

2. **Translate** We translate the problem to mathematical language. The sum of the areas is a sum of products. We find these products and then collect like terms.

area of A plus area of B plus area of C plus area of D

$$4x \;\; + \;\; 5x \;\; + \;\; x \cdot x \;\; + \;\; 2 \cdot 5$$
$$= 4x + 5x + x^2 + 10$$
$$= x^2 + 9x + 10$$

A polynomial for the sum of the areas is $x^2 + 9x + 10$.

EXERCISE SET 4.4

Add.

1. $3x + 2$ and $-4x + 3$
2. $5x^2 + 6x + 1$ and $-7x + 2$
3. $-6x + 2$ and $x^2 + x - 3$
4. $6x^4 + 3x^3 - 1$ and $4x^2 - 3x + 3$
5. $3x^5 + 6x^2 - 1$ and $7x^2 + 6x - 2$
6. $7x^3 + 3x^2 + 6x$ and $-3x^2 - 6$
7. $-4x^4 + 6x^2 - 3x - 5$ and $6x^3 + 5x + 9$
8. $5x^3 + 6x^2 - 3x + 1$ and $5x^4 - 6x^3 + 2x - 5$
9. $(1 + 4x + 6x^2 + 7x^3) + (5 - 4x + 6x^2 - 7x^3)$
10. $(3x^4 - 6x - 5x^2 + 5) + (6x^2 - 4x^3 - 1 + 7x)$
11. $5x^4 - 6x^3 - 7x^2 + x - 1$ and $4x^3 - 6x + 1$
12. $8x^5 - 6x^3 + 6x + 5$ and $-4x^4 + 3x^3 - 7x$
13. $9x^8 - 7x^4 + 2x^2 + 5$ and $8x^7 + 4x^4 - 2x$
14. $4x^5 - 6x^3 - 9x + 1$ and $6x^3 + 9x^2 + 9x$
15. $\frac{1}{4}x^4 + \frac{2}{3}x^3 + \frac{5}{8}x^2 + 7$ and $-\frac{3}{4}x^4 + \frac{3}{8}x^2 - 7$
16. $(\frac{1}{3}x^9 + \frac{1}{5}x^5 - \frac{1}{2}x^2 + 7) + (-\frac{1}{5}x^9 + \frac{1}{4}x^4 - \frac{3}{5}x^5 + \frac{3}{4}x^2 + \frac{1}{2})$

4.4 ADDITION OF POLYNOMIALS

17. $0.02x^5 - 0.2x^3 + x + 0.08$ and $-0.01x^5 + x^4 - 0.8x - 0.02$

18. $(0.03x^6 + 0.05x^3 + 0.22x + 0.05) + (\frac{7}{100}x^6 - \frac{3}{100}x^3 + 0.5)$

19. $\begin{array}{r} -3x^4 + 6x^2 + 2x - 1 \\ -3x^2 + 2x + 1 \\ \hline \end{array}$

20. $\begin{array}{r} -4x^3 + 8x^2 + 3x - 2 \\ -4x^2 + 3x + 2 \\ \hline \end{array}$

21. $\begin{array}{r} 3x^5 - 6x^3 + 3x \\ -3x^4 + 3x^3 + x^2 \\ \hline \end{array}$

22. $\begin{array}{r} 4x^5 - 5x^3 + 2x \\ -4x^4 + 2x^3 + 2x^2 \\ \hline \end{array}$

23. $\begin{array}{r} -3x^2 + x \\ 5x^3 - 6x^2 + 1 \\ 3x - 8 \\ \hline \end{array}$

24. $\begin{array}{r} -4x^2 + 2x \\ 3x^3 - 5x^2 + 3 \\ 5x - 5 \\ \hline \end{array}$

25. $\begin{array}{r} -\frac{1}{2}x^4 - \frac{3}{4}x^3 + 6x \\ \frac{1}{2}x^3 + x^2 + \frac{1}{4}x \\ \frac{3}{4}x^4 + \frac{1}{2}x^2 + \frac{1}{2}x + \frac{1}{4} \\ \hline \end{array}$

26. $\begin{array}{r} -\frac{1}{4}x^4 - \frac{1}{2}x^3 + 2x \\ \frac{3}{4}x^3 - x^2 + \frac{1}{2}x \\ \frac{1}{2}x^4 + \frac{1}{2}x^2 + \frac{1}{2}x + \frac{1}{2} \\ \hline \end{array}$

27. $\begin{array}{r} -4x^2 \\ 4x^4 - 3x^3 + 6x^2 + 5x \\ 6x^3 - 8x^2 + 1 \\ -5x^4 \\ 6x^2 - 3x \\ \hline \end{array}$

28. $\begin{array}{r} 3x^2 \\ 5x^4 - 2x^3 + 4x^2 + 5x \\ 5x^3 - 5x^2 + 2 \\ -7x^4 \\ 3x^2 - 2x \\ \hline \end{array}$

29. $\begin{array}{r} 3x^4 - 6x^2 + 7x \\ 3x^2 - 3x + 1 \\ -2x^4 + 7x^2 + 3x \\ 5x - 2 \\ \hline \end{array}$

30. $\begin{array}{r} 5x^4 - 8x^2 + 4x \\ 5x^2 - 2x + 3 \\ -3x^4 + 3x^2 + 5x \\ 3x - 5 \\ \hline \end{array}$

31. $\begin{array}{r} 3x^5 - 6x^4 + 3x^3 - 1 \\ 6x^4 - 4x^3 + 6x^2 \\ 3x^5 + 2x^3 \\ -6x^4 - 7x^2 \\ -5x^5 + 3x^3 + 2 \\ \hline \end{array}$

32. $\begin{array}{r} 4x^5 - 3x^4 + 2x^3 - 2 \\ 6x^4 + 5x^3 + 3x^2 \\ 5x^5 + 4x^3 \\ -6x^4 - 5x^2 \\ -3x^5 + 2x^3 + 5 \\ \hline \end{array}$

33. $\begin{array}{r} -x^3 + 6x^2 + 3x + 5 \\ x^4 - 3x^2 + 2 \\ -5x + 3 \\ 6x^4 + 4x^2 - 1 \\ -x^3 + 6x \\ \hline \end{array}$

34. $\begin{array}{r} -2x^3 + 3x^2 + 5x + 3 \\ x^4 - 5x^2 + 1 \\ -7x + 4 \\ 4x^4 + 6x^2 - 2 \\ -x^3 + 5x \\ \hline \end{array}$

35. $\begin{array}{r} 1 + 5x - 6x^2 + 6x^3 - 3x^4 \\ -5x - 3x^3 + 5x^5 \\ -8 + 7x^2 + 4x^4 \\ 1 + 3x - 2x^5 \\ \hline \end{array}$

36. $\begin{array}{r} 2 + 3x - 7x^2 + 4x^3 - 5x^4 \\ -6x - 7x^3 + 3x^5 \\ -3 + 5x + 5x^3 + 3x^4 \\ 4 + 10x^2 - 5x^5 \\ \hline \end{array}$

37. $\begin{array}{r} 0.15x^4 + 0.10x^3 - 0.9x^2 \\ -0.01x^3 + 0.01x^2 + x \\ 1.25x^4 + 0.11x^2 + 0.01 \\ 0.27x^3 + 0.99 \\ -0.35x^4 + 15x^2 - 0.03 \\ \hline \end{array}$

38. $\begin{array}{r} 0.05x^4 + 0.12x^3 - 0.5x^2 \\ -0.02x^3 + 0.02x^2 + 2x \\ 1.5x^4 + 0.01x^2 + 0.15 \\ 0.25x^3 + 0.85 \\ -0.25x^4 + 10x^2 - 0.04 \\ \hline \end{array}$

POLYNOMIALS

Problem solving

39. Solve.

a) Find a polynomial for the sum of the areas of these rectangles.

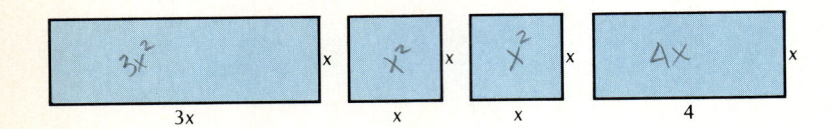

[Rectangles labeled with areas $3x^2$, x^2, x^2, $4x$ with heights x and bases $3x$, x, x, 4]

b) Find the sum of the areas when $x = 3$ and $x = 8$.

40. Solve.

a) Find a polynomial for the sum of the areas of these circles.

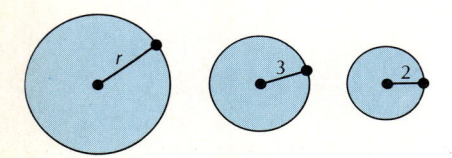

[Three circles with radii r, 3, 2]

b) Find the sum of the areas when $r = 5$ and $r = 11.3$.

Find a polynomial for the perimeter of each figure.

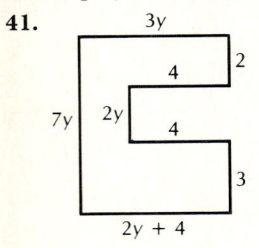

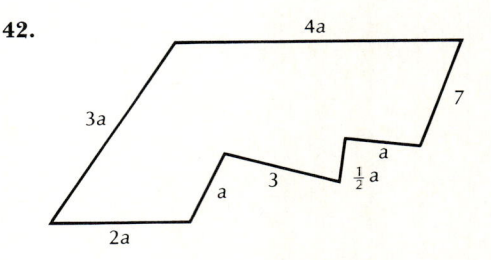

41. [Figure with sides: $3y$, 4, 2, $2y$, 4, 3, $2y + 4$, $7y$]

42. [Figure with sides: $4a$, 7, a, $\frac{1}{2}a$, 3, a, $2a$, $3a$]

For each figure find two algebraic expressions for the area.

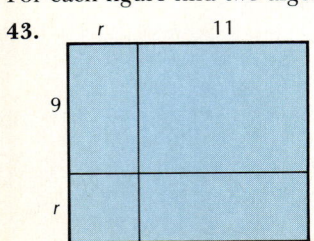

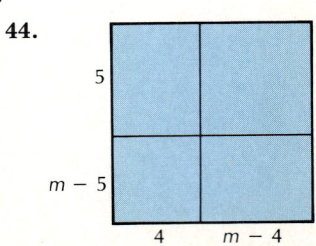

43. [Rectangle divided into 4 parts with dimensions r, 11 across top and 9, r down the side]

44. [Square divided into 4 parts with dimensions 5, $m-5$ vertically and 4, $m-4$ horizontally]

45. Find $(x + 3)^2$ using the four areas of the following square.

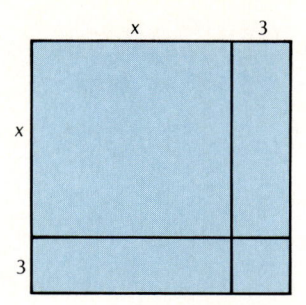

46. Add: $(-20.344x^6 - 70.789x^5 + 890x) + (68.888x^6 + 69.994x^5)$.

47. Addition of real numbers is commutative. That is, $a + b = b + a$, where a and b are any real numbers.

a) Show that addition of binomials such as $(ax + b)$ and $(cx + d)$ is commutative.
b) Show that addition of trinomials such as $(ax^2 + bx + c)$ and $(dx^2 + ex + f)$ is commutative.

48. The sum of a number and 2 is multiplied by the number and then 3 is subtracted from the result. Find a polynomial for the final result.

49. Three brothers have ages that are consecutive multiples of five. The sum of their ages two years ago was 69. Find their ages now.

50. Find four consecutive multiples of four when the sum of the first two is the fourth.

4.5 SUBTRACTION OF POLYNOMIALS

We now consider subtraction of polynomials. To do so we first consider the concept of the additive inverse of a polynomial.

Additive Inverses of Polynomials

We know that two numbers are additive inverses of each other if their sum is zero. The same definition is true of polynomials.

> Two polynomials are *additive inverses* of each other if their sum is zero.

To find a way to determine an additive inverse, look for a pattern in the following examples:

a) $2x + (-2x) = 0$;
b) $-6x^2 + 6x^2 = 0$;

c) $(5t^3 - 2) + (-5t^3 + 2) = 0$;

d) $(7x^3 - 6x^2 - x + 4) + (-7x^3 + 6x^2 + x - 4) = 0$.

Since $(5t^3 - 2) + (-5t^3 + 2) = 0$, we know that the additive inverse of $(5t^3 - 2)$ is $(-5t^3 + 2)$. To say the same thing with purely algebraic symbolism, consider

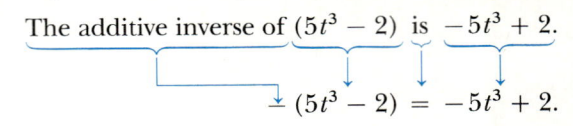

The additive inverse of $(5t^3 - 2)$ is $-5t^3 + 2$.

$-(5t^3 - 2) = -5t^3 + 2$.

> We can find an equivalent polynomial for the additive inverse of a polynomial by replacing each term by its additive inverse (changing the sign of every term).

EXAMPLE 1 Find two equivalent expressions for the additive inverse of $4x^5 - 7x^3 - 8x + \frac{5}{6}$.

a) $-(4x^5 - 7x^3 - 8x + \frac{5}{6})$

b) $-4x^5 + 7x^3 + 8x - \frac{5}{6}$ Changing the sign of every term

EXAMPLE 2 Simplify: $-(-7x^4 - \frac{5}{9}x^3 + 8x^2 - x + 67)$.

$-(-7x^4 - \frac{5}{9}x^3 + 8x^2 - x + 67) = 7x^4 + \frac{5}{9}x^3 - 8x^2 + x - 67$

Subtraction of Polynomials

Recall that we can subtract a rational number by adding its additive inverse: $a - b = a + (-b)$. This allows us to find an equivalent expression for the difference of two polynomials.

EXAMPLE 3 Subtract: $(9x^5 + x^3 - 2x^2 + 4) - (2x^5 + x^4 - 4x^3 - 3x^2)$.

$(9x^5 + x^3 - 2x^2 + 4) - (2x^5 + x^4 - 4x^3 - 3x^2)$
$= (9x^5 + x^3 - 2x^2 + 4) + [-(2x^5 + x^4 - 4x^3 - 3x^2)]$ Adding an inverse
$= (9x^5 + x^3 - 2x^2 + 4) + (-2x^5 - x^4 + 4x^3 + 3x^2)$ Finding the inverse by changing the sign of *every* term
$= 7x^5 - x^4 + 5x^3 + x^2 + 4$ Collecting like terms

After some practice you will be able to subtract mentally.

EXAMPLE 4 Subtract: $(9x^5 + x^3 - 2x) - (-2x^5 + 5x^3 + 6)$.

$(9x^5 + x^3 - 2x) - (-2x^5 + 5x^3 + 6)$
$= (9x^5 + 2x^5) + (x^3 - 5x^3) - 2x - 6$ Subtract the like terms mentally.
$= 11x^5 - 4x^3 - 2x - 6$ Write only this.

We can use columns to subtract. We replace coefficients by their inverses, as shown in Example 3. You can also do it mentally.

EXAMPLE 5 Subtract: $(5x^2 - 3x + 6) - (9x^2 - 5x - 3)$.

a) $5x^2 - 3x + 6$
$9x^2 - 5x - 3$ Writing similar terms in columns

b) $5x^2 - 3x + 6$
$-9x^2 \pm 5x \pm 3$ Changing signs

c) $5x^2 - 3x + 6$
$-9x^2 + 5x + 3$ Adding
$\overline{-4x^2 + 2x + 9}$

If you can do so without error, you should skip step (b). Just write the answer.

EXAMPLE 6 Subtract: $(x^3 + x^2 + 2x - 12) - (2x^3 + x^2 - 3x)$.

$$x^3 + x^2 + 2x - 12$$
$$2x^3 + x^2 - 3x$$
$$\overline{-x^3 + 5x - 12}$$

Problem Solving

EXAMPLE 7 A 4-ft by 4-ft sandbox is placed on a square lawn x ft on a side. Find a polynomial for the remaining area.

1. **Familiarize** We draw a picture of the situation as follows.

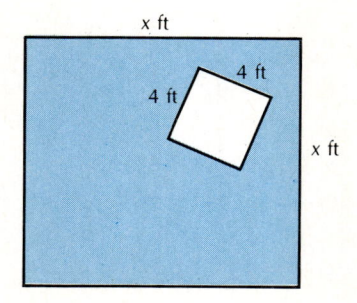

2. **Translate** We reword the problem and translate as follows.

(Area of lawn) − (Area of sandbox) = Area left over

$x \cdot x - 4 \cdot 4 =$ Area left over

3. **Carry out** We carry out the manipulations by multiplying the numbers:

$$x^2 - 16 = \text{Area left over.}$$

4. **Check** We could check the problem by assigning some value to x, say, 10, and carrying out the computation of the area two ways:

a) Area of lawn = $10 \cdot 10 = 100$;
b) Area of sandbox = $4 \cdot 4 = 16$;
c) Area left over = $100 - 16 = 84$.

This is the same as substituting 10 for x in $x^2 - 16$:

$$10^2 - 16 = 100 - 16 = 84.$$

5. **State** The area left over is $x^2 - 16$.

EXERCISE SET 4.5

Find two equivalent expressions for the additive inverse of each polynomial.

1. $-5x$
2. $x^2 - 3x$
3. $-x^2 + 10x - 2$
4. $-4x^3 - x^2 - x$
5. $12x^4 - 3x^3 + 3$
6. $4x^3 - 6x^2 - 8x + 1$

Simplify.

7. $-(3x - 7)$
8. $-(-2x + 4)$
9. $-(4x^2 - 3x + 2)$
10. $-(-6a^3 + 2a^2 - 9a + 1)$
11. $-(-4x^4 - 6x^2 + \frac{3}{4}x - 8)$
12. $-(-5x^4 + 4x^3 - x^2 + 0.9)$

Subtract.

13. $(5x^2 + 6) - (3x^2 - 8)$
14. $(7x^3 - 2x^2 + 6) - (7x^2 + 2x - 4)$
15. $(6x^5 - 3x^4 + x + 1) - (8x^5 + 3x^4 - 1)$
16. $(\frac{1}{2}x^2 - \frac{3}{2}x + 2) - (\frac{3}{2}x^2 + \frac{1}{2}x - 2)$
17. $(6x^2 + 2x) - (-3x^2 - 7x + 8)$
18. $7x^3 - (-3x^2 - 2x + 1)$
19. $(\frac{5}{8}x^3 - \frac{1}{4}x - \frac{1}{3}) - (-\frac{1}{8}x^3 + \frac{1}{4}x - \frac{1}{3})$
20. $(\frac{1}{5}x^3 + 2x^2 - 0.1) - (-\frac{2}{5}x^3 + 2x^2 + 0.01)$
21. $(0.08x^3 - 0.02x^2 + 0.01x) - (0.02x^3 + 0.03x^2 - 1)$
22. $(0.8x^4 + 0.2x - 1) - (\frac{7}{10}x^4 + \frac{1}{5}x - 0.1)$

23. $x^2 + 5x + 6$
    $x^2 + 2x$
    ───────────

24. $x^3 + 1$
    $x^3 + x^2$
    ───────────

25. $x^4 - 3x^2 + x + 1$
    $x^4 - 4x^3$
    ───────────────────

26. $3x^2 - 6x + 1$
    $6x^2 + 8x - 3$
    ───────────

27. $5x^4 + 6x^3 - 9x^2$
    $-6x^4 - 6x^3 + 8x + 9$
    ───────────────────

28. $5x^4 + 6x^2 - 3x + 6$
    $ 6x^3 + 7x^2 - 8x - 9$
    ───────────────────

29. $\begin{array}{r} 3x^4 + 6x^2 + 8x - 1 \\ 4x^5 - 6x^4 - 8x - 7 \\ \hline \end{array}$

30. $\begin{array}{r} 6x^5 + 3x^2 - 7x + 2 \\ 10x^5 + 6x^3 - 5x^2 - 2x + 4 \\ \hline \end{array}$

31. $\begin{array}{r} x^5 - 1 \\ x^5 - x^4 + x^3 - x^2 + x - 1 \\ \hline \end{array}$

32. $\begin{array}{r} x^5 + x^4 - x^3 + x^2 - x + 2 \\ x^5 - x^4 + x^3 - x^2 - x + 2 \\ \hline \end{array}$

Problem solving

Find a polynomial for the color area of each figure.

33.

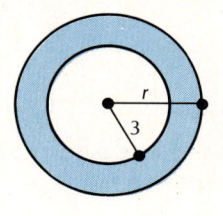

34.

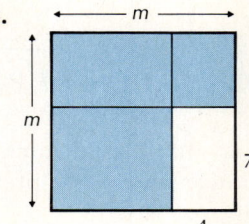

35.

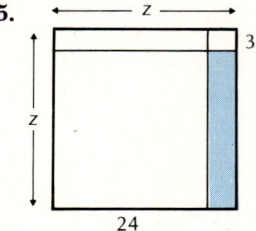

36.

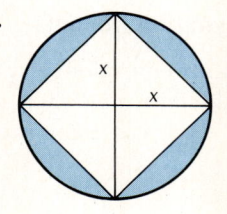

37.

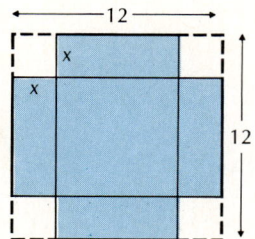

38. Find $(y-2)^2$ using the four parts of this square.

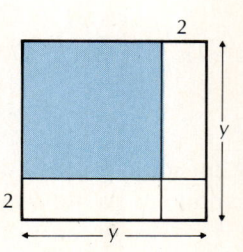

○ ─────────────────────────

Simplify.

39. $(y + 4) + (y - 5) - (y + 8)$
40. $(7y^2 - 5y + 6) - (3y^2 + 8y - 12) + (8y^2 - 10y + 3)$
41. $(4a^2 - 3a) + (7a^2 - 9a - 13) - (6a - 9)$
42. $(3x^2 - 4x + 6) - (-2x^2 + 4) + (-5x - 3)$
43. $(-8y^2 - 4) - (3y + 6) - (2y^2 - y)$

44. $(5x^3 - 4x^2 + 6) - (2x^3 + x^2 - x) + (x^3 - x)$

45. $(-y^4 - 7y^3 + y^2) + (-2y^4 + 5y - 2) - (-6y^3 + y^2)$

46. $(-4 + x^2 + 2x^3) - (-6 - x + 3x^3) - (-x^2 - 5x^3)$

47. 🖩 Subtract: $(345.099x^3 - 6.178x) - (-224.508x^3 + 8.99x)$.

48. Does replacing each occurrence of the variable x in $5x^3 - 3x^2 + 2x$ with its additive inverse result in the additive inverse of the polynomial?

4.6 MULTIPLICATION OF POLYNOMIALS

We now multiply polynomials using techniques based, for the most part, on the distributive laws. As we proceed in this chapter we will develop special ways to find certain products.

Multiplying Monomials

To find an equivalent expression for the product of two monomials, we multiply the coefficients and then use properties of exponents. Recall that a product such as $(3x) \cdot (4x)$ can be written as $(3x)(4x)$.

EXAMPLES Multiply.

1. $(3x)(4x) = (3 \cdot 4)(x \cdot x)$ Multiplying the coefficients
$= 12x^2$ Simplifying

2. $(3x)(-x) = (3x)(-1x)$
$= (3)(-1)(x \cdot x) = -3x^2$

3. $(-7x^5)(4x^3) = (-7 \cdot 4)(x^5 \cdot x^3)$
$= -28x^{5+3}$
$= -28x^8$ Adding exponents and simplifying

After some practice you can do this mentally. Multiply the coefficients and add the exponents. Write only the answer.

Multiplying a Monomial and a Binomial

To find an equivalent expression for the product of a monomial, such as $2x$, by a binomial, such as $5x + 3$, we use a distributive law.

EXAMPLE 4 Multiply: $2x$ and $5x + 3$.

$(2x)(5x + 3) = (2x)(5x) + (2x)(3)$ Using a distributive law
$= 10x^2 + 6x$ Multiplying the monomials

4.6 MULTIPLICATION OF POLYNOMIALS

Multiplying Two Binomials

To find an equivalent expression for the product of two binomials, we use the distributive laws more than once. In Example 5 we use a distributive law three times.

EXAMPLE 5 Multiply: $x + 5$ and $x + 4$.

$$(x + 5)(x + 4) = (x + 5)x + (x + 5)4 \quad \text{Using a distributive law}$$
$$= x \cdot x + 5 \cdot x + x \cdot 4 + 5 \cdot 4 \quad \text{Using a distributive law on each part}$$
$$= x^2 + 5x + 4x + 20 \quad \text{Multiplying the monomials}$$
$$= x^2 + 9x + 20 \quad \text{Collecting like terms}$$

EXAMPLE 6 Multiply: $4x + 3$ and $x - 2$.

$$(4x + 3)(x - 2) = (4x + 3)x + (4x + 3)(-2) \quad \text{Using a distributive law}$$
$$= (4x) \cdot x + (3) \cdot x + (4x)(-2) + (3)(-2) \quad \text{Using a distributive law on each part}$$
$$= 4x^2 + 3x - 8x - 6 \quad \text{Multiplying the monomials}$$
$$= 4x^2 - 5x - 6 \quad \text{Collecting like terms}$$

Multiplying a Binomial and a Trinomial

We again use a distributive law three times.

EXAMPLE 7 Multiply: $(x^2 + 2x - 3)(x^2 + 4)$.

$$(x^2 + 2x - 3)(x^2 + 4) = (x^2 + 2x - 3)x^2 + (x^2 + 2x - 3)4$$
$$= (x^2)(x^2) + 2x(x^2) - 3(x^2) + (x^2) \cdot 4 + (2x) \cdot 4 + (-3) \cdot 4$$
$$= x^4 + 2x^3 - 3x^2 + 4x^2 + 8x - 12$$
$$= x^4 + 2x^3 + x^2 + 8x - 12$$

Multiplying Any Polynomials

Perhaps you have discovered the following in the preceding examples.

> To multiply two polynomials, multiply each term of one by every term of the other. Then add the results.

We can use columns for long multiplications. We multiply each term at the top by every term at the bottom. We locate like terms in columns, and then we add the results.

EXAMPLE 8   Multiply: $(4x^2 - 2x + 3)(x + 2)$.

$$4x^2 - 2x + 3$$
$$\underline{x + 2}$$
$$4x^3 - 2x^2 + 3x \quad \text{Multiplying the top row by } x$$
$$\underline{ 8x^2 - 4x + 6} \quad \text{Multiplying the top row by 2}$$
$$4x^3 + 6x^2 - x + 6 \quad \text{Adding}$$

EXAMPLE 9   Multiply: $(5x^3 - 3x + 4)(-2x^2 - 3)$.

$$5x^3 - 3x + 4$$
$$\underline{-2x^2 - 3}$$
$$-10x^5 + 6x^3 - 8x^2 \quad \text{Multiplying by } -2x^2$$
$$\underline{ - 15x^3 + 9x - 12} \quad \text{Multiplying by } -3$$
$$-10x^5 - 9x^3 - 8x^2 + 9x - 12 \quad \text{Adding}$$

When we multiplied $-2x^2$ by $-3x$, the power dropped from x^5 to x^3 so we left a space for the missing x^4 term.

In addition, we leave spaces for "missing terms."

EXAMPLE 10   Multiply: $(2x^2 + 3x - 4)(2x^2 - x + 3)$.

$$2x^2 + 3x - 4$$
$$\underline{2x^2 - x + 3}$$
$$4x^4 + 6x^3 - 8x^2 \quad \text{Multiplying by } 2x^2$$
$$ - 2x^3 - 3x^2 + 4x \quad \text{Multiplying by } -x$$
$$\underline{ 6x^2 + 9x - 12} \quad \text{Multiplying by 3}$$
$$4x^4 + 4x^3 - 5x^2 + 13x - 12 \quad \text{Adding}$$

EXERCISE SET 4.6

Multiply.

1. $6x^2$ and 7
2. $5x^2$ and -2
3. $-x^3$ and $-x$
4. $-x^4$ and x^2
5. $-x^5$ and x^3
6. $-x^6$ and $-x^2$
7. $3x^4$ and $2x^2$
8. $5x^3$ and $4x^5$
9. $7t^5$ and $4t^3$
10. $10a^2$ and $3a^2$
11. $-0.1x^6$ and $0.2x^4$
12. $0.3x^3$ and $-0.4x^6$
13. $-\frac{1}{5}x^3$ and $-\frac{1}{3}x$
14. $-\frac{1}{4}x^4$ and $\frac{1}{5}x^8$
15. $-4x^2$ and 0
16. $3x$ and $-x + 5$
17. $2x$ and $4x - 6$
18. $4x^2$ and $3x + 6$
19. $5x^2$ and $-2x + 1$
20. $-6x^2$ and $x^2 + x$
21. $-4x^2$ and $x^2 - x$
22. $3y^2$ and $6y^4 + 8y^3$
23. $4y^4$ and $y^3 - 6y^2$
24. $3x^4$ and $14x^{50} + 20x^{11} + 6x^{57} + 60x^{15}$

4.6 MULTIPLICATION OF POLYNOMIALS

25. $5x^6$ and $4x^{32} - 10x^{19} + 5x^8$
26. $-4a^7$ and $20a^{19} + 6a^{15} - 5a^{12} + 14a$
27. $-6y^8$ and $11y^{100} - 7y^{50} + 11y^{41} - 60y^4 + 9$
28. $(x + 6)(x + 3)$
29. $(x + 5)(x + 2)$
30. $(x + 5)(x - 2)$
31. $(x + 6)(x - 2)$
32. $(x - 4)(x - 3)$
33. $(x - 7)(x - 3)$
34. $(x + 3)(x - 3)$
35. $(x + 6)(x - 6)$
36. $(5 - x)(5 - 2x)$
37. $(3 + x)(6 + 2x)$
38. $(2x + 5)(2x + 5)$
39. $(3x - 4)(3x - 4)$
40. $(3y - 4)(3y + 4)$
41. $(2y + 1)(2y - 1)$
42. $(x - \frac{5}{2})(x + \frac{2}{5})$
43. $(x + \frac{4}{3})(x + \frac{3}{2})$
44. $(x^2 + x + 1)(x - 1)$
45. $(x^2 - x + 2)(x + 2)$
46. $(2x^2 + 6x + 1)(2x + 1)$
47. $(4x^2 - 2x - 1)(3x - 1)$
48. $(3y^2 - 6y + 2)(y^2 - 3)$
49. $(y^2 + 6y + 1)(3y^2 - 3)$
50. $(x^3 + x^2 - x)(x^3 + x^2)$
51. $(x^3 - x^2 + x)(x^3 - x^2)$
52. $(-5x^3 - 7x^2 + 1)(2x^2 - x)$
53. $(-4x^3 + 5x^2 - 2)(5x^2 + 1)$
54. $(1 + x + x^2)(-1 - x + x^2)$
55. $(1 - x + x^2)(1 - x + x^2)$
56. $(2x^2 + 3x - 4)(2x^2 + x - 2)$
57. $(2x^2 - x - 3)(2x^2 - 5x - 2)$
58. $(2t^2 - t - 4)(3t^2 + 2t - 1)$
59. $(3a^2 - 5a + 2)(2a^2 - 3a + 4)$
60. $(2x^2 + x - 2)(-2x^2 + 4x - 5)$
61. $(3x^2 - 8x + 1)(-2x^2 - 4x + 2)$
62. $(x - x^3 + x^5)(x^2 - 1 + x^4)$
63. $(x - x^3 + x^5)(3x^2 + 3x^6 + 3x^4)$
64. $(x^3 + x^2 + x + 1)(x - 1)$
65. $(x^3 - x^2 + x - 2)(x - 2)$
66. $(x^3 + x^2 - x - 3)(x - 3)$
67. $(x^3 - x^2 - x + 4)(x + 4)$

○ ─────────────────────────────────

Multiply.

68. $(a + b)^2$
69. $(a - b)^2$
70. $(2x + 3)^2$
71. $(5y + 6)^2$

Problem solving

Find an expression for each shaded area.

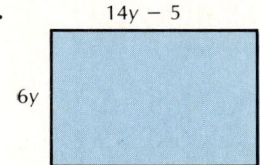

72. [rectangle with width $14y - 5$ and height $6y$]

73. [large rectangle $21t + 8$ by $4t$ with inner white rectangle $3t - 4$ by $2t$]

74. A box with a square bottom is to be made from a 12-inch-square piece of cardboard. Squares with side x are cut out of the corners and the sides are folded up. Find polynomials for the volume and the outside surface area of the box.

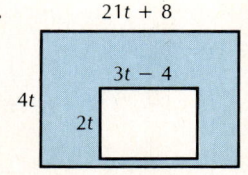

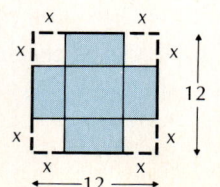

[diagram: 12 by 12 square with squares of side x cut from each corner]

75. An open wooden box is a cube with side x cm. The wood from which the box is made is 1 cm thick. Find a polynomial for the interior volume of the cube.

76. The height of a triangle is 4 ft longer than its base. The area of the triangle is 30 ft^2. Find its height and base.

77. A rectangular garden is twice as long as it is wide. It is surrounded by a sidewalk 4 ft wide. The area of the garden and sidewalk together is 256 ft^2 more than the area of the garden alone. Find the dimensions of the garden.

Compute.

78. a) $(x+3)(x+6) + (x+3)(x+6)$
b) $(x+4)(x+5) - (x+4)(x+5)$

79. a) $(x-2)(x-7) + (x-2)(x-7)$
b) $(x-6)(x-2) - (x-6)(x-2)$

80. a) $(x+5)(x-3) + (x+5)(x-3)$
b) $(x+9)(x-4) - (x+9)(x-4)$

81. a) $(x+7)(x-8) + (x-7)(x+8)$
b) $(x+2)(x-5) - (x-2)(x+5)$

4.7 SPECIAL PRODUCTS OF POLYNOMIALS

We now consider special products of polynomials. These are products that we encounter so often that it is helpful to have methods of computing them that are faster than multiplying each term of one by each term of the other.

Product of a Monomial and Any Polynomial

There is a quick way to multiply a monomial and any polynomial. We use the distributive law mentally, multiplying every term inside by the monomial. Just write the answer.

EXAMPLE 1 Multiply: $5x(2x^2 - 3x + 4)$.

$$5x(2x^2 - 3x + 4) = 10x^3 - 15x^2 + 20x$$

Products of Two Binomials

To multiply two binomials, we multiply each term of one by every term of the other, as shown here:

$$(A + B)(C + D) = AC + AD + BC + BD$$

1. Multiply **F**irst terms: **AC**
2. Multiply **O**utside terms: **AD**
3. Multiply **I**nside terms: **BC**
4. Multiply **L**ast terms: **BD**

FOIL   This will help you remember the rule.

4.7 SPECIAL PRODUCTS OF POLYNOMIALS

EXAMPLE 2 Multiply: $(x + 8)(x^2 + 5)$.

$$\overset{F\quad\quad O\quad\quad I\quad\quad L}{(x + 8)(x^2 + 5) = x^3 + 5x + 8x^2 + 40}$$

Often we can collect like terms after we multiply.

EXAMPLES Multiply.

3. $(x + 6)(x - 6) = x^2 - 6x + 6x - 36$ Using FOIL
$ = x^2 - 36$ Collecting like terms

4. $(x + 3)(x - 2) = x^2 - 2x + 3x - 6$
$ = x^2 + x - 6$

5. $(x^3 + 5)(x^3 - 5) = x^6 - 5x^3 + 5x^3 - 25$
$ = x^6 - 25$

6. $(4x^3 + 5)(3x^2 - 2) = 12x^5 - 8x^3 + 15x^2 - 10$

EXAMPLES Multiply.

7. $(x - \tfrac{2}{3})(x + \tfrac{2}{3}) = x^2 + \tfrac{2}{3}x - \tfrac{2}{3}x - \tfrac{4}{9}$
$\phantom{(x - \tfrac{2}{3})(x + \tfrac{2}{3})} = x^2 - \tfrac{4}{9}$

8. $(x^2 - 0.3)(x^2 - 0.3) = x^4 - 0.3x^2 - 0.3x^2 + 0.09$
$ = x^4 - 0.6x^2 + 0.09$

9. $(3 - 4x)(7 - 5x^3) = 21 - 15x^3 - 28x + 20x^4$
$ = 21 - 28x - 15x^3 + 20x^4$

(*Note:* If the original polynomials are in ascending order, it is natural to write the product in ascending order, but this is not a "must.")

10. $(5x^4 + 2x^3)(3x^2 - 7x) = 15x^6 - 35x^5 + 6x^5 - 14x^4$
$ = 15x^6 - 29x^5 - 14x^4$

EXERCISE SET 4.7

Multiply. Write only the answer.

1. $4x(x + 1)$
2. $3x(x + 2)$
3. $-3x(x - 1)$
4. $-5x(-x - 1)$
5. $x^2(x^3 + 1)$
6. $-2x^3(x^2 - 1)$
7. $3x(2x^2 - 6x + 1)$
8. $-4x(2x^3 - 6x^2 - 5x + 1)$
9. $(x + 1)(x^2 + 3)$
10. $(x^2 - 3)(x - 1)$
11. $(x^3 + 2)(x + 1)$
12. $(x^4 + 2)(x + 12)$
13. $(x + 2)(x - 3)$
14. $(x + 2)(x + 2)$

POLYNOMIALS

15. $(3x + 2)(3x + 3)$
16. $(4x + 1)(2x + 2)$
17. $(5x - 6)(x + 2)$
18. $(x - 8)(x + 8)$
19. $(3x - 1)(3x + 1)$
20. $(2x + 3)(2x + 3)$
21. $(4x - 2)(x - 1)$
22. $(2x - 1)(3x + 1)$
23. $(x - \frac{1}{4})(x + \frac{1}{4})$
24. $(x + \frac{3}{4})(x + \frac{3}{4})$
25. $(x - 0.1)(x + 0.1)$
26. $(3x^2 + 1)(x + 1)$
27. $(2x^2 + 6)(x + 1)$
28. $(2x^2 + 3)(2x - 1)$
29. $(-2x + 1)(x + 6)$
30. $(3x + 4)(2x - 4)$
31. $(x + 7)(x + 7)$
32. $(2x + 5)(2x + 5)$
33. $(1 + 2x)(1 - 3x)$
34. $(-3x - 2)(x + 1)$
35. $(x^2 + 3)(x^3 - 1)$
36. $(x^4 - 3)(2x + 1)$
37. $(x^2 - 2)(x - 1)$
38. $(x^3 + 2)(x - 3)$
39. $(3x^2 - 2)(x^4 - 2)$
40. $(x^{10} + 3)(x^{10} - 3)$
41. $(3x^5 + 2)(2x^2 + 6)$
42. $(1 - 2x)(1 + 3x^2)$
43. $(8x^3 + 1)(x^3 + 8)$
44. $(4 - 2x)(5 - 2x^2)$
45. $(4x^2 + 3)(x - 3)$
46. $(7x - 2)(2x - 7)$
47. $(4x^4 + x^2)(x^2 + x)$
48. $(5x^6 + 3x^3)(2x^6 + 2x^3)$

Multiply.

49. $4y(y + 5)(2y + 8)$
50. $8x(2x - 3)(5x + 9)$
51. $[(x + 1) - x^2][(x - 2) + 2x^2]$
52. $[(2x - 1)(2x + 1)](4x^2 + 1)$

Solve.

53. $(x + 2)(x - 5) = (x + 1)(x - 3)$
54. $(2x + 5)(x - 4) = (x + 5)(2x - 4)$
55. $(x + 1)(x + 2) = (x + 3)(x + 4)$
56. $(x + 3)(x + 1) + (2x - 3)(x - 2) = (3x + 4)(x - 1)$

Problem solving

The height of a box is one more than its length l, and the length is one more than its width w. Find a polynomial for the volume in terms of each of the following.

57. The width w.
58. The length l.
59. The height h.

Find two expressions for each color region.

60. 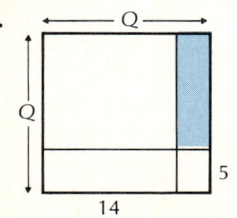
61. 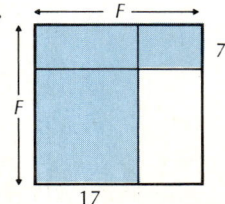
62.

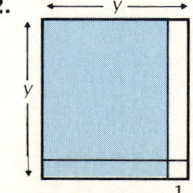

63. A cab company charges 70¢ for the first $\frac{1}{7}$ mile and 10¢ each additional $\frac{1}{7}$ mile per trip. A person takes x number of 4-mile trips every month for 11 months, and 6 trips in August before going on vacation. How much did a year of taxi service cost?

4.8 MORE SPECIAL PRODUCTS

Now we find special products that are sums and differences and those that are squares of binomials.

Multiplying Sums and Differences of Two Expressions

A product of the sum and difference of two expressions, such as $(x + 2)(x - 2)$, occurs quite often. To find a faster way to compute such products, look for a pattern in the following:

a) $(x + 2)(x - 2) = x^2 - 2x + 2x - 4$
$= x^2 - 4;$

b) $(3x - 5)(3x + 5) = 9x^2 + 15x - 15x - 25$
$= 9x^2 - 25.$

Perhaps you discovered the following.

> The product of the sum and difference of two expressions is the square of the first expression minus the square of the second:
>
> $$(A + B)(A - B) = A^2 - B^2.$$

It is helpful to memorize this rule in both words and symbols.

EXAMPLES Multiply. (Carry out the rule, and say the words as you go.)

$(A + B)(A - B) = A^2 - B^2$

1. $(x + 4)(x - 4) = x^2 - 4^2$ "The square of the first expression x^2 minus the square of the second 4^2."
$= x^2 - 16$ Simplifying

2. $(5 + 2w)(5 - 2w) = 5^2 - (2w)^2$
$= 25 - 4w^2$

3. $(3x^2 - 7)(3x^2 + 7) = (3x^2)^2 - 7^2$
$= 9x^4 - 49$

4. $(-4x - 10)(-4x + 10) = (-4x)^2 - 10^2$
$= 16x^2 - 100$

Squaring Binomials

In this special product we multiply a binomial by itself. This is also called "squaring a binomial." Look for a pattern below:

a) $(x + 3)^2 = (x + 3)(x + 3)$
$= x^2 + 3x + 3x + 9 = x^2 + 6x + 9;$

b) $(5 + 3x)^2 = (5 + 3x)(5 + 3x)$
$= 25 + 15x + 15x + 9x^2 = 25 + 30x + 9x^2;$

c) $(x - 3)^2 = (x - 3)(x - 3)$
$= x^2 - 3x - 3x + 9 = x^2 - 6x + 9;$

d) $(3x - 5)^2 = (3x - 5)(3x - 5)$
$= 9x^2 - 15x - 15x + 25 = 9x^2 - 30x + 25.$

Perhaps you discovered a quick way to square a binomial.

> The square of a sum or difference of two expressions is the square of the first expression plus or minus twice the product of the two expressions, plus the square of the last:
>
> $$(A + B)^2 = A^2 + 2AB + B^2;$$
> $$(A - B)^2 = A^2 - 2AB + B^2.$$

EXAMPLES Multiply. (Carry out the rule, and say the words as you go.)

$(A + B)^2 = A^2 + 2 \cdot A \cdot B + B^2$

5. $(x + 3)^2 = x^2 + 2 \cdot x \cdot 3 + 3^2$
$= x^2 + 6x + 9$

6. $(t - 5)^2 = t^2 - 2 \cdot t \cdot 5 + 5^2$
$= t^2 - 10t + 25$

7. $(2x + 7)^2 = (2x)^2 + 2 \cdot 2x \cdot 7 + 7^2 = 4x^2 + 28x + 49$

8. $(5x - 3x^2)^2 = (5x)^2 - 2 \cdot 5x \cdot 3x^2 + (3x^2)^2$
$= 25x^2 - 30x^3 + 9x^4$

Note carefully in these examples that the square of a sum is *not* the sum of squares:

$$(A + B)^2 \neq A^2 + B^2.$$

To see this, note that

$$(20 + 5)^2 = 25^2 = 625,$$

but

$$20^2 + 5^2 = 400 + 25 = 425 \neq 625.$$

4.8 MORE SPECIAL PRODUCTS

Multiplications of Various Types

Now that we have considered how to quickly multiply certain kinds of polynomials, let us try several kinds mixed together so we can learn to sort them out. When you multiply, first see what kind of multiplication you have. Then use the best method. The formulas and methods you have used so far are as follows.

> 1. $(A + B)(A + B) = (A + B)^2 = A^2 + 2AB + B^2$
> 2. $(A - B)(A - B) = (A - B)^2 = A^2 - 2AB + B^2$
> 3. $(A - B)(A + B) = A^2 - B^2$
> 4. FOIL
> 5. The product of a monomial and any polynomial is found by multiplying each term of the polynomial by the monomial.

Note that FOIL will work for any of the first three rules, but it is faster to learn to use them as they are given.

EXAMPLE 9 Multiply: $(x + 3)(x - 3)$.

$(x + 3)(x - 3) = x^2 - 9$ Using method 3 (the product of the sum and difference of two expressions)

EXAMPLE 10 Multiply: $(t + 7)(t - 5)$.

$(t + 7)(t - 5) = t^2 + 2t - 35$ Using method 4 (the product of two binomials, but neither the square of a binomial nor the product of the sum and difference of two expressions)

EXAMPLE 11 Multiply: $(x + 7)(x + 7)$.

$(x + 7)(x + 7) = x^2 + 14x + 49$ Using method 1 (the square of a binomial sum)

EXAMPLE 12 Multiply: $2x^3(9x^2 + x - 7)$.

$2x^3(9x^2 + x - 7) = 18x^5 + 2x^4 - 14x^3$ Using method 5 (the product of a monomial and a trinomial; multiply each term of the trinomial by the monomial)

EXAMPLE 13 Multiply: $(3x^2 - 7x)^2$.

$(3x^2 - 7x)^2 = 9x^4 - 42x^3 + 49x^2$ Using method 2 (the square of a binomial difference)

EXAMPLE 14 Multiply: $(3x + \frac{1}{4})^2$.

$(3x + \frac{1}{4})^2 = 9x^2 + 2(3x)(\frac{1}{4}) + \frac{1}{16}$ Using method 1 (the square of a binomial sum. To get the middle term, we multiply $3x$ by $\frac{1}{4}$ and double.)

$= 9x^2 + \frac{3}{2}x + \frac{1}{16}$

EXAMPLE 15 Multiply: $(4x - \frac{3}{4})^2$.

$$(4x - \tfrac{3}{4})^2 = 16x^2 - 2(4x)(\tfrac{3}{4}) + \tfrac{9}{16}$$
$$= 16x^2 - 6x + \tfrac{9}{16}$$

EXERCISE SET 4.8

Multiply mentally.

1. $(x + 4)(x - 4)$
2. $(x + 1)(x - 1)$
3. $(2x + 1)(2x - 1)$
4. $(x^2 + 1)(x^2 - 1)$
5. $(5m - 2)(5m + 2)$
6. $(3x^4 + 2)(3x^4 - 2)$
7. $(2x^2 + 3)(2x^2 - 3)$
8. $(6x^5 - 5)(6x^5 + 5)$
9. $(3x^4 - 4)(3x^4 + 4)$
10. $(t^2 - 0.2)(t^2 + 0.2)$
11. $(x^6 - x^2)(x^6 + x^2)$
12. $(2x^3 - 0.3)(2x^3 + 0.3)$
13. $(x^4 + 3x)(x^4 - 3x)$
14. $(\tfrac{3}{4} + 2x^3)(\tfrac{3}{4} - 2x^3)$
15. $(x^{12} - 3)(x^{12} + 3)$
16. $(12 - 3x^2)(12 + 3x^2)$
17. $(2x^8 + 3)(2x^8 - 3)$
18. $(x - \tfrac{2}{3})(x + \tfrac{2}{3})$
19. $(x + 2)^2$
20. $(2x - 1)^2$
21. $(3x^2 + 1)^2$
22. $(3x + \tfrac{3}{4})^2$
23. $(x - \tfrac{1}{2})^2$
24. $(2x - \tfrac{1}{5})^2$
25. $(3 + x)^2$
26. $(x^3 - 1)^2$
27. $(x^2 + 1)^2$
28. $(8x - x^2)^2$
29. $(2 - 3x^4)^2$
30. $(6x^3 - 2)^2$
31. $(5 + 6x^2)^2$
32. $(3x^2 - x)^2$
33. $(3 - 2x^3)^2$
34. $(x - 4x^3)^2$
35. $4x(x^2 + 6x - 3)$
36. $8x(-x^5 + 6x^2 + 9)$
37. $(2x^2 - \tfrac{1}{2})(2x^2 - \tfrac{1}{2})$
38. $(-x^2 + 1)^2$
39. $(-1 + 3p)(1 + 3p)$
40. $(-3x + 2)(3x + 2)$
41. $3t^2(5t^3 - t^2 + t)$
42. $-6x^2(x^3 + 8x - 9)$
43. $(6x^4 + 4)^2$
44. $(8a + 5)^2$
45. $(3x + 2)(4x^2 + 5)$
46. $(2x^2 - 7)(3x^2 + 9)$
47. $(8 - 6x^4)^2$
48. $(\tfrac{1}{5}x^2 + 9)(\tfrac{3}{5}x^2 - 7)$

49. ▦ Multiply: $(67.58x + 3.225)^2$.

Calculate as the difference of squares.

50. 18×22 (*Hint:* $(20 - 2)(20 + 2)$.)
51. 93×107

Multiply. (Do not collect like terms before multiplying.)

52. $[(2x - 1)(2x + 1)](4x^2 + 1)$
53. $[(a + 5) + 1][(a + 5) - 1]$
54. $[3a - (2a - 3)][3a + (2a - 3)]$
55. $[(x + 3) + 2]^2$

Solve.

56. $x^2 = (x + 10)^2$
57. $x^2 = (x - 12)^2$
58. $(x + 4)^2 = (x + 8)(x - 8)$
59. $(4x - 1)^2 - (3x + 2)^2 = (7x + 4)(x - 1)$

Problem solving

60. Consider the following rectangle.

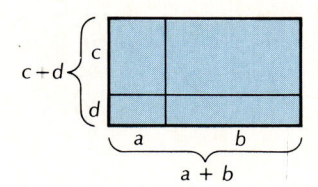

a) Find a polynomial for each of the areas of the four smaller rectangles.
b) Find a polynomial for the sum of the areas in part (a).
c) Find a polynomial for the area of the entire rectangle. Compare your result with the answer to part (b).

61. A polynomial for the color region in this rectangle is $(a + b)(a - b)$.

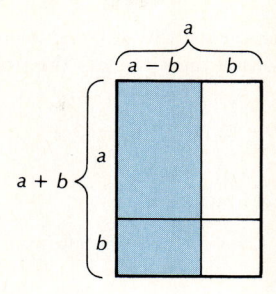

a) Find a polynomial for the area of the entire rectangle.
b) Find a polynomial for the sum of the two small uncolored rectangles.
c) Find a polynomial for the area in part (a) minus the area in part (b).
d) Find a polynomial for the area of the color region and compare this with the polynomial found in part (c).

62. Find three consecutive integers, the sum of whose squares is 65 more than 3 times the square of the smallest.

63. Find $(10x + 5)^2$. Use your result to show how to mentally square any two-digit number ending in 5.

SUMMARY AND REVIEW: CHAPTER 4

The following contains a summary of what you should be able to do after completing this chapter. The review exercises are for practice. Answers are at the back of the book. If you miss an exercise, restudy the section indicated alongside the answer.

You should be able to:

Multiply and divide expressions such as y^8, $3x^5$, and 8^3 and raise a power to a power.

Multiply.

1. $7^2 \cdot 7^4$
2. $y^7 \cdot y^3 \cdot y$
3. $(3x)^5 \cdot (3x)^9$
4. $t^8 \cdot t^0$

Divide.

5. $\dfrac{4^5}{4^2}$
6. $\dfrac{a^5}{a^8}$
7. $\dfrac{(7x)^4}{(7x)^4}$

Simplify.

8. $(x^3)^4$
9. $(3t^4)^2$
10. $(-2xy^2)^3$

Evaluate a polynomial for a given value of a variable.

Evaluate the polynomial for $x = -1$.

11. $7x - 10$

12. $x^2 - 3x + 6$

For a polynomial, identify the terms, the missing terms, the coefficients of each term, the degree of each term, and the degree of the polynomial, and classify the polynomial as a monomial, binomial, trinomial, or none of these.

Identify the terms of the polynomial.

13. $3x^2 + 6x + \frac{1}{2}$

14. $-4y^5 + 7y^2 - 3y - 2$

15. Identify the missing terms in $x^3 + x$.

Identify the coefficient of each term of the polynomial.

16. $6x^2 + 17$

17. $4x^3 + 6x^2 - 5x + \frac{5}{3}$

Identify the degree of each term and the degree of the polynomial.

18. $x^3 + 4x - 6$

19. $3 - 2x^4 + 3x^9 + x^6 - \frac{3}{4}x^3$

Tell whether the polynomial is a monomial, binomial, trinomial, or none of these.

20. $4x^3 - 1$

21. $4 - 9t^3 - 7t^4 + 10t^2$

22. $7y^2$

Collect like terms of a polynomial and arrange the terms in descending order.

Collect like terms.

23. $5x - x^2 + 4x$

24. $\frac{3}{4}x^3 + 4x^2 - x^3 + 7$

25. $-2x^4 + 16 + 2x^4 + 9 - 3x^5$

Collect like terms and then arrange in descending order.

26. $3x^2 - 2x + 3 - 5x^2 - 1 - x$

27. $-x + \frac{1}{2} + 14x^4 - 7x^2 - 1 - 4x^4$

Add and subtract polynomials and find polynomials for certain perimeters and areas of figures.

Add.

28. $(3x^4 - x^3 + x - 4) + (x^5 + 7x^3 - 3x^2 - 5) + (-5x^4 + 6x^2 - x)$

29. $(3x^5 - 4x^4 + x^3 - 3) + (3x^4 - 5x^3 + 3x^2) + (4x^5 + 4x^3) + (-5x^5 - 5x^2) + (-5x^4 + 2x^3 + 5)$

30. $-\frac{3}{4}x^4 + \frac{1}{2}x^3 \qquad\qquad + \frac{7}{8}$
$\qquad\quad -\frac{1}{4}x^3 - \frac{3}{3}x^2 - \frac{7}{4}x$
$\qquad\quad\quad\;\; + \frac{2}{3}x^2 \qquad\quad - \frac{1}{8}$

Subtract.

31. $(5x^2 - 4x + 1) - (3x^2 + 7)$

32. $(3x^5 - 4x^4 + 2x^2 + 3) - (2x^5 - 4x^4 + 3x^3 + 4x^2 - 5)$

33. $2x^5 \quad\;\; - x^3 \quad\;\; + x + 3$
$\;\;\;\;\; 3x^5 - x^4 + 4x^3 + 2x^2 - x + 3$

34. The length of a rectangle is 4 m greater than its width.

a) Express the perimeter as a polynomial.
b) Express the area as a polynomial.

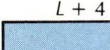

Multiply two monomials, a monomial and a binomial, two binomials, a binomial and a trinomial, and any polynomials. Find special products such as the sum and difference of two expressions and the square of binomials.

Multiply.

35. $3x(-4x^2)$
36. $(7x + 1)^2$
37. $(x + \frac{2}{3})(x + \frac{1}{2})$
38. $(1.5x - 6.5)(0.2x + 1.3)$
39. $(4x^2 - 5x + 1)(3x - 2)$
40. $(x - 9)^2$
41. $5x^4(3x^3 - 8x^2 + 10x + 2)$
42. $(x + 4)(x - 7)$
43. $(x - 0.3)(x - 0.75)$
44. $(x^4 - 2x + 3)(x^3 + x - 1)$
45. $(3x^2 - 2x)^2$
46. $(2x^2 + 3)(x^2 - 7)$
47. $(x^3 - 2x + 3)(4x^2 - 5x)$
48. $(3x^2 + 4)(3x^2 - 4)$
49. $(2 - x)(2 + x)$
50. $(13x - 3)(x - 13)$

○ ──────────────────────────────

51. If a and b are positive, how many terms are there in each of the following?
    a) $(x - a)(x - b) + (x - a)(x - b)$
    b) $(x + a)(x - b) + (x - a)(x + b)$
    c) $(x + a)(x - b) - (x + a)(x - b)$
52. Collect like terms:

    $$-3x^5 \cdot 3x^3 - x^6(2x)^2 + ((3x^4))^2 + (2x^2)^4 - 40x^2(x^3)^2.$$

53. A polynomial has degree 4. The x^2-term is missing. The coefficient of x^4 is 2 times the coefficient of x^3. The coefficient of x is 3 less than the coefficient of x^4. The remaining coefficient is 7 less than the coefficient of x. The sum of the coefficients is 15. Find the polynomial.
54. Multiply: $[(x - 4) - x^3][(x + 4) + 4x^3]$.
55. Solve: $(x - 7)(x + 10) = (x - 4)(x - 6)$.

TEST: CHAPTER 4

Multiply.

1. $6^2 \cdot 6^3$
2. $x^6 \cdot x^2 \cdot x$
3. $(4a)^3 \cdot (4a)^8$

Divide.

4. $\dfrac{3^5}{3^2}$
5. $\dfrac{x^3}{x^8}$
6. $\dfrac{(2x)^5}{(2x)^5}$

Simplify.

7. $(x^3)^2$
8. $(-3y^2)^3$
9. $(2a^3b)^4$

10. Evaluate the polynomial:

    $x^2 + 5x - 1$ for $x = -2$.

11. Identify the coefficient of each term of the polynomial:

    $\frac{1}{3}x^5 - x + 7$.

12. Identify the degree of each term and the degree of the polynomial:
$$2x^3 - 4 + 5x + 3x^6.$$

13. Tell whether the polynomial is a monomial, binomial, trinomial, or none of these:
$$7 - x.$$

Collect like terms.

14. $4a^2 - 6 + a^2$

15. $y^2 - 3y - y + \frac{3}{4}y^2$

16. Collect like terms and then arrange in descending order:
$$3 - x^2 + 2x^3 + 5x^2 - 6x - 2x + x^5.$$

Add.

17. $(3x^5 + 5x^3 - 5x^2 - 3) + (x^5 + x^4 - 3x^3 - 3x^2 + 2x - 4)$

18. $(x^4 + \frac{2}{3}x + 5) + (4x^4 + 5x^2 + \frac{1}{3}x)$

Subtract.

19. $(2x^4 + x^3 - 8x^2 - 6x - 3) - (6x^4 - 8x^2 + 2x)$

20. $(x^3 - 0.4x^2 - 12) - (x^5 + 0.3x^3 + 0.4x^2 + 9)$

Multiply.

21. $-3x^2(4x^2 - 3x - 5)$

22. $(x - \frac{1}{3})^2$

23. $(3x + 10)(3x - 10)$

24. $(3b + 5)(b - 3)$

25. $(x^6 - 4)(x^8 + 4)$

26. $(8 - y)(6 + 5y)$

27. $(2x + 1)(3x^2 - 5x - 3)$

28. $(5t + 2)^2$

○ ─────────────────────────

29. The height of a box is one less than its length, and the length is 2 more than its width. Find the volume in terms of the length.

30. Solve:
$$x^2 + (x - 7)(x + 4) = 2(x - 6)^2.$$

POLYNOMIALS AND FACTORING

5

How can we find the area of a beverage can? A polynomial can be evaluated to solve such a problem.

Factoring is the reverse of multiplication. To *factor* a polynomial, or other algebraic expression, is to find an equivalent expression that is a product. In this chapter we study factoring polynomials. To learn to factor quickly, we study quick methods for multiplication.

At the end of the chapter we get the payoff for learning to factor. We have certain new equations containing second-degree polynomials that we can now solve. This then allows us to solve problems that we could not have solved before.

5.1 FACTORING POLYNOMIALS

In this section we learn to factor polynomials with one term. We also learn how to factor out a common factor.

Factoring Monomials

To factor a monomial we find two monomials whose product is equivalent to that monomial. Compare.

	Multiplying	*Factoring*
a)	$(4x)(5x) = 20x^2$	$20x^2 = (4x)(5x)$
b)	$(2x)(10x) = 20x^2$	$20x^2 = (2x)(10x)$
c)	$(-4x)(-5x) = 20x^2$	$20x^2 = (-4x)(-5x)$
d)	$(x)(20x) = 20x^2$	$20x^2 = (x)(20x)$

The monomial $20x^2$ thus has many factorizations. There are still other ways to factor $20x^2$.

To factor a monomial, factor the coefficient first. Then shift some of the letters to one factor and some to the other.

EXAMPLE 1 Find three factorizations of $15x^3$.

a) $15x^3 = (3 \cdot 5)x^3$
$= (3x)(5x^2)$

b) $15x^3 = (3 \cdot 5)x^3$
$= (3x^2)(5x)$

c) $15x^3 = (-1) \cdot (-15)x^3$
$= (-x)(-15x^2)$

Factoring When Terms Have a Common Factor

To multiply a monomial and a polynomial with more than one term, we multiply each term by the monomial. To factor, we do the reverse. Compare.

Multiply

$3x(x^2 + 2x - 4)$
$= 3x \cdot x^2 + 3x \cdot 2x + 3x(-4)$
$= 3x^3 + 6x^2 - 12x$

Factor

$3x^3 + 6x^2 - 12x$
$= 3x \cdot x^2 + 3x \cdot 2x + 3x \cdot (-4)$
$= 3x(x^2 + 2x - 4)$

Consider the following:

$$3x^3 + 6x^2 - 12x = 3 \cdot x \cdot x \cdot x + 6 \cdot x \cdot x - 4 \cdot 3x.$$

The parts, or terms, of the expression have been factored but the expression itself has not been factored. This is not a correct factorization.

We are finding a factor common to all the terms. There may not always be one other than 1. When there is, we generally use the factor with the largest possible coefficient and the largest exponent. In this way we "factor completely."

EXAMPLE 2 Factor: $3x^2 + 6$.

$$3x^2 + 6 = 3 \cdot x^2 + 3 \cdot 2$$
$$= 3(x^2 + 2) \quad \text{Factoring out the common factor, 3}$$

EXAMPLE 3 Factor: $16x^3 + 20x^2$.

$$16x^3 + 20x^2 = (4x^2)(4x) + (4x^2)(5)$$
$$= 4x^2(4x + 5) \quad \text{Factoring out } 4x^2$$

EXAMPLE 4 Factor: $15x^5 - 12x^4 + 27x^3 - 3x^2$.

$$15x^5 - 12x^4 + 27x^3 - 3x^2 = (3x^2)(5x^3) - (3x^2)(4x^2) + (3x^2)(9x) - (3x^2)(1)$$
$$= 3x^2(5x^3 - 4x^2 + 9x - 1) \quad \text{Factoring out } 3x^2$$

If you can spot the common factor without factoring each term, you should write just the answer.

EXAMPLE 5 Factor: $\tfrac{4}{5}x^2 + \tfrac{1}{5}x + \tfrac{2}{5}$.

$$\tfrac{4}{5}x^2 + \tfrac{1}{5}x + \tfrac{2}{5} = \tfrac{1}{5}(4x^2 + x + 2)$$

Factoring by Grouping

The method that we are about to consider is called *factoring by grouping*. Certain polynomials with four terms can be factored using this method. Consider

$$x^2 + 3x + 4x + 12.$$

There is no factor common to all the terms other than 1. But we can factor $x^2 + 3x$ and $4x + 12$:

$$x^2 + 3x = x(x + 3); \quad \text{Factoring } x^2 + 3x$$
$$4x + 12 = 4(x + 3). \quad \text{Factoring } 4x + 12$$

Then

$$x^2 + 3x + 4x + 12 = x(x + 3) + 4(x + 3).$$

Note the common *binomial* factor $x + 3$. We can use the distributive law again like this:

$$x(x + 3) + 4(x + 3) = (x + 4)(x + 3). \quad \text{Factoring out the common factor, } x + 3$$

EXAMPLES Factor. For purposes of learning this method, do not collect like terms.

6. $x^2 + 7x + 2x + 14 = (x^2 + 7x) + (2x + 14) \quad \text{Separating into two binomials}$
$ = x(x + 7) + 2(x + 7) \quad \text{Factoring each binomial}$
$ = (x + 2)(x + 7) \quad \text{Factoring out the common factor, } x + 7$

7. $5x^2 - 10x + 2x - 4 = (5x^2 - 10x) + (2x - 4)$
$= 5x(x - 2) + 2(x - 2)$
$= (5x + 2)(x - 2)$

8. $x^2 + 3x - x - 3 = (x^2 + 3x) + (-x - 3)$
$= x(x + 3) \quad (x + 3)$

$= (x - 1)(x + 3)$

9. $2x^2 - 12x - 3x + 18 = (2x^2 - 12x) + (-3x + 18)$
$= 2x(x - 6) - 3(x - 6)$
$= (2x - 3)(x - 6)$

10. $12x^5 + 20x^2 - 21x^3 - 35 = (12x^5 + 20x^2) + (-21x^3 - 35)$
$= 4x^2(3x^3 + 5) - 7(3x^3 + 5)$
$= (4x^2 - 7)(3x^3 + 5)$

This method is called *factoring by grouping*. Not all expressions with four terms can be factored by this method.

EXERCISE SET 5.1   E.O.O. to #49

Find three factorizations for each monomial.

1. $6x^3$
2. $9x^4$
3. $-9x^5$
4. $-12x^6$
5. $24x^4$
6. $15x^5$

Factor.

7. $x^2 - 4x$
8. $x^2 + 8x$
9. $2x^2 + 6x$
10. $3x^2 - 3x$
11. $x^3 + 6x^2$
12. $4x^4 + x^2$
13. $8x^4 - 24x^2$
14. $5x^5 + 10x^3$
15. $2x^2 + 2x - 8$
16. $6x^2 + 3x - 15$
17. $17x^5 + 34x^3 + 51x$
18. $16x^6 - 32x^5 - 48x$
19. $6x^4 - 10x^3 + 3x^2$
20. $5x^5 + 10x^2 - 8x$
21. $x^5 + x^4 + x^3 - x^2$
22. $x^9 - x^7 + x^4 + x^3$
23. $2x^7 - 2x^6 - 64x^5 + 4x^3$
24. $10x^3 + 25x^2 + 15x - 20$
25. $1.6x^4 - 2.4x^3 + 3.2x^2 + 6.4x$
26. $2.5x^6 - 0.5x^4 + 5x^3 + 10x^2$
27. $\frac{5}{3}x^6 + \frac{4}{3}x^5 + \frac{1}{3}x^4 + \frac{1}{3}x^3$
28. $\frac{5}{7}x^7 + \frac{3}{7}x^5 - \frac{6}{7}x^3 - \frac{1}{7}x$

Factor by grouping.

29. $y^2 + 4y + y + 4$
30. $x^2 + 5x + 2x + 10$
31. $x^2 - 4x - x + 4$
32. $a^2 + 5a - 2a - 10$
33. $6x^2 + 4x + 9x + 6$
34. $3x^2 - 2x + 3x - 2$

35. $3x^2 - 4x - 12x + 16$
36. $24 - 18y - 20y + 15y^2$
37. $35x^2 - 40x + 21x - 24$
38. $8x^2 - 6x - 28x + 21$
39. $4x^2 + 6x - 6x - 9$
40. $2x^4 - 6x^2 - 5x^2 + 15$
41. $2x^4 + 6x^2 + 5x^2 + 15$
42. $4x^4 - 6x^2 - 6x^2 + 9$
43. $2x^3 + 6x^2 + x + 3$
44. $3x^3 + 2x^2 + 3x + 2$
45. $8x^3 - 12x^2 + 6x - 9$
46. $10x^3 - 25x^2 + 4x - 10$
47. $12x^3 - 16x^2 + 3x - 4$
48. $18x^3 - 21x^2 + 30x - 35$
49. $x^3 + 8x^2 - 3x - 24$
50. $2x^3 + 12x^2 - 5x - 30$

Two polynomials are *relatively prime* if they have no common factors other than constants. Tell which pairs are relatively prime.

51. $5x, x^2$
52. $3x, ax - 3$
53. $x + x^2, 3x^3$
54. $y - 6, y$
55. $7a, a$
56. $2p^2 + 2, 2p$
57. $t^2 - 4t, t^2 - 4$
58. $4x^5 + 8x^3 - 6x, 8x^3 + 12x^2 + 24x - 16$
59. $b^3 + 3b^2, b^3 - a^2$
60. $ax^2 + a^2x, ax - 2a$

Factor.

61. $4x^5 + 6x^3 + 6x^2 + 9$
62. $4x^5 + 6x^4 + 6x^3 + 9x^2$
63. $x^6 + x^4 + x^2 + 1$
64. $x^{13} + x^7 + x^6 + 1$
65. Subtract $(x^2 + 1)^2$ from $x^2(x + 1)^2$ and factor the result.

5.2 DIFFERENCES OF SQUARES

The following polynomials are differences of squares:

$$x^2 - 9, \quad 4t^2 - 49.$$

In this section we learn how to factor differences of squares.

Recognizing Differences of Squares

A difference of squares is an expression like the following:

$$A^2 - B^2.$$

How can we recognize such expressions? Look at $A^2 - B^2$. In order for a binomial to be a difference of squares:

a) There must be two expressions, both squares, such as

$$4x^2, \quad 9, \quad 25t^4, \quad 1, \quad x^6.$$

b) There must be a minus sign between them.

EXAMPLE 1 Is $9x^2 - 64$ a difference of squares?

    a) The first expression is a square: $9x^2 = (3x)^2$.
         The second expression is a square: $64 = (8)^2$.

    b) There is a minus sign between them.

So we have a difference of squares.

Factoring Differences of Squares

To factor a difference of squares we can use the following equation:

$$A^2 - B^2 = (A - B)(A + B).$$

    Where does this equation come from? If you multiply out the two expressions on the right, you do get the expression on the left. So the two expressions are equivalent. In effect, we are going to use the equation $(A - B)(A + B) = A^2 - B^2$ in reverse so we can factor.

    To factor a difference of squares $A^2 - B^2$, we find A and B, the square roots of the expressions A^2 and B^2. We write a plus sign one time and a minus sign one time.

EXAMPLE 2 Factor: $x^2 - 4$.

$$x^2 - 4 = x^2 - 2^2 = (x - 2)(x + 2)$$
$$A^2 - B^2 = (A - B)(A + B)$$

EXAMPLE 3 Factor: $18x^2 - 50x^6$.

Always look first for a factor common to all terms. This time there is one:

$$18x^2 - 50x^6 = 2x^2(9 - 25x^4) = 2x^2[3^2 - (5x^2)^2]$$
$$= 2x^2(3 - 5x^2)(3 + 5x^2)$$

EXAMPLE 4 Factor: $49x^4 - 9x^6$.

$$49x^4 - 9x^6 = x^4(49 - 9x^2) = x^4(7 - 3x)(7 + 3x)$$

Note carefully in these examples that a difference of squares is *not* the square of the difference; that is,

$$A^2 - B^2 \neq (A - B)^2.$$

For example,

$$(45 - 5)^2 = 40^2 = 1600,$$

5.2 DIFFERENCES OF SQUARES

but
$$45^2 - 5^2 = 2025 - 25 = 2000.$$

Factoring Completely

If a factor with more than one term can still be factored, you should do so. When no factor can be factored further, you have *factored completely*. Always factor completely even when the directions do not say so.

EXAMPLE Factor: $1 - 16x^{12}$.

$$\begin{aligned}1 - 16x^{12} &= (1 + 4x^6)(1 - 4x^6) \quad \text{Factoring a difference of squares}\\ &= (1 + 4x^6)(1 - 2x^3)(1 + 2x^3) \quad \text{Factoring further (The factor } 1 - 4x^6 \text{ is a difference of squares.)}\end{aligned}$$

FACTORING HINTS

1. Always look first for a common factor.
2. Always factor completely.
3. Check by multiplying.
4. Never try to factor a sum of squares, $A^2 + B^2$.

EXERCISE SET 5.2

Which of the following are differences of squares?

1. $x^2 - 4$
2. $x^2 - 36$
3. $x^2 + 36$
4. $x^2 + 4$
5. $x^2 - 35$
6. $x^2 - 50$
7. $16x^2 - 25$
8. $36x^2 - 1$
9. $49x^2 - 2$

Factor. Remember to look first for a common factor.

10. $x^2 - 4$
11. $x^2 - 36$
12. $x^2 - 9$
13. $x^2 - 1$
14. $16a^2 - 9$
15. $25x^2 - 4$
16. $4x^2 - 25$
17. $9a^2 - 16$
18. $8x^2 - 98$
19. $24x^2 - 54$
20. $36x - 49x^3$
21. $16x - 81x^3$
22. $16x^2 - 25x^4$
23. $x^{16} - 9x^2$
24. $49a^4 - 81$
25. $25a^4 - 9$
26. $a^{12} - 4a^2$
27. $121a^8 - 100$
28. $81y^6 - 25$
29. $100y^6 - 49$

Factor.

30. $x^4 - 1$
31. $x^4 - 16$
32. $4x^4 - 64$
33. $5x^4 - 80$
34. $1 - y^8$
35. $x^8 - 1$
36. $x^{12} - 16$
37. $x^8 - 81$
38. $\frac{1}{16} - y^2$
39. $\frac{1}{25} - x^2$
40. $25 - \frac{1}{49}x^2$
41. $4 - \frac{1}{9}y^2$
42. $16 - t^4$
43. $1 - a^4$

POLYNOMIALS AND FACTORING

Problem-solving practice

44. In a recent year, 29,090 people were arrested for counterfeiting. This was down 1.2% from the year before. How many were arrested the year before?

45. The first angle of a t[riangle is] as the second. The measure is four times as large greater than that of the se[cond.] The third angle is 30° angles? [H]ow large are the

Factor.

46. $4x^4 - 4x^2$

47. $3x^5 - 12x^3$

48. $3x^2 - \frac{1}{3}$

49. 1

50. $x^2 - 2.25$

51. $x^3 - \dfrac{x}{1.69}$

52. $3.24x^2 - 0.81$

53. 0.64

54. $1.28x^2 - 2$

55. $(x+3)^2 - 9$

56. $(y-5)^2 - 36$

57. $(3a +$

58. $(2y-7)^2 - 1$

59. $y^8 - 256$

60. $x^{16} - 1$

61. $x^2 - \left(\dfrac{1}{x}\right)$

A polynomial is called *irreducible* if it cannot be factored except for removing a common constant facto[r.] coefficient of the leading term is 1, the irreducible polynomial is called *prime*. Which of these polynomi[als are] irreducible? Which are prime?

62. $3x^3 + 9x$

63. $4x^2 + 2y$

64. $4x^2 + 16y^2$

65. $x^2 + y$

66. $16x^3 - 9x$

67. $-25y^2 - 49$

5.3 TRINOMIAL SQUARES

Recall that a trinomial is a polynomial with just three terms. Some trinomials are squares of binomials. For example, the trinomial $x^2 + 10x + 25$ is the square of $(x + 5)$. To see this we can calculate $(x + 5)^2$. It is $x^2 + 2 \cdot 5 \cdot x + 5^2$, or $x^2 + 10x + 25$.

A trinomial that is the square of a binomial is called a *trinomial square*.

Recognizing Trinomial Squares

We use the equations for squaring a binomial in reverse to factor trinomial squares:

$$A^2 + 2AB + B^2 = (A + B)^2;$$

$$A^2 - 2AB + B^2 = (A - B)^2.$$

How can we recognize when an expression to be factored is a trinomial square? Look at $A^2 + 2AB + B^2$ and $A^2 - 2AB + B^2$. In order for an expression to be a

5.2 DIFFERENCES OF SQUARES

but
$$45^2 - 5^2 = 2025 - 25 = 2000.$$

Factoring Completely

If a factor with more than one term can still be factored, you should do so. When no factor can be factored further, you have *factored completely*. Always factor completely even when the directions do not say so.

EXAMPLE 5 Factor: $1 - 16x^{12}$.

$$1 - 16x^{12} = (1 + 4x^6)(1 - 4x^6) \quad \text{Factoring a difference of squares}$$
$$= (1 + 4x^6)(1 - 2x^3)(1 + 2x^3) \quad \text{Factoring further (The factor } 1 - 4x^6 \text{ is a difference of squares.)}$$

FACTORING HINTS

1. Always look first for a common factor.
2. Always factor completely.
3. Check by multiplying.
4. Never try to factor a sum of squares, $A^2 + B^2$.

EXERCISE SET 5.2

Which of the following are differences of squares?

1. $x^2 - 4$
2. $x^2 - 36$
3. $x^2 + 36$
4. $x^2 + 4$
5. $x^2 - 35$
6. $x^2 - 50$
7. $16x^2 - 25$
8. $36x^2 - 1$
9. $49x^2 - 2$

Factor. Remember to look first for a common factor.

10. $x^2 - 4$
11. $x^2 - 36$
12. $x^2 - 9$
13. $x^2 - 1$
14. $16a^2 - 9$
15. $25x^2 - 4$
16. $4x^2 - 25$
17. $9a^2 - 16$
18. $8x^2 - 98$
19. $24x^2 - 54$
20. $36x - 49x^3$
21. $16x - 81x^3$
22. $16x^2 - 25x^4$
23. $x^{16} - 9x^2$
24. $49a^4 - 81$
25. $25a^4 - 9$
26. $a^{12} - 4a^2$
27. $121a^8 - 100$
28. $81y^6 - 25$
29. $100y^6 - 49$

Factor.

30. $x^4 - 1$
31. $x^4 - 16$
32. $4x^4 - 64$
33. $5x^4 - 80$
34. $1 - y^8$
35. $x^8 - 1$
36. $x^{12} - 16$
37. $x^8 - 81$
38. $\frac{1}{16} - y^2$
39. $\frac{1}{25} - x^2$
40. $25 - \frac{1}{49}x^2$
41. $4 - \frac{1}{9}y^2$
42. $16 - t^4$
43. $1 - a^4$

Problem-solving practice

44. In a recent year, 29,090 people were arrested for counterfeiting. This was down 1.2% from the year before. How many were arrested the year before?

45. The first angle of a triangle is four times as large as the second. The measure of the third angle is 30° greater than that of the second. How large are the angles?

Factor.

46. $4x^4 - 4x^2$

47. $3x^5 - 12x^3$

48. $3x^2 - \frac{1}{3}$

49. $18x^3 - \frac{8}{25}x$

50. $x^2 - 2.25$

51. $x^3 - \frac{x}{1.69}$

52. $3.24x^2 - 0.81$

53. $0.64x^2 - 1.21$

54. $1.28x^2 - 2$

55. $(x+3)^2 - 9$

56. $(y-5)^2 - 36$

57. $(3a+4)^2 - 49$

58. $(2y-7)^2 - 1$

59. $y^8 - 256$

60. $x^{16} - 1$

61. $x^2 - \left(\frac{1}{x}\right)^2$

A polynomial is called *irreducible* if it cannot be factored except for removing a common constant factor. If the coefficient of the leading term is 1, the irreducible polynomial is called *prime*. Which of these polynomials are irreducible? Which are prime?

62. $3x^3 + 9x$

63. $4x^2 + 2y$

64. $4x^2 + 16y^2$

65. $x^2 + y$

66. $16x^3 - 9x$

67. $-25y^2 - 49$

5.3 TRINOMIAL SQUARES

Recall that a trinomial is a polynomial with just three terms. Some trinomials are squares of binomials. For example, the trinomial $x^2 + 10x + 25$ is the square of $(x + 5)$. To see this we can calculate $(x + 5)^2$. It is $x^2 + 2 \cdot 5 \cdot x + 5^2$, or $x^2 + 10x + 25$.

A trinomial that is the square of a binomial is called a *trinomial square*.

Recognizing Trinomial Squares

We use the equations for squaring a binomial in reverse to factor trinomial squares:

$$A^2 + 2AB + B^2 = (A + B)^2;$$
$$A^2 - 2AB + B^2 = (A - B)^2.$$

How can we recognize when an expression to be factored is a trinomial square? Look at $A^2 + 2AB + B^2$ and $A^2 - 2AB + B^2$. In order for an expression to be a

5.3 TRINOMIAL SQUARES

trinomial square:

a) Two of the terms, A^2 and B^2, must be squares, such as

$$4, \quad x^2, \quad 25x^4, \quad 16t^2.$$

b) There must be no minus sign before A^2 or B^2.

c) If we multiply A and B (the square roots of these expressions) and double the result, we get the remaining term $2 \cdot A \cdot B$, or its additive inverse, $-2 \cdot A \cdot B$.

EXAMPLE 1 Determine whether $x^2 + 6x + 9$ is a trinomial square.

a) We know that x^2 and 9 are squares.

b) There is no minus sign before x^2 or 9.

c) If we multiply the square roots, x and 3, and double the product, we get the remaining term: $2 \cdot 3 \cdot x = 6x$.

Thus $x^2 + 6x + 9$ is the square of a binomial.

EXAMPLE 2 Determine whether $x^2 + 6x + 11$ is a trinomial square.

The answer is *no*, because only one term is a square.

EXAMPLE 3 Determine whether $16x^2 + 49 - 56x$ is a trinomial square.

a) We know that $16x^2$ and 49 are squares.

b) There is no minus sign before $16x^2$ or 49.

c) If we multiply the square roots, $4x$ and 7, and double the product, we get the additive inverse of the remaining term: $2 \cdot 4x \cdot 7 = 56x$; and $56x$ is the additive inverse of $-56x$.

Thus $16x^2 + 49 - 56x$ is a trinomial square.

Factoring Trinomial Squares

To factor trinomial squares we use the following equations:

$$A^2 + 2AB + B^2 = (A + B)^2;$$
$$A^2 - 2AB + B^2 = (A - B)^2.$$

We use the square roots of the squared terms and the sign of the remaining term.

EXAMPLE 4 Factor: $x^2 + 6x + 9$.

$$x^2 + 6x + 9 = x^2 + 2 \cdot x \cdot 3 + 3^2 = (x + 3)^2 \qquad \text{The sign of the middle term is positive.}$$
$$ A^2 + 2AB + B^2 = (A + B)^2$$

EXAMPLE 5 Factor: $x^2 + 49 - 14x$.

$x^2 + 49 - 14x = x^2 - 14x + 49$ Changing order
$= x^2 - 2 \cdot x \cdot 7 + 7^2$
$= (x - 7)^2$ The sign of the middle term is negative.

EXAMPLE 6 Factor: $16x^2 - 40x + 25$.

$16x^2 - 40x + 25 = (4x)^2 - 2 \cdot 4x \cdot 5 + 5^2 = (4x - 5)^2$
$A^2 - 2AB + B^2 = (A - B)^2$

EXERCISE SET 5.3

13–37

Determine whether each of the following is a trinomial square.

1. $x^2 - 14x + 49$
2. $x^2 - 16x + 64$
3. $x^2 + 16x - 64$
4. $x^2 - 14x - 49$
5. $x^2 - 3x + 9$
6. $x^2 + 2x + 4$
7. $8x^2 + 40x + 25$
8. $9x^2 - 36x + 24$
9. $36x^2 - 24x + 16$

Factor. Remember to look first for a common factor.

10. $x^2 - 14x + 49$
11. $x^2 - 16x + 64$
12. $x^2 + 16x + 64$
13. $x^2 + 14x + 49$
14. $x^2 - 2x + 1$
15. $x^2 + 2x + 1$
16. $4 + 4x + x^2$
17. $4 + x^2 - 4x$
18. $y^2 - 6y + 9$
19. $y^2 + 6y + 9$
20. $2x^2 - 4x + 2$
21. $2x^2 - 40x + 200$
22. $x^3 - 18x^2 + 81x$
23. $x^3 + 24x^2 + 144x$
24. $20x^2 + 100x + 125$
25. $12x^2 + 36x + 27$
26. $49 - 42x + 9x^2$
27. $64 - 112x + 49x^2$
28. $5y^4 + 10y^2 + 5$
29. $a^4 + 14a^2 + 49$
30. $y^6 + 26y^3 + 169$
31. $y^6 - 16y^3 + 64$
32. $16x^{10} - 8x^5 + 1$
33. $9x^{10} + 12y^5 + 4$
34. $1 + 4x^4 + 4x^2$
35. $1 - 2a^3 + a^6$
36. $\frac{1}{81}x^6 + \frac{8}{27}x^3 + \frac{16}{9}$
37. $\frac{1}{9}a^2 + \frac{1}{3}a + \frac{1}{4}$

Problem-solving practice

38. About 5 L of oxygen can be dissolved in 100 L of water at 0°C. This is 1.6 times the amount that can be dissolved in the same volume of water at 20°C. How much oxygen can be dissolved at 20°C?

39. The perimeter of a rectangle is 540 m. The width is 19 m less than the length. Find the width and the length.

Factor, if possible.

40. $49x^2 - 216$
41. $27x^3 - 13x$
42. $x^2 + 22x + 121$
43. $4x^2 + 9$
44. $x^2 - 5x + 25$
45. $18x^3 + 12x^2 + 2x$
46. $63x - 28$
47. $162x^2 - 82$
48. $x^4 - 9$
49. $8.1x^2 - 6.4$
50. $x^8 - 2^8$
51. $3^4 - x^4$

Factor.

52. $(y + 3)^2 + 2(y + 3) + 1$
53. $(a + 4)^2 - 2(a + 4) + 1$
54. $4(a + 5)^2 + 20(a + 5) + 25$
55. $49(x + 1)^2 - 42(x + 1) + 9$
56. $(x + 7)^2 - 4x - 24$
57. $(a + 4)^2 - 6a - 15$
58. Is $(y + 2)^2(y - 2)^2$ a factorization of $y^4 - 8y^2 + 16$? Prove your answer.
59. Is $(x + 3)^2(x - 3)^2$ a factorization of $x^4 + 18x^2 + 81$? Prove your answer.

Factor.

60. $9x^{18} + 48x^9 + 64$
61. $x^{2n} + 10x^n + 25$

Factor the trinomial square, and then the difference of two squares.

62. $a^2 + 2a + 1 - 9$
63. $y^2 + 6y + 9 - x^2 - 8x - 16$

Find c so that the polynomial will be the square of a binomial.

64. $cy^2 + 6y + 1$
65. $cy^2 - 24y + 9$

66. Show that the difference of the squares of two consecutive integers is the sum of the integers. (*Hint:* Use x for the smaller number.)
67. Find the value of a if $x^2 + a^2x + a^2$ factors into $(x + a)^2$.

5.4 FACTORING TRINOMIALS OF THE TYPE $x^2 + bx + c$

Some trinomials are not trinomial squares, as in the following examples:

$$x^2 + 5x + 6 \quad \text{and} \quad x^2 + 3x - 10.$$

To try to factor such trinomials, we use a trial-and-error procedure.

Constant Term Positive

Recall the FOIL method of multiplying two binomials:

$$(x + 2)(x + 5) = x^2 + \underbrace{5x + 2x}_{\text{F O I L}} + 10$$

$$= x^2 + 7x + 10.$$

The product is a trinomial. In the example, the term of highest degree, called the leading term, has a coefficient of 1. The constant term is positive. To factor $x^2 + 7x + 10$, we think of FOIL in reverse. We multiplied x times x to get the first term of the trinomial. So the first term of each binomial factor is x:

$$(x + \underline{})(x + \underline{}).$$

To get the middle term and the last term of the trinomial we look for two numbers whose product is 10 and whose sum is 7. Those numbers are 2 and 5. Thus the factorization is

$$(x + 2)(x + 5).$$

EXAMPLE 1 Factor: $x^2 + 5x + 6$.

Think of FOIL in reverse. The first term of each factor is x:

$$(x + \underline{})(x + \underline{}).$$

Then look for two numbers whose product is 6 and whose sum is 5. Since both 5 and 6 are positive, we need only consider positive factors.

Pairs of factors	Sums of factors
1, 6	7
2, 3	5

The numbers we want are 2 and 3. The factorization is $(x + 2)(x + 3)$. We can check by multiplying to see whether we get the original trinomial.

Consider this multiplication:

$$(x - 3)(x - 4) = x^2 - 4x - 3x + 12$$
$$= x^2 - 7x + 12.$$

When the constant term of a trinomial is positive, we look for two numbers with the same sign. The sign is that of the middle term:

$$(x^2 - 7x + 12) = (x - 3)(x - 4).$$

EXAMPLE 2 Factor: $x^2 - 8x + 12$.

Since the constant term is positive and the coefficient of the middle term is negative, we look for a factorization of 12 in which both factors are negative. Their sum must be -8.

Pairs of factors	Sums of factors
$-1, -12$	-13
$-2, -6$	-8
$-3, -4$	-7

The numbers we want are -2 and -6. The factorization is $(x-2)(x-6)$.

Constant Term Negative

Sometimes when we use FOIL, the product has a negative constant term. Consider these multiplications:

a) $(x-5)(x+2) = x^2 + \underbrace{2x - 5x}_{} - 10$
$ = x^2 - 3x - 10;$

b) $(x+5)(x-2) = x^2 \underbrace{- 2x + 5x}_{} - 10$
$ = x^2 + 3x - 10.$

When the constant term is negative, the middle term may be positive or negative. In these cases, we still look for two factors whose product is -10. One of them must be positive and the other negative. Their sum must still be the coefficient of the middle term.

EXAMPLE 3 Factor: $x^2 - 8x - 20$.

Since the constant term is negative, we look for a factorization of -20 in which one factor is positive and one factor is negative. Their sum must be -8.

Pairs of factors	Sums of factors
$-1, 20$	19
$1, -20$	-19
$-2, 10$	8
$2, -10$	-8
$-5, 4$	-1
$5, -4$	1

The numbers we want are 2 and -10. The factorization is $(x+2)(x-10)$.

EXAMPLE 4 Factor: $x^2 + 5x - 24$.

We look for a factorization of -24 in which one factor is positive and the other is negative. Their sum must be 5.

Pairs of factors	Sums of factors
1, −24	−23
−1, 24	23
2, −12	−10
−2, 12	10
−8, 3	−5
8, −3	5
4, −6	−2
−4, 6	2

The numbers we want are 8 and -3. The factorization is $(x+8)(x-3)$.

EXAMPLE 5 Factor: $x^2 - x - 110$.

Since the constant term is negative, we look for a factorization of -110 in which one factor is positive and one factor is negative. Their sum must be -1. The numbers we want are 10 and -11. The factorization is

$$(x+10)(x-11).$$

Some trinomials cannot be factored. An example is

$$x^2 - x + 5.$$

EXERCISE SET 5.4

Factor.

1. $x^2 + 8x + 15$
2. $x^2 + 5x + 6$
3. $x^2 + 7x + 12$
4. $x^2 + 9x + 8$
5. $x^2 - 6x + 9$
6. $y^2 + 11y + 28$
7. $x^2 + 9x + 14$
8. $a^2 + 11a + 30$
9. $b^2 + 5b + 4$
10. $x^2 - \frac{2}{5}x + \frac{1}{25}$
11. $x^2 + \frac{2}{3}x + \frac{1}{9}$
12. $z^2 - 8z + 7$
13. $d^2 - 7d + 10$
14. $x^2 - 8x + 15$
15. $y^2 - 11y + 10$
16. $x^2 - 2x - 15$
17. $x^2 + x - 42$
18. $x^2 + 2x - 15$
19. $x^2 - 7x - 18$
20. $y^2 - 3y - 28$
21. $x^2 - 6x - 16$
22. $x^2 - x - 42$
23. $y^2 - 4y - 45$
24. $x^2 - 7x - 60$
25. $x^2 - 2x - 99$
26. $x^2 - 72 + 6x$
27. $c - 56 + c^2$
28. $b^2 + 5b - 24$
29. $a^2 + 2a - 35$
30. $2 - x - x^2$

Factor.

31. $x^2 + 20x + 100$
32. $x^2 + 20x + 99$
33. $x^2 - 21x - 100$
34. $x^2 - 20x + 96$
35. $x^2 - 21x - 72$
36. $4x^2 + 40x + 100$
37. $x^2 - 25x + 144$
38. $y^2 - 21y + 108$
39. $a^2 + a - 132$
40. $a^2 + 9a - 90$
41. $120 - 23x + x^2$
42. $96 + 22d + d^2$
43. $108 - 3x - x^2$
44. $112 + 9y - y^2$
45. Find all integers m for which $y^2 + my + 50$ can be factored.
46. Find all integers b for which $a^2 + ba - 50$ can be factored.

Factor.

47. $x^2 - \frac{1}{2}x - \frac{3}{16}$
48. $x^2 - \frac{1}{4}x - \frac{1}{8}$
49. $x^2 + \frac{30}{7}x - \frac{25}{7}$
50. $\frac{1}{3}x^3 + \frac{1}{3}x^2 - 2x$

Find a polynomial in factored form for each shaded region. (Leave answers in terms of π.)

51.

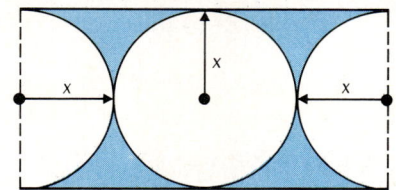

52.

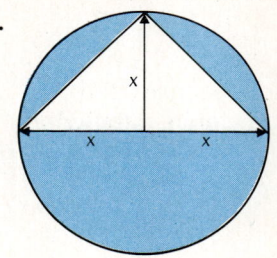

5.5 FACTORING TRINOMIALS OF THE TYPE $ax^2 + bx + c$, $a \neq 1$

In Section 5.4 we learned to factor trinomials of the type $x^2 + bx + c$. In this section we consider two methods of factoring trinomials where the leading, or x^2, coefficient is not 1. You may choose the one that works best for you, or you may use the one that your instructor chooses for you. Both methods involve trial and error, but the first requires trial and error in only one step.

Method 1

We know how to factor the trinomial $x^2 + 5x + 6$. We look for factors of the constant term, 6, whose sum is the coefficient, 5, of the middle term:

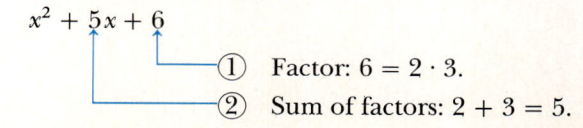

$$x^2 + 5x + 6$$

① Factor: $6 = 2 \cdot 3$.
② Sum of factors: $2 + 3 = 5$.

What happens when the coefficient of the first, or x^2, term is not 1? Consider the trinomial $6x^2 + 23x + 20$. The method we use is similar to what we used for the preceding trinomial, but we need two more steps. We first multiply the leading coefficient 6 and the constant 20, and get 120. Then we look for a factorization of 120 in which the sum of the factors is the coefficient, 23, of the middle term. Next we write the middle term as a sum using these factors.

$6x^2 + 23x + 20$

① Multiply 6 and 20: $6 \cdot 20 = 120$.
② Factor 120: $120 = 8 \cdot 15$ and $8 + 15 = 23$.
③ Write the middle term as a sum: $23x = 8x + 15x$.
④ Then factor by grouping.

We factor by grouping as follows:

$$6x^2 + 23x + 20 = 6x^2 + 8x + 15x + 20$$
$$= 2x(3x + 4) + 5(3x + 4) \quad \text{Factoring by grouping (see Section 5.1)}$$
$$= (2x + 5)(3x + 4).$$

It does not matter which way we write the middle term as a sum. We still get the same factorization, although the factors may be in a different order. Note the following:

$$6x^2 + 23x + 20 = 6x^2 + 15x + 8x + 20$$
$$= 3x(2x + 5) + 4(2x + 5)$$
$$= (3x + 4)(2x + 5).$$

The method we used to factor trinomials of the type $x^2 + bx + c$ was based on FOIL. So too is the method we have just introduced. Before we show why, we state the method more formally and consider other examples.

To factor $ax^2 + bx + c$:

a) First look for a common factor.

b) Multiply the leading coefficient a and the constant c.

c) Try to factor the product ac so that the sum of the factors is b. That is, find p and q such that $ac = pq$ and $p + q = b$.

d) Write the middle term, bx, as a sum: $bx = px + qx$. (We can think of this as "splitting" the middle term.)

e) Then factor by grouping. Regroup $ax^2 + px$ and $qx + c$.

EXAMPLE 1 Factor: $3x^2 - 10x - 8$.

a) First look for a common factor. There is none (other than 1).

b) Multiply the leading coefficient and the constant, 3 and -8:
$$3(-8) = -24.$$

c) Try to factor -24 so that the sum of the factors is -10.

Pairs of factors	Sums of factors
4, −6	−2
−4, 6	2
12, −2	10
−12, 2	−10

d) Write $-10x$ as a sum using the results of part (c). That is, split the middle term as follows:
$$-10x = -12x + 2x.$$

e) Factor by grouping:
$$3x^2 - 10x - 8 = 3x^2 - 12x + 2x - 8 \quad \text{Substituting } -12x + 2x \text{ for } -10x$$
$$= 3x(x - 4) + 2(x - 4)$$
$$= (3x + 2)(x - 4)$$

EXAMPLE 2 Factor: $8x^2 + 8x - 6$.

a) First look for a common factor. The number 2 is common to all three terms, so we factor it out:
$$2(4x^2 + 4x - 3).$$

b) Now factor the trinomial $4x^2 + 4x - 3$. Multiply the leading coefficient and the constant, 4 and -3:
$$4(-3) = -12.$$

c) Try to factor -12 so that the sum of the factors is 4.

Pairs of factors	Sums of factors
−3, 4	1
3, −4	−1
12, −1	11
−6, 2	−4
6, −2	4

d) Split the middle term $4x$, as follows:
$$4x = 6x - 2x.$$

e) Then factor by grouping:

$$4x^2 + 4x - 3 = 4x^2 + 6x - 2x - 3$$
$$= 2x(2x + 3) - (2x + 3)$$
$$= 2x(2x + 3) - 1(2x + 3)$$
$$= (2x - 1)(2x + 3).$$

The factorization of $4x^2 + 4x - 3$ is $(2x - 1)(2x + 3)$. But, don't forget the common factor! We must include it to get a factorization of the original trinomial:

$$8x^2 + 8x - 6 = 2(2x - 1)(2x + 3).$$

This method of factoring is based on FOIL as shown here:

$$(ax + b)(cx + d) = acx^2 + adx + bcx + bd$$
$$= acx^2 + (ad + bc)x + bd$$

$$(ac)(bd) = (ad)(bc)$$

We multiply the outside coefficients and factor them in such a way that we can write the middle term as a sum.

Method 2

We now consider an alternative method for factoring trinomials of the type $ax^2 + bx + c$. Consider the following multiplication:

$$\begin{matrix} & \text{F} & \text{O} & \text{I} & \text{L} \\ (2x + 5)(3x + 4) = & 6x^2 & + 8x & + 15x & + 20 \\ & = 6x^2 & + & 23x & + 20 \end{matrix}$$

F	O + I	L
$2 \cdot 3$	$2 \cdot 4 + 5 \cdot 3$	$5 \cdot 4$

Now to factor $6x^2 + 23x + 20$, we do the reverse of what we just did:

$$\begin{matrix} \text{F} & \text{O + I} & \text{L} \\ 6x^2 & + 23x & + 20 \\ = (2x + 5) & (3x + 4) \end{matrix}$$

We look for numbers (x-coefficients) whose product is 6—in this case, 2 and 3—and numbers whose product is 20—in this case, 4 and 5. The product of the outside terms plus the product of the inside terms must, of course, be 23.

To factor $ax^2 + bx + c$, we look for two binomials like this:

$$(\_\_x + \_\_)(\_\_x + \_\_),$$

where products of numbers in the blanks are as follows.

> 1. The numbers in the *first* blanks of each binomial have product a.
> 2. The *outside* product and the *inside* product add up to b.
> 3. The numbers in the *last* blanks of each binomial have product c.

EXAMPLE 3 Factor: $3x^2 + 5x + 2$.

We look for two numbers whose product is 3. These are

$$1, 3 \quad \text{and} \quad -1, -3.$$

We have these possibilities:

$$(x + \_\_)(3x + \_\_)$$

or

$$(-x + \_\_)(-3x + \_\_).$$

Now we look for numbers whose product is 2. These are

$$1, 2 \quad \text{and} \quad -1, -2.$$

Here are some possibilities for factorizations. There are eight possibilities, but we have not listed all of them here:

$$(x + 1)(3x + 2), \quad (-x + 1)(-3x + 2), \quad (x + 2)(3x + 1),$$
$$(x - 1)(3x - 2), \quad (-x - 1)(-3x - 2), \quad (x - 2)(3x - 1).$$

When we multiply, we must get $3x^2 + 5x + 2$. When we multiply, we find that both of the expressions in color above are factorizations. We usually choose the one in which the first coefficients are positive. Thus the factorization is

$$(x + 1)(3x + 2).$$

We always look first for a common factor. If there is one, we remove that common factor before proceeding.

EXAMPLE 4 Factor: $8x^2 + 8x - 6$.

We first look for a factor common to all three terms. The number 2 is a common factor, so we factor it out:

$$2(4x^2 + 4x - 3).$$

Now we factor the trinomial $4x^2 + 4x - 3$. We look for pairs of numbers whose product is 4. The pairs that are positive are

$$4, 1 \quad \text{and} \quad 2, 2. \qquad \text{Both positive}$$

We then have these possibilities:

$$(4x + \underline{})(x + \underline{}) \quad \text{and} \quad (2x + \underline{})(2x + \underline{}).$$

Next we look for pairs of numbers whose product is -3. They are

$$3, -1 \quad \text{and} \quad -3, 1.$$

Then we have these possibilities for factorizations:

$$(4x + 3)(x - 1), \quad (2x + 3)(2x - 1),$$
$$(4x - 1)(x + 3), \quad (2x - 3)(2x + 1).$$
$$(4x + 1)(x - 3),$$
$$(4x - 3)(x + 1),$$

We usually do not write all of these. We multiply until we find the factors that give the product $4x^2 + 4x - 3$. We find that the factorization is

$$(2x + 3)(2x - 1).$$

But don't forget the common factor. We must include it in order to get a factorization of the original polynomial:

$$8x^2 + 8x - 6 = 2(2x + 3)(2x - 1).$$

Keep in mind that no matter which of the two methods you use to factor trinomials of the type $ax^2 + bx + c$, you involve trial and error. This is the way such factoring is done. As you practice you will find that you can make better and better guesses. Don't forget: When factoring any polynomials, always look first for a common factor. Failure to do so is a common error.

EXERCISE SET 5.5

Factor.

1. $2x^2 - 7x - 4$
2. $3x^2 - x - 4$
3. $5x^2 + x - 18$
4. $3x^2 - 4x - 15$
5. $6x^2 + 23x + 7$
6. $6x^2 + 13x + 6$
7. $3x^2 + 4x + 1$
8. $7x^2 + 15x + 2$
9. $4x^2 + 4x - 15$
10. $9x^2 + 6x - 8$
11. $2x^2 - x - 1$
12. $15x^2 - 19x - 10$
13. $9x^2 + 18x - 16$
14. $2x^2 + 5x + 2$
15. $3x^2 - 5x - 2$
16. $18x^2 - 3x - 10$
17. $12x^2 + 31x + 20$
18. $15x^2 + 19x - 10$

19. $14x^2 + 19x - 3$
20. $35x^2 + 34x + 8$
21. $9x^2 + 18x + 8$
22. $6 - 13x + 6x^2$
23. $49 - 42x + 9x^2$
24. $25x^2 + 40x + 16$
25. $24x^2 + 47x - 2$
26. $16a^2 + 78a + 27$
27. $35x^2 - 57x - 44$
28. $9a^2 + 12a - 5$
29. $20 + 6x - 2x^2$
30. $15 + x - 2x^2$
31. $12x^2 + 28x - 24$
32. $6x^2 + 33x + 15$
33. $30x^2 - 24x - 54$
34. $20x^2 - 25x + 5$
35. $4x + 6x^2 - 10$
36. $-9 + 18x^2 - 21x$
37. $3x^2 - 4x + 1$
38. $6x^2 - 13x + 6$
39. $12x^2 - 28x - 24$
40. $6x^2 - 33x + 15$
41. $-1 + 2x^2 - x$
42. $-19x + 15x^2 + 6$
43. $9x^2 - 18x - 16$
44. $14x^2 + 35x + 14$
45. $15x^2 - 25x - 10$
46. $18x^2 + 3x - 10$
47. $12x^3 + 31x^2 + 20x$
48. $15x^3 + 19x^2 - 10x$
49. $14x^4 + 19x^3 - 3x^2$
50. $70x^4 + 68x^3 + 16x^2$
51. $168x^3 - 45x^2 + 3x$

Problem-solving practice

52. The earth is a sphere (or ball) that is about 40,000 km in circumference. Find the radius of the earth, in kilometers and in miles. (*Hint:* 1 km ≈ 0.62 mi.)

53. In an apartment, lamps, an air conditioner, and a television set are all operating at the same time. The lamps take 10 times as many watts as the television set, and the air conditioner takes 40 times as many watts as the television set. The total wattage used in the apartment is 2550 watts. How many watts are used by each appliance?

○ ─────────────────────────

Factor, if possible.

54. $9x^4 + 18x^2 + 8$
55. $6 - 13x - 6x^2$
56. $9x^2 - 42x + 49$
57. $15x^4 - 19x^2 + 6$
58. $6x^3 + 4x^2 - 10x$
59. $18x^3 - 21x^2 - 9x$
60. $x^2 + 3x - 7$
61. $x^2 + 13x - 12$
62. $x^5 + 2x^4 + 2x + 1$
63. $27x^3 - 63x^2 - 147x + 343$

Factor.

64. $20x^{2n} + 16x^n + 3$
65. $-15x^{2m} + 26x^m - 8$
66. $3x^{6a} - 2x^{3a} - 1$
67. $x^{2n+1} - 2x^{n+1} + x$
68. $3(a+1)^{n+1}(a+3)^2 - 5(a+1)^n(a+3)^3$

5.6 FACTORING: A GENERAL STRATEGY

We now try to put all of our factoring techniques together and consider a general strategy for factoring polynomials. Here we will encounter polynomials of all the types we have considered, mixed up, so you will have to determine which method to use.

> **To factor a polynomial:**
>
> **A.** Always look first for a common factor.
>
> **B.** Then look at the number of terms.
>
> *Two terms:* Determine whether you have a difference of squares. Do not try to factor a sum of squares: $A^2 + B^2$.
>
> *Three terms:* Determine whether the trinomial is a square. If so, you know how to factor. If not, try trial and error.
>
> *Four terms:* Try factoring by grouping.
>
> **C.** Always *factor completely*. If a factor with more than one term can still be factored, you should do so. When no factor can be factored further, you have finished.

EXAMPLE 1 Factor: $10x^3 - 40x$.

    A. We look first for a common factor:
$$10x^3 - 40x = 10x(x^2 - 4). \qquad \text{Factoring out the largest common factor}$$

    B. The factor $x^2 - 4$ has only two terms. It is a difference of squares. We factor it, being careful to include the common factor:
$$10x(x - 2)(x + 2).$$

    C. Have we factored completely? Yes, because no factor with more than one term can be factored further.

EXAMPLE 2 Factor: $t^4 - 16$.

    A. We look for a common factor. There isn't one.

    B. There are only two terms. It is a difference of squares: $(t^2)^2 - 4^2$. We factor it:
$$(t^2 + 4)(t^2 - 4).$$

We see that one of the factors is still a difference of squares. We factor it:

$(t^2 + 4)(t - 2)(t + 2).$

        ↑   This is a sum of squares. It cannot be factored!

    C. We have factored completely because no factors with more than one term can be factored further.

EXAMPLE 3 Factor: $2x^3 + 10x^2 + x + 5$.

    A. We look for a common factor. There isn't one.

    B. There are four terms. We try factoring by grouping:
$$\begin{aligned} 2x^3 + 10x^2 + x + 5 &= (2x^3 + 10x^2) + (x + 5) &&\text{Separating into two binomials} \\ &= 2x^2(x + 5) + 1(x + 5) &&\text{Factoring each binomial} \\ &= (2x^2 + 1)(x + 5). &&\text{Factoring out the common factor, } x + 5 \end{aligned}$$

C. No factor with more than one term can be factored further, so we have factored completely.

EXAMPLE 4 Factor: $x^5 - 2x^4 - 35x^3$.

A. We look first for a common factor. This time there is one:
$$x^5 - 2x^4 - 35x^3 = x^3(x^2 - 2x - 35).$$

B. The factor $x^2 - 2x - 35$ has three terms, but it is not a trinomial square. We factor it using trial and error:
$$x^5 - 2x^4 - 35x^3 = x^3(x^2 - 2x - 35) = x^3(x - 7)(x + 5).$$

Don't forget to include the common factor in your final answer!

C. No factor with more than one term can be factored further, so we have factored completely.

EXAMPLE 5 Factor: $x^4 - 10x^2 + 25$.

A. We look first for a common factor. There isn't one.

B. There are three terms. We see that this is a trinomial square. We factor it:
$$x^4 - 10x^2 + 25 = (x^2)^2 - 2 \cdot 5 \cdot x^2 + 5^2 = (x^2 - 5)^2.$$

C. No factor with more than one term can be factored further, so we have factored completely.

EXERCISE SET 5.6

Factor.

1. $2x^2 - 128$
2. $3t^2 - 27$
3. $a^2 + 25 - 10a$
4. $y^2 + 49 + 14y$
5. $2x^2 - 11x + 12$
6. $8y^2 - 18y - 5$
7. $x^3 + 24x^2 + 144x$
8. $x^3 - 18x^2 + 81x$
9. $x^3 + 3x^2 - 4x - 12$
10. $x^3 - 5x^2 - 25x + 125$
11. $24x^2 - 54$
12. $8x^2 - 98$
13. $20x^3 - 4x^2 - 72x$
14. $9x^3 + 12x^2 - 45x$
15. $x^2 + 4$
16. $t^2 + 25$
17. $x^4 + 7x^2 - 3x^3 - 21x$
18. $m^4 + 8m^3 + 8m^2 + 64m$
19. $x^5 - 14x^4 + 49x^3$
20. $2x^6 + 8x^5 + 8x^4$

21. $20 - 6x - 2x^2$
22. $45 - 3x - 6x^2$
23. $x^2 + 3x + 1$
24. $x^2 + 5x + 2$
25. $4x^4 - 64$
26. $5x^5 - 80x$
27. $1 - y^8$
28. $t^8 - 1$
29. $x^5 - 4x^4 + 3x^3$
30. $x^6 - 2x^5 + 7x^4$
31. $36a^2 - 15a + \frac{25}{16}$
32. $\frac{1}{81}x^6 - \frac{8}{27}x^3 + \frac{16}{9}$

Factor completely.
33. $a^4 - 2a^2 + 1$
34. $x^4 + 9$
35. $12.25x^2 - 7x + 1$
36. $\frac{1}{5}x^2 - x + \frac{4}{5}$
37. $5x^2 + 13x + 7.2$
38. $x^3 - (x - 3x^2) - 3$
39. $18 + y^3 - 9y - 2y^2$
40. $-(x^4 - 7x^2 - 18)$
41. $a^3 + 4a^2 + a + 4$
42. $x^3 + x^2 - (4x + 4)$
43. $x^4 - 7x^2 - 18$
44. $3x^4 - 15x^2 + 12$
45. $x^3 - x^2 - 4x + 4$
46. $y^2(y + 1) - 4y(y + 1) - 21(y + 1)$
47. $y^2(y - 1) - 2y(y - 1) + (y - 1)$
48. Factor $x^{2k} - 2^{2k}$ when $k = 4$.
49. Factor $a^4 - 81$.
50. Factor $acx^{m+n} + adx^n + bcx^m + bd$, where a, b, c, and d are constants.

5.7 SOLVING EQUATIONS BY FACTORING

In this section we introduce a new equation-solving method and use it along with factoring to solve certain equations.

The Principle of Zero Products

The product of two numbers is 0 if one of the numbers is 0. Furthermore, *if any product is 0, then a factor must be 0.* For example, if $7x = 0$, then we know that $x = 0$. If $x(2x - 9) = 0$, we know that $x = 0$ or $2x - 9 = 0$. If $(x + 3)(x - 2) = 0$, we know that $x + 3 = 0$ or $x - 2 = 0$.

EXAMPLE 1 Solve: $(x + 3)(x - 2) = 0$.

We have a product of 0. This equation will be true when either factor is 0. Hence it is true when

$$x + 3 = 0 \quad \text{or} \quad x - 2 = 0.$$

5.7 SOLVING EQUATIONS BY FACTORING

Here we have two simple equations, which we know how to solve:

$$x = -3 \quad \text{or} \quad x = 2.$$

There are two solutions, -3 and 2.

We have another principle to help in solving equations.

THE PRINCIPLE OF ZERO PRODUCTS

An equation with 0 on one side and with factors on the other can be solved by finding those numbers that make the factors 0.

EXAMPLE 2 Solve: $(5x + 1)(x - 7) = 0$.

$5x + 1 = 0 \quad \text{or} \quad x - 7 = 0$ Using the principle of zero products

$5x = -1 \quad \text{or} \quad x = 7$

$x = -\frac{1}{5} \quad \text{or} \quad x = 7$ Solving the two equations separately

Check: For $-\frac{1}{5}$:   For 7:

$$\begin{array}{c|c} (5x + 1)(x - 7) = 0 \\ \hline (5(-\tfrac{1}{5}) + 1)(-\tfrac{1}{5} - 7) & 0 \\ (-1 + 1)(-7\tfrac{1}{5}) \\ 0(-7\tfrac{1}{5}) \\ 0 \end{array} \qquad \begin{array}{c|c} (5x + 1)(x - 7) = 0 \\ \hline (5 \cdot 7 + 1)(7 - 7) & 0 \\ (35 + 1) \cdot 0 \\ 0 \end{array}$$

The solutions are $-\frac{1}{5}$ and 7.

    The "possible solutions" we get by using the principle of zero products are actually always solutions, unless we have made an error in solving. Thus, when we use this principle, a check is not necessary, except to detect errors. Keep in mind that you must have 0 on one side of an equation before you can apply the principle of zero products.

    When some factors have only one term, you can still use the principle of zero products in the same way.

EXAMPLE 3 Solve: $x(2x - 9) = 0$.

$x = 0 \quad \text{or} \quad 2x - 9 = 0$ Using the principle of zero products

$x = 0 \quad \text{or} \quad 2x = 9$

$x = 0 \quad \text{or} \quad x = \frac{9}{2}$

The solutions are 0 and $\frac{9}{2}$.

Using Factoring to Solve Equations

Using factoring and the principle of zero products, we can solve some new kinds of equations. Thus we have extended our equation-solving abilities.

EXAMPLE 4 Solve: $x^2 + 5x + 6 = 0$.

We first factor the polynomial. Then we use the principle of zero products:

$$x^2 + 5x + 6 = 0$$
$$(x + 2)(x + 3) = 0 \qquad \text{Factoring}$$
$$x + 2 = 0 \quad \text{or} \quad x + 3 = 0 \qquad \text{Using the principle of zero products}$$
$$x = -2 \quad \text{or} \quad x = -3.$$

Check:

$$\begin{array}{c|c} x^2 + 5x + 6 = 0 \\ \hline (-2)^2 + 5(-2) + 6 \;\big|\; 0 \\ 4 - 10 + 6 \\ -6 + 6 \\ 0 \end{array} \qquad \begin{array}{c|c} x^2 + 5x + 6 = 0 \\ \hline (-3)^2 + 5(-3) + 6 \;\big|\; 0 \\ 9 - 15 + 6 \\ -6 + 6 \\ 0 \end{array}$$

The solutions are -2 and -3.

Again, keep in mind that you *must* have 0 on one side before you can use the principle of zero products.

EXAMPLE 5 Solve: $x^2 - 8x = -16$.

We first add 16 to get 0 on one side:

$$x^2 - 8x + 16 = 0 \qquad \text{Adding 16}$$
$$(x - 4)(x - 4) = 0 \qquad \text{Factoring}$$
$$x - 4 = 0 \quad \text{or} \quad x - 4 = 0 \qquad \text{Using the principle of zero products}$$
$$x = 4 \quad \text{or} \quad x = 4.$$

There is only one solution, 4. The check is left to the student.

EXAMPLE 6 Solve: $x^2 + 5x = 0$.

$$x(x + 5) = 0 \qquad \text{Factoring out a common factor}$$
$$x = 0 \quad \text{or} \quad x + 5 = 0 \qquad \text{Using the principle of zero products}$$
$$x = 0 \quad \text{or} \quad x = -5$$

The solutions are 0 and -5. The check is left to the student.

5.7 SOLVING EQUATIONS BY FACTORING

EXAMPLE 7 Solve: $4x^2 - 25 = 0$.

$$(2x - 5)(2x + 5) = 0 \quad \text{Factoring a difference of squares}$$
$$2x - 5 = 0 \quad \text{or} \quad 2x + 5 = 0$$
$$2x = 5 \quad \text{or} \quad 2x = -5$$
$$x = \tfrac{5}{2} \quad \text{or} \quad x = -\tfrac{5}{2}$$

The solutions are $\tfrac{5}{2}$ and $-\tfrac{5}{2}$. The check is left to the student.

EXERCISE SET 5.7

Solve.

1. $(x + 8)(x + 6) = 0$
2. $(x + 3)(x + 2) = 0$
3. $(x - 3)(x + 5) = 0$
4. $(x + 9)(x - 3) = 0$
5. $(x + 12)(x - 11) = 0$
6. $(x - 13)(x + 53) = 0$
7. $x(x + 5) = 0$
8. $y(y + 7) = 0$
9. $y(y - 13) = 0$
10. $v(v - 4) = 0$
11. $0 = y(y + 10)$
12. $0 = x(x - 21)$
13. $(2x + 5)(x + 4) = 0$
14. $(2x + 9)(x + 8) = 0$
15. $(3x - 1)(x + 2) = 0$
16. $(3x - 9)(x + 3) = 0$
17. $(5x + 1)(4x - 12) = 0$
18. $(4x + 9)(14x - 7) = 0$
19. $(7x - 28)(28x - 7) = 0$
20. $(12x - 11)(8x - 5) = 0$
21. $2x(3x - 2) = 0$
22. $75x(8x - 9) = 0$
23. $\tfrac{1}{2}x(\tfrac{2}{3}x - 12) = 0$
24. $\tfrac{5}{7}x(\tfrac{3}{4}x - 6) = 0$
25. $(\tfrac{1}{3} - 3x)(\tfrac{1}{5} - 2x) = 0$
26. $(\tfrac{1}{5} + 2x)(\tfrac{1}{9} - 3x) = 0$
27. $(\tfrac{1}{3}y - \tfrac{2}{3})(\tfrac{1}{4}y - \tfrac{3}{2}) = 0$
28. $(\tfrac{7}{4}x - \tfrac{1}{12})(\tfrac{2}{3}x - \tfrac{12}{11}) = 0$
29. $(0.3x - 0.1)(0.05x - 1) = 0$
30. $(0.1x - 0.3)(0.4x - 20) = 0$
31. $9x(3x - 2)(2x - 1) = 0$
32. $(x - 5)(x + 55)(5x - 1) = 0$
33. $x^2 + 6x + 5 = 0$
34. $x^2 + 7x + 6 = 0$
35. $x^2 + 7x - 18 = 0$
36. $x^2 + 4x - 21 = 0$
37. $x^2 - 8x + 15 = 0$
38. $x^2 - 9x + 14 = 0$
39. $x^2 - 8x = 0$
40. $x^2 - 3x = 0$
41. $x^2 + 19x = 0$
42. $x^2 + 12x = 0$
43. $x^2 = 16$
44. $100 = x^2$
45. $9x^2 - 4 = 0$
46. $4x^2 - 9 = 0$
47. $0 = 6x + x^2 + 9$
48. $0 = 25 + x^2 + 10x$
49. $x^2 + 16 = 8x$
50. $1 + x^2 = 2x$

51. $5x^2 = 6x$
52. $7x^2 = 8x$
53. $6x^2 - 4x = 10$
54. $3x^2 - 7x = 20$
55. $12y^2 - 5y = 2$
56. $2y^2 + 12y = -10$
57. $x(x - 5) = 14$
58. $t(3t + 1) = 2$
59. $64m^2 = 81$
60. $100t^2 = 49$
61. $3x^2 + 8x = 9 + 2x$
62. $x^2 - 5x = 18 + 2x$
63. $10x^2 - 23x + 12 = 0$
64. $12x^2 - 17x - 5 = 0$

Solve.

65. $b(b + 9) = 4(5 + 2b)$
66. $y(y + 8) = 16(y - 1)$
67. $(t - 3)^2 = 36$
68. $(t - 5)^2 = 2(5 - t)$
69. $x^2 - \frac{1}{64} = 0$
70. $x^2 - \frac{25}{36} = 0$
71. $\frac{5}{16}x^2 = 5$
72. $\frac{27}{25}x^2 = \frac{1}{3}$

73. Find an equation that has the given numbers as solutions. For example, 3 and -2 are solutions to $x^2 - x - 6 = 0$.

a) $1, -3$
b) $3, -1$
c) $2, 2$
d) $3, 4$
e) $3, -4$
f) $-3, 4$
g) $-3, -4$
h) $\frac{1}{2}, \frac{1}{2}$
i) $5, -5$
j) $0, 0.1, \frac{1}{4}$

74. Check the numbers found below. What's wrong with the methods used? Try to find the solutions.

a) $(x - 3)(x + 4) = 8$
   $x - 3 = 0$ or $x + 4 = 8$
   $x = 3$ or $x = 4$

b) $(x - 3)(x + 4) = 8$
   $x - 3 = 2$ or $x + 4 = 4$
   $x = 5$ or $x = 0$

75. Solve: $(0.00005x + 0.1)(0.0097x + 0.5) = 0$.

76. For each equation on the left, find an equivalent equation on the right.

a) $3x^2 - 4x + 8 = 0$
b) $(x - 6)(x + 3) = 0$
c) $x^2 + 2x + 9 = 0$
d) $(2x - 5)(x + 4) = 0$
e) $5x^2 - 5 = 0$
f) $x^2 + 10x - 2 = 0$
g) $4x^2 + 8x + 36 = 0$
h) $(2x + 8)(2x - 5) = 0$
i) $9x^2 - 12x + 24 = 0$
j) $(x + 1)(5x - 5) = 0$
k) $x^2 - 3x - 18 = 0$
l) $2x^2 + 20x - 4 = 0$

5.8 PROBLEM SOLVING

We can use our five-step problem-solving process and our new methods for solving equations to solve problems.

EXAMPLE 1 One more than a number times one less than a number is 8. Find the number.

1. **Familiarize** The problem is stated explicitly enough that we can go right to the translation.

5.8 PROBLEM SOLVING

2. **Translate** We translate as follows:

   One more than a number times one less than that number is 8.
   $(x+1)$ $(x-1)$ $= 8$

3. **Carry out** We solve the equation as follows:

   $(x+1)(x-1) = 8$
   $x^2 - 1 = 8$ Multiplying
   $x^2 - 1 - 8 = 0$ Adding -8 to get 0 on one side
   $x^2 - 9 = 0$
   $(x-3)(x+3) = 0$ Factoring
   $x - 3 = 0$ or $x + 3 = 0$ Using the principle of zero products
   $x = 3$ or $x = -3$

4. **Check** One more than 3 (this is 4) times one less than 3 (this is 2) is 8. Thus, 3 checks. The check for -3 is left to the student.

5. **State** There are two such numbers, 3 and -3.

EXAMPLE 2 The square of a number minus twice the number is 48. Find the number.

1. **Familiarize** Again, the problem is stated explicitly enough that we can go right to the translation.

2. **Translate** We translate as follows:

   The square of a number minus twice the number is 48.
   x^2 $-$ $2x$ $= 48$

3. **Carry out** We solve the equation as follows:

   $x^2 - 2x = 48$
   $x^2 - 2x - 48 = 0$ Adding -48 to get 0 on one side
   $(x-8)(x+6) = 0$
   $x - 8 = 0$ or $x + 6 = 0$ Using the principle of zero products
   $x = 8$ or $x = -6$

4. **Check** The square of 8 is 64, and twice the number 8 is 16. Then $64 - 16$ is 48, so 8 checks. The check for -6 is left to the student.

5. **State** There are two such numbers, 8 and -6.

POLYNOMIALS AND FACTORING

EXAMPLE 3 The height of a triangular sail is 7 ft more than the base. The area of the triangle is 30 ft². Find the height and the base.

1. **Familiarize** We first make a drawing. If you don't remember the formula for the area of a triangle, look it up, either in this book or a geometry book. The area is

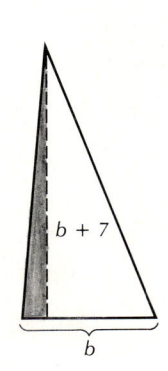

$$\tfrac{1}{2} \cdot \text{base} \cdot \text{height}.$$

2. **Translate** It helps to reword this problem before translating:

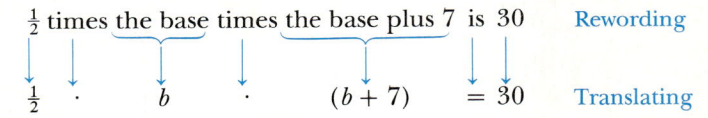

$\tfrac{1}{2}$ times the base times the base plus 7 is 30   Rewording

$\tfrac{1}{2} \cdot b \cdot (b + 7) = 30$ Translating

3. **Carry out** We solve the equation as follows:

$\tfrac{1}{2} \cdot b \cdot (b + 7) = 30$
$\tfrac{1}{2}(b^2 + 7b) = 30$ Multiplying
$b^2 + 7b = 60$ Multiplying by 2
$b^2 + 7b - 60 = 0$ Adding -60 to get 0 on one side
$(b + 12)(b - 5) = 0$ Factoring
$b + 12 = 0$ or $b - 5 = 0$ Using the principle of zero products
$b = -12$ or $b = 5$

4. **Check** The solutions of the equation are -12 and 5. The base of a triangle cannot have a negative length, so -12 cannot be a solution. Suppose the base is 5 ft. Then the height is 7 ft more than the base, so the height is 12 ft and the area is $\tfrac{1}{2}(5)(12)$ or 30 ft². These numbers check in the original problem.

5. **State** The height is 12 ft and the base is 5 ft.

5.8 PROBLEM SOLVING

EXAMPLE 4 In a sports league of n teams in which each team plays every other team twice, the total number N of games to be played is given by

$$N = n^2 - n.$$

A basketball league plays a total of 240 games. How many teams are in the league?

1. **Familiarize** To familiarize yourself with this problem, reread Example 4 in Section 4.2 where we first considered it.

2. **Translate** We are trying to find the number of teams n in a league when 240 games are played. We substitute 240 for N in order to solve for n:

$$n^2 - n = 240. \quad \text{Substituting 240 for } N$$

3. **Carry out** We solve the equation as follows:

$$n^2 - n = 240$$
$$n^2 - n - 240 = 0 \quad \text{Adding } -240 \text{ to get 0 on one side}$$
$$(n - 16)(n + 15) = 0 \quad \text{Factoring}$$
$$n - 16 = 0 \quad \text{or} \quad n + 15 = 0 \quad \text{Using the principle of zero products}$$
$$n = 16 \quad \text{or} \quad n = -15$$

4. **Check** The solutions of the equation are 16 and -15. Since the number of teams cannot be negative, -15 cannot be a solution. But 16 checks, since $16^2 - 16 = 256 - 16 = 240$.

5. **State** There are 16 teams in the league.

EXAMPLE 5 The product of two consecutive integers is 156. Find the integers.

1. **Familiarize** Recall that *consecutive* integers are next to each other, such as 49 and 50, or -6 and -5. If x represents the smaller integer, then $x + 1$ represents the larger integer.

2. **Translate** It helps to reword the problem before translating:

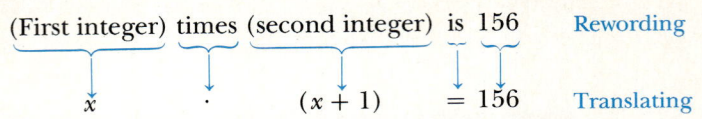

(First integer) times (second integer) is 156   Rewording

$$x \cdot (x+1) = 156 \quad \text{Translating}$$

We have let x represent the first integer. Then $x+1$ represents the second.

3. **Carry out** We solve the equation as follows:

$$x(x+1) = 156$$
$$x^2 + x = 156 \quad \text{Multiplying}$$
$$x^2 + x - 156 = 0 \quad \text{Adding } -156 \text{ to get 0 on one side}$$
$$(x-12)(x+13) = 0 \quad \text{Factoring}$$
$$x - 12 = 0 \quad \text{or} \quad x + 13 = 0 \quad \text{Using the principle of zero products}$$
$$x = 12 \quad \text{or} \quad x = -13.$$

4. **Check** The solutions of the equation are 12 and -13. When x is 12, then $x+1$ is 13, and $12 \cdot 13 = 156$. The numbers 12 and 13 are consecutive integers that are solutions to the problem. When x is -13, then $x+1$ is -12, and $(-13)(-12) = 156$. The numbers -13 and -12 are also consecutive integers that are solutions to the problem.

5. **State** We have two solutions, each of which consists of a pair of numbers: 12 and 13, and -13 and -12.

EXERCISE SET 5.8

Solve.

1. If you subtract a number from 4 times its square, the result is 3. Find the number.

2. If 7 is added to the square of a number, the result is 32. Find the number.

3. Eight more than the square of a number is 6 times the number. Find the number.

4. Fifteen more than the square of a number is 8 times the number. Find the number.

5. The product of two consecutive integers is 182. Find the integers.

6. The product of two consecutive integers is 56. Find the integers.

7. The product of two consecutive even integers is 168. Find the integers.

8. The product of two consecutive even integers is 224. Find the integers.

9. The product of two consecutive odd integers is 255. Find the integers.

10. The product of two consecutive odd integers is 143. Find the integers.

11. The length of a rectangular garden is 4 m greater than the width. The area of the rectangle is 96 m². Find the length and width.

12. The length of a rectangular calculator is 5 cm greater than the width. The area of the rectangle is 84 cm². Find the length and width.

13. The area of a square bookcase is 5 ft² more than the perimeter. Find the length of a side.

14. The perimeter of a square porch is 3 yd more than the area. Find the length of a side.

15. The base of a triangle is 10 cm greater than the height. The area is 28 cm². Find the height and base.

16. The height of a triangle is 8 m less than the base. The area is 10 m². Find the height and base.

17. If the sides of a square are lengthened by 3 m, the area becomes 81 m². Find the length of a side of the original square.

18. If the sides of a square are lengthened by 7 km, the area becomes 121 km². Find the length of a side of the original square.

19. The sum of the squares of two consecutive odd positive integers is 74. Find the integers.

20. The sum of the squares of two consecutive odd positive integers is 130. Find the integers.

Use $N = n^2 - n$ for Exercises 21–24.

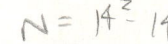

21. A slow-pitch softball league has 23 teams. What is the total number of games to be played?

22. A basketball league has 14 teams. What is the total number of games to be played?

23. A slow-pitch softball league plays a total of 132 games. How many teams are in the league?

24. A basketball league plays a total of 90 games. How many teams are in the league?

The number of possible handshakes within a group of n people is given by $N = \frac{1}{2}(n^2 - n)$.

25. At a meeting there are 40 people. How many handshakes are possible?

26. At a party there are 100 people. How many handshakes are possible?

27. Everyone shook hands at a party. There were 190 handshakes in all. How many were at the party?

28. Everyone shook hands at a meeting. There were 300 handshakes in all. How many were at the meeting?

---

29. A cement walk of constant width is built around a 20-ft × 40-ft rectangular pool. The total area of the pool and walk is 1500 ft². Find the width of the walk.

30. A model rocket is launched using an engine that will generate a speed of 180 feet per second. The formula $h = rt - 16t^2$ gives the height of an object projected upward at a rate of r feet per second after t seconds. After how many seconds will the rocket reach a height of 464 feet? After how many seconds will it be at that height again?

31. When the speed of an object is measured in meters per second and distance in meters, the formula of Exercise 30 becomes $h = rt - 4.9t^2$. A baseball is thrown upward with a speed of 20.6 meters per second.

a) After how many seconds will the ball reach a height of 21.6 meters?

b) How long after it is thrown will it hit the ground?

32. The one's digit of a number less than 100 is 4 greater than the ten's digit. The sum of the number and the product of the digits is 58. Find the number.

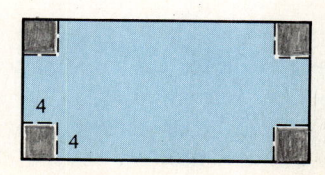

33. The total surface area of a box is 350 m². The box is 9 m high and has a square base. Find the length of the side of the base.

34. A rectangular piece of cardboard is twice as long as it is wide. A 4-cm square is cut out of each corner, and the sides are turned up to make a box. The volume of the box is 616 cm³. Find the original dimensions of the cardboard.

35. An open rectangular gutter is made by turning up the sides of a piece of metal 20 in. wide. The area of the cross-section of the gutter is 50 in². Find the depth of the gutter.

36. The length of each side of a square is increased by 5 cm to form a new square. The area of the new square is $2\frac{1}{4}$ times the area of the original square. Find the area of each square.

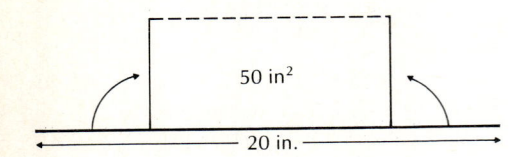

50 in²

20 in.

5.9 POLYNOMIALS IN SEVERAL VARIABLES

The polynomials we have been studying have only one variable. A *polynomial in several variables* is an expression like those you have already seen, but we allow that there can be more than one variable. Here are some examples:

$$3x + xy^2 + 5y + 4, \qquad 8xy^2z - 2x^3z - 13x^4y^2 + 15.$$

We now learn how to add, subtract, multiply, and factor polynomials in several variables.

Evaluating Polynomials

EXAMPLE 1 Evaluate the polynomial $4 + 3x + xy^2 + 8x^3y^3$ for $x = -2$ and $y = 5$.

We replace x by -2 and y by 5:

$$4 + 3(-2) + (-2) \cdot 5^2 + 8(-2)^3 \cdot 5^3 = 4 - 6 - 50 - 8000 = -8052.$$

EXAMPLE 2 (*The magic number.*) The Boston Red Sox are leading the New York Yankees for the Eastern Division championship of the American League. The magic number is 8. This means that any combination of Red Sox wins and Yankee losses that totals 8 will ensure the championship for the Red Sox. The magic number is given by the polynomial

$$G - P - L + 1,$$

where G is the number of games in the season, P is the number of games the leading team has played, and L is the number of games ahead in the loss column.

Given the situation shown in the table and assuming a 162-game season, what is the magic number for the Philadelphia Phillies?

5.9 POLYNOMIALS IN SEVERAL VARIABLES

EASTERN DIVISION				
	W	L	Pct.	GB
Philadelphia	77	40	.658	—
Pittsburgh	65	53	.551	$12\frac{1}{2}$
New York	61	60	.504	18
Chicago	55	67	.451	$24\frac{1}{2}$
St. Louis	51	65	.440	$25\frac{1}{2}$
Montreal	41	73	.360	$34\frac{1}{2}$

We evaluate the polynomial for $G = 162$, $P = 77 + 40$, or 117, and $L = 53 - 40$, or 13:

$$162 - 117 - 13 + 1 = 33.$$

Coefficients and Degrees

The *degree* of a term is the sum of the exponents of the variables. The *degree of a polynomial* is the degree of the term of highest degree.

EXAMPLE 3 Identify the coefficient and degree of each term of

$$9x^2y^3 - 14xy^2z^3 + xy + 4y + 5x^2 + 7.$$

Term	Coefficient	Degree	
$9x^2y^3$	9	5	
$-14xy^2z^3$	-14	6	
xy	1	2	
$4y$	4	1	Think: $4y = 4y^1$
$5x^2$	5	2	
7	7	0	Think: $7 = 7x^0$

EXAMPLE 4 What is the degree of $5x^3y + 9xy^4 - 8x^3y^3$?

The term of highest degree is $-8x^3y^3$. Its degree is 6. Thus the degree of the polynomial is 6.

Collecting Like Terms

Like terms (or *similar terms*) have exactly the same variables with exactly the same exponents. For example,

$3x^2y^3$ and $-7x^2y^3$ are like terms;
$9x^4z^7$ and $12x^4z^7$ are like terms.

But

$13xy^2$ and $-2x^2y$ are *not* like terms;

$3xyz^2$ and $4xy$ are *not* like terms.

Collecting like terms is based on the distributive law.

EXAMPLES Collect like terms.

5. $5x^2y + 3xy^2 - 5x^2y - xy^2 = (5-5)x^2y + (3-1)xy^2 = 2xy^2$

6. $3xy - 5xy^2 + 3xy^2 + 9xy = -2xy^2 + 12xy$

Addition

The sum of two polynomials in several variables can be found by writing a plus sign between them and then collecting like terms.

EXAMPLE 7 Add: $-5x^3 + 3y - 5y^2$ and $8x^3 + 4x^2 + 7y^2$.

$(-5x^3 + 3y - 5y^2) + (8x^3 + 4x^2 + 7y^2) = (-5+8)x^3 + 4x^2 + 3y + (-5+7)y^2$
$= 3x^3 + 4x^2 + 3y + 2y^2$

EXAMPLE 8 Add: $(5xy^2 - 4x^2y + 5x^3 + 2) + (3xy^2 - 2x^2y + 3x^3y - 5)$.

We look for like terms. The like terms are $5xy^2$ and $3xy^2$, $-4x^2y$ and $-2x^2y$, and 2 and -5. We collect these. There are no more like terms. Thus the answer is

$$8xy^2 - 6x^2y + 5x^3 + 3x^3y - 3.$$

Subtraction

We subtract a polynomial by adding its inverse. An equivalent expression for the additive inverse of a polynomial is found by replacing each coefficient by its additive inverse, or by changing the sign of each term. For example, the additive inverse of the polynomial

$$4x^2y - 6x^3y^2 + x^2y^2 - 5y$$

can be represented by

$$-(4x^2y - 6x^3y^2 + x^2y^2 - 5y).$$

An equivalent expression can be found by replacing each coefficient by its additive inverse. Thus

$$-(4x^2y - 6x^3y^2 + x^2y^2 - 5y) = -4x^2y + 6x^3y^2 - x^2y^2 + 5y.$$

5.9 POLYNOMIALS IN SEVERAL VARIABLES

EXAMPLE 9 Subtract: $(4x^2y + x^3y^2 + 3x^2y^3 + 6y) - (4x^2y - 6x^3y^2 + x^2y^2 - 5y)$.

$(4x^2y + x^3y^2 + 3x^2y^3 + 6y) - (4x^2y - 6x^3y^2 + x^2y^2 - 5y)$
$= 4x^2y + x^3y^2 + 3x^2y^3 + 6y - 4x^2y + 6x^3y^2 - x^2y^2 + 5y$ Adding the inverse
$= 7x^3y^2 + 3x^2y^3 - x^2y^2 + 11y$ Collecting like terms (Try to write just the answer!)

Multiplication

To multiply polynomials in several variables, we can multiply each term of one by every term of the other. Where appropriate, we use special products.

EXAMPLE 10 Multiply: $(3x^2y - 2xy + 3y)(xy + 2y)$.

$$\begin{array}{r} 3x^2y - 2xy + 3y \\ xy + 2y \\ \hline 3x^3y^2 - 2x^2y^2 + 3xy^2 \\ 6x^2y^2 - 4xy^2 + 6y^2 \\ \hline 3x^3y^2 + 4x^2y^2 - xy^2 + 6y^2 \end{array}$$

Multiplying by xy
Multiplying by $2y$
Adding

EXAMPLES Multiply.

FOIL
11. $(x^2y + 2x)(xy^2 + y^2) = x^3y^3 + x^2y^3 + 2x^2y^2 + 2xy^2$
12. $(p + 5q)(2p - 3q) = 2p^2 - 3pq + 10pq - 15q^2$
$ = 2p^2 + 7pq - 15q^2$

$(A + B)^2 = A^2 + 2\ A\ \ B + B^2$
13. $(3x + 2y)^2 = (3x)^2 + 2(3x)(2y) + (2y)^2$
$ = 9x^2 + 12xy + 4y^2$
14. $(2y^2 - 5x^2y)^2 = (2y^2)^2 - 2(2y^2)(5x^2y) + (5x^2y)^2$
$ = 4y^4 - 20x^2y^3 + 25x^4y^2$

$(A + B)(A - B) = A^2 - B^2$
15. $(3x^2y + 2y)(3x^2y - 2y) = (3x^2y)^2 - (2y)^2$
$ = 9x^4y^2 - 4y^2$
16. $(-2x^3y^2 + 5t)(2x^3y^2 + 5t) = (5t - 2x^3y^2)(5t + 2x^3y^2)$
$ = (5t)^2 - (2x^3y^2)^2 = 25t^2 - 4x^6y^4$

$(A - B)(A + B) = A^2 - B^2$
17. $(2x + 3 - 2y)(2x + 3 + 2y) = (2x + 3)^2 - (2y)^2$
$ = 4x^2 + 12x + 9 - 4y^2$

Factoring

To factor polynomials in several variables, we can use the same general strategy that we considered in Section 5.6, which you might review before studying the following examples.

EXAMPLE 18 Factor: $20x^3y + 12x^2y$.

A. We look first for a common factor:

$$20x^3y + 12x^2y = (4x^2y)(5x) + (4x^2y) \cdot 3$$
$$= 4x^2y(5x + 3). \quad \text{Factoring out the largest common factor}$$

B. Then we look at the number of terms. There are only two terms, but the binomial $5x + 3$ is not a difference of squares. It cannot be factored further.

C. We have factored completely because no factors with more than one term can be factored further.

EXAMPLE 19 Factor: $6x^2y - 21x^3y^2 + 3x^2y^3$.

A. We look first for a common factor:

$$6x^2y - 21x^3y^2 + 3x^2y^3 = 3x^2y(2 - 7xy + y^2).$$

B. There are three terms in $2 - 7xy + y^2$. Determine whether the trinomial is a square. Since only y^2 is a square, we do not have a trinomial square. Can the trinomial be factored by trial? A key to the answer is that x is only in the term $-7xy$. The polynomial might be in a form like $(1 - y)(2 + y)$, but there would be no x in the middle term.

C. Have we factored completely? Yes, because no factor with more than one term can be factored further.

EXAMPLE 20 Factor: $(p + q)(x + 2) + (p + q)(x + y)$.

A. We look first for a common factor:

$$(p + q)(x + 2) + (p + q)(x + y) = (p + q)[(x + 2) + (x + y)]$$
$$= (p + q)(2x + y + 2).$$

B. There are three terms in $2x + y + 2$, but this trinomial cannot be factored further.

C. No factor with more than one term can be factored further, so we have factored completely.

EXAMPLE 21 Factor: $px + py + qx + qy$.

A. We look first for a common factor. There isn't one.

B. There are four terms. We try factoring by grouping:

$$px + py + qx + qy = p(x + y) + q(x + y)$$
$$= (p + q)(x + y).$$

C. Have we factored completely? Since no factor with more than one term can be factored further, we have factored completely.

EXAMPLE 22 Factor: $25x^2 + 20xy + 4y^2$.

A. We look first for a common factor. There isn't one.

B. There are three terms. Determine whether the trinomial is a square. The first term and the last term are squares:

$$25x^2 = (5x)^2 \quad \text{and} \quad 4y^2 = (2y)^2.$$

Twice the product of $5x$ and $2y$ should be the other term:

$$2 \cdot 5x \cdot 2y = 20xy.$$

Thus the trinomial is a perfect square.
    We factor by writing the square roots of the square terms and the sign of the other term:

$$25x^2 + 20xy + 4y^2 = (5x + 2y)^2.$$

We can check by squaring $5x + 2y$.

C. No factor with more than one term can be factored further, so we have factored completely.

EXAMPLE 23 Factor: $p^2q^2 + 7pq + 12$.

A. We look first for a common factor. There isn't one.

B. There are three terms. Determine whether the trinomial is a square. The first term is a square, but neither of the other terms is a square, so we do not have a trinomial square. We use trial and error:

$$p^2q^2 + 7pq + 12 = (pq)^2 + (3 + 4)pq + 3 \cdot 4$$
$$= (pq + 3)(pq + 4).$$

C. No factor with more than one term can be factored further, so we have factored completely.

EXAMPLE 24 Factor: $8x^4 - 20x^2y - 12y^2$.

A. We look first for a common factor:

$$8x^4 - 20x^2y - 12y^2 = 4(2x^4 - 5x^2y - 3y^2).$$

B. There are three terms in $2x^4 - 5x^2y - 3y^2$. Determine whether the trinomial is a square. Since none of the terms is a square, we do not have a trinomial square. We use trial and error:

$$8x^4 - 20x^2y - 12y^2 = 4(2x^4 - 5x^2y - 3y^2)$$
$$= 4[(2x^2)(x^2) + (-6 + 1)x^2y + (-3y)y]$$
$$= 4(2x^2 + y)(x^2 - 3y).$$

C. No factor with more than one term can be factored further, so we have factored completely.

EXAMPLE 25 Factor: $a^4 + a^3b - 6a^2b^2$.

A. We look first for a common factor:
$$a^4 + a^3b - 6a^2b^2 = a^2(a^2 + ab - 6b^2).$$

B. There are three terms in $a^2 + ab - 6b^2$. Determine whether the trinomial is a square. Since neither ab nor $-6b^2$ is a square, we do not have a trinomial square. We use trial and error:
$$a^4 + a^3b - 6a^2b^2 = a^2(a^2 + ab - 6b^2)$$
$$= a^2(a - 2b)(a + 3b).$$

C. No factor with more than one term can be factored further, so we have factored completely.

EXAMPLE 26 Factor: $a^4 - 16b^4$.

$$a^4 - 16b^4 = (a^2 - 4b^2)(a^2 + 4b^2) = (a - 2b)(a + 2b)(a^2 + 4b^2)$$

EXERCISE SET 5.9

Evaluate each polynomial for $x = 3$ and $y = -2$.

1. $x^2 - y^2 + xy$
2. $x^2 + y^2 - xy$

Evaluate each polynomial for $x = 2$, $y = -3$, and $z = -1$.

3. $xyz^2 + z$
4. $xy - xz + yz$

An amount of money P is invested at interest rate r. In 3 years it will grow to an amount given by the polynomial
$$P + 3rP + 3r^2P + r^3P.$$

5. Evaluate the polynomial for $P = 10{,}000$ and $r = 0.08$ to find the amount to which $10{,}000 will grow at 8% interest for 3 years.

6. Evaluate the polynomial for $P = 10{,}000$ and $r = 0.07$ to find the amount to which $10{,}000 will grow at 7% interest for 3 years.

The area of a right circular cylinder is given by the polynomial

$$2\pi rh + 2\pi r^2,$$

where h is the height and r is the radius of the base.

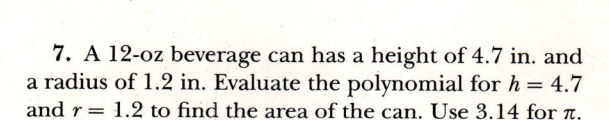

7. A 12-oz beverage can has a height of 4.7 in. and a radius of 1.2 in. Evaluate the polynomial for $h = 4.7$ and $r = 1.2$ to find the area of the can. Use 3.14 for π.

8. A 16-oz beverage can has a height of 6.3 in. and a radius of 1.2 in. Evaluate the polynomial for $h = 6.3$ and $r = 1.2$ to find the area of the can. Use 3.14 for π.

Identify the coefficient and degree of each term of the following polynomials. Then find the degree of the polynomial.

9. $x^3y - 2xy + 3x^2 - 5$
10. $5y^3 - y^2 + 15y + 1$
11. $17x^2y^3 - 3x^3yz - 7$
12. $6 - xy + 8x^2y^2 - y^5$

Collect like terms.

13. $a + b - 2a - 3b$
14. $y^2 - 1 + y - 6 - y^2$
15. $3x^2y - 2xy^2 + x^2$
16. $m^3 + 2m^2n - 3m^2 + 3mn^2$
17. $2u^2v - 3uv^2 + 6u^2v - 2uv^2$
18. $3x^2 + 6xy + 3y^2 - 5x^2 - 10xy - 5y^2$
19. $6au + 3av - 14au + 7av$
20. $3x^2y - 2z^2y + 3xy^2 + 5z^2y$

Add.

21. $(2x^2 - xy + y^2) + (-x^2 - 3xy + 2y^2)$
22. $(2z - z^2 + 5) + (z^2 - 3z + 1)$
23. $(r - 2s + 3) + (2r + s) + (s + 4)$
24. $(b^3a^2 - 2b^2a^3 + 3ba + 4) + (b^2a^3 - 4b^3a^2 + 2ba - 1)$
25. $(2x^2 - 3xy + y^2) + (-4x^2 - 6xy - y^2) + (x^2 + xy - y^2)$

Subtract.

26. $(x^3 - y^3) - (-2x^3 + x^2y - xy^2 + 2y^3)$
27. $(xy - ab) - (xy - 3ab)$
28. $(3y^4x^2 + 2y^3x - 3y) - (2y^4x^2 + 2y^3x - 4y - 2x)$
29. $(-2a + 7b - c) - (-3b + 4c + 8d)$
30. Find the sum of $2a + b$ and $3a - 4b$. Then subtract $5a + 2b$.

Multiply.

31. $(3z - u)(2z + 3u)$
32. $(a - b)(a^2 + b^2 + 2ab)$
33. $(a^2b - 2)(a^2b - 5)$
34. $(xy + 7)(xy - 4)$
35. $(a + a^2 - 1)(a^2 + 1 - y)$
36. $(r + tx)(vx + s)$
37. $(a^3 + bc)(a^3 - bc)$
38. $(m^2 + n^2 - mn)(m^2 + mn + n^2)$
39. $(y^4x + y^2 + 1)(y^2 + 1)$
40. $(a - b)(a^2 + ab + b^2)$

41. $(3xy - 1)(4xy + 2)$
42. $(m^3n + 8)(m^3n - 6)$
43. $(3 - c^2d^2)(4 + c^2d^2)$
44. $(6x - 2y)(5x - 3y)$
45. $(m^2 - n^2)(m + n)$
46. $(pq + 0.2)(0.4pq - 0.1)$
47. $(xy + x^5y^5)(x^4y^4 - xy)$
48. $(x - y^3)(2y^3 + x)$
49. $(x + h)^2$
50. $(3a + 2b)^2$
51. $(r^3t^2 - 4)^2$
52. $(3a^2b - b^2)^2$
53. $(p^4 + m^2n^2)^2$
54. $(ab + cd)^2$
55. $(2a^3 - \frac{1}{2}b^3)^2$
56. $-5x(x + 3y)^2$
57. $3a(a - 2b)^2$
58. $(a^2 + b + 2)^2$
59. $(2a - b)(2a + b)$
60. $(x - y)(x + y)$
61. $(c^2 - d)(c^2 + d)$
62. $(p^3 - 5q)(p^3 + 5q)$
63. $(ab + cd^2)(ab - cd^2)$
64. $(xy + pq)(xy - pq)$
65. $(x + y - 3)(x + y + 3)$
66. $(p + q + 4)(p + q - 4)$
67. $[x + y + z][x - (y + z)]$
68. $[a + b + c][a - (b + c)]$
69. $(a + b + c)(a - b - c)$
70. $(3x + 2 - 5y)(3x + 2 + 5y)$

Factor.

71. $12n^2 + 24n^3$
72. $ax^2 + ay^2$
73. $9x^2y^2 - 36xy$
74. $x^2y - xy^2$
75. $2\pi rh + 2\pi r^2$
76. $10p^4q^4 + 35p^3q^3 + 10p^2q^2$
77. $(a + b)(x - 3) + (a + b)(x + 4)$
78. $5c(a^3 + b) - (a^3 + b)$
79. $(x - 1)(x + 1) - y(x + 1)$
80. $x^2 + x + xy + y$
81. $n^2 + 2n + np + 2p$
82. $a^2 - 3a + ay - 3y$
83. $2x^2 - 4x + xz - 2z$
84. $6y^2 - 3y + 2py - p$
85. $x^2 + y^2 - 2xy$
86. $4b^2 + a^2 - 4ab$
87. $9c^2 + 6cd + d^2$
88. $16x^2 + 24xy + 9y^2$
89. $49m^4 - 112m^2n + 64n^2$
90. $4x^2y^2 + 12xyz + 9z^2$
91. $y^4 + 10y^2z^2 + 25z^4$
92. $0.01x^4 - 0.1x^2y^2 + 0.25y^4$
93. $\frac{1}{4}a^2 + \frac{1}{3}ab + \frac{1}{9}b^2$
94. $4p^2q + pq^2 + 4p^3$
95. $a^2 - ab - 2b^2$
96. $3b^2 - 17ab - 6a^2$
97. $2mn - 360n^2 + m^2$
98. $15 + x^2y^2 + 8xy$
99. $m^2n^2 - 4mn - 32$
100. $p^2q^2 + 7pq + 6$
101. $a^5b^2 + 3a^4b - 10a^3$
102. $m^2n^6 + 4mn^5 - 32n^4$
103. $a^5 + 4a^4b - 5a^3b^2$
104. $2s^6t^2 + 10s^3t^3 + 12t^4$
105. $x^6 + x^3y - 2y^2$
106. $a^4 + a^2bc - 2b^2c^2$
107. $x^2 - y^2$
108. $p^2q^2 - r^2$
109. $7p^4 - 7q^4$
110. $a^4b^4 - 16$
111. $81a^4 - b^4$
112. $1 - 16x^{12}y^{12}$

Find a polynomial for the area of each shaded region. (Leave results in terms of π where appropriate.)

113.

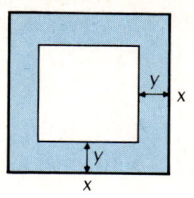

114.

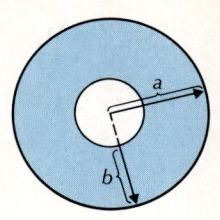

115.

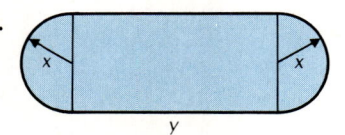

117. Find a formula for $(A + B)^3$.

Factor.

118. $6(x - 1)^2 + 7y(x - 1) - 3y^2$

119. $(y + 4)^2 + 2x(y + 4) + x^2$

120. $2(a + 3)^2 - (a + 3)(b - 2) - (b - 2)^2$

SUMMARY AND REVIEW: CHAPTER 5

The following contains a summary of what you should be able to do after completing this chapter. The review exercises are for practice. Answers are at the back of the book. If you miss an exercise, restudy the section indicated alongside the answer.

You should be able to:

Factor monomials.

Find three factorizations of each monomial.

1. $-10x^2$ **2.** $36x^5$

Factor polynomials when the terms have a common factor; differences of two squares; trinomial squares; trinomials of the type $x^2 + bx + c$; trinomials of the type $ax^2 + bx + c$, $a \neq 1$; and polynomials with four terms by grouping; and apply the general strategy for factoring.

Factor.

3. $5 - 20x^6$ **4.** $x^2 - 3x$ **5.** $9x^2 - 4$ **6.** $x^2 + 4x - 12$

7. $x^2 + 14x + 49$ **8.** $6x^3 + 12x^2 + 3x$ **9.** $x^3 + x^2 + 3x + 3$ **10.** $6x^2 - 5x + 1$

11. $x^4 - 81$
12. $9x^3 + 12x^2 - 45x$
13. $2x^2 - 50$
14. $x^4 + 4x^3 - 2x - 8$
15. $16x^4 - 1$
16. $8x^6 - 32x^5 + 4x^4$
17. $75 + 12x^2 + 60x$
18. $x^2 + 9$
19. $x^3 - x^2 - 30x$
20. $4x^2 - 25$
21. $9x^2 + 25 - 30x$
22. $6x^2 - 28x - 48$
23. $x^2 - 6x + 9$
24. $2x^2 - 7x - 4$
25. $18x^2 - 12x + 2$
26. $3x^2 - 27$
27. $15 - 8x + x^2$
28. $25x^2 - 20x + 4$

Solve equations by factoring and then by using the principle of zero products.

Solve.
29. $(x - 1)(x + 3) = 0$
30. $x^2 + 2x - 35 = 0$
31. $x^2 + x - 12 = 0$
32. $3x^2 + 2 = 5x$
33. $2x^2 + 5x = 12$
34. $16 = x(x - 6)$

Solve problems involving equations that can be factored.

35. The square of a number is 6 more than the number. Find the number.
36. The product of two consecutive even integers is 288. Find the integers.
37. The product of two consecutive odd integers is 323. Find the integers.
38. Twice the square of a number is 10 more than the number. Find the number.

Evaluate a polynomial in several variables for given values of the variables and identify the coefficients and the degrees of the terms and the degree of the polynomial. Also collect like terms of a polynomial in several variables.

39. Evaluate the polynomial $2 - 5xy + y^2 - 4xy^3 + x^6$ for $x = -1$ and $y = 2$.

Identify the coefficient and degree of each term of the following polynomials. Then find the degree of the polynomial.
40. $x^5y - 7xy + 9x^2 - 8$
41. $x^2y^5z^9 - y^{40} + x^{13}z^{10}$

Collect like terms.
42. $y + w - 2y + 8w - 5$
43. $m^6 - 2m^2n + m^2n^2 + n^2m - 6m^3 + m^2n^2 + 7n^2m$

Add, subtract, multiply, and factor polynomials in several variables.

44. Add: $(5x^2 - 7xy + y^2) + (-6x^2 - 3xy - y^2) + (x^2 + xy - 2y^2)$.
45. Subtract: $(6x^3y^2 - 4x^2y - 6x) - (-5x^3y^2 + 4x^2y + 6x^2 - 6)$.

Multiply.
46. $(p - q)(p^2 + pq + q^2)$
47. $(3a^4 - \frac{1}{3}b^3)^2$

Factor.
48. $x^2y^2 + xy - 12$
49. $12a^2 + 84ab + 147b^2$
50. $m^2 + 5m + mt + 5t$

Solve.
51. The pages of a book measure 15 cm by 20 cm. Margins of equal width surround the printing on each page and constitute one half of the area of the page. Find the width of the margins.
52. The cube of a number is the same as twice the square of the number. Find the number.

53. The length of a rectangle is 2 times its width. When the length is increased by 20 and the width decreased by 1, the area is 160. Find the original length and width.

Solve.

54. $x^2 + 25 = 0$

55. $(x-2)(x+3)(2x-5) = 0$

TEST: CHAPTER 5

1. Find three factorizations of $4x^3$.

Factor.

2. $x^2 - 7x + 10$
3. $x^2 + 25 - 10x$
4. $6y^2 - 8y^3 + 4y^4$
5. $x^3 + x^2 + 2x + 2$
6. $x^2 - 5x$
7. $x^3 + 2x^2 - 3x$
8. $28x - 48 + 10x^2$
9. $4x^2 - 9$
10. $x^2 - x - 12$
11. $6m^3 + 9m^2 + 3m$
12. $3w^2 - 75$
13. $60x + 45x^2 + 20$
14. $3x^4 - 48$
15. $49x^2 - 84x + 36$
16. $5x^2 - 26x + 5$
17. $x^4 + 2x^3 - 3x - 6$
18. $80 - 5x^4$
19. $4x^2 - 4x - 15$
20. $6t^3 + 9t^2 - 15t$

Solve.

21. $x^2 - x - 20 = 0$
22. $2x^2 + 7x = 15$
23. $x(x-3) = 28$

24. The square of a number is 24 more than 5 times the number. Find the number.

25. The length of a rectangle is 6 m more than the width. The area of the rectangle is 40 m². Find the length and the width.

26. Collect like terms:

$$x^3y - y^3 + xy^3 + 8 - 6x^3y - x^2y^2 + 11.$$

27. Subtract:

$$(8a^2b^2 - ab + b^3) - (-6ab^2 - 7ab - ab^3 + 5b^3).$$

28. Multiply:

$$(3x^5 - 4y^5)(3x^5 + 4y^5).$$

29. Factor:

$$3m^2 - 9m - 30n^2.$$

30. Solve: The length of a rectangle is 5 times its width. When the length is decreased by 3 and the width is increased by 2, the area of the new rectangle is 60. Find the original length and width.

31. Factor:

$$(a+3)^2 - 2(a+3) - 35.$$

GRAPHS, SYSTEMS OF EQUATIONS, AND PROBLEM SOLVING

6

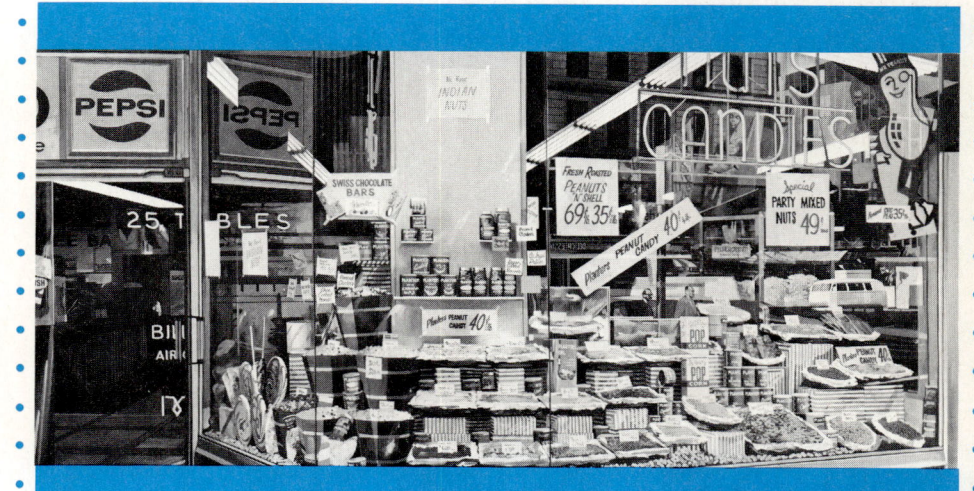

How can candies of different costs be mixed in order to create a mixture that has a cost between the two costs? Systems of equations can be used to find an answer.

We now study equations that have two variables. These variables will be raised to the first power only, and there will be no products or quotients involving variables. Such equations are called *linear*. We first learn to graph such equations. The graph of an equation is a geometric picture of its solutions. The graphs of linear equations are straight lines. Then we consider where two such graphs might intersect. By doing this we can solve what is called a *system of equations*.

Systems of equations have extensive applications to many fields such as sociology, psychology, business, education, engineering, and science.

Richard Estes. The Candy Store. 1969. Detail. Oil and synthetic polymer. $47\frac{3}{4} \times 68\frac{3}{4}$ inches. Collection of Whitney Museum of American Art. Gift of the Friends of the Whitney Museum of American Art. Acq. #69.21.

6.1 GRAPHS AND EQUATIONS

We have graphed numbers on a line. We now learn to graph number pairs on a plane, to enable us to graph an equation that contains two variables.

Points and Ordered Pairs

On a number line each point is the graph of a number. On a plane each point is the graph of a number pair. We use two perpendicular number lines called *axes*. They cross at a point called the *origin*. It has coordinates (0, 0) but is usually labeled 0. The arrows show the positive directions.

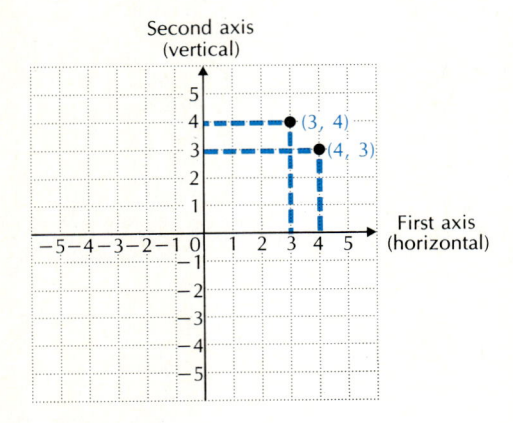

Plotting Points

Note that (4, 3) and (3, 4) give different points. They are called *ordered pairs* of numbers because it makes a difference which number comes first.

EXAMPLE 1 Plot the point $(-3, 4)$.

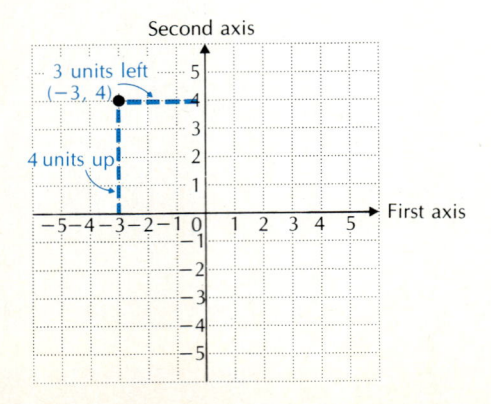

The first number, −3, is negative. We go −3 units in the first direction (3 units to the left). The second number, 4, is positive. We go 4 units in the second direction (up).

The numbers in an ordered pair are called *coordinates*. In (−3, 4), the *first coordinate* is −3 and the *second coordinate* is 4.

Quadrants

This figure shows some points and their coordinates. In region I (the first *quadrant*) both coordinates of any point are positive. In region II (the second *quadrant*) the first coordinate is negative and the second positive, and so on.

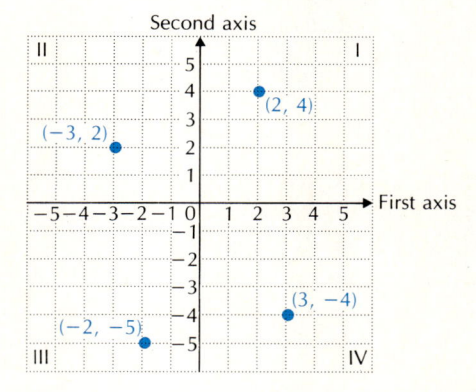

Finding Coordinates

To find coordinates of a point, we see how far to the right or left of zero it is located and how far up or down.

EXAMPLE 2 Find the coordinates of points A, B, C, D, E, F, and G.

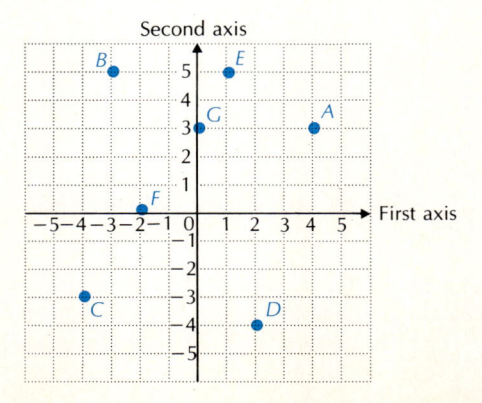

Point A is 4 units to the right (first direction) and 3 units up (second direction). Its coordinates are (4, 3). The coordinates of the other points are as follows:

$$B: (-3, 5); \quad C: (-4, -3); \quad D: (2, -4);$$
$$E: (1, 5); \quad F: (-2, 0); \quad G: (0, 3).$$

EXERCISE SET 6.1

1. Plot these points.

(2, 5)   (−1, 3)   (3, −2)   (−2, −4)
(0, 4)   (0, −5)   (5, 0)   (−5, 0)

In which quadrant is each point located?

3. (−5, 3)

4. (−12, 1)

7. (−6, −29)

8. (−3.6, −105.9)

11. In quadrant III, first coordinates are always _____ and second coordinates are always _____.

13. Find the coordinates of points A, B, C, D, and E.

2. Plot these points.

(4, 4)   (−2, 4)   (5, −3)   (−5, −5)
(0, 4)   (0, −4)   (3, 0)   (−4, 0)

5. (100, −1)

6. (35.6, −2.5)

9. (3.8, 9.2)

10. (1895, 1492)

12. In quadrant II, _____ coordinates are always positive and _____ coordinates are always negative.

14. Find the coordinates of points A, B, C, D, and E.

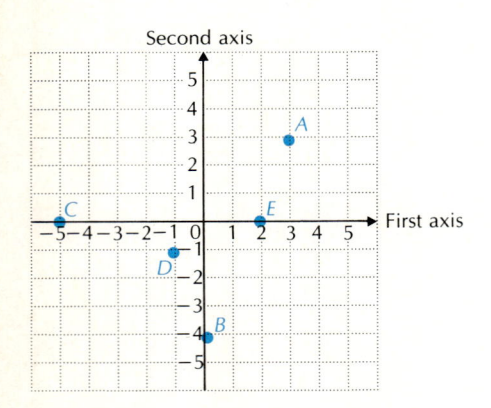

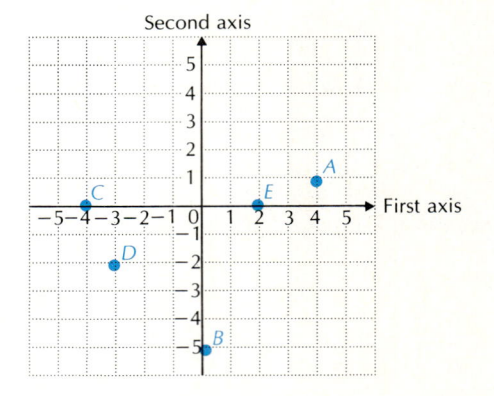

Use graph paper. Draw a first and second axis. Then plot these points.

15. (0, −3),   (−1, −5),   (1, −1),   (2, 1)

16. (0, 1),   (1, 4),   (−1, −2),   (−2, −5)

Problem-solving practice

17. The radius of Jupiter is about 11 times that of the earth. The radius of the earth is about 4030 mi. What is the radius of Jupiter?

18. The length of a rectangle is 4 in. greater than the width. The area of the rectangle is 21 in^2. Find the perimeter of the rectangle.

In Exercises 19–24, tell in which quadrant(s) each of the following points could be located.

19. The first coordinate is positive.

20. The second coordinate is negative.

21. The first and second coordinates are equal.

22. The first coordinate is the additive inverse of the second coordinate.

23. The points $(-1, 1)$, $(4, 1)$, and $(4, -5)$ are three vertices of a rectangle. Find the coordinates of the fourth vertex.

24. Three parallelograms share the vertices $(-2, -3)$, $(-1, 2)$, and $(4, -3)$. Find the fourth vertex of each parallelogram.

25. Graph eight points such that the sum of the coordinates is 6.

26. Graph eight points such that the first coordinate minus the second coordinate is 1.

27. Find the perimeter of a rectangle whose vertices have coordinates $(5, 3)$, $(5, -2)$, $(-3, -2)$, and $(-3, 3)$.

28. Find the area of a triangle whose vertices have coordinates $(0, 9)$, $(0, -4)$, and $(5, -4)$.

6.2 GRAPHING EQUATIONS

We now see how to graph equations in two variables on a plane.

Solutions of Equations

An equation with two variables has *pairs* of numbers for solutions. We usually take the variables in alphabetical order. Then we get *ordered* pairs for solutions.

EXAMPLE 1 Determine whether $(3, 7)$ is a solution of $y = 2x + 1$.

$$\begin{array}{c|c} y = 2x + 1 \\ \hline 7 & 2 \cdot 3 + 1 \\ & 6 + 1 \\ & 7 \end{array}$$

We substitute 3 for x and 7 for y (alphabetical order of variables).

The equation becomes true: $(3, 7)$ is a solution.

EXAMPLE 2 Determine whether $(-2, 3)$ is a solution of $2t = 4s - 8$.

$$\begin{array}{c|c} 2t = 4s - 8 \\ \hline 2 \cdot 3 & 4(-2) - 8 \\ 6 & -8 - 8 \\ & -16 \end{array}$$

We substitute -2 for s and 3 for t.

The equation becomes false: $(-2, 3)$ is not a solution.

Graphing Equations $y = mx$ and $y = mx + b$

The graph of an equation is a drawing of its solutions.

> To *graph* an equation means to make a drawing of its solutions.

If an equation has a graph that is a line, we can graph it by plotting a few points and then drawing a line through them.

EXAMPLE 3 Graph: $y = x$.

We will use alphabetical order. Thus the first axis will be the *x*-axis and the second axis will be the *y*-axis. Next, we find some solutions of the equation. In this case it is easy. Here are a few:

$$(0, 0), \quad (1, 1), \quad (5, 5), \quad (-2, -2), \quad (-4, -4).$$

Now we plot these points. We can see that if we were to plot a million solutions, the dots we draw would resemble a solid line. Once we see the pattern, we can draw the line with a ruler. The line is the graph of the equation $y = x$. We label the line $y = x$ on the graph paper.

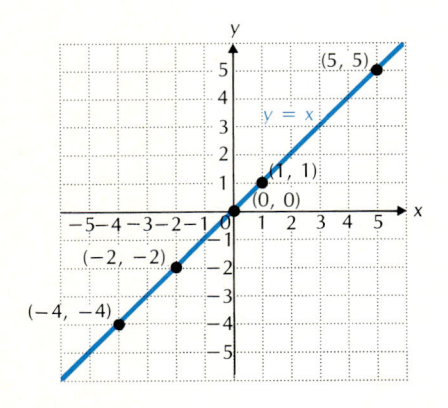

A graph of an equation is a picture of its solutions. Each point of the picture gives an ordered pair (a, b) that is a solution. No other points give solutions.

EXAMPLE 4 Graph: $y = 2x$.

We find some ordered pairs that are solutions, keeping the results in a table. We choose *any* number for *x* and then determine *y* by substitution. Suppose we choose 0 for *x*. Then

$$y = 2x = 2 \cdot 0 = 0.$$

We get a solution: the ordered pair $(0, 0)$. Suppose we choose 3 for x. Then

$$y = 2x = 2 \cdot 3 = 6.$$

We get a solution: the ordered pair $(3, 6)$. We make some negative choices for x as well as some positive ones. If a number takes us off the graph paper, we generally do not use it. Continuing in this manner we get a table like the one shown below. In this case, since $y = 2x$, we get y by doubling x.

Now we plot these points. If we had enough of them, they would make a line. We draw it with a ruler and label it $y = 2x$.

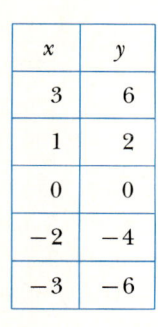

x	y
3	6
1	2
0	0
-2	-4
-3	-6

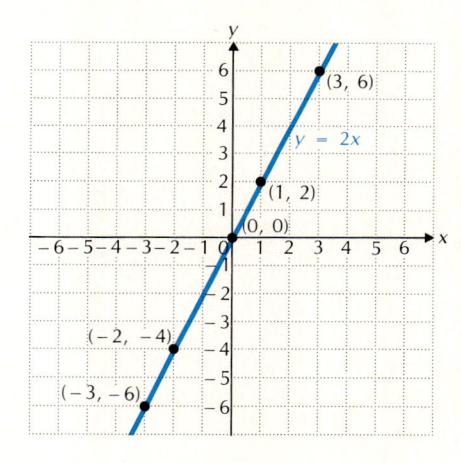

EXAMPLE 5 Graph: $y = -3x$.

We make a table of solutions. Then we plot the points. If we had enough of them, they would make a line. We draw it with a ruler and label it $y = -3x$.

x	y
0	0
1	-3
-1	3
2	-6
-2	6

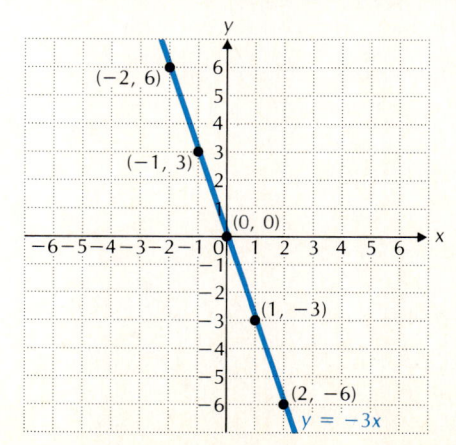

EXAMPLE 6 Graph: $y = -\frac{5}{3}x$.

We make a table of solutions.

When $x = 0$, $\quad y = -\frac{5}{3} \cdot 0 = 0$.
When $x = 3$, $\quad y = -\frac{5}{3} \cdot 3 = -5$.
When $x = -3$, $\quad y = -\frac{5}{3}(-3) = 5$.
When $x = 1$, $\quad y = -\frac{5}{3} \cdot 1 = -\frac{5}{3}$.

Note that if we substitute multiples of 3, we can avoid fractions.

Next we plot the points. If we had enough of them, they would make a line.

x	y
0	0
3	−5
−3	5
1	$-\frac{5}{3}$

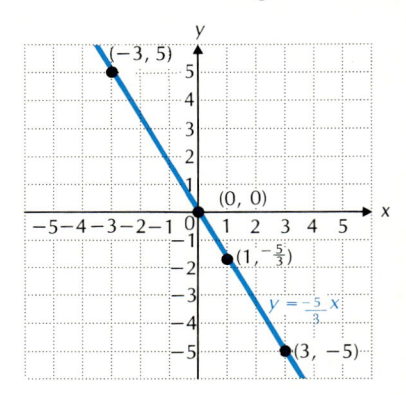

Every equation $y = mx$ has a graph that is a straight line. It contains the origin. The number m, called the *slope*, tells us how the line slants. For a positive slope a line slants up from left to right, as in Examples 3 and 4. For a negative slope a line slants down from left to right, as in Examples 5 and 6.

We know that the graph of any equation $y = mx$ is a straight line through the origin, with slope m. What will happen if we add a number b on the right side to get an equation $y = mx + b$?

EXAMPLE 7 Graph $y = x + 2$ and compare it with $y = x$.

We first make a table of values.

x	y (or x + 2)
0	2
1	3
−1	1
2	4
−2	0
3	5

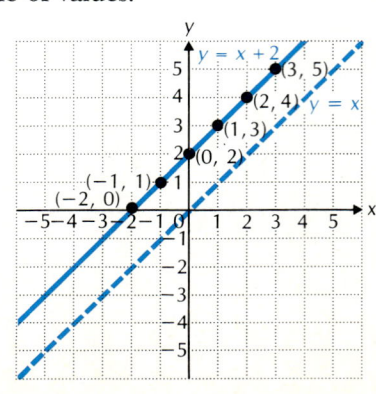

We then plot these points. If we had enough of them, they would make a line. We draw this line with a ruler and label it $y = x + 2$. The graph of $y = x$ is drawn for comparison. Note that the graph of $y = x + 2$ looks just like the graph of $y = x$, but it is moved up 2 units.

EXAMPLE 8 Graph $y = 2x - 3$ and compare it with $y = 2x$.

We first make a table of values.

x	y (or $2x - 3$)
0	-3
1	-1
2	1
-1	-5

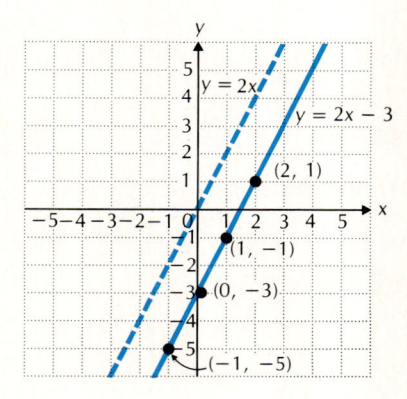

We draw the graph of $y = 2x - 3$. It looks just like the graph of $y = 2x$, but it is moved down 3 units.

The graph of $y = mx$ goes through the origin $(0, 0)$. The graph of any equation $y = mx + b$ is also a line. It is parallel to $y = mx$, but moved up or down. It goes through the point $(0, b)$. That point is called the *y-intercept*. We may also refer to the number b as the *y*-intercept. The number m is still called the *slope*. It tells us how steeply the line slants.

EXAMPLE 9 Graph: $y = \frac{2}{5}x + 4$.

We first make a table of values. Using multiples of 5 avoids fractions.

When $x = 0$, $\quad y = \frac{2}{5} \cdot 0 + 4 = 0 + 4 = 4.$
When $x = 5$, $\quad y = \frac{2}{5} \cdot 5 + 4 = 2 + 4 = 6.$
When $x = -5$, $\quad y = \frac{2}{5}(-5) + 4 = -2 + 4 = 2.$

Since two points determine a line, that is all you really need to graph a line, but you should always plot a third point as a check.

x	y
0	4
5	6
-5	2

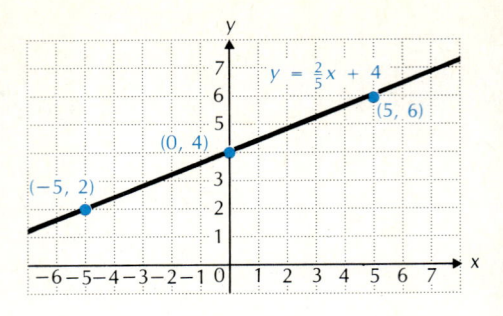

We draw the graph of $y = \frac{2}{5}x + 4$.

EXAMPLE 10 Graph: $y = -\frac{3}{4}x - 2$.

We first make a table of values.

When $x = 0$, $y = -\frac{3}{4} \cdot 0 - 2 = 0 - 2 = -2$.
When $x = 4$, $y = -\frac{3}{4} \cdot 4 - 2 = -3 - 2 = -5$.
When $x = -4$, $y = -\frac{3}{4}(-4) - 2 = 3 - 2 = 1$.

x	y
0	-2
4	-5
-4	1

We plot these points and draw a line through them.

We plot this point for a check to see whether it is on the line.

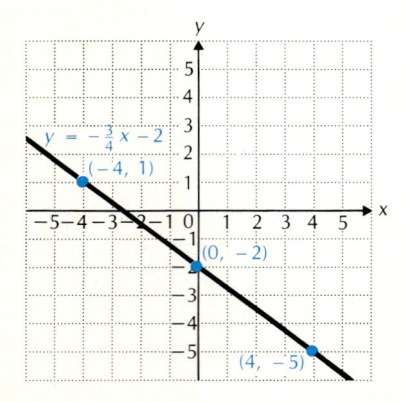

We draw the graph of $y = -\frac{3}{4}x - 2$. Every graph should be labeled.

6.2 GRAPHING EQUATIONS

Any equation $y = mx + b$ has a graph that is a straight line. It goes through the point $(0, b)$ and the y-intercept, and it has slope m.

EXERCISE SET 6.2

Determine whether the given point is a solution of the equation.

1. $(2, 5)$; $y = 3x - 1$
2. $(1, 7)$; $y = 2x + 5$
3. $(2, -3)$; $3x - y = 4$
4. $(-1, 4)$; $2x + y = 6$
5. $(-2, -1)$; $2a + 2b = -7$
6. $(0, -4)$; $4m + 2n = -9$

Graph.

7. $y = 4x$
8. $y = 2x$
9. $y = -2x$
10. $y = -4x$
11. $y = \frac{1}{3}x$
12. $y = \frac{1}{4}x$
13. $y = -\frac{3}{2}x$
14. $y = -\frac{5}{4}x$
15. $y = x + 1$
16. $y = -x + 1$
17. $y = 2x + 2$
18. $y = 3x - 2$
19. $y = \frac{1}{3}x - 1$
20. $y = \frac{1}{2}x + 1$
21. $y = -x - 3$
22. $y = -x - 2$
23. $y = \frac{5}{2}x + 3$
24. $y = \frac{5}{3}x - 2$
25. $y = -\frac{5}{2}x - 2$
26. $y = -\frac{5}{3}x + 3$
27. $y = x$
28. $y = -x$
29. $y = 3 - 2x$
30. $y = 7 - 5x$
31. $y = \frac{4}{3} - \frac{1}{3}x$
32. $y = -\frac{1}{4}x - \frac{1}{2}$

Problem-solving practice

33. A post is placed through some water into the mud at the bottom of the lake. Half of the post is in the mud and $\frac{1}{3}$ is in the water, and the part above water is $5\frac{1}{2}$ ft long. How long is the post?

34. The sum of two consecutive even integers is 130. Find the product of the integers.

○ ─────────────

35. Complete the table for $y = x^2 + 1$. Plot the points on graph paper and draw the graph.

x	0	-1	1	-2	2	-3	3
y							

36. Find all whole-number solutions of $x + y = 6$.
37. Find all whole-number solutions of $x + 3y = 15$.
38. Translate to an equation: n nickels and d dimes total $1.95. Find three solutions.
39. Translate to an equation: n nickels and q quarters total $2.35. Find three solutions.
40. Find three solutions of $y = |x|$.
41. Find three solutions of $y = |x| + 1$.
42. Two machines A and B produce rivets. Machine A produces 68 rivets per hour, while machine B produces 76 rivets per hour. Let x represent the number of hours machine A runs and y represent the number of hours machine B runs. Translate to an equation: The combined production of A and B on a given day is 864. Find a solution to the equation. Explain your solution.

6.3 LINEAR EQUATIONS

We now develop faster procedures for graphing equations whose graphs are straight lines. Such equations are called *linear equations*.

Graphing Using Intercepts

The fastest method for graphing equations whose graphs are straight lines involves the use of intercepts. Look at the graph of $y - 2x = 4$ shown below. We could graph this equation by solving for y to get $y = 2x + 4$ and proceed as before, but we want to develop a faster method.

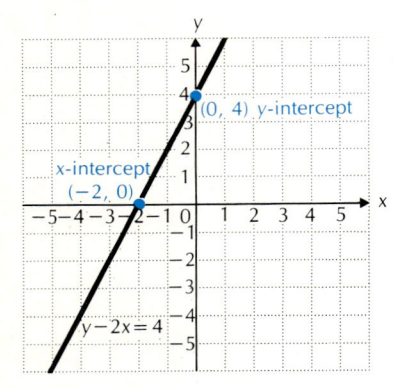

The y-intercept is $(0, 4)$. It occurs where the line crosses the y-axis and always has 0 as the first coordinate. The x-intercept is $(-2, 0)$. It occurs where the line crosses the x-axis and always has 0 as the second coordinate.

> The x-intercept is $(a, 0)$. To find a, let $y = 0$.
> The y-intercept is $(0, b)$. To find b, let $x = 0$.

Now let us draw a graph using intercepts.

EXAMPLE 1   Graph: $4x + 3y = 12$.

To find the x-intercept, let $y = 0$. Then

$$4x + 3 \cdot 0 = 12$$
$$4x = 12$$
$$x = 3.$$

Thus (3, 0) is the *x*-intercept. Note that this amounts to covering up the *y*-term and looking at the rest of the equation.

To find the *y*-intercept, let $x = 0$. Then

$$4 \cdot 0 + 3y = 12$$
$$3y = 12$$
$$y = 4.$$

Thus (0, 4) is the *y*-intercept.

We plot these points and draw the line. A third point should be used as a check. We substitute any arbitrary value for *x* and solve for *y*.

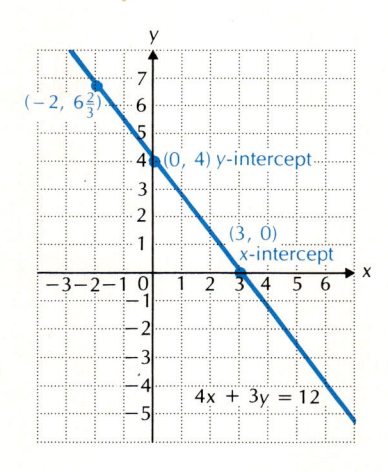

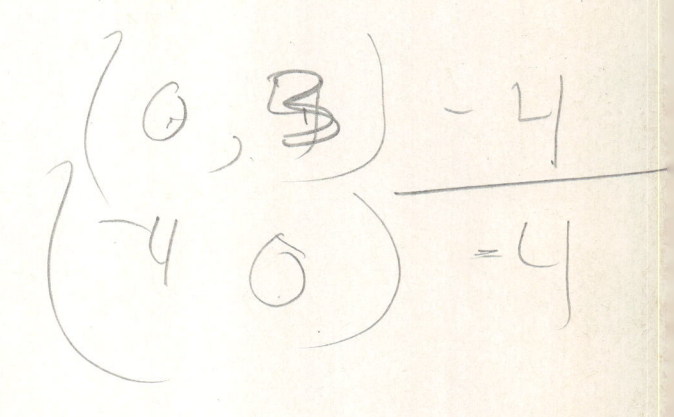

If we let $x = -2$, then

$$4(-2) + 3y = 12 \qquad \text{Substituting } -2 \text{ for } x$$
$$-8 + 3y = 12$$
$$3y = 12 + 8 = 20$$
$$y = \tfrac{20}{3}, \text{ or } 6\tfrac{2}{3}. \qquad \text{Solving for } y$$

We see that the point $(-2, 6\tfrac{2}{3})$ is on the graph, so the graph is probably correct.

Equations with a Missing Variable

Consider the equation $y = 3$. We can think of it as $y = 0 \cdot x + 3$. No matter what number we choose for *x*, we find that *y* is 3.

EXAMPLE 2 Graph: $y = 3$.

Any ordered pair $(x, 3)$ is a solution. So the line is parallel to the x-axis with y-intercept $(0, 3)$.

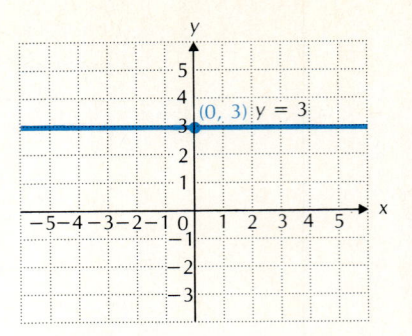

EXAMPLE 3 Graph: $x = -4$.

Any ordered pair $(-4, y)$ is a solution. So the line is parallel to the y-axis with x-intercept $(-4, 0)$.

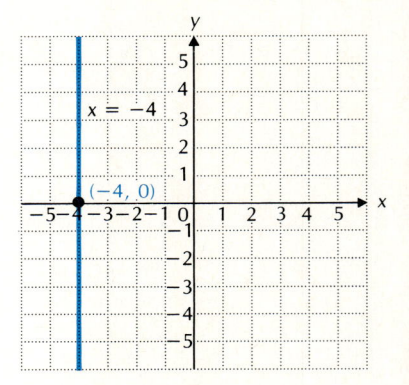

• •

The graph of $y = b$ is a horizontal line. The graph of $x = a$ is a vertical line.

Equations whose graphs are straight lines are linear equations. We summarize the best procedure for graphing linear equations.

TO GRAPH LINEAR EQUATIONS

1. Is the equation of the type $x = a$ or $y = b$? If so, the graph will be a line parallel to an axis.
2. If the line is not of the type $x = a$ or $y = b$, find the intercepts. Graph using the intercepts if this is feasible.
3. If the intercepts are too close together, choose another point farther from the origin.
4. In any case, use a third point as a check.

If you have trouble remembering whether a graph such as $y = 3$ or $x = -4$ is horizontal or vertical, the following may help.

Consider $y = 3$. Make up a table with all 3's in the y-column.

x	y
	3
	3
	3

Choose any numbers for x.

(y must be 3)

x	y
-2	3
0	3
4	3

Now when you plot the ordered pairs $(-2, 3)$, $(0, 3)$, and $(4, 3)$ and connect the points, you will obtain a horizontal line. Similarly, consider $x = -4$. Make up a table with all -4's in the x-column.

x	y
-4	
-4	
-4	

Choose any numbers for y.

x	y
-4	-5
-4	1
-4	3

Now when you plot the ordered pairs $(-4, -5)$, $(-4, 1)$, and $(-4, 3)$ and connect them, you will obtain a vertical line.

EXERCISE SET 6.3

Find the intercepts. Then graph.

1. $x + 3y = 6$
2. $x + 2y = 8$
3. $-x + 2y = 4$
4. $-x + 3y = 9$
5. $3x + y = 9$
6. $2x + y = 6$
7. $2y - 2 = 6x$
8. $3y - 6 = 9x$
9. $3x - 9 = 3y$
10. $5x - 10 = 5y$
11. $2x - 3y = 6$
12. $2x - 5y = 10$
13. $4x + 5y = 20$
14. $2x + 6y = 12$
15. $2x + 3y = 8$
16. $x - 1 = y$
17. $x - 3 = y$
18. $2x - 1 = y$
19. $3x - 2 = y$
20. $4x - 3y = 12$
21. $6x - 2y = 18$
22. $7x + 2y = 6$
23. $3x + 4y = 5$
24. $y = -4 - 4x$
25. $y = -3 - 3x$
26. $-3x = 6y - 2$
27. $-4x = 8y - 5$

Graph.

28. $x = -2$
29. $x = -1$
30. $y = 2$
31. $y = 4$
32. $x = 2$
33. $x = 3$
34. $y = 0$
35. $y = -1$
36. $x = \frac{3}{2}$
37. $x = -\frac{5}{2}$

Problem-solving practice

38. A salesperson gets a weekly salary of $235 plus a $2 commission for each tire that is sold. How much did the salesperson make in a four-week period in which the salesperson sold 84 tires?

39. The base of a triangle is 5 m greater than the height. The area is 7 m². Find the base and the height.

40. Write an equation for the *y*-axis.

41. Write an equation for the *x*-axis.

42. Find the coordinates of the point of intersection of the graphs of the equations $x = -3$ and $y = 6$.

43. Write an equation of a line parallel to the *x*-axis and 5 units below it.

44. Write an equation of a line parallel to the *y*-axis and 13 units to the right of it.

45. Write an equation of a line parallel to the *x*-axis and intersecting the *y*-axis at (0, 2.8).

46. Find the value of m in $y = mx + 3$ so that the *x*-intercept of its graph will be (2, 0).

47. Find the value of b in $2y = -5x + 3b$ so that the *y*-intercept of its graph will be (0, −12).

6.4 TRANSLATING PROBLEMS TO EQUATIONS

As you have probably already noted, in the five-step problem-solving process the most difficult and time-consuming part is translating to mathematical language. In this chapter we consider situations in which the translation can be done by writing more than one equation. That is usually much easier than translating to a single equation. In this section we just practice translating a problem to two equations.

EXAMPLE 1 Translate the following problem situation to mathematical language, using two equations.

The sum of two numbers is 15. One number is four times the other. Find the numbers.

2. **Translate** There are two statements in this problem. We translate the first one.

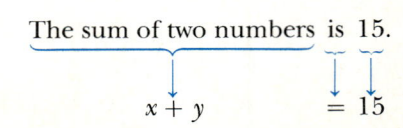

The sum of two numbers is 15.

$x + y \qquad = 15$

6.4 TRANSLATING PROBLEMS TO EQUATIONS

We have used x and y for the numbers. Now we translate the second statement, remembering to use x and y.

$$\underbrace{\text{One number}}_{y} \underbrace{\text{is}}_{=} \underbrace{\text{four}}_{4} \underbrace{\text{times}}_{\cdot} \underbrace{\text{the other.}}_{x}$$

For the second statement we could have also translated to $x = 4y$. That would also have been correct. The problem has been translated to a pair or *system of equations*. We list what the variables represent and then list the equations:

$$x + y = 15,$$
$$y = 4x.$$

EXAMPLE 2 Translate the following problem situation to mathematical language, using two equations.

Badger Rent-A-Car rents compact cars at a daily rate of $43.95 plus 40¢ per mile. Thirsty Rent-A-Car rents compact cars at a daily rate of $42.95 plus 42¢ per mile. For what mileage is the cost the same?

2. **Translate** We translate the first statement, using $0.40 for 40¢.

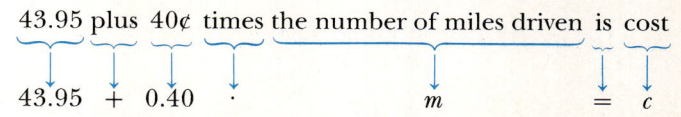

$$\underbrace{43.95}_{43.95} \underbrace{\text{plus}}_{+} \underbrace{40¢}_{0.40} \underbrace{\text{times}}_{\cdot} \underbrace{\text{the number of miles driven}}_{m} \underbrace{\text{is}}_{=} \underbrace{\text{cost}}_{c}$$

We have let m represent the mileage and c the cost. We translate the second statement, but again it helps to reword it first.

$$\underbrace{42.95}_{42.95} \underbrace{\text{plus}}_{+} \underbrace{42¢}_{0.42} \underbrace{\text{times}}_{\cdot} \underbrace{\text{the number of miles driven}}_{m} \underbrace{\text{is cost.}}_{= c}$$

We have now translated to a system of equations:

$$43.95 + 0.40m = c,$$
$$42.95 + 0.42m = c.$$

The familiarization step often aids translating. In particular, making a drawing is often helpful.

EXAMPLE 3 Translate the following problem situation to mathematical language, using two equations.

The perimeter of a rectangle is 90 cm. The length is 20 cm greater than the width. Find the length and the width.

1. **Familiarize** We make a drawing and label it. We have called the length l and the width w.

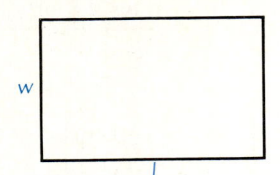

2. **Translate** From the drawing we see that the perimeter (the distance around) of the rectangle is $l + l + w + w$, or $2l + 2w$. We translate the first statement.

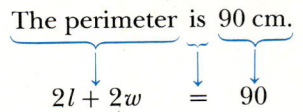

$$\underbrace{\text{The perimeter}}_{2l + 2w} \underbrace{\text{is}}_{=} \underbrace{\text{90 cm.}}_{90}$$

We translate the second statement.

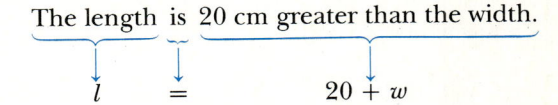

$$\underbrace{\text{The length}}_{l} \underbrace{\text{is}}_{=} \underbrace{\text{20 cm greater than the width.}}_{20 + w}$$

We have translated to a system of equations:

$$2w + 2l = 90,$$
$$l = 20 + w.$$

EXERCISE SET 6.4

Translate to a system of equations. Do not attempt to solve. Save for later use.

1. The sum of two numbers is 58. The difference is 16. Find the numbers.

2. The sum of two numbers is 26.4. One number is five times the other. Find the numbers.

3. The perimeter of a rectangle is 400 m. The width is 40 m less than the length. Find the length and width.

4. The perimeter of a rectangle is 76 cm. The width is 17 cm less than the length. Find the length and width.

5. Acme Rent-A-Car rents an intermediate-size car at a daily rate of $53.95 plus 30¢ per mile. Hartz Rent-A-Car rents an intermediate-size car at a daily rate of $54.95 plus 20¢ per mile. For what mileage is the cost the same?

6. Badger rents a basic car at a daily rate of $45.95 plus 40¢ per mile. Hartz rents a basic car at a daily rate of $46.95 plus 20¢ per mile. For what mileage is the cost the same?

7. The difference between two numbers is 16. Three times the larger number is seven times the smaller. What are the numbers?

8. The difference between two numbers is 18. Twice the smaller number plus three times the larger is 74. What are the numbers?

9. Two angles are supplementary. One is 8° more than three times the other. Find the angles. (Supplementary angles are angles whose sum is 180°.)

10. Two angles are supplementary. One is 30° more than two times the other. Find the angles.

11. Two angles are complementary. Their difference is 34°. Find the angles. (Complementary angles are angles whose sum is 90°.)

12. Two angles are complementary. One angle is 42° more than $\frac{1}{2}$ the other. Find the angles.

13. In a vineyard a vintner uses 820 hectares to plant Chardonnay and Riesling grapes. The vintner knows that profits will be greatest by planting 140 hectares more of Chardonnay than Riesling. How many hectares of each grape should be planted? (A *hectare* is a unit of area, about 2.47 arces.)

14. The Hayburner Horse Farm allots 650 hectares to plant hay and oats. The owners know that their needs are best met if they plant 180 hectares more of hay than oats. How many hectares of each should they plant?

15. The difference between two numbers is 18. Twice the smaller number plus three times the larger is 74. What are the numbers?

16. The perimeter of a rectangle is 400 m. The length is 40 m more than the width. Find the length and width.

17. The perimeter of a rectangle is 76 cm. The length is 17 cm more than the width. Find the length and width.

18. The perimeter of a football field (excluding the end zones) is $306\frac{2}{3}$ yd. The length is $46\frac{2}{3}$ yd longer than the width. Find the length and width.

Translate to a system of equations. Do not attempt to solve. Save for later use.

19. Patrick's age is 20% of his father's age. Twenty years from now, Patrick's age will be 52% of his father's age. How old are Patrick and his father?

20. If 5 is added to a man's age and the total is divided by 5, the result will be his daughter's age. Five years ago the father's age was eight times his daughter's age. Find their present ages.

21. When the base of a triangle is increased by 2 ft and the height is decreased by 1 ft, the height becomes $\frac{1}{3}$ of the base, and the area becomes 24 ft². Find the original dimensions of the triangle.

6.5 SYSTEMS OF EQUATIONS

When a problem has been translated, as in Section 6.4, we have a system of equations. We now learn how to solve such systems.

Identifying Solutions

Consider the system of equations

$$x + y = 8,$$
$$2x - y = 1.$$

A *solution* of a system of two equations is an ordered pair that makes both equations true. Consider the system listed above. Look at the graphs. Recall that a graph of an equation is a picture of its solution set. Each point on a graph corresponds to a solution. Which points (ordered pairs) are solutions of *both* equations? The graph shows that there is only one. It is the point P where the graphs cross. This point looks as if its coordinates are (3, 5). We check:

$$\begin{array}{c|c} x + y = 8 \\ \hline 3 + 5 & 8 \\ 8 & \end{array} \qquad \begin{array}{c|c} 2x - y = 1 \\ \hline 2 \cdot 3 - 5 & 1 \\ 6 - 5 & \\ 1 & \end{array}$$

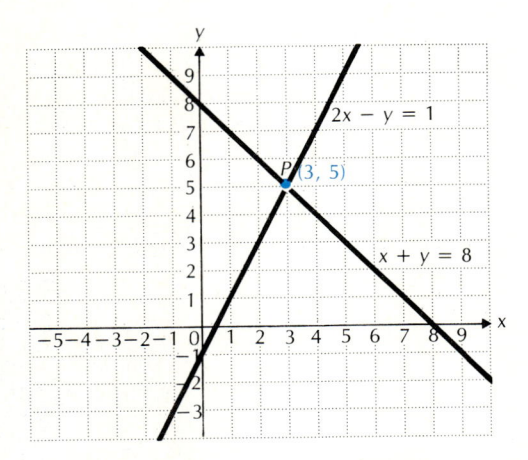

There is just one solution of the system of equations. It is (3, 5). In other words, $x = 3$ and $y = 5$.

EXAMPLE 1 Determine whether (1, 2) is a solution of the system

$$y = x + 1,$$
$$2x + y = 4.$$

We check:

$$\begin{array}{c|c} y = x + 1 \\ \hline 2 & 1 + 1 \\ 2 & 2 \end{array} \qquad \begin{array}{c|c} 2x + y = 4 \\ \hline 2 \cdot 1 + 2 & 4 \\ 2 + 2 & \\ 4 & \end{array}$$

Thus (1, 2) is a solution of the system.

EXAMPLE 2 Determine whether $(-3, 2)$ is a solution of the system

$$a + b = -1,$$
$$b + 3a = 4.$$

6.5 SYSTEMS OF EQUATIONS

We check:

$$\begin{array}{c|c} a+b=-1 \\ \hline -3+2 & -1 \\ -1 & \end{array} \qquad \begin{array}{c|c} b+3a=4 \\ \hline 2+3(-3) & 4 \\ 2-9 & \\ -7 & \end{array}$$

The point $(-3, 2)$ is not a solution of $b + 3a = 4$. Thus it is not a solution of the system.

Solving Systems by Graphing

> To solve a system of equations by graphing, we graph both equations and find coordinates of the point(s) of intersection. Then we check. If the lines are parallel, there is no solution.

EXAMPLE 3 Solve by graphing:

$$x + 2y = 7,$$
$$x = y + 4.$$

We graph the equations. Point P looks as if it has coordinates $(5, 1)$.

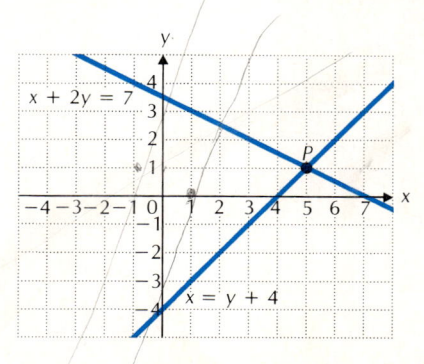

Check:

$$\begin{array}{c|c} x+2y=7 \\ \hline 5+2\cdot 1 & 7 \\ 5+2 & \\ 7 & \end{array} \qquad \begin{array}{c|c} x=y+4 \\ \hline 5 & 1+4 \\ 5 & 5 \end{array}$$

The solution is $(5, 1)$.

EXAMPLE 4 Solve by graphing:

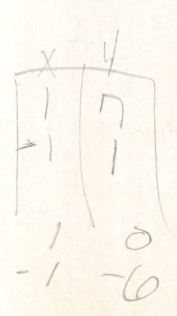

$$y = 3x + 4,$$
$$y = 3x - 3.$$

The graphs are parallel. There is no point where they cross, so the system has no solution.

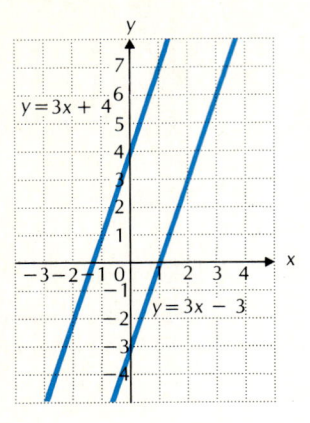

When we graph a system of two equations, one of the following three things can happen.

> 1. The lines have one point of intersection, as in Example 3. The point of intersection is the only solution of the system.
> 2. The lines are parallel, as in Example 4. If so, there is no point that satisfies both equations. The system has no solution.
> 3. The lines coincide. Thus the equations have the same graph, and every solution of one equation is a solution of the other. There is an unlimited number of solutions.

EXERCISE SET 6.5

Determine whether the given ordered pair is a solution of the system of equations. Remember to use alphabetical order of variables.

1. $(3, 2)$; $2x + 3y = 12,$
    $x - 4y = -5$
2. $(1, 5)$; $5x - 2y = -5,$
    $3x - 7y = -32$
3. $(3, 2)$; $3t - 2s = 0,$
    $t + 2s = 15$
4. $(2, -2)$; $b + 2a = 2,$
    $b - a = -4$
5. $(15, 20)$; $3x - 2y = 5,$
    $6x - 5y = -10$
6. $(-1, -3)$; $3r + s = -6,$
    $2r = 1 + s$
7. $(-1, 1)$; $x = -1,$
    $x - y = -2$
8. $(-3, 4)$; $2x = -y - 2,$
    $y = -4$
9. $(12, 3)$; $y = \frac{1}{4}x,$
    $3x - y = 33$
10. $(-3, 1)$; $y = -\frac{1}{3}x,$
    $3y = -5x - 12$

Use graph paper. Solve each system graphically and check.

11. $x + y = 3$,
    $x - y = 1$
12. $x - y = 2$,
    $x + y = 6$
13. $x + 2y = 10$,
    $3x + 4y = 8$
14. $x - 2y = 6$,
    $2x - 3y = 5$
15. $8x - y = 29$,
    $2x + y = 11$
16. $4x - y = 10$,
    $3x + 5y = 19$
17. $u = v$,
    $4u = 2v - 6$
18. $x = 3y$,
    $3y - 6 = 2x$
19. $x = -y$,
    $x + y = 4$
20. $-3x = 5 - y$,
    $2y = 6x + 10$
21. $a = \frac{1}{2}b + 1$,
    $a - 2b = -2$
22. $x = \frac{1}{3}y + 2$,
    $-2x - y = 1$
23. $y = 3$,
    $x = 5$
24. $y = 3x$,
    $y = -3x + 2$
25. $x + y = 9$,
    $3x + 3y = 27$
26. $x + y = 4$,
    $x + y = -4$

○

27. The solution of the following system is $(2, -3)$. Find A and B.

$$Ax - 3y = 13,$$
$$x - By = 8$$

Determine whether the given ordered pair is a solution of each system of three equations.

28. $(2, -3)$; $x + 3y = -7$,
    $-x + y = -5$,
    $2x - y = 1$
29. $(-1, -5)$; $4a - b = 1$,
    $-a + b = -4$,
    $2a + 3b = -17$

30. Describe in words the graph of the system in Exercise 29. Describe the solution.
31. Solve this system by graphing. What happens when you check your possible solution?

$$3x + 7y = 5,$$
$$6x - 7y = 1$$

6.6 SOLVING BY SUBSTITUTION: PROBLEM SOLVING

Graphing helps picture the solution of a system of equations, but solving by graphing is not fast or accurate. Let us learn better ways using algebra.

The Substitution Method

One method for solving systems is known as the *substitution method*.

EXAMPLE 1 Solve the system:

$$x + y = 6,$$
$$x = y + 2.$$

The second equation says that x and $y + 2$ name the same thing. Thus in the first equation, we can substitute $y + 2$ for x:

$$x + y = 6$$
$$(y + 2) + y = 6. \quad \text{Substituting } y + 2 \text{ for } x$$

This last equation has only one variable. We solve it:

$$(y + 2) + y = 6$$
$$2y + 2 = 6 \quad \text{Collecting like terms}$$
$$2y = 4$$
$$y = 2.$$

We return to the original pair of equations. We substitute 2 for y in either of them. We use the first equation:

$$x + y = 6$$
$$x + 2 = 6 \quad \text{Substituting 2 for } y$$
$$x = 4.$$

The ordered pair (4, 2) may be a solution. We check.

Check:
$$\begin{array}{c|c} x + y = 6 \\ \hline 4 + 2 & 6 \\ 6 & \end{array} \qquad \begin{array}{c|c} x = y + 2 \\ \hline 4 & 2 + 2 \\ & 4 \end{array}$$

Since (4, 2) checks, we have the solution. We could also express the answer as $x = 4$, $y = 2$.

EXAMPLE 2 Solve:

$$s = 13 - 3t,$$
$$s + t = 5.$$

We substitute $13 - 3t$ for s in the second equation:

$$s + t = 5$$
$$(13 - 3t) + t = 5. \quad \text{Substituting } 13 - 3t \text{ for } s$$

Now we solve for t:

$$13 - 2t = 5 \quad \text{Collecting like terms}$$
$$-2t = -8 \quad \text{Adding } -13$$
$$t = \frac{-8}{-2}, \text{ or } 4 \quad \text{Multiplying by } \frac{1}{-2}$$

Next we substitute 4 for t in the second equation of the original system:
$$s + t = 5$$
$$s + 4 = 5 \quad \text{Substituting 4 for } t$$
$$s = 1.$$

We check the ordered pair $(1, 4)$.

Check:
$$\begin{array}{c|c} s = 13 - 3t \\ \hline 1 & 13 - 3 \cdot 4 \\ & 13 - 12 \\ & 1 \end{array} \qquad \begin{array}{c|c} s + t = 5 \\ \hline 1 + 4 & 5 \\ 5 & \end{array}$$

Since $(1, 4)$ checks, it is the solution.

Solving for the Variable First

Sometimes neither equation of a pair has a variable alone on one side. Then we solve one equation for one of the variables and proceed as before.

EXAMPLE 3 Solve:
$$x - 2y = 6,$$
$$3x + 2y = 4.$$

We solve one equation for one variable. Since the coefficient of x is 1 in the first equation, it is easier to solve it for x:
$$x - 2y = 6$$
$$x = 6 + 2y.$$

We substitute $6 + 2y$ for x in the second equation of the original pair and solve:
$$3x + 2y = 4$$
$$3(6 + 2y) + 2y = 4 \quad \text{Substituting } 6 + 2y \text{ for } x$$
$$18 + 6y + 2y = 4$$
$$18 + 8y = 4$$
$$8y = -14$$
$$y = \frac{-14}{8}, \text{ or } -\frac{7}{4}.$$

We go back to either of the original equations and substitute $-\frac{7}{4}$ for y. It will be easier to solve for x in the first equation:
$$x - 2y = 6$$
$$x - 2(-\tfrac{7}{4}) = 6$$
$$x + \tfrac{7}{2} = 6$$
$$x = 6 - \tfrac{7}{2}$$
$$x = \tfrac{5}{2}$$

Check:

$$\begin{array}{c|c} x - 2y = 6 \\ \hline \frac{5}{2} - 2(-\frac{7}{4}) \mid 6 \\ \frac{5}{2} + \frac{7}{2} \\ \frac{12}{2} \\ 6 \end{array} \qquad \begin{array}{c|c} 3x + 2y = 4 \\ \hline 3 \cdot \frac{5}{2} + 2(-\frac{7}{4}) \mid 4 \\ \frac{15}{2} - \frac{7}{2} \\ \frac{8}{2} \\ 4 \end{array}$$

Since $(\frac{5}{2}, -\frac{7}{4})$ checks, it is the solution.

Problem Solving

Now let us use the substitution method in problem solving.

EXAMPLE 4 The perimeter of a rectangle is 90 cm. The length is 20 cm greater than the width. Find the length and the width.

The **Familiarize** and **Translate** steps have been done in Example 3 of Section 6.4. The resulting system of equations is

$$2l + 2w = 90,$$
$$l = 20 + w;$$

where l represents the length and w the width.

3. **Carry out** We solve the system. We substitute $20 + w$ for l in the first equation and solve:

$$2(20 + w) + 2w = 90 \qquad \text{Substituting } 20 + w \text{ for } l$$
$$40 + 2w + 2w = 90$$
$$40 + 4w = 90$$
$$4w = 50$$
$$w = \tfrac{50}{4}, \text{ or } 12.5.$$

We go back to the original equations and substitute 12.5 for w in the second equation:

$$l = 20 + w$$
$$l = 20 + 12.5$$
$$l = 32.5.$$

4. **Check** A possible solution is a length of 32.5 cm and a width of 12.5 cm. The perimeter would be $2(32.5) + 2(12.5)$, or $65 + 25$, or 90. Also, the length is 20 cm greater than the width. These check.

5. **State** The length is 32.5 cm and the width is 12.5 cm.

EXERCISE SET 6.6

Solve by the substitution method.

1. $x + y = 4,$
   $y = 2x + 1$
2. $x + y = 10,$
   $y = x + 8$
3. $y = x + 1,$
   $2x + y = 4$
4. $y = x - 6,$
   $x + y = -2$
5. $y = 2x - 5,$
   $3y - x = 5$
6. $y = 2x + 1,$
   $x + y = -2$
7. $x = -2y,$
   $x + 4y = 2$
8. $r = -3s,$
   $r + 4s = 10$

Solve by the substitution method. Get one variable alone first.

9. $s + t = -4,$
   $s - t = 2$
10. $x - y = 6,$
    $x + y = -2$
11. $y - 2x = -6,$
    $2y - x = 5$
12. $x - y = 5,$
    $x + 2y = 7$
13. $2x + 3y = -2,$
    $2x - y = 9$
14. $x + 2y = 10,$
    $3x + 4y = 8$
15. $x - y = -3,$
    $2x + 3y = -6$
16. $3b + 2a = 2,$
    $-2b + a = 8$
17. $r - 2s = 0,$
    $4r - 3s = 15$
18. $y - 2x = 0,$
    $3x + 7y = 17$
19. $x - 3y = 7,$
    $-4x + 12y = 28$
20. $8x + 2y = 6,$
    $4x = 3 - y$

Problem solving

21. The sum of two numbers is 27. One number is 3 more than the other. Find the numbers.

22. The sum of two numbers is 36. One number is 2 more than the other. Find the numbers.

23. Find two numbers whose sum is 58 and whose difference is 16.

24. Find two numbers whose sum is 66 and whose difference is 8.

25. The difference between two numbers is 16. Three times the larger number is seven times the smaller. What are the numbers?

26. The difference between two numbers is 18. Twice the smaller number plus three times the larger is 74. What are the numbers?

27. The perimeter of a rectangle is 400 m. The length is 40 m more than the width. Find the length and width.

28. The perimeter of a rectangle is 76 cm. The length is 17 cm more than the width. Find the length and width.

29. The perimeter of a football field (excluding the end zones) is $306\frac{2}{3}$ yards. The length is $46\frac{2}{3}$ yards longer than the width. Find the length and width.

○ ─────────────────────

Solve by the substitution method.

30. $y - 2.35x = -5.97$
    $2.14y - x = 4.88$
31. $\frac{1}{4}(a - b) = 2$
    $\frac{1}{6}(a + b) = 1$
32. $\frac{x}{2} + \frac{3y}{2} = 2$
    $\frac{x}{5} - \frac{y}{2} = 3$
33. $0.4x + 0.7y = 0.1$
    $0.5x - 0.1y = 1.1$

34. Determine whether $(2, -3)$ is a solution of this system of three equations:

$$x + 3y = 7,$$
$$-x + y = -5,$$
$$2x - y = 1.$$

35. A rectangle has a perimeter of P feet. The width is 5 feet less than the length. Find the length in terms of P.

36. A rectangle has a perimeter of P meters. The length is 8 meters longer than the width. Find the width in terms of P.

Exercises 37 and 38 contain systems of three equations in three variables. A solution is an ordered triple, listed in alphabetical order. Use the substitution method to solve.

37. $x + y + z = 4,$
$x - 2y - z = 1,$
$y = -1$

38. $x + y + z = 180,$
$x = z - 70,$
$2y - z = 0$

39. Consider this system of equations:
$$3y + 3x = 14,$$
$$y = -x + 4.$$
Try to solve by the substitution method. Can you explain your results?

40. Consider this system of equations:
$$y = x + 5,$$
$$-3x + 3y = 15.$$
Try to solve by the substitution method. Can you explain your results?

41. Why is there no solution to the following system? (*Hint:* Use substitution more than once.)
$$x + y = 10,$$
$$y + z = 10,$$
$$x + z = 10,$$
$$x + y + z = 10$$

6.7 THE ADDITION METHOD: PROBLEM SOLVING

The *addition method* for solving systems of equations makes use of the *addition principle*. Some systems are much easier to solve using the addition method.

Solving by the Addition Method

Another method of solving systems of equations is called the *addition method*.

EXAMPLE 1 Solve:
$$x + y = 5,$$
$$x - y = 1.$$

We will use the addition principle for equations. According to the second equation, $x - y$ and 1 are the same thing. Thus we can add $x - y$ to the left side of the first equation and 1 to the right side:

$$x + y = 5$$
$$\underline{x - y = 1}$$
$$2x + 0y = 6.$$

6.7 THE ADDITION METHOD: PROBLEM SOLVING

We have made one variable "disappear." We have an equation with just one variable:
$$2x = 6.$$
We solve for x: $x = 3$. Next we substitute 3 for x in either of the original equations:

$x + y = 5$
$3 + y = 5$ Substituting 3 for x in the first equation
$y = 2.$ Solving for y

Check:

$x + y = 5$	$x - y = 1$
$3 + 2$ | 5	$3 - 2$ | 1
5	1

Since (3, 2) checks, it is the solution.

Using the Multiplication Principle First

The addition method allows us to eliminate a variable. We may need to multiply by -1 to make this happen.

EXAMPLE 2   Solve:
$$2x + 3y = 8,$$
$$x + 3y = 7.$$

If we add, we do not eliminate a variable. However, if the $3y$ were $-3y$ in one equation we could. We multiply on both sides of the second equation by -1 and then add:

$2x + 3y = 8$
$\underline{-x - 3y = -7}$ Multiplying by -1
$x = 1.$ Adding

Now we substitute 1 for x in one of the original equations:

$x + 3y = 7$
$1 + 3y = 7$ Substituting 1 for x in the second equation
$3y = 6$
$y = 2$ Solving for y

Check:

$2x + 3y = 8$	$x + 3y = 7$
$2 \cdot 1 + 3 \cdot 2$ | 8	$1 + 3 \cdot 2$ | 7
$2 + 6$	$1 + 6$
8	7

Since (1, 2) checks, it is the solution.

In Example 2 we used the multiplication principle, multiplying by -1. We often need to multiply by something other than -1.

EXAMPLE 3 Solve:
$$3x + 6y = -6,$$
$$5x - 2y = 14.$$

This time we multiply by 3 on both sides of the second equation. Then we add:

$$\begin{array}{rl} 3x + 6y = & -6 \\ 15x - 6y = & 42 \quad \text{Multiplying by 3} \\ \hline 18x = & 36 \quad \text{Adding} \\ x = & 2 \quad \text{Solving for } x \end{array}$$

We go back to the first equation and substitute 2 for x:

$$\begin{array}{rl} 3 \cdot 2 + 6y = -6 & \text{Substituting} \\ 6 + 6y = -6 & \\ 6y = -12 & \\ y = -2 & \text{Solving for } y \end{array}$$

Check:
$$\begin{array}{c|c} 3x + 6y = -6 & 5x - 2y = 14 \\ \hline 3 \cdot 2 + 6 \cdot (-2) \;\big|\; -6 & 5 \cdot 2 - 2 \cdot (-2) \;\big|\; 14 \\ 6 + (-12) & 10 - (-4) \\ -6 & 14 \end{array}$$

Since $(2, -2)$ checks, it is the solution.

Solving a *system* of equations in two variables requires finding an ordered *pair* of numbers. Once you have solved for one variable, don't forget the other.

EXAMPLE 4 Solve:
$$3x + 5y = 6,$$
$$5x + 3y = 4.$$

We use the multiplication principle with both equations:
$$3x + 5y = 6,$$
$$5x + 3y = 4$$

Thus we have

$$\begin{array}{rl} 15x + 25y = 30 & \text{Multiplying on both sides of the first equation by 5} \\ -15x - 9y = -12 & \text{Multiplying on both sides of the second equation by } -3 \\ \hline 16y = 18 & \text{Adding} \\ y = \tfrac{18}{16}, \text{ or } \tfrac{9}{8}. \end{array}$$

6.7 THE ADDITION METHOD: PROBLEM SOLVING

We substitute $\frac{9}{8}$ for y in one of the original equations:

$$3x + 5y = 6$$
$$3x + 5 \cdot \tfrac{9}{8} = 6 \qquad \text{Substituting } \tfrac{9}{8} \text{ for } y \text{ in the first equation}$$
$$3x + \tfrac{45}{8} = 6$$
$$3x = 6 - \tfrac{45}{8}$$
$$3x = \tfrac{48}{8} - \tfrac{45}{8}$$
$$3x = \tfrac{3}{8}$$
$$x = \tfrac{3}{8} \cdot \tfrac{1}{3}, \quad \text{or } \tfrac{1}{8} \qquad \text{Solving for } x$$

Check:

$3x + 5y = 6$		$5x + 3y = 4$	
$3 \cdot \tfrac{1}{8} + 5 \cdot \tfrac{9}{8}$	6	$5 \cdot \tfrac{1}{8} + 3 \cdot \tfrac{9}{8}$	4
$\tfrac{3}{8} + \tfrac{45}{8}$		$\tfrac{5}{8} + \tfrac{27}{8}$	
$\tfrac{48}{8}$		$\tfrac{32}{8}$	
6		4	

The solution is $(\tfrac{1}{8}, \tfrac{9}{8})$.

Some systems have no solution.

EXAMPLE 5 Solve:
$$y - 3x = 2,$$
$$y - 3x = 1.$$

We multiply by -1 on both sides of the second equation:

$$y - 3x = 2$$
$$\underline{-y + 3x = -1} \qquad \text{Multiplying by} -1$$
$$0 = -1. \qquad \text{Adding}$$

We obtain a *false* equation, $0 = 1$, so there is no solution. The graphs of the equations are parallel lines. They do not intersect.

When decimals or fractions appear, multiply to clear them. Then proceed as before.

EXAMPLE 6 Solve:
$$\tfrac{1}{4}x + \tfrac{5}{12}y = \tfrac{1}{2},$$
$$\tfrac{5}{12}x + \tfrac{1}{4}y = \tfrac{1}{3}.$$

The number 12 is a multiple of all of the denominators. We multiply on both sides of each equation by 12:

$$12(\tfrac{1}{4}x + \tfrac{5}{12}y) = 12 \cdot \tfrac{1}{2} \qquad\qquad 12(\tfrac{5}{12}x + \tfrac{1}{4}y) = 12 \cdot \tfrac{1}{3}$$
$$12 \cdot \tfrac{1}{4}x + 12 \cdot \tfrac{5}{12}y = 6 \qquad\qquad 12 \cdot \tfrac{5}{12}x + 12 \cdot \tfrac{1}{4}y = 4$$
$$3x + 5y = 6; \qquad\qquad 5x + 3y = 4.$$

The resulting system is

$$3x + 5y = 6,$$
$$5x + 3y = 4.$$

The solution of this system is given in Example 4.

Problem Solving

Now let us use the addition method in problem solving.

EXAMPLE 7 Badger Rent-A-Car rents compact cars at a daily rate of $43.95 plus 40¢ per mile. Thirsty Rent-A-Car rents compact cars at a daily rate of $42.95 plus 42¢ per mile. For what mileage is the cost the same?

The *Familiarize* and *Translate* steps have been done in Example 2 of Section 6.4. The resulting system of equations is

$$43.95 + 0.40m = c,$$
$$42.95 + 0.42m = c;$$

where m represents the mileage and c the cost.

3. *Carry out* We solve the system of equations. We clear the system of decimals by multiplying on both sides by 100. Then we multiply the second equation by -1 and use the addition method.

$$4395 + 40m = 100c$$
$$\underline{-4295 - 42m = -100c}$$
$$100 - 2m = 0$$
$$100 = 2m$$
$$50 = m$$

4. *Check* For 50 mi, the cost of the Badger car is $43.95 + 0.40(50)$, or $43.95 + 20$, or $63.95, and the cost of the other car is $42.95 + 0.42(50)$, or $42.95 + 21$, or $63.95, so the costs are the same when the mileage is 50.

5. *State* When the cars are driven 50 miles, the costs will be the same.

EXERCISE SET 6.7

Solve using the addition method.

1. $x + y = 10,$
   $x - y = 8$

2. $x - y = 7,$
   $x + y = 3$

3. $x + y = 8,$
   $-x + 2y = 7$

4. $x + y = 6,$
   $-x + 3y = -2$

6.7 THE ADDITION METHOD: PROBLEM SOLVING

5. $3x - y = 9,$
   $2x + y = 6$

6. $4x - y = 1,$
   $3x + y = 13$

7. $4a + 3b = 7,$
   $-4a + b = 5$

8. $7c + 5d = 18,$
   $c - 5d = -2$

9. $8x - 5y = -9,$
   $3x + 5y = -2$

10. $3a - 3b = -15,$
    $-3a - 3b = -3$

11. $4x - 5y = 7,$
    $-4x + 5y = 7$

12. $2x + 3y = 4,$
    $-2x - 3y = -4$

Solve using the multiplication principle first. Then add.

13. $-x - y = 8,$
    $2x - y = -1$

14. $x + y = -7,$
    $3x + y = -9$

15. $x + 3y = 19,$
    $x - y = -1$

16. $3x - y = 8,$
    $x + 2y = 5$

17. $x + y = 5,$
    $5x - 3y = 17$

18. $x - y = 7,$
    $4x - 5y = 25$

19. $2w - 3z = -1,$
    $3w + 4z = 24$

20. $7p + 5q = 2,$
    $8p - 9q = 17$

21. $2a + 3b = -1,$
    $3a + 5b = -2$

22. $3x - 4y = 16,$
    $5x + 6y = 14$

23. $x - 3y = 0,$
    $5x - y = -14$

24. $5a - 2b = 0,$
    $2a - 3b = -11$

25. $3x - 2y = 10,$
    $5x + 3y = 4$

26. $2p + 5q = 9,$
    $3p - 2q = 4$

27. $3x - 8y = 11,$
    $x + 6y - 8 = 0$

28. $m - n = 32,$
    $3m - 8n - 6 = 0$

29. $0.06x + 0.05y = 0.07,$
    $0.04x - 0.03y = 0.11$

30. $x - \frac{3}{2}y = 13,$
    $\frac{3}{2}x - y = 17$

Problem solving

Many of these problems have already been translated in Exercise Set 6.4.

31. Acme rents an intermediate-size car at a daily rate of $53.95 plus 30¢ per mile. Another company rents an intermediate-size car for $54.95 plus 20¢ per mile. For what mileage is the cost the same?

32. Badger rents a basic car at a daily rate of $45.95 plus 40¢ per mile. Another company rents a basic car for $46.95 plus 20¢ per mile. For what mileage is the cost the same?

33. Two angles are supplementary. One is 8° more than three times the other. Find the angles. (Supplementary angles are angles whose sum is 180°.)

34. Two angles are supplementary. One is 30° more than two times the other. Find the angles.

35. Two angles are complementary. Their difference is 34°. Find the angles. (Complementary angles are angles whose sum is 90°.)

36. Two angles are complementary. One angle is 42° more than $\frac{1}{2}$ the other. Find the angles.

37. In a vineyard a vintner uses 820 hectares to plant Chardonnay and Riesling grapes. The vintner knows that profits will be greatest by planting 140 hectares more of Chardonnay than Riesling. How many hectares of each grape should be planted?

38. The Hayburner Horse Farm allots 650 hectares to plant hay and oats. The owners know that their needs are best met if they plant 180 hectares more of hay than oats. How many hectares of each should they plant?

39. Several ancient Chinese books included problems that can be solved by translating to systems of equations. *Arithmetical Rules in Nine Sections* is a book of 246 problems compiled by a Chinese mathematician, Chang Tsang, who died in 152 B.C. One of the problems is: Suppose there are a number of rabbits and pheasants confined in a cage. In all there are 35 heads and 94 feet. How many rabbits and how many pheasants are there? Solve the problem.

40–42. Solve the systems of equations you set up in Exercises 19–21 of Exercise Set 6.4.

Solve.

43. $3(x - y) = 9,$
    $x + y = 7$

44. $5(a - b) = 10,$
    $a + b = 2$

45. $2(x - y) = 3 + x,$
    $x = 3y + 4$

46. $2(5a - 5b) = 10,$
$-5(6a + 2b) = 10$

47. $1.5x + 0.85y = 1637.5,$
$0.01(x + y) = 15.25$

48. $\dfrac{x}{3} + \dfrac{y}{2} = 1\dfrac{1}{3},$
$x + 0.05y = 4$

Solve for x and y.

49. $y = ax + b,$
$y = x + c$

50. $ax + by + c = 0,$
$ax + cy + b = 0$

Solve.

51. $3(7 - a) - 2(1 + 2b) + 5 = 0,$
$3a + 2b - 18 = 0$

52. $\dfrac{2}{x} - \dfrac{3}{y} = -\dfrac{1}{2},$
$\dfrac{1}{x} + \dfrac{2}{y} = \dfrac{11}{12}$

53. Suppose we can get a system into the form

$$ax + by = c$$
$$dx + ey = f,$$

where a, b, c, d, e, and f are positive or negative rational numbers.

a) Solve the system for x and y.

b) Use the pattern of your solution to part (a) to solve the following system.

$$1.425x - 5695y = 1000,$$
$$0.875x + 275y = 2500.$$

6.8 MORE PROBLEM SOLVING

We continue solving problems using the five-step process and our methods for solving systems of equations. The question may arise, after we have translated to a system, as to which method to use to solve the system: the substitution method or the addition method.

Although there are exceptions, the substitution method is probably better when a variable has a coefficient of 1, as in the following:

$$8x + 4y = 10, \qquad x + y = 61,$$
$$-x + y = 3; \qquad x = 2y.$$

Otherwise, the addition method is better. Both methods work. When in doubt, use the addition method.

EXAMPLE 1 Major league baseball teams play 162 games in a regular season. In a recent year, a team won 34 more games than it lost. How many games did the team win and how many did the team lose?

6.8 MORE PROBLEM SOLVING

1. **Familiarize** Let's familiarize ourselves with the problem by making a guess. Suppose the team won 103 games. Since they play 162 games, and we know that

    Number of games won plus Number of games lost = 162,

    it follows that they lost $162 - 103$, or 59 games. The difference

    Games won − Games lost

    is $103 - 59$, or 44. The results do not quite check in the problem, but at least we have become more familiar with the problem.

2. **Translate** We let x = the number of games won and y = the number of games lost. Then, the first two statements of the problem can be reworded and translated as follows.

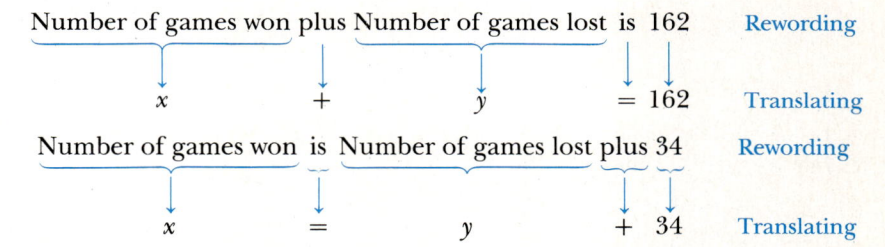

Number of games won	plus	Number of games lost	is 162	Rewording
x	$+$	y	$= 162$	Translating
Number of games won	is	Number of games lost	plus 34	Rewording
x	$=$	y	$+\ 34$	Translating

3. **Carry out** We solve the system of equations using the substitution method:

    $$x + y = 162,$$
    $$x = y + 34.$$

    We substitute $y + 34$ for x in the first equation:

    $$x + y = 162$$
    $$(y + 34) + y = 162$$
    $$34 + 2y = 162$$
    $$2y = 128$$
    $$y = 64.$$

    We substitute 64 for y in the second equation and solve for x:

    $$x = y + 34$$
    $$x = 64 + 34$$
    $$x = 98.$$

4. **Check** If the team won 98 games and lost 64, then the number of games won plus the number of games lost is $98 + 64$, or 162. Since games lost plus 34 is $64 + 34$, or 98, we know that it is true that the team won 34 more games than it lost.

5. **State** The team won 98 games and lost 64.

EXAMPLE 2 Howie is 21 years older than Izzi. In six years, Howie will be twice as old as Izzi. How old are they now?

1. **Familiarize** Let us consider some conditions of the problem. We let x represent Howie's age now and y Izzi's age now. We know that now Howie is 21 years older than Izzi. We make a table to organize our information. How do the ages relate in 6 years? In 6 years Izzi will be $y + 6$ and Howie will be $x + 6$, or $2(y + 6)$.

	Age now	Age in 6 years
Howie	x, or $y + 21$	$x + 6$, or $2(y + 6)$
Izzi	y	$y + 6$

2. **Translate** From the present ages we get the following rewording and translation.

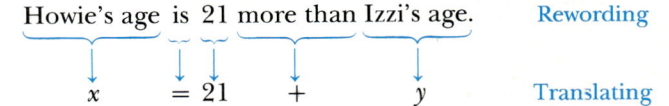

Howie's age is 21 more than Izzi's age.   *Rewording*

$$x = 21 + y$$ *Translating*

From their ages in 6 years we get the following rewording and translation.

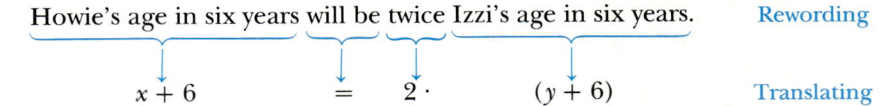

Howie's age in six years will be twice Izzi's age in six years.   *Rewording*

$$x + 6 = 2 \cdot (y + 6)$$ *Translating*

3. **Carry out** We solve the system of equations:

$$x = 21 + y,$$
$$x + 6 = 2(y + 6).$$

We use the addition method. We first add $-y$ on both sides of the first equation:

$$x - y = 21.$$

We also simplify the second equation:

$$x + 6 = 2y + 12$$
$$x - 2y = 6.$$

We solve the system

$$x - y = 21,$$
$$x - 2y = 6.$$

We multiply on both sides of the second equation by -1 and add:

$$\begin{array}{rl} x - y = 21 & \\ -x + 2y = -6 & \text{Multiplying by } -1 \\ \hline y = 15 & \text{Adding} \end{array}$$

We find x by substituting 15 for y in $x - y = 21$:
$$x - 15 = 21$$
$$x = 36.$$

4. **Check** Howie's age is 36, which is 21 more than 15, Izzi's age. In six years when Howie will be 42 and Izzi 21, Howie's age will be twice Izzi's age.

5. **State** Howie is now 36 and Izzi is 15.

EXAMPLE 3 There were 411 people at a movie. Admission was $1.00 for adults and $0.75 for children. The receipts were $395.75. How many adults and how many children attended?

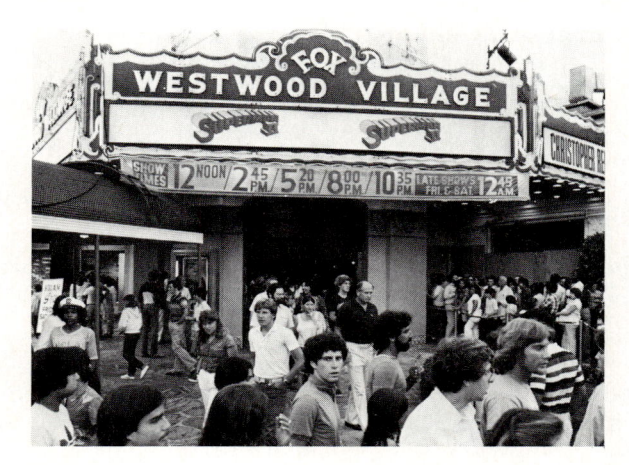

1. **Familiarize** There are many ways to familiarize ourselves with a problem situation. This time, let us make a guess and do some calculations. We guess:

   240 adults and

   180 children.

   Does the guess check? Does it make sense?
   The total number of people attending was supposed to be 411, so our guess, which totaled 420, cannot be right. Let's try another guess.

   240 adults and

   171 children.

   Now the total number sold is 411, so our guess is right with respect to the number of people attending.
   How much money was taken in? The problem says that adults paid $1.00, so the total amount of money collected from the adults was

   $$240(\$1), \text{ or } \$240.$$

The children paid $0.75, so the total amount of money collected from the children was

$$171(\$0.75), \text{ or } \$128.25.$$

This makes the total receipts

$$\$240 + \$128.25, \text{ or } \$368.25.$$

Our guess is not the answer to the problem because the total taken in, according to the problem, was $395.75. However, we are at least more familiar with the problem situation.

Let us list the information in a table. That usually helps a great deal in the familiarization process.

People	Paid	Number attending	Money taken in
Adults	$1.00	x	$1.00x$
Children	$0.75	y	$0.75y$
Totals		411	$395.75

2. **Translate** We let $x =$ the number of adults and $y =$ the number of children. The total number of people attending was 411, so

$$x + y = 411.$$

The amount taken in from the adults was $1.00x$, and the amount taken in from the children was $0.75y$. These amounts are in dollars. The total was $395.75, so we have

$$1.00x + 0.75y = 395.75.$$

We can multiply on both sides by 100 to clear of decimals. Thus we have the translation, a system of equations:

$$x + y = 411,$$
$$100x + 75y = 39{,}575.$$

3. **Carry out** We solve the system of equations. We use the addition method. We multiply on both sides of the first equation by -100 and then add:

$$-100x - 100y = -41{,}100 \quad \text{Multiplying by } -100$$
$$\underline{100x + 75y = 39{,}575}$$
$$-25y = -1{,}525 \quad \text{Adding}$$

$$y = \frac{-1525}{-25} \quad \text{Dividing by } -25$$

$$y = 61.$$

We go back to the first equation and substitute 61 for y:

$$x + y = 411$$
$$x + 61 = 411$$
$$x = 350.$$

4. **Check** We leave the check to the student. It is similar to what we did in the familiarization step.

5. **State** 350 adults and 61 children attended.

EXAMPLE 4 A chemist has one solution that is 80% acid (and the rest water), and another solution that is 30% acid. What is needed is 200 liters of a solution that is 62% acid. The chemist will prepare it by mixing the two solutions on hand. How much of each should be used?

1. **Familiarize** We can draw a picture of the situation. The chemist uses x liters of the first solution and y liters of the second solution.

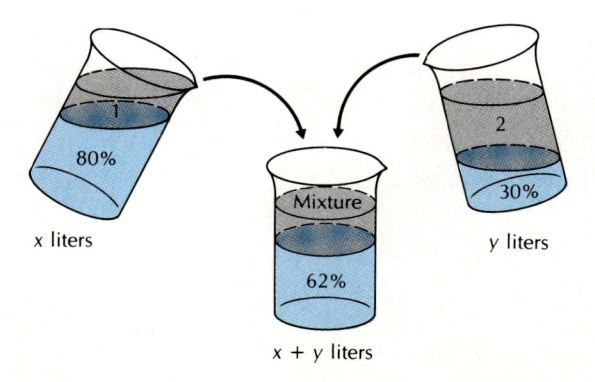

We can arrange the information in a table.

Type of solution	Amount of solution	Percent of acid	Amount of acid in solution
1	x	80%	80%x
2	y	30%	30%y
Mixture	200 liters	62%	62% × 200, or 124 liters

2. **Translate** The chemist uses x liters of the first solution and y liters of the second. Since the total is to be 200 liters, we have

$$x + y = 200.$$

The *amount* of acid in the new mixture is to be 62% of 200 liters, or 124 liters. The amounts of acid from the two solutions are 80%x and 30%y. Thus,

$$80\%x + 30\%y = 124,$$

or

$$0.8x + 0.3y = 124.$$

We clear the decimals by multiplying on both sides by 10:

$$10(0.8x + 0.3y) = 10 \cdot 124$$
$$8x + 3y = 1240.$$

3. **Carry out** We solve the system:

$$x + y = 200,$$
$$8x + 3y = 1240.$$

We use the addition method. We multiply on both sides of the first equation by -3 and then add:

$$\begin{aligned}
-3x - 3y &= -600 \quad &&\text{Multiplying by } -3 \\
\underline{8x + 3y} &= \underline{1240} \\
5x &= 640 \quad &&\text{Adding} \\
x &= \tfrac{640}{5} \quad &&\text{Multiplying by } \tfrac{1}{5} \\
x &= 128. \quad &&\text{Solving for } x
\end{aligned}$$

We go back to the first equation and substitute 128 for x:

$$x + y = 200$$
$$128 + y = 200$$
$$y = 72.$$

The solution is $x = 128$ and $y = 72$.

6.8 MORE PROBLEM SOLVING

4. **Check** The sum of 128 and 72 is 200. Now 80% of 128 is 102.4 and 30% of 72 is 21.6. These add up to 124.

5. **State** The chemist should use 128 liters of the 80%-acid solution and 72 liters of the 30%-acid solution.

EXAMPLE 5 A grocer wishes to mix some candy worth 45¢ per lb and some worth 80¢ per lb to make 350 lb of a mixture worth 65¢ per lb. How much of each should be used?

1. **Familiarize** Arranging information in a table will help.

	Cost of candy	Amount (lb)	Value (¢)
1	45¢	x	$45x$
2	80¢	y	$80y$
Mixture	65¢	350	65¢ (350), or 22,750

2. **Translate** We use x for the amount of 45¢ candy and y for the amount of 80¢ candy. Then
$$x + y = 350.$$

Our second equation will come from the values. The value of the first candy, in cents, is $45x$ (x lb at 45¢ per lb). The value of the second is $80y$, and the value of the mixture is 350×65. Thus we have
$$45x + 80y = 350 \times 65.$$

3. **Carry out** We solve the system:
$$x + y = 350,$$
$$45x + 80y = 22{,}750.$$

We use the addition method. We multiply on both sides of the first equation by -45 and then add:

$$-45x - 45y = -15{,}750 \quad \text{Multiplying by } -45$$
$$\underline{45x + 80y = 22{,}750}$$
$$35y = 7{,}000 \quad \text{Adding}$$
$$y = \frac{7{,}000}{35}$$
$$y = 200.$$

We go back to the first equation and substitute 200 for y:

$$x + y = 350$$
$$x + 200 = 350$$
$$x = 150.$$

4. **Check** We consider $x = 150$ lb and $y = 200$ lb. The sum is 350 lb. The value of the candy is $45(150) + 80(200)$, or $22{,}750$¢. These values check.

5. **State** The grocer should mix 150 lb of the 45¢ candy with 200 lb of the 80¢ candy.

EXERCISE SET 6.8

1. A firm sells cars and trucks. There is room on its lot for 510 vehicles. From experience they know that profits will be greatest if there are 190 more cars than trucks on the lot. How many of each vehicle should the firm have for the greatest profit?

2. A family went camping at a park 45 km from town. They drove 13 km more than they walked to get to the campsite. How far did they walk?

3. Sammy is twice as old as his daughter. In four years Sammy's age will be three times what his daughter's age was six years ago. How old is each at present?

4. Ann is eighteen years older than her son. She was three times as old one year ago. How old is each?

5. Marge is twice as old as Consuelo. The sum of their ages seven years ago was 13. How old are they now?

6. Andy is four times as old as Wendy. In twelve years, Wendy's age will be half of Andy's. Find their ages now.

7. A collection of dimes and quarters is worth $15.25. There are 103 coins in all. How many of each are there?

8. A collection of quarters and nickels is worth $1.25. There are 13 coins in all. How many of each are there?

9. A collection of nickels and dimes is worth $25. There are three times as many nickels as dimes. How many of each are there?

10. A collection of nickels and dimes is worth $2.90. There are 19 more nickels then dimes. How many of each are there?

11. There were 429 people at a play. Admission was $1 for adults and 75¢ for children. The receipts were $372.50. How many adults and how many children attended?

12. The attendance at a school concert was 578. Admission was $2 for adults and $1.50 for children. The receipts were $985. How many adults and how many children attended?

13. There were 200 tickets sold for a college basketball game. Tickets for students were $0.50 and for adults were $0.75. The total amount of money collected was $132.50. How many of each type of ticket were sold?

14. There were 203 tickets sold for a wrestling match. For activity card holders the price was $1.25 and for noncard holders the price was $2. The total amount of money collected was $310. How many of each type of ticket were sold?

15. Solution A is 50% acid and solution B is 80% acid. How much of each should be used to make 100 grams of a solution that is 68% acid? (*Hint:* 68% of what is acid?)

16. Solution A is 30% alcohol and solution B is 75% alcohol. How much of each should be used to make 100 liters of a solution that is 50% alcohol?

17. Farmer Jones has 100 liters of milk that is 4.6% butterfat. How much skim milk (no butterfat) should be mixed with it to make milk that is 3.2% butterfat?

18. A tank contains 8000 liters of a solution that is 40% acid. How much water should be added to make a solution that is 30% acid?

19. A solution containing 30% insecticide is to be mixed with a solution containing 50% insecticide to make 200 liters of a solution containing 42% insecticide. How much of each solution should be used?

20. A solution containing 28% fungicide is to be mixed with a solution containing 40% fungicide to make 300 liters of a solution containing 36% fungicide. How much of each solution should be used?

21. The Nuthouse has 10 kg of mixed cashews and pecans worth $8.40 per kg. Cashews alone sell for $8.00 per kg and pecans sell for $9.00 per kg. How many kg of each are in the mixture?

22. A coffee shop mixes Brazilian coffee worth $5 per kg with Turkish coffee worth $8 per kg. The mixture is to sell for $7 per kg. How much of each type of coffee should be used to make a mixture of 300 kg?

23. A total of $27,000 is invested, part of it at 12% and part of it at 13%. The total yield after one year is $3385. How much was invested at each rate?

24. A two-digit number is six times the sum of its digits. The ten's digit is one more than the unit's digit. Find the number.

25. The sum of the digits of a two-digit number is 12. When the digits are reversed, the number is decreased by 18. Find the original number.

26. An automobile radiator contains 16 liters of antifreeze and water. This mixture is 30% antifreeze. How much of this mixture should be drained and replaced with pure antifreeze so that the mixture will be 50% antifreeze?

27. An employer has a daily payroll of $325 when employing some workers at $20 per day and others at $25 per day. When the number of $20 workers is increased by 50% and the number of $25 workers is decreased by $\frac{1}{5}$, the new daily payroll is $400. Find how many were originally employed at each rate.

28. A student earned $288 on investments. If $1100 was invested at one yearly rate and $1800 at a rate that was 1.5% higher, find the two rates of interest.

29. In a two-digit number, the sum of the unit's digit and the number is 43 more than five times the ten's digit. The sum of the digits is 11. Find the number.

30. The sum of the digits of a three-digit number is 9. If the digits are reversed, the number increases by 495. The sum of the ten's and hundred's digit is half the unit's digit. Find the number.

31. Together, a bat, ball, and glove cost $99.00. The bat costs $9.95 more than the ball, and the glove costs $65.45 more than the bat. How much does each cost?

32. In Lewis Carroll's "Through the Looking Glass" Tweedledum says to Tweedledee, "The sum of your weight and twice mine is 361 pounds." Then Tweedledee says to Tweedledum, "Contrariwise, the sum of your weight and twice mine is 362 pounds." Find the weight of Tweedledum and Tweedledee.

6.9 PROBLEMS INVOLVING MOTION

Many problems deal with distance, time, and speed. A basic formula comes from the definition of speed.

$$\text{Speed} = \frac{\text{Distance}}{\text{Time}}, \quad r = \frac{d}{t}.$$

From $r = d/t$, we can obtain two other formulas by solving. They are

$$d = rt \quad \text{Multiplying by } t$$

and

$$t = \frac{d}{r}. \quad \text{Multiplying by } t \text{ and dividing by } r$$

In most problems involving motion, you will use one of these formulas. It is probably easiest to remember the definition of speed, $r = d/t$. You can easily obtain either of the other formulas as you need them.

We have a five-step process for problem solving. The following steps are also helpful when solving motion problems.

1. **Organize the information in a chart.**
2. **Look for as many things as you can that are the same, so you can write equations.**

EXAMPLE 1 A train leaves Podunk traveling east at 35 kilometers per hour (km/h). An hour later another train leaves Podunk on a parallel track at 40 km/h. How far from Podunk will the trains meet?

1. **Familiarize** First make a drawing.

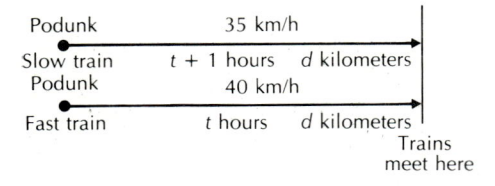

```
Podunk          35 km/h
•───────────────────────────→
Slow train   t + 1 hours   d kilometers
Podunk          40 km/h
•───────────────────────────→
Fast train    t hours      d kilometers
                              Trains
                              meet here
```

From the drawing we see that the distances are the same. Let's call the distance d. We don't know the times. Let t represent the time for the faster train. Then the time for the slower train will be $t + 1$. We can organize the information in a chart. In this case the distances are the same, so we shall use the formula $d = rt$.

6.9 PROBLEMS INVOLVING MOTION

	Distance	Speed	Time
Slow train	d	35	$t+1$
Fast train	d	40	t

2. **Translate** In these problems we look for things that are the same, so we can find equations. From each row of the chart we get an equation, $d = rt$. Thus we have two equations:

$$d = 35(t + 1),$$
$$d = 40t.$$

3. **Carry out** We solve the system using the substitution method:

$35(t + 1) = 40t$ Using the substitution method (substituting $35(t + 1)$ for d in the second equation)

$35t + 35 = 40t$ Multiplying

$35 = 5t$ Adding $-35t$

$\frac{35}{5} = t$ Multiplying by $\frac{1}{5}$

$7 = t.$

The problem asks us to find how far from Podunk the trains meet. Thus we need to find d. We can do this by substituting 7 for t in the equation $d = 40t$:

$$d = 40(7)$$
$$= 280.$$

4. **Check** If the time is 7 hr, then the distance the slow train travels is $35(7 + 1)$, or 280 km. The fast train travels $40(7)$, or 280 km. Since the distances are the same, we know how far from Podunk the trains will meet.

5. **State** The trains meet 280 km from Podunk.

EXAMPLE 2 A motorboat took 3 hr to make a downstream trip with a 6-km/h current. The return trip against the same current took 5 hr. Find the speed of the boat in still water.

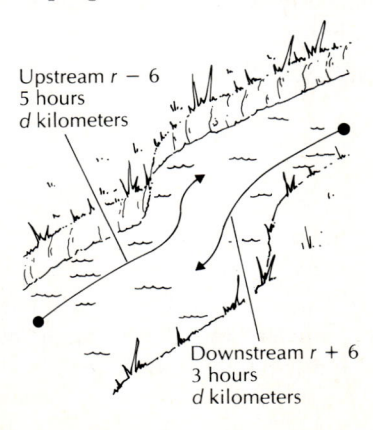

Upstream $r - 6$
5 hours
d kilometers

Downstream $r + 6$
3 hours
d kilometers

1. **Familiarize** We first make a drawing. From the drawing we see that the distances are the same. Let's call the distance d. Let r represent the speed of the boat in still water. Then, when the boat is traveling downstream, its speed is $r + 6$ (the current helps the boat along). When it is traveling upstream, its speed is $r - 6$ (the current holds the boat back some). We can organize the information in a chart. In this case the distances are the same, so we shall use the formula $d = rt$.

	Distance	Speed	Time
Downstream	d	$r + 6$	3
Upstream	d	$r - 6$	5

2. **Translate** From each row of the chart, we get an equation, $d = rt$:

$$d = (r + 6)3,$$
$$d = (r - 6)5.$$

3. **Carry out** We solve the system using substitution:

$(r + 6)3 = (r - 6)5$ Using substitution (substituting $(r + 6)3$ for d in the second equation)

$3r + 18 = 5r - 30$ Multiplying

$-2r + 18 = -30$ Adding $-5r$

$-2r = -48$ Adding -18

$r = \dfrac{-48}{-2}$, or 24. Multiplying by $-\frac{1}{2}$

4. **Check** When $r = 24$, $r + 6 = 30$, and $30 \cdot 3 = 90$, the distance. When $r = 24$, $r - 6 = 18$, and $18 \cdot 5 = 90$. In both cases we get the same distance.

5. **State** The speed in still water is 24 km/h.

EXAMPLE 3 Two cars leave town at the same time going in opposite directions. One of them travels 60 mph and the other 30 mph. In how many hours will they be 150 miles apart?

1. **Familiarize** We first make a drawing.

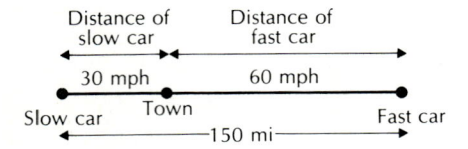

```
     Distance of      Distance of
     slow car         fast car
     ←——————→ ←——————————————→
         30 mph           60 mph
     •—————————•—————————————————•
     Slow car  Town              Fast car
               ←————— 150 mi —————→
```

From the wording of the problem and the drawing, we see that the distances may *not* be the same. But the times the cars travel are the same, so we can just use t for time. We can organize the information in a chart.

	Distance	Speed	Time
Fast car	Distance of fast car	60	t
Slow car	Distance of slow car	30	t

2. **Translate** From the drawing we see that

$$(\text{Distance of fast car}) + (\text{Distance of slow car}) = 150.$$

Then using $d = rt$ in each row of the table we get

$$60t + 30t = 150.$$

3. **Carry out** We solve the equation:

$$90t = 150 \quad \text{Collecting like terms}$$
$$t = \tfrac{150}{90}, \text{ or } \tfrac{5}{3}, \text{ or } 1\tfrac{2}{3} \text{ hours.} \quad \text{Multiplying by } \tfrac{1}{90}$$

4. **Check** When $t = \tfrac{5}{3}$ hr,

$$(\text{Distance of fast car}) + (\text{Distance of slow car}) = 60(\tfrac{5}{3}) + 30(\tfrac{5}{3})$$
$$= 100 + 50, \text{ or } 150 \text{ miles.}$$

Thus the time of $\tfrac{5}{3}$ hr, or $1\tfrac{2}{3}$ hr checks.

5. **State** In $1\tfrac{2}{3}$ hr the cars will be 150 miles apart.

EXERCISE SET 6.9

1. Two cars leave town at the same time going in opposite directions. One travels 55 mph and the other travels 48 mph. In how many hours will they be 206 miles apart?

2. Two cars leave town at the same time going in opposite directions. One travels 44 mph and the other travels 55 mph. In how many hours will they be 297 miles apart?

3. Two cars leave town at the same time going in the same direction. One travels 30 mph and the other travels 46 mph. In how many hours will they be 72 miles apart?

4. Two cars leave town at the same time going in the same direction. One travels 32 mph and the other travels 47 mph. In how many hours will they be 69 miles apart?

5. A train leaves a station and travels east at 72 km/h. Three hours later a second train leaves on a parallel track and travels east at 120 km/h. When will it overtake the first train?

6. A private airplane leaves an airport and flies due south at 192 km/h. Two hours later a jet leaves the same airport and flies due south at 960 km/h. When will the jet overtake the plane?

7. A canoeist paddled for 4 hours with a 6-km/h current to reach a campsite. The return trip against the same current took 10 hours. Find the speed of the canoe in still water.

8. An airplane flew for 4 hours with a 20-km/h tail wind. The return flight against the same wind took 5 hours. Find the speed of the plane in still air.

9. It takes a passenger train 2 hours less time than it takes a freight train to make the trip from Central City to Clear Creek. The passenger train averages 96 km/h, whereas the freight train averages 64 km/h. How far is it from Central City to Clear Creek?

10. It takes a small jet 4 hours less time than it takes a propeller-driven plane to travel from Glen Rock to Oakville. The jet averages 637 km/h, whereas the propeller plane averages 273 km/h. How far is it from Glen Rock to Oakville?

11. An airplane took 2 hours to fly 600 km against a head wind. The return trip with the wind took $1\frac{2}{3}$ hours. Find the speed of the plane in still air.

12. It took 3 hours to row a boat 18 km against the current. The return trip with the current took $1\frac{1}{2}$ hours. Find the speed of the rowboat in still water.

13. A motorcycle breaks down and the rider has to walk the rest of the way to work. The motorcycle was being driven at 45 mph and the rider walks at a speed of 6 mph. The distance from home to work is 25 miles and the total time for the trip was 2 hours. How far did the motorcycle go before it broke down?

14. A student walks and jogs to college each day. The student averages 5 km/h walking and 9 km/h jogging. The distance from home to college is 8 km and the student makes the trip in 1 hour. How far does the student jog?

15. An airplane flew for 4.23 hours with a 25.5-km/h tail wind. The return flight against the same wind took 4.97 hours. Find the speed of the plane in still air.

16. An airplane took $2\frac{1}{2}$ hours to fly 625 miles with the wind. It took 4 hours and 10 minutes to make the return trip against the same wind. Find the wind speed and the speed of the plane in still air.

17. To deliver a package, a messenger must travel at a speed of 60 mph on land and then use a motorboat whose speed is 20 mph in still water. While delivering the package, the messenger goes by land to a dock and then travels on a river against a current of 4 mph. The messenger reaches the destination in 4.5 hours and then returns to the starting point in 3.5 hours. How far did the messenger travel by land and how far by water?

18. Against a headwind, Gary computes his flight time for a trip of 2900 miles at 5 hours. The flight would take 4 hours and 50 minutes if the headwind were half as much. Find the headwind and the plane's air speed.

19. A car travels from one town to another at a speed of 32 mph. If it had gone 4 mph faster, it could have made the trip in $\frac{1}{2}$ hour less time. How far apart are the towns?

20. Two airplanes start at the same time and fly toward each other from points 1000 km apart at rates of 420 km/h and 330 km/h. When will they meet?

21. A truck and a car leave a service station at the same time and travel in the same direction. The truck travels at 55 mph and the car at 40 mph. They can maintain CB radio contact within a range of 10 miles. When will they lose contact?

22. Charles Lindbergh flew the Spirit of St. Louis in 1927 from New York to Paris at an average speed of 107.4 mph. Eleven years later, Howard Hughes flew the same route, averaged 217.1 mph and took 16 hours, 57 minutes less time. Find the length of their route.

6.10 EQUATIONS OF LINES AND SLOPE*

Slope

Graphs of some linear equations slant upward from left to right. Others slant downward. Some are vertical and some are horizontal. Some slant more steeply than others. We are looking for a way to describe such possibilities with numbers.

Consider a line with two points marked P and Q. As we move from P to Q, the x-coordinate changes from 2 to 6 and the y-coordinate changes from 1 to 3. The change in x is $6 - 2$, or 4. The change in y is $3 - 1$, or 2.

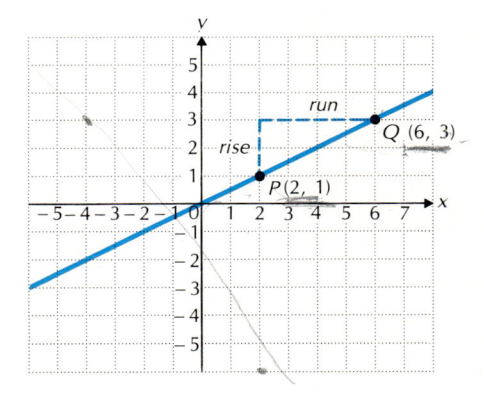

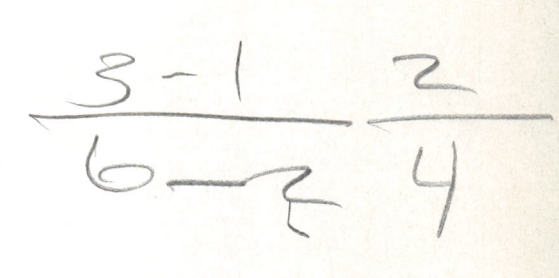

We call the change in y the *rise*, and the change in x the *run*. The ratio rise/run is the same for any two points on a line. We call this ratio the *slope*. Slope describes the slant of a line. The slope of the line in the graph is given by

$$\frac{\text{rise}}{\text{run}}, \text{ or } \frac{2}{4}, \text{ or } \frac{1}{2}.$$

- - -

The *slope* of a line containing points (x_1, y_1) and (x_2, y_2) is given by

$$m = \frac{\text{rise}}{\text{run}} = \frac{\text{the change in } y}{\text{the change in } x} = \frac{y_2 - y_1}{x_2 - x_1}.$$

- - -

EXAMPLE 1 Graph the line containing the points $(-4, 3)$ and $(2, -6)$ and find the slope.

The graph is shown on the following page. From $(-4, 3)$ to $(2, -6)$ the change in y, or rise, is $3 - (-6)$, or 9. The change in x, or run, is $-4 - 2$, or -6.

---

* To the instructor: this section is optional.

GRAPHS, SYSTEMS OF EQUATIONS, AND PROBLEM SOLVING

$$\text{Slope} = \frac{\text{rise}}{\text{run}} = \frac{\text{change in } y}{\text{change in } x} = \frac{3 - (-6)}{-4 - 2}$$

$$= \frac{9}{-6} = -\frac{9}{6}, \quad \text{or} \quad -\frac{3}{2}.$$

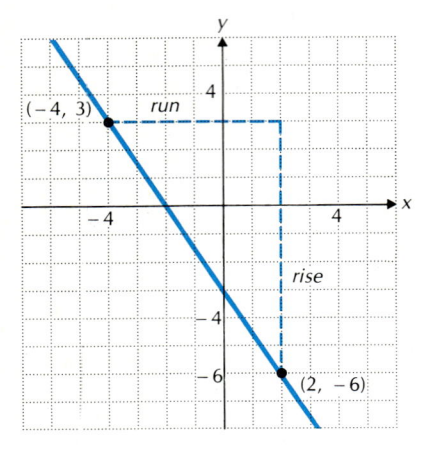

When we use the formula

$$m = \frac{y_2 - y_1}{x_2 - x_1},$$

we subtract in two ways. We just have to remember to subtract the y-coordinates in the same order that we subtract the x-coordinates. Let's do Example 1 again.

$$\text{Slope} = \frac{\text{change in } y}{\text{change in } x} = \frac{-6 - 3}{2 - (-4)} = \frac{-9}{6} = -\frac{3}{2}.$$

The slope of a line tells how it slants. A line with a positive slope slants up from left to right. The larger the positive slope, the steeper the slant. A line with negative slope slants downward from left to right.

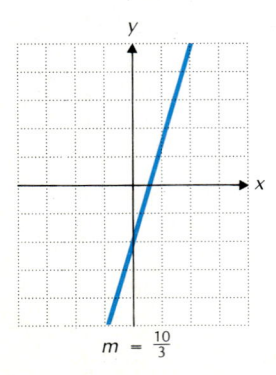

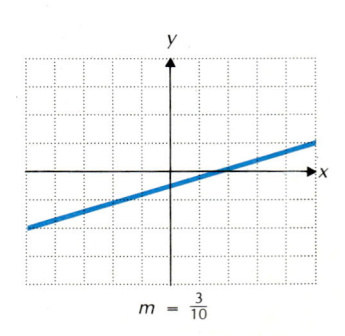

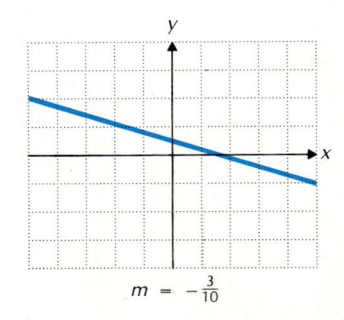

$m = \frac{10}{3}$

$m = \frac{3}{10}$

$m = -\frac{3}{10}$

Horizontal and Vertical Lines

What about the slope of a horizontal or a vertical line?

EXAMPLE 2 Find the slope of the line $y = 4$.

Consider the points $(-3, 4)$ and $(2, 4)$, which are on the line.

The change in $y = 4 - 4$, or 0.

The change in $x = -3 - 2$, or -5.

$$m = \frac{4 - 4}{-3 - 2}$$

$$= \frac{0}{-5}$$

$$= 0$$

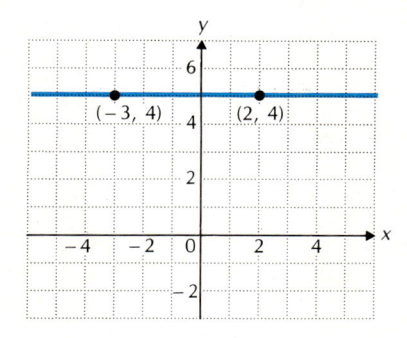

Any two points on a horizontal line have the same y-coordinate. Thus the change in y is 0, so the slope is 0.

EXAMPLE 3 Find the slope of the line $x = -3$.

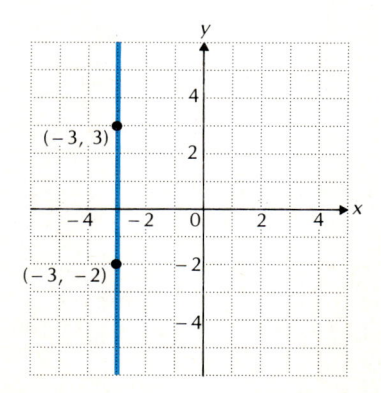

Consider the points $(-3, 3)$ and $(-3, -2)$, which are on the line.

The change in $y = 3 - (-2)$, or 5.

The change in $x = -3 - (-3)$, or 0.

$$m = \frac{3 - (-2)}{-3 - (-3)}$$

$$= \frac{5}{0}$$

Since division by 0 is not defined, this line has no slope.

- -

A horizontal line has slope 0. A vertical line has *no* slope.

Finding Slope from an Equation

It is possible to find the slope of a line from its equation. Let us consider the equation
$$y = 2x + 3.$$
Two points on the line are $(0, 3)$ and $(1, 5)$. We can find such points by picking arbitrary values for x, say 0 and 1, and substituting to find corresponding y-values. The slope of the line is found as follows.

$$m = \frac{\text{change in } y}{\text{change in } x} = \frac{5 - 3}{1 - 0} = \frac{2}{1} = 2$$

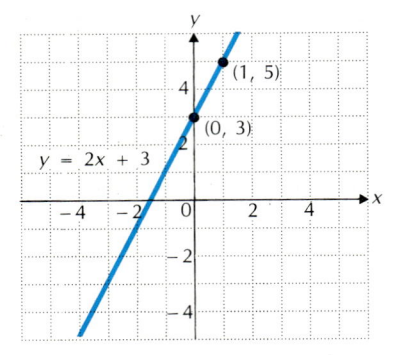

The slope is 2. This is also the coefficient of the x-term in the equation $y = 2x + 3$.

> The slope of the line $y = mx + b$ is m. To find the slope of a nonvertical linear equation in x and y, first solve the equation for y, and get the resulting equation in the form $y = mx + b$. The coefficient of the x-term, m, is the slope of the line.

6.10 EQUATIONS OF LINES AND SLOPE

EXAMPLE 4 Find the slope of the line $2x + 3y = 7$.

We solve for y:

$$2x + 3y = 7$$
$$3y = -2x + 7$$
$$y = \tfrac{1}{3}(-2x + 7)$$
$$y = \tfrac{1}{3}(-2x) + \tfrac{1}{3}(7)$$
$$y = -\tfrac{2}{3}x + \tfrac{7}{3}.$$

The slope is $-\tfrac{2}{3}$.

The Slope–Intercept Equation of a Line

In the equation $y = mx + b$, we know that m is the slope. What is the y-intercept? To find out we let $x = 0$, and solve for y:

$$y = mx + b$$
$$y = m(0) + b$$
$$y = b.$$

Thus the y-intercept is $(0, b)$.

- -

The equation $y = mx + b$ is called the *slope–intercept* equation. The slope is m and the y-intercept is $(0, b)$.

EXAMPLE 5 Find the slope and y-intercept of $y = 3x - 4$.

Since the equation is already in the form $y = mx + b$, we simply read the slope and y-intercept from the equation.

$$y = 3x - 4$$

The slope is 3.   The y-intercept is $(0, -4)$.

EXAMPLE 6 Find the slope and y-intercept of $2x + 3y = 8$.

We first solve for y:

$$2x + 3y = 8$$
$$3y = -2x + 8$$
$$y = -\tfrac{2}{3}x + \tfrac{8}{3}.$$

The slope is $-\tfrac{2}{3}$ and the y-intercept is $(0, \tfrac{8}{3})$.

The Point–Slope Equation of a Line

Suppose we know the slope of a line and that it contains a certain point. We can use the slope–intercept equation to find an equation of the line.

EXAMPLE 7 Find an equation of the line with slope 3 that contains the point (4, 1).

Step 1. Use the point (4, 1) and substitute 4 for x and 1 for y in $y = mx + b$. We also substitute 3 for m, the slope. Then we solve for b:

$$y = mx + b$$
$$1 = 3 \cdot 4 + b \quad \text{Substituting}$$
$$-11 = b. \quad \text{Solving for } b, \text{ the } y\text{-intercept}$$

Step 2. Write the equation $y = mx + b$ by substituting 3 for m and -11 for b:

$$y = mx + b$$
$$y = 3x - 11.$$

Now consider a line with slope 2 and containing the point (1, 3) as shown. Suppose (x, y) is any point on this line. Using the definition of slope and the two points (1, 3) and (x, y), we get

$$m = \frac{\text{change in } y}{\text{change in } x}$$
$$= \frac{y - 3}{x - 1}.$$

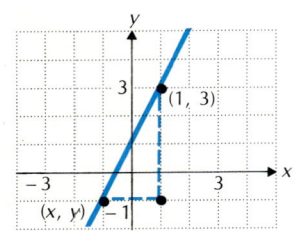

We know that the slope is 2, so

$$2 = \frac{y - 3}{x - 1}, \quad \text{or} \quad \frac{y - 3}{x - 1} = 2.$$

Multiplying on both sides by $x - 1$, we get

$$y - 3 = 2(x - 1).$$

Solving for y, we obtain

$$y - 3 = 2x - 2$$
$$y = 2x + 1.$$

THE POINT–SLOPE EQUATION

A nonvertical line that contains a point (x_1, y_1) with slope m has an equation

$$y - y_1 = m(x - x_1).$$

EXAMPLE 8 Find an equation of the line with slope 5 that contains the point $(-2, -3)$.

We use the point–slope equation. We substitute 5 for m, -2 for x_1, and -3 for y_1:

$$y - y_1 = m(x - x_1)$$
$$y - (-3) = 5[x - (-2)]$$
$$y + 3 = 5(x + 2)$$
$$y + 3 = 5x + 10$$
$$y = 5x + 7.$$

We can also use the point–slope equation to find an equation of a line that contains two given points.

EXAMPLE 9 Find an equation of the line containing $(2, 3)$ and $(-6, 1)$.

First we find the slope:

$$m = \frac{3 - 1}{2 - (-6)}$$
$$= \frac{1}{4}.$$

Then we use the point–slope equation:

$$y - y_1 = m(x - x_1)$$
$$y - 3 = \tfrac{1}{4}(x - 2) \qquad \text{Substituting 2 for } x_1, \text{ 3 for } y_1, \text{ and } \tfrac{1}{4} \text{ for } m$$
$$y - 3 = \tfrac{1}{4}x - \tfrac{1}{2}$$
$$y = \tfrac{1}{4}x + \tfrac{5}{2}.$$

Applications of Slope

Slope has many real-world applications. For example, numbers like 2%, 3%, and 6% are often used to represent the *grade* of a road. Such a number is meant to tell how steep a road up a hill or mountain is. For example, a 3% grade means that for every horizontal distance of 100 ft, the road rises 3 ft. The concept of grade also occurs in cardiology when a person runs on a treadmill. A physician may change the steepness of the treadmill to measure its effect on heartbeat.

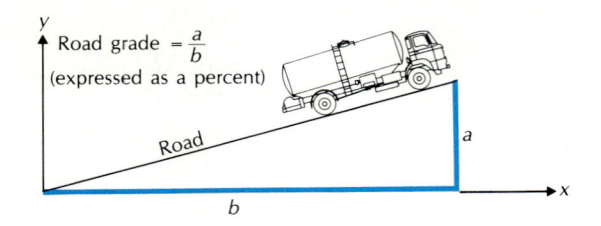

Another example occurs in hydrology. When a river flows, the strength or force of the river depends on how much the river falls vertically compared to how much it flows horizontally.

EXERCISE SET 6.10

Find the slope, if it exists, of the line containing each pair of points.

1. $(3, 2)$ and $(-1, 2)$
2. $(4, 1)$ and $(-2, -3)$
3. $(-2, 4)$ and $(3, 0)$
4. $(-4, 2)$ and $(2, -3)$
5. $(4, 0)$ and $(5, 7)$
6. $(3, 0)$ and $(6, 2)$
7. $(0, 8)$ and $(-3, 10)$
8. $(0, 9)$ and $(4, 7)$
9. $(3, -2)$ and $(5, -6)$
10. $(-2, 4)$ and $(6, -7)$
11. $(-2, \frac{1}{2})$ and $(-5, \frac{1}{2})$
12. $(8, -3)$ and $(10, -3)$
13. $(9, -4)$ and $(9, -7)$
14. $(-10, 3)$ and $(-10, 4)$

Find the slope, if it exists, of each line.

15. $x = -8$
16. $x = -4$
17. $y = 2$
18. $y = 17$
19. $x = 9$
20. $x = 6$
21. $y = -9$
22. $y = -4$

Find the slope of each line.

23. $3x + 2y = 6$
24. $4x - y = 5$
25. $x + 4y = 8$
26. $x + 3y = 6$
27. $-2x + y = 4$
28. $-5x + y = 5$
29. $4x - 3y = -12$
30. $x - 2y = 9$
31. $x - 3y = -2$
32. $x + y = 7$
33. $-2x + 4y = 8$
34. $-5x + 7y = 2$

Find the slope and y-intercept of each line.

35. $y = -4x - 9$
36. $y = -3x - 5$
37. $y = 1.8x$
38. $y = -27.4x$
39. $2x + 3y = 9$
40. $5x + 4y = 12$
41. $-8x - 7y = 21$
42. $-2x - 9y = 13$
43. $9x = 3y + 5$
44. $4x = 9y + 7$
45. $-6x = 4y + 2$
46. $y = -17$

Find an equation for the line containing the given point and having the given slope.

47. $(2, 5)$, $m = 5$
48. $(-3, 0)$, $m = -2$
49. $(2, 4)$, $m = \frac{3}{4}$
50. $(\frac{1}{2}, 2)$, $m = -1$
51. $(2, -6)$, $m = 1$
52. $(4, -2)$, $m = 6$
53. $(-3, 0)$, $m = -3$
54. $(0, 3)$, $m = -3$
55. $(5, 6)$, $m = \frac{2}{3}$
56. $(2, 7)$, $m = \frac{5}{6}$

Find an equation of the line that contains the given pair of points.

57. $(-6, 1)$ and $(2, 3)$
58. $(12, 16)$ and $(1, 5)$
59. $(0, 4)$ and $(4, 2)$
60. $(0, 0)$ and $(4, 2)$
61. $(3, 2)$ and $(1, 5)$
62. $(-4, 1)$ and $(-1, 4)$
63. $(5, 0)$ and $(0, -2)$
64. $(-2, -2)$ and $(1, 3)$
65. $(-2, -4)$ and $(2, -1)$
66. $(-3, 5)$ and $(-1, -3)$

67. Find an equation of the line that contains the point $(2, -3)$ and has the same slope as the line $3x - y + 4 = 0$.

68. Find an equation of the line that has the same y-intercept as the line $x - 3y = 6$ and contains the point $(5, -1)$.

69. Find an equation of the line with the same slope as $3x - 2y = 8$ and the same y-intercept as $2y + 3x = -4$.

70. Graph several equations that have the same slope. How are these lines related?

SUMMARY AND REVIEW: CHAPTER 6

The following contains a summary of what you should be able to do after completing this chapter. The review exercises are for practice. Answers are at the back of the book. If you miss an exercise, restudy the section indicated alongside the answer.

You should be able to:

Plot points associated with ordered pairs of numbers, determine the quadrant in which a point lies, and find coordinates of a point on a graph.

Use graph paper. Plot these points.

1. $(2, 5)$
2. $(0, -3)$
3. $(-4, -2)$

In which quadrant is each point located?

4. $(3, -8)$
5. $(-20, -14)$
6. $(4.9, 1.3)$

Find the coordinates of each point.

7. A
8. B
9. C

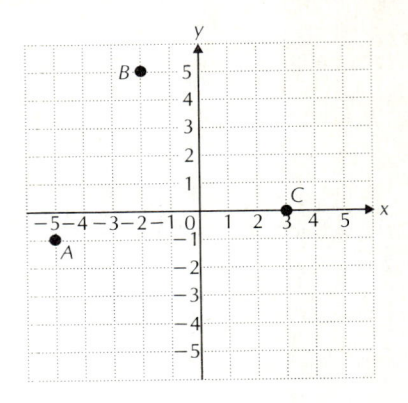

Determine whether an ordered pair of numbers is a solution of an equation in two variables.

Determine whether the given point is a solution of the equation $2y - x = 10$.

10. $(2, -6)$
11. $(0, 5)$

Graph linear equations.

Graph.

12. $y = 2x - 5$
13. $y = -\frac{3}{4}x$
14. $y = -x + 4$
15. $y = 3 - 4x$
16. $5x - 2y = 10$
17. $y = 3$
18. $x = -\frac{3}{4}$
19. $x - 2y = 6$

Determine whether an ordered pair is a solution of a system of equations.

Determine whether the given ordered pair is a solution of the system of equations.

20. $(6, -1)$; $x - y = 3$,
    $2x + 5y = 6$
21. $(2, -3)$; $2x + y = 1$,
    $x - y = 5$
22. $(-2, 1)$; $x + 3y = 1$,
    $2x - y = -5$
23. $(-4, -1)$; $x - y = 3$,
    $x + y = -5$

Solve systems of equations using graphing and the substitution and addition methods.

Use graph paper. Solve each system graphically.

24. $x + y = 4$,
    $x - y = 8$
25. $x + 3y = 12$,
    $2x - 4y = 4$
26. $2x + y = 1$,
    $x = 2y + 8$
27. $3x - 2y = -4$,
    $2y - 3x = -2$

Solve by the substitution method.

28. $y = 5 - x$,
    $3x - 4y = -20$
29. $x + 2y = 6$,
    $2x + 3y = 8$
30. $3x + y = 1$,
    $x - 2y = 5$

31. $x + y = 6,$
    $y = 3 - 2x$

32. $s + t = 5,$
    $s = 13 - 3t$

33. $x - y = 4,$
    $y = 2 - x$

Solve by the addition method.

34. $x + y = 4,$
    $2x - y = 5$

35. $x + 2y = 9,$
    $3x - 2y = -5$

36. $x - y = 8,$
    $2x + y = 7$

37. $\frac{2}{3}x + y = -\frac{5}{3},$
    $x - \frac{1}{3}y = -\frac{13}{3}$

38. $2x + 3y = 8,$
    $5x + 2y = -2$

39. $5x - 2y = 2,$
    $3x - 7y = 36$

40. $-x - y = -5,$
    $2x - y = 4$

41. $6x + 2y = 4,$
    $10x + 7y = -8$

Solve problems using the five-step process and the substitution and addition methods for solving systems of equations.

42. The sum of two numbers is 8. Their difference is 12. Find the numbers.

43. The sum of two numbers is 27. One-half of the first number plus one-third of the second number is 11. Find the numbers.

44. The perimeter of a rectangle is 96 cm. The length is 27 cm more than the width. Find the length and width.

45. An airplane flew for 4 hours with a 15-km/h tail wind. The return flight against the same wind took 5 hours. Find the speed of the airplane in still air.

46. There were 508 people at an organ recital. Orchestra seats cost $5.00 per person with balcony seats costing $3.00. The total receipts were $2118. Find the number of orchestra and the number of balcony seats sold.

47. Solution A is 30% alcohol and solution B is 60% alcohol. How much of each is needed to make 80 liters of a solution that is 45% alcohol?

48. Jeff is three times as old as his son. In 9 years, Jeff will be twice as old as his son. How old is each now?

---

49. The solution of the following system is (6, 2). Find C and D.

$$2x - Dy = 6,$$
$$Cx + 4y = 14$$

50. Solve using the substitution method.

$$x - y + 2z = -3,$$
$$2x + y - 3z = 11,$$
$$z = -2$$

51. Solve.

$$3(x - y) = 4 + x,$$
$$x = 5y + 2$$

52. For a two-digit number, the sum of the unit's digit and the ten's digit is 6. When the digits are reversed, the new number is 18 more than the original number. Find the original number.

53. A stable boy agreed to work for one year. At the end of that time he was to receive $240 and one horse. After 7 months the boy quit the job, but still received the horse and $100. What was the value of the horse?

TEST: CHAPTER 6

In which quadrant is each point located?

1. $(-\frac{1}{2}, 7)$

2. $(-5, -6)$

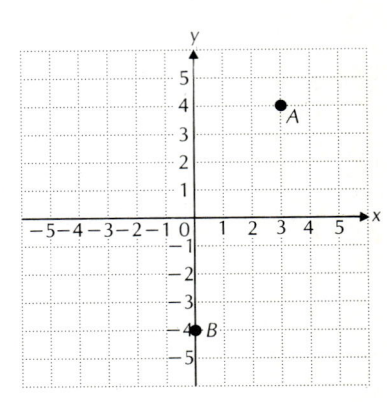

Find the coordinates of each point.

3. A

4. B

5. Determine whether the ordered pair $(2, -4)$ is a solution of the equation $y - 3x = -10$.

Graph.

6. $y = 2x - 1$

7. $2x - 4y = -8$

8. $y = 5$

9. $y = -\frac{3}{2}x$

10. Determine whether the given ordered pair is a solution of the system of equations.

$(-2, -1);\quad x = 4 + 2y,$
$ 2y - 3x = 4$

11. Use graph paper. Solve graphically.

$x - y = 3,$
$x - 2y = 4$

Solve by the substitution method.

12. $y = 6 - x,$
$2x - 3y = 22$

13. $x + 2y = 5,$
$x + y = 2$

Solve by the addition method.

14. $x - y = 6,$
$3x + y = -2$

15. $\frac{1}{2}x - \frac{1}{3}y = 8,$
$\frac{2}{3}x + \frac{1}{2}y = 5$

16. $4x + 5y = 5,$
$6x + 7y = 7$

17. $2x + 3y = 13,$
$3x - 5y = 10$

18. The difference of two numbers is 12. One-fourth of the larger number plus one-half of the smaller is 9. Find the numbers.

19. A motorboat traveled for 2 hours with an 8-km/h current. The return trip against the same current took 3 hours. Find the speed of the motorboat in still water.

20. Solution A is 25% acid and solution B is 40% acid. How much of each is needed to make 60 liters of a solution that is 30% acid?

21. Find the area of a rectangle whose vertices have coordinates $(-3, 1)$, $(5, 1)$, $(5, 8)$, and $(-3, 8)$.

22. You are in line at a ticket window. There are two more people ahead of you in line than there are behind you. In the entire line there are three times as many people as there are behind you. How many are ahead of you in line?

CUMULATIVE REVIEW: CHAPTERS 1–6

1. Simplify: $\dfrac{144ab}{108abc}$.

2. Evaluate: $(y - 1)^2$ for $y = -6$.

3. Write a true sentence using $<$ or $>$: $\quad -3.1 \quad -3.15$.

Compute and simplify.

4. $-\frac{1}{2} + \frac{3}{8} + (-6) + \frac{3}{4}$

5. $-\frac{2}{3}\left(\frac{18}{15}\right)$

6. $-6.262 \div 1.01$

7. $-\frac{72}{108} \div \left(-\frac{2}{3}\right)$

8. Remove parentheses and simplify: $\quad -4[2(x - 3) - 1]$.

Solve.

9. $3(x - 2) = 24$

10. $6y + 3 = -15$

11. $-4x = -18$

12. $5x + 7 = -3x - 9$

13. $4(y - 5) = -2(y + 2)$

14. $-6x - 2(x - 4) = 10$

15. $\frac{1}{3}x - \frac{2}{9} = \frac{2}{3} + \frac{4}{9}x$

16. $-\frac{5}{6} = x - \frac{1}{3}$

Collect like terms.

17. $x + 2y - 2z + \frac{1}{2}x - z$

18. $2x^3 - 7 + \frac{3}{7}x^2 - 6x^3 - \frac{4}{7}x^2 + 5$

Simplify.

19. $x^8 \cdot x^2$

20. $\dfrac{z^4}{z^7}$

21. $(-3x^2y)^3$

22. Subtract: $\quad (-8y^2 - y + 2) - (y^3 - 6y^2 + y - 5)$.

Multiply.

23. $4(3x + 4y + z)$

24. $(2.5a + 7.5)(0.4a - 1.2)$

25. $(2x^2 - 1)(x^3 + x - 3)$

26. $(1 - 3x^2)(2 - 4x^2)$

27. $(2x^5 + 3)(3x^2 - 6)$

28. $(2x^3 + 1)(2x^3 - 1)$

29. $(6x - 5)^2$

30. $(8 - \frac{1}{3}x)(8 + \frac{1}{3}x)$

Factor.

31. $36 - 81y$

32. $-6 - 2x - 12y$

33. $x^2 - 10x + 24$

34. $8x^2 + 10x + 3$

35. $6x^5 - 36x^3 + 9x^2$

36. $2x^2 - 18$

37. $16x^2 + 40x + 25$

38. $3x^2 + 10x - 8$

39. $x^4 + 2x^3 - 3x - 6$
40. $12x - 4x^2 - 48x^4$
41. $16x^4 - 81$
42. $6x^2 - 28x + 16$
43. $3 - 12x^6$
44. Solve: $2x^2 + 7x = 4$.

Simplify.

45. $(5xy^2 - 6x^2y^2 - 3xy^3) - (-4xy^3 + 7xy^2 - 2x^2y^2)$
46. $(3x^2 + 4y)(3x^2 - 4y)$
47. $(2a^2b - 5ab^2)^2$
48. Factor: $4x^4 - 12x^2y + 9y^2$.

Graph.

49. $y = \frac{1}{2}x$
50. $3x - 5y = 15$
51. $y = -x - 2$
52. $y = 1$
53. Solve by the substitution method.
$$y = 1 - x,$$
$$2x - 5y = 9$$

Solve by the addition method.

54. $3x - 4y = 32,$
$4x + 3y = 1$

55. $-3x - y = 0,$
$2x - y = -5$

56. $\frac{1}{4}x - \frac{1}{5}y = -1,$
$\frac{1}{3}x - \frac{1}{4}y = -\frac{5}{4}$

57. A scale drawing shows a bolt enlarged 5 times. If the drawing is 3.7 cm long, how long is the actual bolt?

58. The product of two even consecutive integers is 224. Find the integers.

59. The second angle of a triangle is twice as large as the first. The third angle is 48° less than the sum of the other two angles. Find the measures of the angles.

60. Six hamburgers and 4 milkshakes cost $11.40. Three hamburgers and one milkshake cost $4.80. Find the cost of a hamburger and the cost of a milkshake.

61. The perimeter of a rectangle is 220 ft. One-fourth the length is 10 less than the width. Find the length and the width.

62. Simplify: $-|0.875 - (-\frac{1}{8}) - 8|$.
63. Solve: $\frac{1}{3}|n| + 8 = 56$.
64. Multiply: $[4y^3 - (y^2 - 3)][4y^3 + (y^2 - 3)]$.
65. Factor: $2a^{32} - 13{,}122b^{40}$.

INEQUALITIES AND SETS

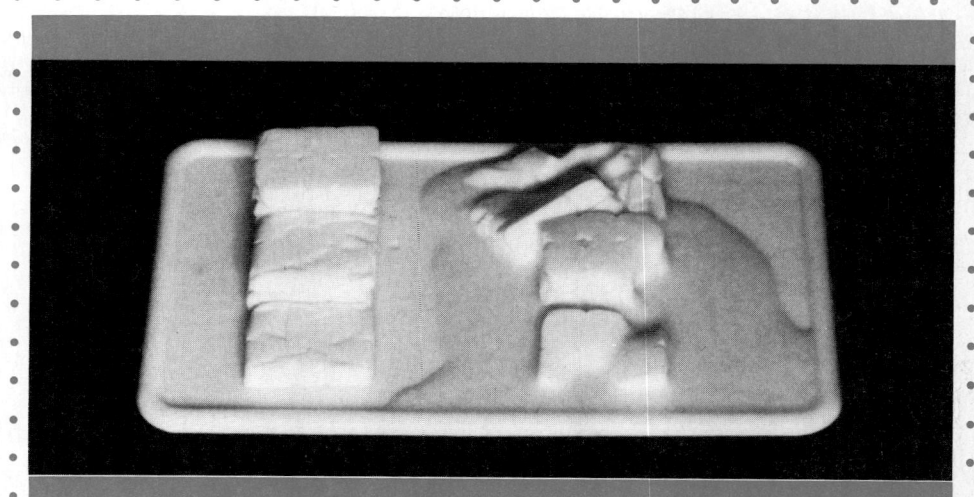

7

How can we find those Celsius temperatures for which butter stays solid? Inequalities can be a help in solving such a problem.

An inequality is a sentence like $2x + 3 < -9$. There are principles for solving inequalities very much like the addition and multiplication principles that we have used to solve equations. We learn how to solve inequalities in one variable and how to graph them. Then we use our new skills for problem solving.

Another goal of the chapter is to graph inequalities in two variables in the plane.

7.1 USING THE ADDITION PRINCIPLE

In this section we learn an addition principle for solving inequalities that is similar to the one we learned for equations.

Solutions of Inequalities

In Section 2.1 we defined the symbols $>$ (greater than) and $<$ (less than). The symbol $\geq$ means *greater than or equal to*. The symbol $\leq$ means *less than or equal to*. For example, $3 \leq 4$ and $3 \leq 3$ are both true, but $-3 \leq -4$ and $0 \geq 2$ are both false. An *inequality* is a number sentence with $>$, $<$, $\geq$, or $\leq$ for its verb—for example,

$$-4 < 5, \quad x < 3, \quad 2x + 5 \geq 0, \quad \text{and} \quad -3y + 7 \leq -8.$$

Some replacements for a variable in an inequality make it true and some make it false.

EXAMPLE 1 Determine whether each number is a solution of $x < 3$.

-2 Since $-2 < 3$ is true, -2 is a solution.

5 Since $5 < 3$ is false, 5 is not a solution.

$\frac{1}{4}$ Since $\frac{1}{4} < 3$ is true, $\frac{1}{4}$ is a solution.

When we have found *all* the numbers that make an inequality true, we have *solved* the inequality.

> The replacements that make an inequality true are called its *solutions*. To *solve* an inequality means to find all of its solutions.

The Addition Principle

Consider the true inequality

$$3 < 7.$$

Add 2 on both sides and we get another true inequality:

$$3 + 2 < 7 + 2 \quad \text{or} \quad 5 < 9.$$

Similarly, if we add -3 to both numbers we get another true inequality:

$$3 + (-3) < 7 + (-3) \quad \text{or} \quad 0 < 4.$$

> **THE ADDITION PRINCIPLE FOR INEQUALITIES**
>
> If any number is added on both sides of a true inequality, we get another true inequality.

7.1 USING THE ADDITION PRINCIPLE

The addition principle holds whether the inequality contains $<$, $>$, $\leq$, or $\geq$. Let's see how we use the addition principle to solve inequalities.

EXAMPLE 2 Solve: $x + 2 > 8$.

We use the addition principle, adding -2:

$$x + 2 + (-2) > 8 + (-2)$$
$$x > 6.$$

By using the addition principle we get an inequality for which we can determine the solutions easily.

Any number greater than 6 makes the last sentence true, hence is a solution of that sentence. Any such number is also a solution of the original sentence. Thus we have it solved.

We cannot check all the solutions of an inequality by substitution, as we can check solutions of equations. There are too many of them.

A partial check could be done by substituting a number greater than 6, say, 7, into the original inequality:

$$\begin{array}{c|c} x + 2 > 8 \\ \hline 7 + 2 & 8 \\ 9 & \end{array}$$

Since $9 > 8$ is true, 7 is a solution.

However, we don't really need to check. Let us see why. Consider the first and last inequalities

$$x + 2 > 8 \quad \text{and} \quad x > 6.$$

Any number that makes the first one true must make the last one true. We know this by the addition principle. Now the question is, will any number that makes the last one true also be a solution of the first one? Let us use the addition principle again, adding 2:

$$x > 6$$
$$x + 2 > 6 + 2$$
$$x + 2 > 8.$$

Now we know that any number that makes $x > 6$ true also makes $x + 2 > 8$ true. Therefore the sentences $x > 6$ and $x + 2 > 8$ have the same solutions. When two equations or inequalities have the same solutions, we say that they are *equivalent*. Any time we use the addition principle a similar thing happens. Thus, whenever we use the principle with inequalities, the first and last sentences will be equivalent.

EXAMPLE 3 Solve: $3x + 1 < 2x - 3$.

$$3x + 1 < 2x - 3$$
$$3x + 1 - 1 < 2x - 3 - 1 \quad \text{Adding } -1$$
$$3x < 2x - 4 \quad \text{Simplifying}$$
$$3x - 2x < 2x - 4 - 2x \quad \text{Adding } -2x$$
$$x < -4 \quad \text{Simplifying}$$

Any number less than -4 is a solution. The following are some of the solutions:

$$-5, \quad -6, \quad -4.1, \quad -2045, \quad -18\pi.$$

To describe all the solutions, we use the set notation

$$\{x \mid x < -4\},$$

which is read:

The set of all x such that x is less than -4.

EXAMPLE 4 Solve: $x + \frac{1}{3} \geq \frac{5}{4}$.

$$x + \frac{1}{3} \geq \frac{5}{4}$$
$$x + \frac{1}{3} - \frac{1}{3} \geq \frac{5}{4} - \frac{1}{3} \quad \text{Adding } -\frac{1}{3}$$
$$x \geq \frac{5}{4} \cdot \frac{3}{3} - \frac{1}{3} \cdot \frac{4}{4} \quad \text{Finding a common denominator}$$
$$x \geq \frac{15}{12} - \frac{4}{12}$$
$$x \geq \frac{11}{12}$$

Any number greater than or equal to $\frac{11}{12}$ is a solution. The *solution set* is

$$\{x \mid x \geq \tfrac{11}{12}\},$$

which is read:

The set of all x such that x is greater than or equal to $\frac{11}{12}$.

EXERCISE SET 7.1

Determine whether each number is a solution of the given inequality.

1. $x > 4$
   a) 4
   b) 0
   c) -4
   d) 6

2. $y < 5$
   a) 0
   b) 5
   c) -1
   d) -5

3. $x \geq 6$
   a) -6
   b) 0
   c) 6
   d) 8

4. $x \leq 10$
   a) 4
   b) -10
   c) 0
   d) 11

5. $x < -8$
   a) 0
   b) -8
   c) -9
   d) -7

6. $x \geq 0$
   a) 2
   b) -3
   c) 0
   d) 3

7. $y \geq -5$
   a) 0
   b) -4
   c) -5
   d) -6

8. $y \leq -\frac{1}{2}$
   a) -1
   b) $-\frac{2}{3}$
   c) 0
   d) -0.5

Solve using the addition principle. Write set notation for answers.

9. $x + 7 > 2$
10. $x + 6 > 3$
11. $y + 5 > 8$
12. $y + 7 > 9$
13. $x + 8 \leq -10$
14. $x + 9 \leq -12$
15. $a + 12 < 6$
16. $a + 20 < 8$
17. $x - 7 \leq 9$
18. $x - 3 \leq 14$
19. $x - 6 > 2$
20. $x - 9 > 4$
21. $y - 7 > -12$
22. $y - 10 > -16$
23. $2x + 3 > x + 5$
24. $2x + 4 > x + 7$
25. $3x + 9 \leq 2x + 6$
26. $3x + 10 \leq 2x + 8$
27. $3x - 6 \geq 2x + 7$
28. $3x - 9 \geq 2x + 11$
29. $5x - 6 < 4x - 2$
30. $6x - 8 < 5x - 9$
31. $3y + 4 \geq 2y - 7$
32. $4y + 5 \leq 3y - 8$
33. $7 + c > 7$
34. $-9 + c > 9$
35. $y + \frac{1}{4} \leq \frac{1}{2}$
36. $y + \frac{1}{3} \leq \frac{5}{6}$
37. $x - \frac{1}{3} > \frac{1}{4}$
38. $x - \frac{1}{8} > \frac{1}{2}$
39. $-14x + 21 > 21 - 15x$
40. $-10x + 15 > 18 - 11x$

Problem-solving practice

41. When the sides of a square are lengthened by 0.2 km, the area becomes 0.64 km². Find the length of a side of the original square.

42. In one year 43,500 hunting and fishing licenses were sold in the United States. There were 5500 more fishing licenses sold than hunting licenses. How many of each were sold that year?

Solve.

43. $3(r + 2) < 2r + 4$
44. $4(r + 5) \geq 3r + 7$
45. $0.8x + 5 \geq 6 - 0.2x$
46. $0.7x + 6 \leq 7 - 0.3x$
47. $2x + 2.4 > x - 9.4$
48. $5x + 2.5 > 4x - 1.5$
49. $12x + 1.2 \leq 11x$
50. $x + 0.8 \leq 7.8 - 6$
51. Solve: $17x + 9,479,756 \leq 16x - 8,579,243$.
52. Suppose that $2x - 5 \geq 9$ is true for some value of x. Determine whether $2x - 5 \geq 8$ is true for that same value of x.

7.2 USING THE MULTIPLICATION PRINCIPLE

There is a multiplication principle for inequalities similar to that for equations, but it must be qualified when multiplying on both sides by a negative number. We consider the multiplication principle and then use it with the addition principle to solve inequalities.

The Multiplication Principle

Consider the true inequality
$$3 < 7.$$

Multiply both numbers by 2 and we get another true inequality:
$$3 \cdot 2 < 7 \cdot 2 \quad \text{or} \quad 6 < 14.$$

Multiply both numbers by -3 and we get the false inequality:
$$3 \cdot (-3) < 7 \cdot (-3)$$

or
$$-9 < -21. \qquad \text{False}$$

However, if we reverse the inequality symbol we get a true inequality:
$$-9 > -21. \qquad \text{True}$$

> **THE MULTIPLICATION PRINCIPLE FOR INEQUALITIES**
>
> If we multiply on both sides of a true inequality by a positive number, we get another true inequality. If we multiply by a negative number and the inequality symbol is reversed, we get another true inequality.

The multiplication principle holds whether the inequality contains $<$, $>$, $\geq$, or $\leq$.

EXAMPLE 1 Solve: $4x < 28$.

$$\tfrac{1}{4} \cdot 4x < \tfrac{1}{4} \cdot 28 \qquad \text{Multiplying by } \tfrac{1}{4}$$
$$\text{The symbol stays the same.}$$
$$x < 7 \qquad \text{Simplifying}$$

The solution set is $\{x \mid x < 7\}$.

EXAMPLE 2 Solve: $-2y < 18$.

$$-\tfrac{1}{2}(-2y) > -\tfrac{1}{2} \cdot 18 \qquad \text{Multiplying by } -\tfrac{1}{2}$$
$$\text{The symbol has to be reversed!}$$
$$y > -9 \qquad \text{Simplifying}$$

The solution set is $\{y \mid y > -9\}$.

Using the Principles Together

We use the addition and multiplication principles together in solving inequalities in much the same way as in solving equations. We generally use the addition principle first.

EXAMPLE 3 Solve: $6 - 5y > 7$.

$$-6 + 6 - 5y > -6 + 7 \quad \text{Adding } -6$$
$$-5y > 1 \quad \text{Simplifying}$$
$$-\tfrac{1}{5} \cdot (-5y) < -\tfrac{1}{5} \cdot 1 \quad \text{Multiplying by } -\tfrac{1}{5}$$
$$\text{The symbol has to be reversed!}$$
$$y < -\tfrac{1}{5}$$

The solution set is $\{y \mid y < -\tfrac{1}{5}\}$.

EXAMPLE 4 Solve: $5x + 9 \leqslant 4x + 3$.

$$5x + 9 - 9 \leqslant 4x + 3 - 9 \quad \text{Adding } -9$$
$$5x \leqslant 4x - 6 \quad \text{Simplifying}$$
$$5x - 4x \leqslant 4x - 6 - 4x \quad \text{Adding } -4x$$
$$x \leqslant -6 \quad \text{Simplifying}$$

The solution set is $\{x \mid x \leqslant -6\}$.

EXAMPLE 5 Solve: $8y - 5 > 17 - 5y$.

$$-17 + 8y - 5 > -17 + 17 - 5y \quad \text{Adding } -17$$
$$8y - 22 > -5y \quad \text{Simplifying}$$
$$-8y + 8y - 22 > -8y - 5y \quad \text{Adding } -8y$$
$$-22 > -13y \quad \text{Simplifying}$$
$$-\tfrac{1}{13} \cdot (-22) < -\tfrac{1}{13} \cdot (-13y) \quad \text{Multiplying by } -\tfrac{1}{13}$$
$$\text{The symbol has to be reversed.}$$
$$\tfrac{22}{13} < y$$

The solution set is $\{y \mid \tfrac{22}{13} < y\}$. Since $\tfrac{22}{13} < y$ has the same meaning as $y > \tfrac{22}{13}$, we can also describe the solution set as $\{y \mid y > \tfrac{22}{13}\}$.

EXERCISE SET 7.2

Solve using the multiplication principle.

1. $5x < 35$
2. $8x \geqslant 32$
3. $9y \leqslant 81$
4. $10x > 240$
5. $7x < 13$
6. $8y < 17$
7. $12x > -36$
8. $16x < -64$
9. $5y \geqslant -2$
10. $7x > -4$
11. $-2x \leqslant 12$
12. $-3y \leqslant 15$

13. $-4y \geq -16$
14. $-7x < -21$
15. $-3x < -17$
16. $-5y > -23$
17. $-2y > \frac{1}{7}$
18. $-4x \leq \frac{1}{9}$
19. $-\frac{6}{5} \leq -4x$
20. $-\frac{7}{8} > -56t$

Solve using the addition and multiplication principles.

21. $4 + 3x < 28$
22. $5 + 4y < 37$
23. $6 + 5y \geq 36$
24. $7 + 8x \geq 71$
25. $3x - 5 \leq 13$
26. $5y - 9 \leq 21$
27. $10y - 9 > 31$
28. $12y - 6 > 42$
29. $13x - 7 < -46$
30. $8y - 4 < -52$
31. $5x + 3 \geq -7$
32. $7y + 4 \geq -10$
33. $4 - 3y > 13$
34. $6 - 8x > 22$
35. $3 - 9x < 30$
36. $5 - 7y < 40$
37. $3 - 6y > 23$
38. $8 - 2y > 14$
39. $4x + 2 - 3x \leq 9$
40. $15x + 3 - 14x \leq 7$
41. $8x + 7 - 7x > -3$
42. $9x + 8 - 8x > -5$
43. $6 - 4y > 4 - 3y$
44. $7 - 8y > 5 - 7y$
45. $5 - 9y \leq 2 - 8y$
46. $6 - 13y \leq 4 - 12y$
47. $19 - 7y - 3y < 39$
48. $18 - 6y - 9y < 63$
49. $21 - 8y < 6y + 49$
50. $33 - 12x < 4x + 97$
51. $14 - 5y - 2y \geq -19$
52. $17 - 6y - 7y \leq -13$
53. $27 - 11x > 14x - 18$
54. $42 - 13y > 15y - 19$

Problem-solving practice

55. The first angle of a triangle is five times as large as the second. The measure of the third angle is 16° less than that of the second. How large are the angles?

56. A number subtracted from twice its square is six. Find the number.

Solve.

57. $5(12 - 3t) \geq 15(t + 4)$
58. $6(z - 5) < 5(7 - 2z)$
59. $4(0.5 - y) + y > 4y - 0.2$
60. $3 + 3(0.6 + y) > 2y + 6.6$
61. $\frac{x}{3} - 2 \leq 1$
62. $\frac{2}{3} - \frac{x}{5} < \frac{4}{15}$
63. $\frac{y}{5} + 1 \leq \frac{2}{5}$
64. $\frac{3x}{5} \geq -15$
65. $\frac{-x}{4} - \frac{3x}{8} + 2 > 3 - x$
66. $11 - x > 5 + \frac{2x}{5}$

Solve for x.

67. $-(x + 5) \geq 4a - 5$
68. $\frac{1}{2}(2x + 2b) > \frac{1}{3}(21 + 3b)$
69. $-6(x + 3) \leq -9(y + 2)$
70. $y < ax + b$
71. $x^2 > 0$
72. $x^2 + 1 > 0$

Determine whether each of the following statements is true for all rational numbers a, b, and c.

73. If $a + c < b + c$, then $a < b$.

74. If $a - c \leqslant b - c$, then $a \leqslant b$.

75. If $ac \leqslant bc$, then $a \leqslant b$.

76. If $a^2 < b^2$, then $a < b$.

77. Suppose we are considering *only* integer solutions to $x > 5$. Find an equivalent inequality involving $\geqslant$.

Translate to an inequality.

78. a is at least b

79. x is at most y

7.3 PROBLEM SOLVING USING INEQUALITIES

We can use inequalities to solve certain kinds of problems. We continue to use our five-step problem-solving process, but we translate to inequalities rather than equations.

EXAMPLE 1 A student is taking an introductory algebra course in which four tests are to be given. To get an A, the student must average at least 90 on the four tests. The student got scores of 91, 86, and 89 on the first three tests. Determine (in terms of an inequality) what scores on the last test will allow the student to get an A.

1. **Familiarize** Suppose the student gets a 96 on the last test. The average of the four scores is their sum divided by the number of tests, 4, and is given by

$$\frac{91 + 86 + 89 + 96}{4}.$$

For this average to be *at least* 90 means it must be greater than or equal to 90. Thus we must have

$$\frac{91 + 86 + 89 + 96}{4} \geqslant 90.$$

Since the average is $\frac{362}{4}$, or 90.5, the test score of 96 will give the student an A. But there are other possible scores. To find all of them, we translate to an inequality and solve.

2. **Translate** We let x represent the student's score on the last test. The average of the four scores is given by

$$\frac{91 + 86 + 89 + x}{4}.$$

Since this average must be *at least* 90, this means that it must be greater than or equal to 90. Thus we can translate the problem to the inequality

$$\frac{91 + 86 + 89 + x}{4} \geqslant 90.$$

3. **Carry out** We solve the inequality. We first multiply by 4 to clear of fractions.

$$\frac{91 + 86 + 89 + x}{4} \geq 90$$

$$4\left(\frac{91 + 86 + 89 + x}{4}\right) \geq 4 \cdot 90 \qquad \text{Multiplying by 4}$$

$$91 + 86 + 89 + x \geq 360$$

$$266 + x \geq 360 \qquad \text{Collecting like terms}$$

$$x \geq 94$$

4. **Check** Suppose x is a score greater than or equal to 94. Then by successively adding 91, 86, and 89 on both sides of the inequality we get

$$91 + 86 + 89 + x \geq 360,$$

so

$$\frac{91 + 86 + 89 + x}{4} \geq \frac{360}{4}, \quad \text{or 90.}$$

From what we did in the familiarization, we know that 96 is a score that checks. This score is greater than or equal to 94. This also gives us a partial check.

5. **State** Any score that is at least 94 will give the student an A in the course.

EXAMPLE 2 A formula for converting Celsius temperatures C to Fahrenheit temperatures F is

$$F = \tfrac{9}{5}C + 32.$$

Butter stays solid at Fahrenheit temperatures below 88°. Determine (in terms of an inequality) those Celsius temperatures for which butter stays solid.

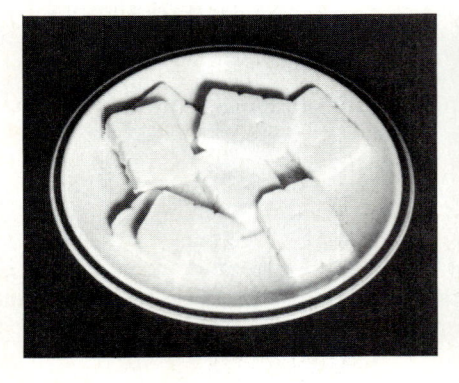

1. **Familiarize** We can draw a picture of the situation as follows.

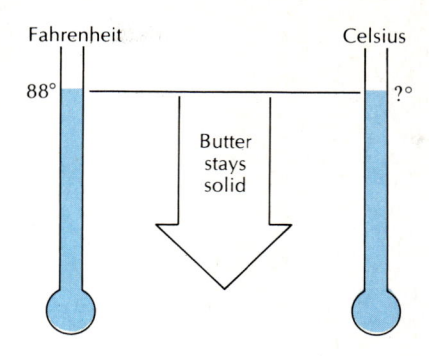

Fahrenheit 88° / Celsius ?°
Butter stays solid

2. **Translate** The Fahrenheit temperature F is to be less than 88. We have the inequality

$$F < 88.$$

To find Celsius temperatures C that satisfy this condition, we substitute $\frac{9}{5}C + 32$ for F:

$$\tfrac{9}{5}C + 32 < 88.$$

Thus we have translated the problem to an inequality.

3. **Carry out** We solve the inequality:

$$\tfrac{9}{5}C + 32 < 88$$
$$\tfrac{9}{5}C < 56 \qquad \text{Adding } -32$$
$$C < \tfrac{5}{9} \cdot 56 \qquad \text{Multiplying by } \tfrac{5}{9}$$
$$C < 31.1. \qquad \text{Rounding to the nearest tenth}$$

4. **Check** The check is left to the student.

5. **State** Butter stays solid at Celsius temperatures below 31.1°.

EXERCISE SET 7.3

1. Your quiz grades are 73, 75, 89, and 91. Determine (in terms of an inequality) those scores you can obtain on the last quiz in order to receive an average quiz grade of 85.

2. A human body is considered to be fevered when its temperature is higher than 98.6°F. Using the formula in Example 2, determine (in terms of an inequality) those Celsius temperatures for which the body is fevered.

3. The formula
$$R = -0.075t + 3.85$$
can be used to predict the world record in the 1500-meter run t years after 1930. Determine (in terms of an inequality) those years for which the world record will be less than 3.5 minutes.

4. Find all numbers such that the sum of the number and 15 is less than four times the number.

5. Find all numbers such that three times the number minus ten times the number is greater than or equal to eight times the number.

6. Acme rents station wagons at a daily rate of $42.95 plus $0.46 per mile. A family wants to rent a wagon one day while on vacation, but must stay within a budget of $200. Determine (in terms of an inequality) those mileages that will allow the family to stay within budget. Round to the nearest tenth of a mile.

7. Atlas rents an intermediate-size car at a daily rate of $44.95 plus $0.39 per mile. A businessperson is not to exceed a daily car rental budget of $250. Determine (in terms of an inequality) those mileages that will allow the businessperson to stay within budget. Round to the nearest tenth of a mile.

8. The length of a rectangle is 4 cm. Determine (in terms of an inequality) those widths for which the area will be less than 86 cm^2.

9. The width of a rectangle is 16 yd. Determine (in terms of an inequality) those lengths for which the area will be greater than or equal to 264 yd^2.

10. The length of a rectangle is 26 cm. What widths will make the perimeter greater than 80 cm?

11. The width of a rectangle is 8 ft. What lengths will make the perimeter at least 200 ft? at most 200 ft?

12. One side of a triangle is 2 cm shorter than the base. The other side is 3 cm longer than the base. What lengths of the base will allow the perimeter to be greater than 19 cm?

13. A salesperson made 18 customer calls last week and 22 calls this week. How many calls must be made next week to maintain an average of at least 20 for the three-week period?

14. George and Joan do volunteer work at a hospital. Joan worked 3 more hours than George and together they worked more than 27 hours. What possible hours did each work?

15. A student is shopping for a new pair of jeans and a sweater. The student is determined to spend no more than $40.00 for the outfit. The student buys jeans for $21.95. What is the most the student can spend for the sweater?

16. The width of a rectangle is 32 km. What lengths will make the area at least 2048 km²?

17. The height of a triangle is 20 cm. What lengths of the base will make the area at most 40 cm²?

18. The area of a square can be no more than 64 cm². What lengths of a side will allow this?

19. The sum of two consecutive odd integers is less than 100. What is the largest possible pair of such integers?

20. The sum of three consecutive odd integers is less than 30. What are the three largest such integers?

21. A salesperson can choose to be paid in one of two ways.

    Plan A: A salary of $600 per month, plus a commission of 4% of gross sales.
    Plan B: A salary of $800 per month, plus a commission of 6% of gross sales over $10,000.

For what gross sales is plan A better than plan B, assuming that gross sales are always more than $10,000?

22. A mason can be paid in one of two ways.

    Plan A: $300 plus $3.00 per hour.
    Plan B: $8.50 per hour.

If a job takes n hours, for what values of n is plan B better for the mason?

7.4 GRAPHS OF INEQUALITIES

A *graph* of an inequality is a drawing of its solutions. An inequality in one variable can be graphed on a number line. An inequality in two variables can be graphed on a coordinate plane.

Inequalities in One Variable

We graph inequalities in one variable on a number line.

EXAMPLE 1 Graph: $x < 2$.

The solutions of $x < 2$ are those numbers less than 2. They are shown on the graph by shading all points to the left of 2. The open circle at 2 indicates that 2 is not part of the graph.

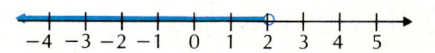

EXAMPLE 2 Graph: $y \geq -3$.

The solutions of $y \geq -3$ are shown by shading the point for -3 and all points to the right of -3. The closed circle at -3 indicates that -3 *is* part of the graph.

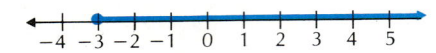

EXAMPLE 3 Graph: $3x + 2 < 5x - 1$.

We first solve:

$$3x + 2 < 5x - 1$$
$$2 < 2x - 1 \quad \text{Adding } -3x$$
$$3 < 2x \quad \text{Adding 1}$$
$$\tfrac{3}{2} < x \quad \text{Multiplying by } \tfrac{1}{2}$$

We shade all points to the right of $\tfrac{3}{2}$.

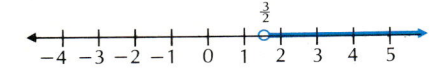

Graphing Inequalities with Absolute Value

EXAMPLE 4 Graph: $|x| < 3$.

The absolute value of a number is its distance from 0 on a number line. For the absolute value of a number to be less than 3 it must be between 3 and -3. Therefore we use open circles at 3 and -3 and shade the points between these two numbers.

EXAMPLE 5 Graph: $|x| \geq 2$.

For the absolute value of a number to be greater than or equal to 2 its distance from 0 must be 2 or more. Thus the number must be 2 or greater, or it must be less than or equal to -2. Therefore we shade the point for 2 and all points to its right. We also shade the point for -2 and all points to its left.

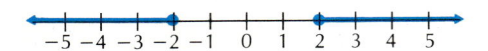

Inequalities in Two Variables

The solutions of inequalities in two variables are ordered pairs.

7.4 GRAPHS OF INEQUALITIES

EXAMPLE 6 Determine whether $(-3, 2)$ is a solution of $5x - 4y < 13$.

We use alphabetical order of variables. We replace x by -3 and y by 2.

$$\begin{array}{c|c} 5x - 4y < 13 & \\ \hline 5(-3) - 4 \cdot 2 & 13 \\ -15 - 8 & \\ -23 & \end{array}$$

Since $-23 < 13$ is true, $(-3, 2)$ is a solution.

EXAMPLE 7 Graph: $y > x$.

We first graph the line $y = x$ for comparison. Every solution of $y = x$ is an ordered pair such as $(3, 3)$. The first and second coordinates are the same. The graph of $y = x$ is shown to the left below.

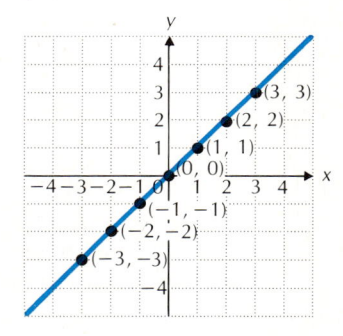

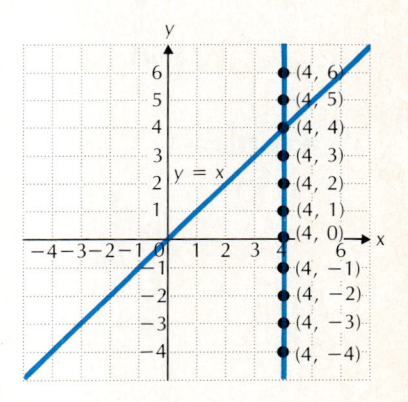

Now look at the graph to the right above. We consider a vertical line and ordered pairs on it. For all points above $y = x$, the second coordinate is greater than the first, $y > x$. For all points below the line, $y < x$. The same thing happens for any vertical line. Then for all points above $y = x$, the ordered pairs are solutions. We shade the half-plane above $y = x$. This is the graph of $y > x$. Points on $y = x$ are not in the graph, so we draw it dashed.

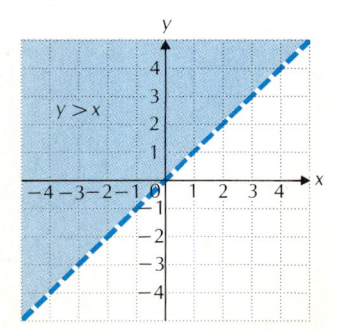

EXAMPLE 8 Graph: $y \leq x - 1$.

First we sketch the line $y = x - 1$. Points on the line $y = x - 1$ are also in the graph of $y \leq x - 1$, so we draw the line solid. For points above the line, $y > x - 1$. These points are not in the graph. For points below the line, $y < x - 1$. These are in the graph, so we shade the lower half-plane.

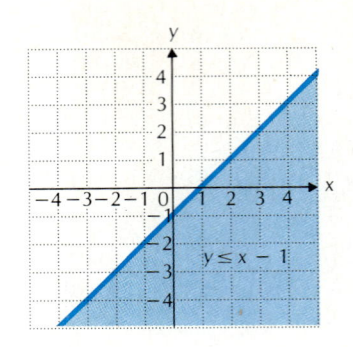

A *linear inequality* is one that we can get from a linear equation by changing the equals symbol to an inequality symbol. Every linear equation has a graph that is a straight line. The graph of a linear inequality is a half-plane, sometimes including the line along the edge. In the following example we show a different method of graphing. We graph the line using intercepts. Then we determine which side to shade by substituting a point from either half-plane.

EXAMPLE 9 Graph: $6x - 2y < 10$.

We first graph the line $y = 3x - 5$. The intercepts are $(0, -5)$ and $(\frac{5}{3}, 0)$. The point $(3, 4)$ is also on the graph. This line forms the boundary of the solutions of the inequality. In this case points on the line are not solutions of the inequality. We shade the half-plane above the line. We know to do this by picking a test point on the line. We try $(3, -2)$ and substitute:

$$6(3) - 2(-2) < 10, \quad \text{or} \quad 22 < 10.$$

Since this inequality is *false*, the point $(3, -2)$ is *not* a solution; no point in the half-plane containing $(3, -2)$ is a solution. Thus the points in the opposite half-plane are solutions.

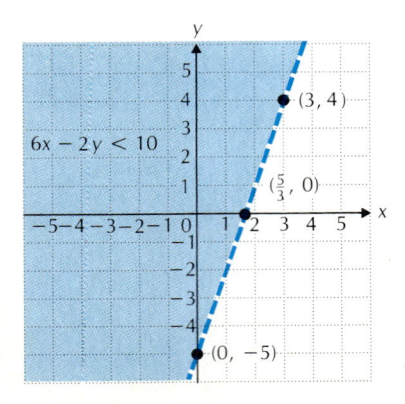

EXAMPLE 10 Graph: $2x + 3y < 6$.

a) First we graph the line $2x + 3y = 6$. The intercepts are $(3, 0)$ and $(0, 2)$. We use a dashed line for the graph since we have $<$.

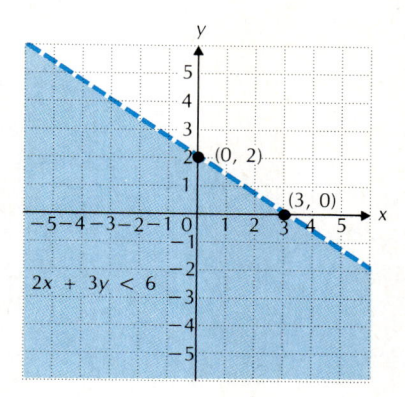

b) Let's pick a point that does not belong to the line. Substitute to determine whether this point is a solution. The origin $(0, 0)$ is usually an easy one to use: $2 \cdot 0 + 3 \cdot 0 < 6$ is true, so the origin is a solution. This means we shade the lower half-plane. Had the substitution given us a false inequality, we would have shaded the other half-plane.

If the line goes through the origin, we must test some other point not on the line. The point $(1, 1)$ is often a good one to try.

EXERCISE SET 7.4

Graph on a number line.

1. $x < 5$
2. $y < 0$
3. $t < -3$
4. $y > 5$
5. $x \geqslant 6$
6. $y \geqslant -4$
7. $x + 2 > 7$
8. $x + 3 > 9$
9. $z - 3 < 4$
10. $x - 2 < 6$
11. $t - 3 \leqslant -7$
12. $x - 4 \leqslant -8$
13. $x - 8 \geqslant 0$
14. $2x + 6 < 14$
15. $3y + 5 < 26$
16. $4x - 8 \geqslant 12$
17. $5y - 4 > 11$
18. $3x + 7 < 8x - 3$
19. $4y + 9 > 11y - 12$
20. $5t + 8 \geqslant 12t - 27$
21. $6x + 11 \leqslant 14x + 7$
22. $|x| < 2$
23. $|y| < 6$
24. $|t| \leqslant 1$
25. $|x| \leqslant 7$
26. $|y| > 3$
27. $|t| > 4$
28. $|x| \geqslant 7$
29. $|y| \geqslant 9$

30. Determine whether $(-3, -5)$ is a solution of $-x - 3y < 18$.

31. Determine whether $(5, -3)$ is a solution of $-2x + 4y \leq -2$.

32. Determine whether $(\frac{1}{2}, -\frac{1}{4})$ is a solution of $7y - 9x > -3$.

33. Determine whether $(-6, 5)$ is a solution of $x + 0 \cdot y < 3$.

Graph on a plane.

34. $x > 2y$
35. $x > 3y$
36. $y \leq x - 3$
37. $y \leq x - 5$
38. $y < x + 1$
39. $y < x + 4$
40. $y \geq x - 2$
41. $y \geq x - 1$
42. $y \leq 2x - 1$
43. $y \leq 3x + 2$
44. $x + y \leq 3$
45. $x + y \leq 4$
46. $x - y > 7$
47. $x - y > -2$
48. $x - 3y < 6$
49. $x - y < -10$
50. $2x + 3y \leq 12$
51. $5x + 4y \geq 20$
52. $y \geq 1 - 2x$
53. $y - 2x \leq -1$
54. $y + 4x > 0$
55. $y - x < 0$
56. $y > -3x$
57. $y < -5x$

○

Graph on a plane.

58. $y > 2$
(*Hint:* Think of this as $0 \cdot x + y > 2$.)
59. $x \geq 3$
60. $x > 0$
61. $y \leq 0$

Graph the following pair of inequalities using the same set of axes. Then determine whether each point satisfies *both* inequalities:

$$x + y \leq 1, \quad x - y < 1.$$

62. $(0, 0)$
63. $(1, 1)$
64. $(1, 0)$
65. $(0, 1)$
66. $(-1, -1)$
67. $(1, -3)$
68. $(5, 2)$
69. $(-4, 2)$

7.5 SETS

The concept of set is used frequently in more advanced mathematics. We give a basic introduction to sets in this section.

Naming Sets

We have discussed notation like the following:

$$\{x \mid x < 6\},$$

where we were considering the real numbers. If we were just considering whole numbers, the above set could also be named, or symbolized,

$$\{0, 1, 2, 3, 4, 5\}.$$

7.5 SETS

This way of naming a set is known as the *roster* method. In words, it is

The set consisting of 0, 1, 2, 3, 4, and 5.

Membership

The symbol ∈ means *is a member of* or *belongs to*. Thus

$$x \in A$$

means

$$x \text{ is a member of } A$$

or

$$x \text{ belongs to } A.$$

EXAMPLES Determine whether true or false.

1. $1 \in \{1, 2, 3\}$ True
2. $1 \in \{2, 3\}$ False
3. $4 \in \{x \mid x \text{ is an even whole number}\}$ True
4. $5 \in \{x \mid x \text{ is an even whole number}\}$ False

Set membership can be illustrated with a diagram as shown below.

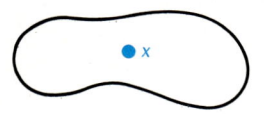

Intersections

The *intersection* of two sets A and B is the set of members common to both sets and is indicated by the symbol

$$A \cap B.$$

Thus,

$$\{0, 1, 3, 5, 25\} \cap \{2, 3, 4, 5, 6, 7, 9\}$$

represents the set

$$\{3, 5\}.$$

Set intersection is illustrated by the diagram below.

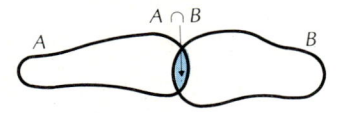

The solution of the system of equations

$$x + 2y = 7,$$
$$x - y = 4$$

is the ordered pair (5, 1). It is the intersection of the graphs of the two lines.

The set without members is known as the *empty set*, and is often named ∅. Each of the following is a description of the empty set:

The set of all six-eyed algebra teachers;

$\{2, 3\} \cap \{5, 6, 7\}$;

$\{x | x$ is an even whole number$\} \cap \{x | x$ is an odd whole number$\}$.

The system of equations

$$x + 2y = 7,$$
$$x + 2y = 9$$

has no solution. The lines are parallel. Their intersection is empty.

Unions

Two sets A and B may be combined to form a new set. It contains the members of A as well as those of B. The new set is called the *union* of A and B, and is represented by the symbol

$$A \cup B.$$

Thus,

$$\{0, 5, 7, 13, 27\} \cup \{0, 2, 3, 4, 5\}$$

represents the set

$$\{0, 2, 3, 4, 5, 7, 13, 27\}.$$

Set union is illustrated by the diagram below.

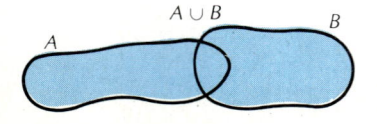

The solution set of the equation $(x - 3)(x + 2) = 0$ is $\{3, -2\}$. This set is the union of the solution sets of the equations $x - 3 = 0$ and $x + 2 = 0$, which are $\{3\}$ and $\{-2\}$.

EXERCISE SET 7.5

Name each set using the roster method.

1. The set of whole numbers 3 through 8
2. The set of whole numbers 101 through 107
3. The set of odd numbers between 40 and 50
4. The set of multiples of 5 between 10 and 40
5. $\{x | \text{the square of } x \text{ is } 9\}$
6. $\{x | x \text{ is the cube of } 0.2\}$

Determine whether true or false.

7. $2 \in \{x | x \text{ is an odd number}\}$
8. $7 \in \{x | x \text{ is an odd number}\}$
9. Elton John $\in$ The set of all rock stars
10. Apple $\in$ The set of all fruit
11. $-3 \in \{-4, -3, 0, 1\}$
12. $0 \in \{-4, -3, 0, 1\}$

Find each intersection.

13. $\{a, b, c, d, e\} \cap \{c, d, e, f, g\}$
14. $\{a, e, i, o, u\} \cap \{q, u, i, c, k\}$
15. $\{1, 2, 5, 10\} \cap \{0, 1, 7, 10\}$
16. $\{0, 1, 7, 10\} \cap \{0, 1, 2, 5\}$
17. $\{1, 2, 5, 10\} \cap \{3, 4, 7, 8\}$
18. $\{a, e, i, o, u\} \cap \{m, n, f, g, h\}$

Find each union.

19. $\{a, e, i, o, u\} \cup \{q, u, i, c, k\}$
20. $\{a, b, c, d, e\} \cup \{c, d, e, f, g\}$
21. $\{0, 1, 7, 10\} \cup \{0, 1, 2, 5\}$
22. $\{1, 2, 5, 10\} \cup \{0, 1, 7, 10\}$
23. $\{a, e, i, o, u\} \cup \{m, n, f, g, h\}$
24. $\{1, 2, 5, 10\} \cup \{a, b\}$

25. Find the union of the set of integers and the set of whole numbers.
26. Find the intersection of the odd integers and the even integers.
27. Find the union of the rational numbers and the irrational numbers.
28. Find the intersection of the even integers and the positive rational numbers.

29. For a set A, find the following.

   a) $A \cup \emptyset$ b) $A \cup A$ c) $A \cap A$ d) $A \cap \emptyset$

30. A set is *closed* under an operation if, when the operation is performed on its members, the result is in the set. For example, the set of real numbers is closed under the operation of addition since the sum of two real numbers is a real number.

   a) Is the set of even numbers closed under addition?
   b) Is the set of odd numbers closed under addition?
   c) Is the set $\{0, 1\}$ closed under addition?
   d) Is the set $\{0, 1\}$ closed under multiplication?
   e) Is the set of real numbers closed under multiplication?
   f) Is the set of integers closed under division?

SUMMARY AND REVIEW: CHAPTER 7

The following contains a summary of what you should be able to do after completing this chapter. The review exercises are for practice. Answers are at the back of the book. If you miss an exercise, restudy the section indicated alongside the answer.

You should be able to:

Determine whether a number is a solution of an inequality.

Determine whether the given number is a solution of the inequality $x \leqslant 4$.

1. -3
2. 7
3. 4

Solve inequalities using the addition principle, the multiplication principle, and both principles together.

Solve. Write set notation for the answers.

4. $y + \frac{2}{3} \geqslant \frac{1}{6}$
5. $9x \geqslant 63$
6. $2 + 6y > 14$
7. $7 - 3y \geqslant 27 + 2y$
8. $3x + 5 < 2x - 6$
9. $-4y < 28$
10. $3 - 4x < 27$
11. $4 - 8x < 13 + 3x$
12. $-3y \geqslant -21$
13. $-4x \leqslant \frac{1}{3}$

Solve problems involving inequalities using the five-step problem-solving process.

14. Your quiz grades are 71, 75, 82, and 86. What is the lowest grade you can get on the last quiz and still have an average of at least 80?

15. The length of a rectangle is 43 cm. What widths will make the perimeter greater than 120 cm?

Determine whether a point is a solution of an inequality.

Determine whether the given point is a solution of the inequality $x - 2y > 1$.

16. $(0, 0)$
17. $(1, 3)$
18. $(4, -1)$

Graph inequalities on a number line and a plane.

Graph on a number line.

19. $y \leqslant 9$
20. $6x - 3 < x + 2$
21. $|x| \leqslant 2$

Graph on a plane.

22. $x < y$
23. $x + 2y \geqslant 4$

Name sets using the roster method and determine whether a given object is a member of a set.

24. Name the set of multiples of 4 between 30 and 60 using the roster method.

Determine whether true or false.

25. $91 \in \{x \mid x \text{ is a prime number}\}$
26. $-4 \in \{-6, -5, -3, -1, 0\}$

Find the intersection and union of sets.

Find each intersection.

27. $\{2, 4, 12, 14\} \cap \{-4, -2, 0, 2, 4\}$
28. $\{a, f, g\} \cap \{b, h, q, r, s\}$

Find each union.

29. $\{0, 1\} \cup \{1, 2, 3\}$

30. $\{A, W, R\} \cup \{E, F, A, B\}$

Solve.

31. $3[2 - 4(y - 3)] < 6(y - 1)$

32. $-\dfrac{x}{5} - \dfrac{4x}{15} + 6 \leqslant 10 - x$

33. The sum of three consecutive odd integers is less than 198. What are the largest three such integers?

34. Find the intersection of odd integers greater than 6 and odd integers less than 15.

TEST: CHAPTER 7

Solve. Write set notation for the answers.

1. $x + 6 \leqslant 2$
2. $14x + 9 > 13x - 4$
3. $12x \leqslant 60$
4. $-2y \geqslant 26$
5. $-4y \leqslant -32$
6. $-5x \geqslant \frac{1}{4}$
7. $4 - 6x > 40$
8. $5 - 9x \geqslant 19 + 5x$

Problem solving

9. The sum of three consecutive integers is greater than 29. What are the least possible values of these integers?

10. Find all numbers such that six times the number is greater than the number plus 30.

Graph on a number line.

11. $4x - 6 < x + 3$

12. $|x| \geqslant 5$

Graph on a plane.

13. $y > x - 1$

14. $2x - y \leqslant 4$

Determine whether the given point is a solution of the inequality $3y - 2x < -2$.

15. $(0, 0)$

16. $(-4, -10)$

17. Name the set of prime numbers between 15 and 29 using the roster method.

18. $\{0, 1\} \cap \{-1, 0\}$

19. $\{q, r, t, s, w\} \cap \{r, s, t, u, v\}$

Find each union.

20. $\{-2, -1, 0, 1, 2\} \cup \{-4, -2, 0, 2, 4\}$

21. $\{a\} \cup \{b\}$

22. Give all integer solutions for $|2x| \leqslant 5$.

23. Solve: $x^2 + 7 > 0$.

24. Graph on a plane: $x < -4$.

FUNCTIONS AND VARIATION

8

How can the length of the human bone be used to find the height of the entire body? A linear function can give the answer.

A function is a special kind of correspondence between sets. The notion of function is basic to most of mathematics. This chapter contains a brief introduction to functions. Our goal is to give you an idea of what a function is and to extend your graphing skills. If time is short, Sections 8.1 and 8.2 can be omitted.

The last two sections of this chapter cover two classes of functions—those involving direct and inverse variation. These have applications in problem solving, as we shall see.

8.1 FUNCTIONS

Identifying Functions

Consider a set of years. To each year there corresponds the cost of a first-class postage stamp that year.

Year	Cost of first-class stamp
1948	3¢
1958	4¢
1964	5¢
1974	10¢
1978	15¢
1983	20¢
1984	20¢
1985	22¢

To each year there corresponds *exactly* one cost. Such a correspondence is called a *function*.

> A *function* is a correspondence (or rule) that assigns to each member of some set (called the *domain*) exactly one member of a set (called the *range*).

The members of the domain are also called *inputs* and the members of the range are called *outputs*. Arrows can be used to describe functions.

Domain (set of inputs)	Range (set of outputs)
1948 →	3
1958 →	4
1964 →	5
1974 →	10
1978 →	15
1983 →	20
1984 ↗	
1985 →	22

Note that each input has exactly one output, even though in the case of 1983 and 1984 those outputs are the same.

EXAMPLE 1 Determine whether or not each of the following correspondences is a function.

8.1 FUNCTIONS

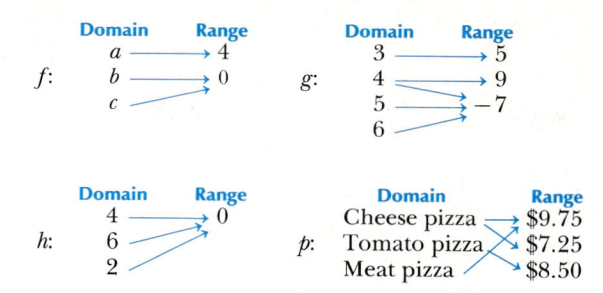

f:
Domain	Range
a	4
b	0
c	

g:
Domain	Range
3	5
4	9
5	−7
6	

h:
Domain	Range
4	0
6	
2	

p:
Domain	Range
Cheese pizza	$9.75
Tomato pizza	$7.25
Meat pizza	$8.50

The correspondence f is a function because each member of the domain is matched to only one member of the range. The correspondence g is not a function because the member 4 of the domain is matched to more than one member of the range. The correspondence h is a function. The correspondence p is not a function because the member cheese pizza of the domain is matched to two members of the range. In a function, a member of the domain can be matched with only one member of the range.

For functions we refer to a member of the domain as an *input* and the corresponding member of the range as its *output*.

Formulas for Functions

Functions are also described by formulas. Such formulas are "recipes" for finding outputs.

EXAMPLE 2 You see a flash of lightning. After a few seconds you hear the thunder associated with that flash. How far away was the lightning?

Your distance from the storm, M (in miles), is a function of n, the number of seconds it takes the sound of the thunder to reach you. We can approximate M by the formula

$$M = \tfrac{1}{5}n.$$

Complete the following table for this function.

n (sec)	0	1	2	3	4	5	6	10
M (mi)	0	$\tfrac{1}{5}$						

To complete the table we successively substitute values of n, and compute M:

For $n = 2$: $M = \tfrac{1}{5}(2) = \tfrac{2}{5}$;   For $n = 3$: $M = \tfrac{1}{5}(3) = \tfrac{3}{5}$;

For $n = 4$: $M = \tfrac{1}{5}(4) = \tfrac{4}{5}$;   For $n = 5$: $M = \tfrac{1}{5}(5) = 1$;

For $n = 6$: $M = \tfrac{1}{5}(6) = \tfrac{6}{5}$, or $1\tfrac{1}{5}$;   For $n = 10$: $M = \tfrac{1}{5}(10) = 2$.

Function Notation

Finding outputs can be thought of in terms of a "function machine." Inputs are entered into the machine. The machine then gives the outputs.

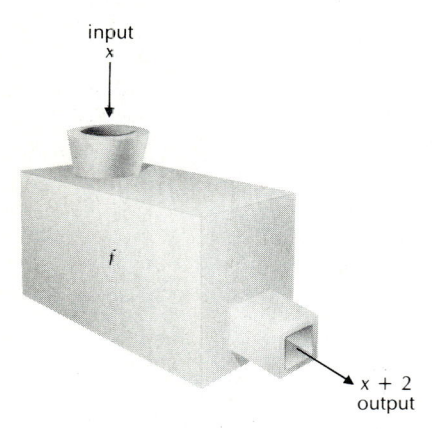

The symbol $f(x)$, read "f of x," denotes the number assigned to x by the correspondence. The preceding function f assigns to each input x the output $x + 2$. We describe this with the formula $f(x) = x + 2$, and then we can calculate as follows:

For the input 8, the output is 10:

$$f(8) = 8 + 2, \quad \text{or } 10.$$

For the input -3, the output is -1:

$$f(-3) = -3 + 2, \quad \text{or } -1.$$

For the input 0, the output is 2:
$$f(0) = 0 + 2, \text{ or } 2.$$
For the input 5, the output is 7:
$$f(5) = 5 + 2, \text{ or } 7.$$

EXAMPLE 3 A function H assigns to each input x the output 5. Find (a) $H(x)$, (b) $H(0)$, and (c) $H(2)$.

a) $H(x) = 5$
b) $H(0) = 5$
c) $H(2) = 5$

A function such as H is known as a *constant* function.

EXAMPLE 4 For the function $f(t) = 2t^2 + 5$, find (a) $f(-2)$, (b) $f(0)$, and (c) $f(3)$.

a) $f(-2) = 2(-2)^2 + 5$
$= 2 \cdot 4 + 5$
$= 13$
b) $f(0) = 2(0)^2 + 5$
$= 5$
c) $f(3) = 2(3)^2 + 5$
$= 2 \cdot 9 + 5$
$= 23$

Outputs are also called *function values*. In Example 4, $f(-2) = 13$. We could say that the "function value at -2 is 13," or "when x is -2, the value of the function is 13."

EXAMPLE 5 (*An application: Predicting heights in anthropology*). An anthropologist can use certain functions to estimate the height of a male or female, given the lengths of certain bones. A *humerus* is the bone from the elbow to the shoulder. Let x = the length

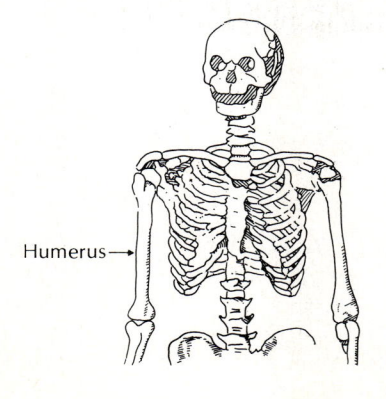

Humerus→

of the humerus in cm. Then the height, in cm, of a male with a humerus of length x is given by

$$M(x) = 2.89x + 70.64.$$

The height, in cm, of a female with a humerus of length x is given by

$$F(x) = 2.75x + 71.48.$$

A 45-cm humerus was uncovered in a ruin.

a) If it was from a male, how tall was he?

b) If it was from a female, how tall was she?

a) We find $M(45)$:

$$M(45) = 2.89(45) + 70.64$$
$$= 130.05 + 70.64$$
$$= 200.69.$$

The male's height was about 200.69 cm.

b) We find $F(45)$:

$$F(45) = 2.75(45) + 71.48$$
$$= 123.75 + 71.48$$
$$= 195.23.$$

The female's height was about 195.23 cm.

EXERCISE SET 8.1

Determine whether each of the following correspondences is a function.

1. Domain → Range: 2 → 9, 5 → 8, 19

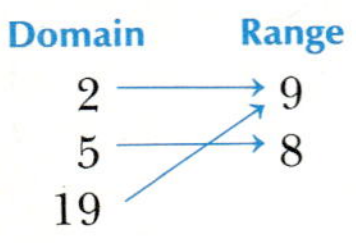

2. Domain → Range: 5 → 3, −3 → 7, 7, −7

3. Domain → Range: −5 → 1, 5, 8

4. Domain → Range: 6 → −6, 7 → −7, 3 → −3

5. Domain → Range: Los Angeles, New York → Mets, Lakers, Dodgers, Yankees

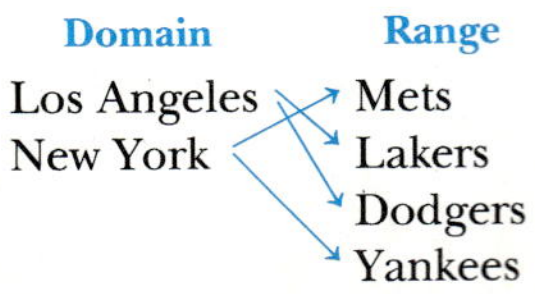

6. Domain → Range: (3, 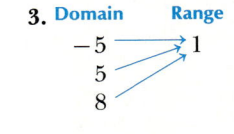 4) → 12, (8, 10) → −11, (4, −2) → 18, (−3, −8) → 2

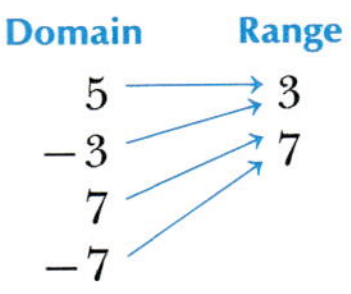

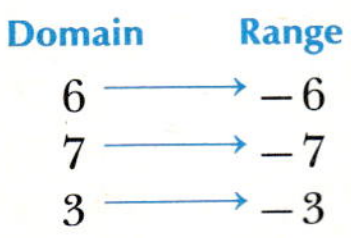

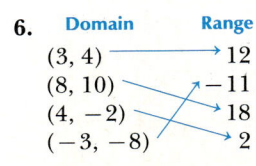

Find the indicated outputs.

7. $f(x) = x + 5$

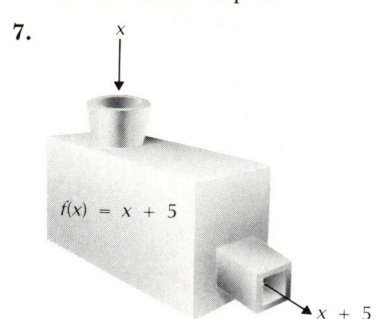

Find $f(3)$, $f(7)$, and $f(-9)$.

8. $g(t) = t - 6$

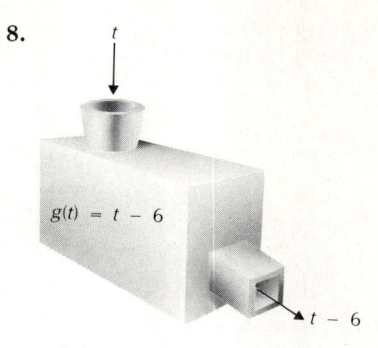

Find $g(0)$, $g(6)$, and $g(18)$.

9. $h(p) = 3p$

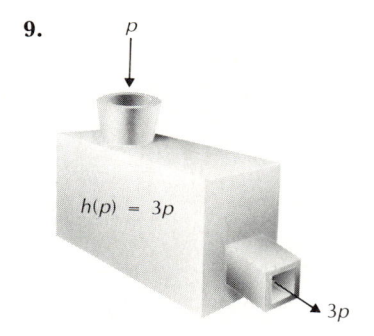

Find $h(-2)$, $h(5)$, and $h(24)$.

10. $f(x) = -4x$

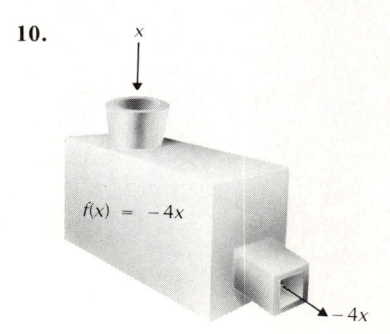

Find $f(6)$, $f(-\frac{1}{2})$, and $f(20)$.

Find the indicated outputs.

11. $g(s) = 2s + 4$; find $g(1)$, $g(-7)$, and $g(6)$.
12. $h(x) = 19$; find $h(4)$, $h(-6)$, and $h(12)$.
13. $F(x) = 2x^2 - 3x + 2$; find $F(0)$, $F(-1)$, and $F(2)$.
14. $P(x) = 3x^2 - 2x + 5$; find $P(0)$, $P(-2)$, and $P(3)$.
15. $h(x) = |x|$; find $h(-4)$, $h(5)$, and $h(-3)$.
16. $f(t) = |t| + 1$; find $f(-5)$, $f(0)$, and $f(-9)$.
17. $f(x) = |x| - 2$; find $f(-3)$, $f(93)$, and $f(-100)$.
18. $g(t) = t^3 + 3$; find $g(1)$, $g(-5)$, and $g(0)$.
19. $h(x) = x^4 - 3$; find $h(0)$, $h(-1)$, and $h(3)$.
20. $f(x) = 2/x$; find $f(-3)$, $f(-2)$, $f(-1)$, $f(\frac{1}{2})$, $f(1)$, $f(2)$, $f(3)$, and $f(10)$.

21. The function $P(d) = 1 + (d/33)$ gives the pressure, in *atmospheres* (atm), at a depth d in the sea (d is in feet). Note that $P(0) = 1$ atm, $P(33) = 2$ atm, and so on. Find the pressure at 20 ft, 30 ft, and 100 ft.

22. The function $R(t) = 33\frac{1}{3}t$ gives the number of revolutions of a $33\frac{1}{3}$ RPM record as a function of t, the time (in minutes) it turns. Find the number of revolutions at 5 min, 20 min, and 25 min.

23. The function $T(d) = 10d + 20$ gives the temperature (in degrees Celsius) inside the earth as a function of d, the depth in kilometers. Find the temperature at 5 km, 20 km, 1000 km.

24. Snow is melted to water in a cylindrical tube. The function $W(d) = 0.112d$ approximates the amount (in centimeters) of water W that will melt from snow that is d cm deep. Find the amount of water that results from snow melting from depths of 16 cm, 25 cm, and 100 cm.

In many physical situations we speak of one quantity being "a function of" another. For instance, the cost of replacing a defective tire is a function of the tread depth. For Exercises 25–28, the following chart gives a rule for this function for a tire with original tread depth of 9 millimeters (mm).

Tread depth (mm)	9	8	7	6	5	4	3	2	1
% charged	No charge	20%	30%	40%	55%	70%	80%	90%	100%

25. Find the cost of replacing a tire whose regular price is $64.50 and whose tread depth is 4 mm.

26. Find the cost of replacing a tire whose regular price is $78.50 and whose tread depth is 7 mm.

27. Find the cost of replacing a tire whose regular price is $67.80 and whose tread depth is 3 mm.

28. Find the cost of replacing a tire whose regular price is $72.40 and whose tread depth is 5 mm.

Find the range of each function for the given domain.

29. $f(x) = 3x + 5$, when the domain is the set of whole numbers less than 4

30. $g(t) = t^2 - 5$, when the domain is the set of integers between -4 and 2

31. $h(x) = |x| - x$, when the domain is the set of integers between -2 and 20.

32. $f(m) = m^3 + 1$, when the domain is the set of integers between -3 and 3

Suppose that $f(x) = 3x$ and $g(x) = -4x^2$. Find each of the following.

33. $f(8) - g(2)$

34. $f(0) - g(-5)$

35. $2f(1) + 3g(4)$

36. $g(-3) \cdot f(-8) + 16$

37. If $f(-1) = -7$ and $f(3) = 8$, find a linear equation for $f(x)$.

38. $H(x - 1) = 5x$; find $H(6)$.

39. When you flip a coin n times, determine whether or not the number of "heads" you get is a function of the number (n) of flips.

8.2 FUNCTIONS AND GRAPHS

To graph a function, such as $f(x) = x + 2$, we do just what we do to graph $y = x + 2$. We find ordered pairs (x, y), or $(x, f(x))$; plot them; and connect the points.

EXAMPLE 1 Graph: $f(x) = x + 2$.

A list of function values is shown in this table.

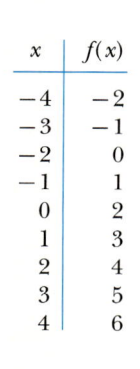

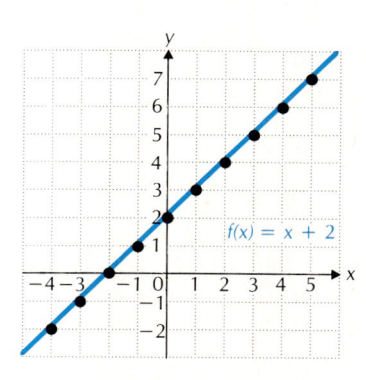

x	$f(x)$
-4	-2
-3	-1
-2	0
-1	1
0	2
1	3
2	4
3	5
4	6

We plot the points and connect them. The graph is a straight line.

EXAMPLE 2 Graph: $g(x) = |x|$.

A list of function values is shown in this table.

x	$g(x)$
-3	3
-2	2
-1	1
0	0
1	1
2	2
3	3

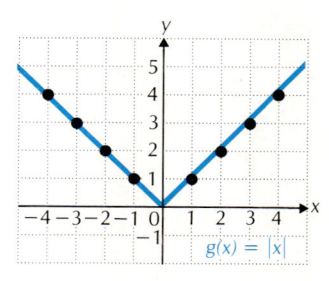

We plot the points and connect them. The graph is V-shaped. Note that as x increases through positive values, the absolute value of x increases. As x decreases through negative values, the absolute value of x increases. Thus the graph is a V-shaped curve that rises on either side of the vertical axis.

Recognizing Graphs of Functions

Consider the function f described by $f(x) = x - 2$. Its graph is shown below. It is also the graph of the equation $y = x - 2$.

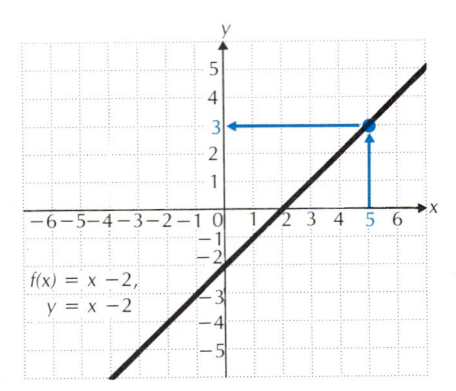

$f(x) = x - 2,$
$y = x - 2$

The domain, the set of all possible inputs, is represented by a set of points on the horizontal axis. The range, the set of all possible outputs, is represented by a set of points on the vertical axis. To find a function value, we locate the input on the horizontal axis, move as shown to the graph of the function, and move horizontally to find the output on the vertical axis.

Some graphs are graphs of functions and some are not. If a vertical line drawn anywhere on the graph can intersect the graph at more than one point, there will be more than one number in the range corresponding to a member of the domain. This graph is not the graph of a function.

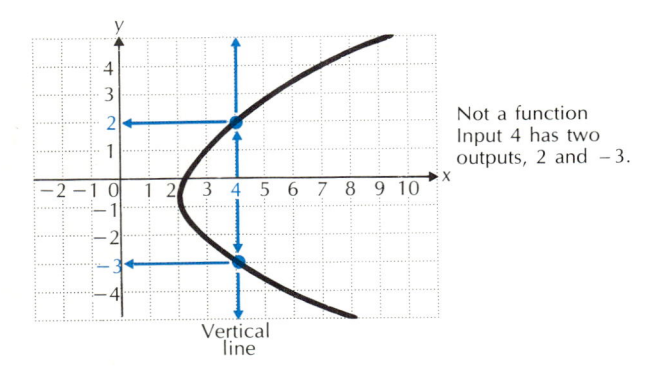

Not a function
Input 4 has two outputs, 2 and −3.

Vertical line

To recognize a graph of a function we can use a vertical-line test. If it is possible for any vertical line to intersect the graph at more than one point, the graph is not the graph of a function.

8.2 FUNCTIONS AND GRAPHS

EXAMPLE 3 Which of the following are graphs of functions?

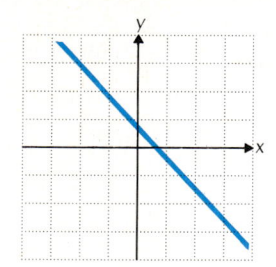

A function. No vertical line could cross the graph at more than one point.

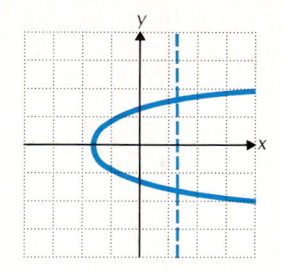

Not a function. A vertical line does cross the graph at more than one point.

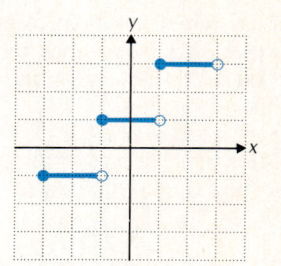

A function. No vertical line can cross the graph at more than one point.

EXERCISE SET 8.2

Graph each function.

1. $f(x) = x + 4$
2. $g(x) = x + 3$
3. $h(x) = 2x - 3$
4. $f(x) = 3x - 1$
5. $g(x) = x - 6$
6. $h(x) = x - 5$
7. $f(x) = 2x - 7$
8. $g(x) = 4x - 13$
9. $f(x) = \frac{1}{2}x + 1$
10. $f(x) = -\frac{3}{4}x - 2$
11. $g(x) = 2|x|$
12. $h(x) = -|x|$
13. $g(x) = x^2$
14. $f(x) = x^2 - 1$
15. $f(x) = \dfrac{2}{x}$
16. $f(x) = -\dfrac{1}{x}$

Which of the following are graphs of functions?

17.

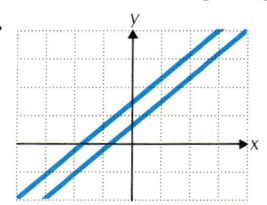

18.

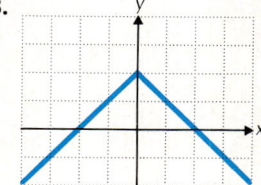

19.

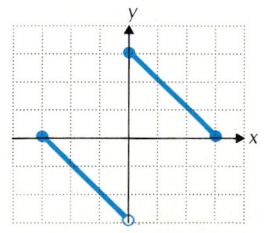

20.

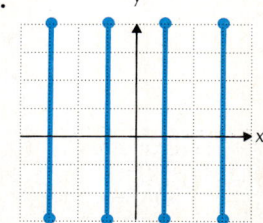

Problem-solving practice

21. A student is 150 km from home. The student drives farther away from home in a straight line at a speed of 60 km/h. How far is the student from home after 4 hr?

22. The length of a rectangle is 2 m greater than the width. The area is 35 m². Find the perimeter.

23. Sketch a graph that is not a function.
24. Draw the graph of $|y| = x$. Is this the graph of a function?
25. Draw the graph of $y^2 = x$. Is this the graph of a function?

8.3 DIRECT VARIATION

A bicycle is traveling at 10 km/h. In 1 hr it goes 10 km. In 2 hr it goes 20 km. In 3 hr it goes 30 km, and so on. We will use the number of hours as the first coordinate and the number of kilometers traveled as the second coordinate (1, 10), (2, 20), (3, 30), (4, 40), and so on. Note that as the first number gets larger, so does the second. Note also that the ratio of distance to time for each of these ordered pairs is $\frac{10}{1}$, or 10.

Whenever a situation produces pairs of numbers in which the ratio is constant, we say that there is *direct variation*. Here the distance varies directly as the time:

$$\frac{d}{t} = 10 \text{ (a constant)}, \quad \text{or } d = 10t.$$

• •

If a situation translates to a function described by $y = kx$, where k is a positive constant, $y = kx$ is called an *equation of direct variation*, and k is called the *variation constant*.

8.3 DIRECT VARIATION

The graph of $y = kx$, $k > 0$, always goes through the origin and rises from left to right. Note that as x increases, y increases.

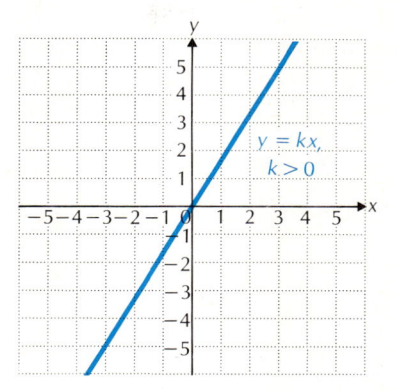

When there is direct variation $y = kx$, the variation constant can be found if one pair of values of x and y is known. Then other values can be found.

EXAMPLE 1 Find an equation of variation where y varies directly as x, and $y = 7$ when $x = 25$.

We substitute to find k:

$$y = kx$$
$$7 = k \cdot 25$$
$$\tfrac{7}{25} = k, \quad \text{or } k = 0.28.$$

Then the equation of variation is $y = 0.28x$. Note that the answer is an *equation*.

EXAMPLE 2 It is known that the karat rating K of a gold object varies directly as the actual percentage of gold in the object. A 14-karat gold object is 58.25% gold. What is the percentage of gold in a 24-karat object?

1. **Familiarize** The problem states that we have direct variation between the variables K and P. Thus an equation $K = kP$, $k > 0$, applies. As the percentage of gold increases, the karat rating increases.

2. **Translate** We write an equation of variation:
$$K = kP.$$
Karat rating varies directly as percentage of gold.

3. **Carry out** The mathematical manipulation has two steps: First, find the equation of variation. Second, compute the percentage of gold in a 24-karat object.

a) First, find an equation of variation:
$$K = kP$$
$$14 = k \cdot 0.5825 \qquad \text{Substituting 14 for } K \text{ and 0.5825 for } P$$
$$\frac{14}{0.5825} = k$$
$$24.03 \approx k. \qquad \text{Dividing and rounding to the nearest hundredth}$$

The equation of variation is $K = 24.03P$.

b) Use the equation to find the percentage of gold in a 24-karat object:
$$K = 24.03P$$
$$24 = 24.03P$$
$$\frac{24}{24.03} = P$$
$$0.999 \approx P$$
$$99.9\% \approx P.$$

4. **Check** The check might be done by repeating the computations. You might also do some reasoning about the answer. The karat rating increased from 14 to 24. Similarly, the percentage increased from 58% to 99.9%.

5. **State** A 24-karat object is 99.9% gold.

EXERCISE SET 8.3

Find an equation of variation where y varies directly as x and the following are true.

1. $y = 28$, when $x = 7$
2. $y = 30$, when $x = 8$
3. $y = 0.7$, when $x = 0.4$
4. $y = 0.8$, when $x = 0.5$
5. $y = 400$, when $x = 125$
6. $y = 630$, when $x = 175$
7. $y = 200$, when $x = 300$
8. $y = 500$, when $x = 60$

Problem solving

9. A person's paycheck P varies directly as the number H of hours worked. For working 15 hours, the pay is $78.75. Find the pay for 35 hours of work.

10. The number B of bolts a machine can make varies directly as the time it operates. It can make 6578 bolts in 2 hours. How many can it make in 5 hours?

11. The number of servings S of meat that can be obtained from a turkey varies directly as its weight W. From a turkey weighing 14 kg one can get 40 servings of meat. How many servings can be obtained from an 8-kg turkey?

12. The number of servings S of meat that can be obtained from round steak varies directly as the weight W. From 9 kg of round steak one can get 70 servings of meat. How many servings can one get from 12 kg of round steak?

13. The weight M of an object on the moon varies directly as its weight E on earth. A person who weighs 78 kg on earth weighs 13 kg on the moon. How much would a 100-kg person weigh on the moon?

14. The weight M of an object on Mars varies directly as its weight E on earth. A person who weighs 95 kg on earth weighs 36.1 kg on Mars. How much would an 80-kg person weigh on Mars?

15. The number of kilograms of water W in a human body varies directly as the total body weight B. A person weighing 75 kg contains 54 kg of water. How many kilograms of water are in a person weighing 95 kg?

16. The amount C that a family spends on car expenses varies directly as its income I. A family making $21,760 a year will spend $3264 a year for car expenses. How much will a family making $30,000 a year spend for car expenses?

Write an equation of direct variation for each situation in Exercises 17–20. If possible, give a value for k and graph the equation.

17. The perimeter P of an equilateral polygon varies directly as the length S of a side.

18. The circumference C of a circle varies directly as the radius r.

19. The number of bags B of peanuts sold at a baseball game varies directly as the number N of people in attendance.

20. The cost C of building a new house varies directly as the area A of the floor space of the house.

21. Show that if p varies directly as q, then q varies directly as p.

22. The area of a circle varies directly as the square of the length of the radius. What is the variation constant?

Write an equation of variation to describe these situations.

23. In a stream, the amount S of salt carried varies directly as the sixth power of the speed V of the stream.

24. The square of the pitch P of a vibrating string varies directly as the tension t on the string.

The volume of a box varies directly as its length. It also varies directly as the height. We then say that the volume varies *jointly* as the length and the height. An equation of variation can be written using the *product* of variables $V = k \cdot l \cdot h$. Write an equation of variation for the following.

25. The power P required in an electric circuit varies jointly as the resistance R and the square of the current I.

8.4 INVERSE VARIATION

A car is traveling a distance of 10 km. At a speed of 10 km/h it will take 1 hr. At 20 km/h, it will take $\frac{1}{2}$ hr. At 30 km/h it will take $\frac{1}{3}$ hr, and so on. This determines a set of pairs of numbers, all having the same product:

$$(10, 1), \quad (20, \tfrac{1}{2}), \quad (30, \tfrac{1}{3}), \quad (40, \tfrac{1}{4}), \quad \text{and so on.}$$

Note that as the first number gets larger, the second number gets smaller. Whenever a situation produces pairs of numbers whose product is constant, we say that there is *inverse variation*. Here the time varies inversely as the speed:

$$rt = 10 \text{ (a constant)}, \quad \text{or } t = \frac{10}{r}.$$

> If a situation translates to a function described by $y = k/x$, where k is a positive constant, $y = k/x$ is called an *equation of inverse variation*. We say that y varies inversely as x.

The graph of $y = k/x$, $k > 0$, is shaped like the graph shown here for positive values of x. Note that as x increases, y decreases; and as x decreases, y increases.

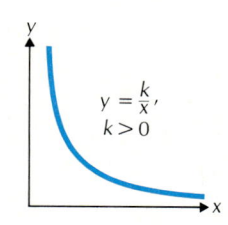

$y = \frac{k}{x}$, $k > 0$

EXAMPLE 1 Find an equation of variation where y varies inversely as x, and $y = 145$ when $x = 0.8$.

We substitute to find k:

$$y = \frac{k}{x}$$

$$145 = \frac{k}{0.8}$$

$$(0.8)145 = k$$

$$116 = k.$$

The equation of variation is $y = 116/x$.

EXAMPLE 2 The time T to do a certain job varies inversely as the number N of people working (assuming all work at the same rate). It takes 4 hr for 20 people to wash and wax the floors in a building. How long would it then take 25 people to do the job?

1. **Familiarize** The problem states that we have inverse variation between the variables T and N. Thus an equation $T = k/N$, $k > 0$, applies. As the number of people increases, the time it takes to do the job decreases.

2. **Translate** We write an equation of variation:

$$T = \frac{k}{N}.$$

Time varies inversely as the number of people.

3. **Carry out** The mathematical manipulation has two steps: First, find the equation of variation. Second, compute the amount of time it would take 25 people to do the job.

   a) First, find an equation of variation:

   $$T = \frac{k}{N}$$

   $$4 = \frac{k}{20} \quad \text{Substituting 4 for } T \text{ and 20 for } N$$

   $$20 \cdot 4 = k$$

   $$80 = k.$$

   The equation of variation is $T = 80/N$.

   b) Use the equation to find the amount of time that it takes 25 people to do the job:

   $$T = \frac{80}{N}$$

   $$T = \frac{80}{25}$$

   $$= 3.2.$$

4. **Check** The check might be done by repeating the computations. We might also analyze the results. The number of people increased from 20 to 25. Did the time decrease? It did, and this confirms what we expect with inverse variation.

5. **State** It takes 3.2 hr for 25 people to do the job.

EXERCISE SET 8.4

Find an equation of variation where y varies inversely as x and the following are true.

1. $y = 25$, when $x = 3$
2. $y = 45$, when $x = 2$
3. $y = 8$, when $x = 10$
4. $y = 7$, when $x = 10$
5. $y = 0.125$, when $x = 8$
6. $y = 6.25$, when $x = 0.16$
7. $y = 42$, when $x = 25$
8. $y = 42$, when $x = 50$
9. $y = 0.2$, when $x = 0.3$
10. $y = 0.4$, when $x = 0.6$

Problem solving

11. It takes 16 hr for 2 people to resurface a gym floor. How long would it take 6 people to do the job?

12. It takes 4 hr for 9 cooks to prepare a school lunch. How long would it take 8 cooks to prepare the lunch?

13. The volume V of a gas varies inversely as the pressure P on it. The volume of a gas is 200 cubic centimeters (cm^3) under a pressure of 32 kg/cm^2. What will be its volume under a pressure of 20 kg/cm^2?

14. The current I in an electrical conductor varies inversely as the resistance R of the conductor. The current is 2 amperes when the resistance is 960 ohms. What is the current when the resistance is 540 ohms?

15. The time t required to empty a tank varies inversely as the rate r of pumping. A pump can empty a tank in 90 minutes at the rate of 1200 L/min. How long will it take the pump to empty the tank at 2000 L/min?

16. The height H of triangles of fixed area varies inversely as the base B. Suppose the height is 50 cm when the base is 40 cm. Find the height when the base is 8 cm. What is the fixed area?

17. The pitch P of a musical tone varies inversely as its wavelength W. One tone has a pitch of 660 vibrations per second and a wavelength of 1.6 feet. Find the wavelength of another tone that has a pitch of 440 vibrations per second.

18. The time t required to drive a fixed distance varies inversely as the speed r. It takes 5 hr at 60 km/h to drive a fixed distance. How long would it take at 40 km/h?

Write an equation of inverse variation for each situation.

19. The cost C per person of chartering a fishing boat varies inversely as the number N of persons sharing the cost.

20. The number N of revolutions of a tire rolling over a given distance varies inversely as the circumference C of the tire.

21. The amount of current I flowing in an electrical circuit varies inversely with the resistance R of the circuit.

22. The density D of a given mass varies inversely as its volume V.

23. The intensity of illumination I from a light source varies inversely as the square of the distance d from the source.

Which of the following vary inversely?

24. The cost of mailing a letter in the United States and the distance it travels.

25. A runner's speed in a race and the time it takes to run it.

26. The number of plays to go 80 yards for a touchdown and the average gain per play.

27. The weight of a turkey and the cooking time.

28. Graph the equation of inverse variation $y = 6/x$.

The time it takes n people to do s jobs varies directly as the number of jobs and inversely as the number of people. An equation of variation is

$$T = ks \cdot \frac{1}{n}.$$

This is *combined variation*. Write an equation of variation for each of the following.

29. The force F needed to keep a car from skidding on a curve varies directly as the square of the car's speed S and its mass m and inversely as the radius of the curve r.

30. For a horizontal beam supported at both ends, the maximum safe load L varies directly as its width w and the square of its thickness t and inversely as the distance d between the supports.

SUMMARY AND REVIEW: CHAPTER 8

The following contains a summary of what you should be able to do after completing this chapter. The review exercises are for practice. Answers are at the back of the book. If you miss an exercise, restudy the section indicated alongside the answer.

You should be able to:

Determine whether a correspondence is a function.

Determine whether each of the following correspondences is a function.

1. Domain   Range
   −1 → 3
   0 → 4
   1 → 5

2. Domain   Range
   −2 → 0
   5 → 1
   7 → 4

Given a function defined by a formula, find function values (outputs) for specified inputs.

Find the indicated outputs.

3. $f(x) = 3x - 4$; find $f(2)$, $f(0)$, and $f(-1)$.

4. $g(t) = |t| - 3$; find $g(3)$, $g(-5)$, and $g(0)$.

5. $h(x) = x^3 + 1$; find $h(-2)$, $h(0)$, and $h(1)$.

6. If you are moderately active, you need each day about 15 calories per pound of body weight. The function $C(p) = 15p$ approximates the number of calories C that are needed to maintain body weight, p, in lb. How many calories would be needed to maintain a body weight of 180 lb?

Graph a function and recognize the graph of a function.

Graph.

7. $g(x) = x + 7$

8. $f(x) = x^2 - 3$

9. $h(x) = 3|x|$

Which of the following are graphs of functions?

10.

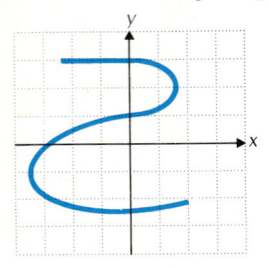

11.

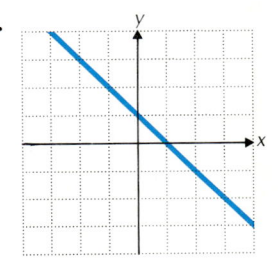

Find an equation of variation where y varies directly as x and y varies inversely as x.

Find an equation of variation where y varies directly as x, and the following are true.

12. $y = 12$, when $x = 4$

13. $y = 4$, when $x = 8$

14. $y = 0.4$, when $x = 0.5$

Find an equation of variation where y varies inversely as x and the following are true.

15. $y = 5$, when $x = 6$

16. $y = 0.5$, when $x = 2$

17. $y = 1.3$, when $x = 0.5$

Solve a problem involving variation using the five-step problem-solving process.

18. A person's paycheck P varies directly as the number H of hours worked. The pay is $165.00 for working 20 hr. Find the pay for 30 hr of work.

19. It takes 5 hr for 2 washing machines to wash a fixed amount. How long would it take 10 washing machines? (The number of hours varies inversely as the number of washing machines.)

20. The stopping distance (at some fixed speed) of regular tires on glare ice is given by a linear function of the air temperature F,

$$D(F) = 2F + 115,$$

where $D(F)$ = the stopping distance, in feet, when the air temperature is F, in degrees Fahrenheit.

a) Find the stopping distance when the air temperature is 0°F.
b) Find the stopping distance when the air temperature is 32°F.

21. Suppose that $f(x) = 2x$ and $g(x) = -3x^2$. Find $3f(2) - f(-1) \cdot g(4)$.

22. Draw the graph of $y^2 + 1 = x$. Is this the graph of a function?

TEST: CHAPTER 8

1. Determine whether the following correspondence is a function.

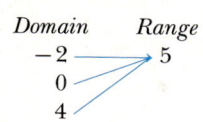

   Domain   Range
   $-2 \longrightarrow 5$
   0
   4

Find the indicated outputs.

2. $f(x) = \frac{1}{2}x + 1$; find $f(0)$, $f(1)$, and $f(2)$.

3. $g(t) = -2|t| + 3$; find $g(-1)$, $g(0)$, and $g(3)$.

4. The world record for the 10,000-m run has been decreasing steadily since 1940. The record is 30.18 min minus 0.12 times the number of years since 1940. The function $R(t) = 30.18 - 0.12t$ gives the record R in min as a function of t, the time in yr since 1940. What will the record be in 1986?

Graph each function.

5. $h(x) = x - 4$

6. $g(x) = x^2 - 1$

Which of the following are graphs of functions?

7.

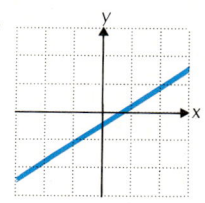

8.

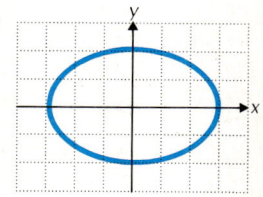

Find an equation of variation where y varies directly as x and the following are true.

9. $y = 6$, when $x = 3$

10. $y = 1.5$, when $x = 3$

Find an equation of variation where y varies inversely as x and the following are true.

11. $y = 6$, when $x = 2$

12. $y = \frac{1}{3}$, when $x = 3$

13. The distance d traveled by a train varies directly as the time t it travels. The train travels 60 km in $\frac{1}{2}$ hr. How far will it travel in 2 hr?

14. It takes 3 hr for 2 cement mixers to mix a certain amount. The number of hours varies inversely as the number of cement mixers. How long would it take 5 cement mixers to do the job?

15. The population of a small town, in thousands, is estimated by a linear function

$$P(t) = 1.25t + 15,$$

where $P(t) =$ the population the tth year after 1940; and $P(0)$ is the population in 1940, $P(45)$ is the population in 1985, and so on.

a) Find the population in 1940.
b) Find the population in 1985.

16. Suppose that $f(x) = x - 4$ and $g(x) = -2x^2$. Find $f(-2) - f(0) \cdot g(3) + g(-1)$.

FRACTIONAL EXPRESSIONS AND EQUATIONS

9

One car travels 20 km/h faster than another. While one of them goes 240 km, the other goes 160 km. How can we use fractional equations to find their speeds?

An expression that is a quotient of two expressions is called a *fractional expression*. The following are fractional expressions:

$$\frac{3}{4}, \quad \frac{5}{x+2}, \quad \frac{t^2+3t-10}{7t^2-4}.$$

In most of the fractional expressions that we shall consider, the numerator and denominator will be polynomials.

Fractional expressions indicate division. For example,

$$\frac{t^2+3t-10}{7t^2-4} \quad \text{means} \quad (t^2+3t-10) \div (7t^2-4).$$

In this chapter we learn how to simplify fractional expressions. We also learn how to add, subtract, multiply, and divide. Then we use these skills to solve equations, formulas, and problems.

9.1 MULTIPLYING AND SIMPLIFYING

We first study multiplying and simplifying fractional expressions. What we do is similar to multiplying and simplifying fractional expressions in arithmetic. But, instead of simplifying an expression like

$$\frac{16}{64},$$

we may simplify an expression like

$$\frac{x^2 - 16}{x + 4}.$$

Just as factoring is important in simplifying in arithmetic, so too is factoring important in factoring fractional expressions. But, the factoring we use most is the factoring of polynomials, which we studied in Chapter 5.

Multiplying

For fractional expressions, multiplication is done as in arithmetic.

> To multiply two fractional expressions, multiply numerators and multiply denominators.

EXAMPLES Multiply.

1. $\dfrac{x-2}{3} \cdot \dfrac{x+2}{x+3} = \dfrac{(x-2)(x+2)}{3(x+3)}$ Multiplying numerators and multiplying denominators

   We could multiply out the numerator and the denominator to get

   $$\frac{x^2 - 4}{3x + 9},$$

   but it is best *not* to do it yet. We will see why later.

2. $\dfrac{-2}{2y+3} \cdot \dfrac{3}{y-5} = \dfrac{-2 \cdot 3}{(2y+3)(y-5)}$

Multiplying by 1

Any fractional expression with the same numerator and denominator is a symbol for 1:

$$\frac{x+2}{x+2} = 1, \qquad \frac{3x^2 - 4}{3x^2 - 4} = 1, \qquad \frac{-1}{-1} = 1.$$

Note that certain replacements are not sensible. For example, in

$$\frac{x+2}{x+2}$$

we should not substitute -2 for x, because we would get 0 for the denominator.

Expressions that have the same value for all sensible replacements are called *equivalent expressions*. We can multiply by 1 to obtain an equivalent expression.

EXAMPLES Multiply.

3. $\dfrac{3x+2}{x+1} \cdot \dfrac{2x}{2x} = \dfrac{(3x+2)2x}{(x+1)2x}$

4. $\dfrac{x+2}{x-1} \cdot \dfrac{x+1}{x+1} = \dfrac{(x+2)(x+1)}{(x-1)(x+1)}$

5. $\dfrac{2+x}{2-x} \cdot \dfrac{-1}{-1} = \dfrac{(2+x)(-1)}{(2-x)(-1)}$

Note in Example 4 that the original expression

$$\frac{x+2}{x-1}$$

has 1 as a replacement that is not sensible. The resulting expression

$$\frac{(x+2)(x+1)}{(x-1)(x+1)}$$

has 1 and -1 as replacements that are not sensible. The two expressions are still equivalent. They have the same value for all the sensible replacements.

Simplifying Fractional Expressions

To simplify, we can do the reverse of multiplying. We factor numerator and denominator and "remove" a factor of 1.

EXAMPLE 6 Simplify by removing a factor of 1: $\dfrac{3x}{x}$.

$\dfrac{3x}{x} = \dfrac{3 \cdot x}{1 \cdot x}$ Factoring the numerator and denominator

$\phantom{\dfrac{3x}{x}} = \dfrac{3}{1} \cdot \dfrac{x}{x}$ Factoring the fractional expression

$\phantom{\dfrac{3x}{x}} = \dfrac{3}{1} \cdot 1 \quad \dfrac{x}{x} = 1$

$\phantom{\dfrac{3x}{x}} = 3$ We "removed" a factor of 1.

In Example 6 we supplied a 1 in the denominator. This can always be done whenever it is helpful.

EXAMPLES Simplify by removing a factor of 1.

7. $\dfrac{6a + 12}{7(a + 2)} = \dfrac{6(a + 2)}{7(a + 2)}$

$= \dfrac{6}{7} \cdot \dfrac{a + 2}{a + 2}$

$= \dfrac{6}{7}$ "Removing" the factor $\dfrac{a + 2}{a + 2}$

8. $\dfrac{6x^2 + 4x}{2x^2 + 2x} = \dfrac{2x(3x + 2)}{2x(x + 1)}$ Factoring numerator and denominator

$= \dfrac{2x}{2x} \cdot \dfrac{3x + 2}{x + 1}$ Factoring the fractional expression

$= \dfrac{3x + 2}{x + 1}$ "Removing" a factor of 1

9. $\dfrac{x^2 + 3x + 2}{x^2 - 1} = \dfrac{(x + 2)(x + 1)}{(x + 1)(x - 1)}$

$= \dfrac{x + 1}{x + 1} \cdot \dfrac{x + 2}{x - 1}$

$= \dfrac{x + 2}{x - 1}$

10. $\dfrac{5a + 15}{10} = \dfrac{5(a + 3)}{5 \cdot 2}$

$= \dfrac{5}{5} \cdot \dfrac{a + 3}{2}$

$= \dfrac{a + 3}{2}$

Multiplying and Simplifying

We usually simplify after we multiply. That is why we do not multiply out the numerator and denominator too soon. We would need to factor them again anyway in order to simplify.

EXAMPLE 11 Multiply and simplify: $\dfrac{x^2 + 6x + 9}{x^2 - 4} \cdot \dfrac{x - 2}{x + 3}$.

$$\frac{x^2 + 6x + 9}{x^2 - 4} \cdot \frac{x-2}{x+3} = \frac{(x^2 + 6x + 9)(x-2)}{(x^2 - 4)(x+3)} \quad \text{Multiplying the numerators and denominators}$$

$$= \frac{(x+3)(x+3)(x-2)}{(x+2)(x-2)(x+3)} \quad \text{Factoring the numerator and denominator}$$

$$= \frac{(x+3)(x-2)}{(x+3)(x-2)} \cdot \frac{x+3}{x+2} \quad \text{Factoring the fractional expression}$$

$$= \frac{x+3}{x+2} \quad \text{Simplifying by removing a factor of 1}$$

EXAMPLE 12 Multiply and simplify: $\dfrac{x^2 + x - 2}{15} \cdot \dfrac{5}{2x^2 - 3x + 1}$.

$$\frac{x^2 + x - 2}{15} \cdot \frac{5}{2x^2 - 3x + 1} = \frac{(x^2 + x - 2)5}{15(2x^2 - 3x + 1)} \quad \text{Multiplying the numerators and denominators}$$

$$= \frac{(x+2)(x-1)5}{(5)(3)(x-1)(2x-1)} \quad \text{Factoring the numerator and denominator}$$

$$= \frac{(x-1)5}{(x-1)5} \cdot \frac{x+2}{3(2x-1)} \quad \text{Factoring the fractional expression}$$

$$= \underbrace{\frac{x+2}{3(2x-1)}}_{\uparrow} \quad \text{Simplifying by removing a factor of 1}$$

You need not carry out this multiplication.

EXERCISE SET 9.1

Multiply. Do not carry out multiplications in the numerator and denominator.

1. $\dfrac{3x}{2} \cdot \dfrac{x+4}{x-1}$
2. $\dfrac{4x}{5} \cdot \dfrac{x-3}{x+2}$
3. $\dfrac{x-1}{x+2} \cdot \dfrac{x+1}{x+2}$
4. $\dfrac{x-2}{x-5} \cdot \dfrac{x-2}{x+5}$
5. $\dfrac{2x+3}{4} \cdot \dfrac{x+1}{x-5}$
6. $\dfrac{-5}{3x-4} \cdot \dfrac{-6}{5x+6}$
7. $\dfrac{a-5}{a^2+1} \cdot \dfrac{a+2}{a^2-1}$
8. $\dfrac{t+3}{t^2-2} \cdot \dfrac{t+3}{t^2-2}$
9. $\dfrac{x+1}{2+x} \cdot \dfrac{x-1}{x+1}$
10. $\dfrac{2x}{2x} \cdot \dfrac{x-1}{x+4}$
11. $\dfrac{3y-1}{2y+1} \cdot \dfrac{y}{y}$
12. $\dfrac{-1}{-1} \cdot \dfrac{3-x}{4-x}$

Simplify.

13. $\dfrac{a^2 - 9}{a^2 + 5a + 6}$

14. $\dfrac{t^2 - 25}{t^2 + t - 20}$

15. $\dfrac{2t^2 + 6t + 4}{4t^2 - 12t - 16}$

16. $\dfrac{3a^2 - 9a - 12}{6a^2 + 30a + 24}$

17. $\dfrac{x^2 - 25}{x^2 - 10x + 25}$

18. $\dfrac{x^2 + 8x + 16}{x^2 - 16}$

19. $\dfrac{a^2 - 1}{a - 1}$

20. $\dfrac{t^2 - 1}{t + 1}$

21. $\dfrac{x^2 + 1}{x + 1}$

22. $\dfrac{y^2 + 4}{y + 2}$

23. $\dfrac{6x^2 - 54}{4x^2 - 36}$

24. $\dfrac{8x^2 - 32}{4x^2 - 16}$

25. $\dfrac{6t + 12}{t^2 - t - 6}$

26. $\dfrac{5y + 5}{y^2 + 7y + 6}$

27. $\dfrac{a^2 - 10a + 21}{a^2 - 11a + 28}$

28. $\dfrac{y^2 - 3y - 18}{y^2 - 2y - 15}$

29. $\dfrac{t^2 - 4}{(t + 2)^2}$

30. $\dfrac{(a - 3)^2}{a^2 - 9}$

Multiply and simplify.

31. $\dfrac{x^2 - 3x - 10}{(x - 2)^2} \cdot \dfrac{x - 2}{x - 5}$

32. $\dfrac{t^2}{t^2 - 4} \cdot \dfrac{t^2 - 5t + 6}{t^2 - 3t}$

33. $\dfrac{a^2 - 9}{a^2} \cdot \dfrac{a^2 - 3a}{a^2 + a - 12}$

34. $\dfrac{x^2 + 10x - 11}{x^2 - 1} \cdot \dfrac{x + 1}{x + 11}$

35. $\dfrac{4a^2}{3a^2 - 12a + 12} \cdot \dfrac{3a - 6}{2a}$

36. $\dfrac{5v + 5}{v - 2} \cdot \dfrac{v^2 - 4v + 4}{v^2 - 1}$

37. $\dfrac{x^4 - 16}{x^4 - 1} \cdot \dfrac{x^2 + 1}{x^2 + 4}$

38. $\dfrac{t^4 - 1}{t^4 - 81} \cdot \dfrac{t^2 + 9}{t^2 + 1}$

39. $\dfrac{(t - 2)^3}{(t - 1)^3} \cdot \dfrac{t^2 - 2t + 1}{t^2 - 4t + 4}$

40. $\dfrac{(y + 4)^3}{(y + 2)^3} \cdot \dfrac{y^2 + 4y + 4}{y^2 + 8y + 16}$

Problem-solving practice

41. The product of two consecutive even integers is 360. Find the integers.

42. A passenger drove from Denver, Colorado, to Cheyenne, Wyoming, for a vacation, and then took the bus back. The car trip took 0.3 hr more than the bus. The car averaged 66 km/h, and the bus averaged 75 km/h. How far is it from Denver to Cheyenne?

○ ─────────────────────────────

Simplify.

43. $\dfrac{x^4 - 16y^4}{(x^2 + 4y^2)(x - 2y)}$

44. $\dfrac{(a - b)^2}{b^2 - a^2}$

45. $\dfrac{t^4 - 1}{t^4 - 81} \cdot \dfrac{t^2 - 9}{t^2 + 1} \cdot \dfrac{(t - 9)^2}{(t + 1)^2}$

46. $\dfrac{(t + 2)^3}{(t + 1)^3} \cdot \dfrac{t^2 + 2t + 1}{t^2 + 4t + 4} \cdot \dfrac{t + 1}{t + 2}$

47. $\dfrac{x^2 - y^2}{(x - y)^2} \cdot \dfrac{x^2 - 2xy + y^2}{x^2 - 4xy - 5y^2}$

48. $\dfrac{x - 1}{x^2 + 1} \cdot \dfrac{x^4 - 1}{(x - 1)^2} \cdot \dfrac{x^2 - 1}{x^4 - 2x^2 + 1}$

Determine those replacements in the following expressions that are not sensible.

49. $\dfrac{x+1}{x^2+4x+4}$

50. $\dfrac{x^2-16}{x^2+2x-3}$

51. $\dfrac{x-7}{x^3-9x^2+14x}$

52. $\dfrac{x^3+47}{x^2-49}$

53. Select any number x, multiply by 2, add 5, multiply by 5, subtract 25, and divide by 10. What do you get? Explain how this procedure can be used for a number trick.

9.2 DIVISION AND RECIPROCALS

In all our work in this chapter there is a similarity with what we have done for rational numbers. In fact, after a replacement, a fractional expression represents a rational number. We now see a similarity with finding reciprocals of fractional expressions and division.

Finding Reciprocals

Two expressions are reciprocals of each other if their product is 1. The reciprocal of a fractional expression is found by interchanging numerator and denominator.

EXAMPLES

1. The reciprocal of $\frac{2}{5}$ is $\frac{5}{2}$. (This is because $\frac{2}{5} \cdot \frac{5}{2} = \frac{10}{10} = 1$.)

2. The reciprocal of $\dfrac{2x^2-3}{x+4}$ is $\dfrac{x+4}{2x^2-3}$.

3. The reciprocal of $x+2$ is $\dfrac{1}{x+2}$. $\left(\text{Think of } x+2 \text{ as } \dfrac{x+2}{1}.\right)$

Division

> To divide, we multiply by a reciprocal and simplify the result.

(To review the reason for this, see Section 1.4.)

EXAMPLES Divide.

4. $\dfrac{3}{4} \div \dfrac{2}{5} = \dfrac{3}{4} \cdot \dfrac{5}{2}$ Multiplying by the reciprocal

$= \dfrac{3 \cdot 5}{4 \cdot 2} = \dfrac{15}{8}$

5. $\dfrac{x+1}{x+2} \div \dfrac{x-1}{x+3} = \dfrac{x+1}{x+2} \cdot \dfrac{x+3}{x-1}$ Multiplying by the reciprocal

$= \dfrac{(x+1)(x+3)}{(x+2)(x-1)}$ ⟵ You need not carry out the multiplications in the numerator and denominator.

EXAMPLE 6 Divide and simplify: $\dfrac{x+1}{x^2-1} \div \dfrac{x+1}{x^2-2x+1}$.

$\dfrac{x+1}{x^2-1} \div \dfrac{x+1}{x^2-2x+1} = \dfrac{x+1}{x^2-1} \cdot \dfrac{x^2-2x+1}{x+1}$ Multiplying by the reciprocal

$= \dfrac{(x+1)(x^2-2x+1)}{(x^2-1)(x+1)}$

$= \dfrac{(x+1)(x-1)(x-1)}{(x-1)(x+1)(x+1)}$ Factoring the numerator and denominator

$= \dfrac{(x+1)(x-1)}{(x+1)(x-1)} \cdot \dfrac{x-1}{x+1}$ Factoring the fractional expression

$= \dfrac{x-1}{x+1}$ Simplifying

EXAMPLE 7 Divide and simplify: $\dfrac{x^2-2x-3}{x^2-4} \div \dfrac{x+1}{x+5}$.

$\dfrac{x^2-2x-3}{x^2-4} \div \dfrac{x+1}{x+5} = \dfrac{x^2-2x-3}{x^2-4} \cdot \dfrac{x+5}{x+1}$ Multiplying by the reciprocal

$= \dfrac{(x^2-2x-3)(x+5)}{(x^2-4)(x+1)}$

$= \dfrac{(x-3)(x+1)(x+5)}{(x-2)(x+2)(x+1)}$ Factoring the numerator and denominator

$= \dfrac{x+1}{x+1} \cdot \dfrac{(x-3)(x+5)}{(x-2)(x+2)}$ Factoring the fractional expression

$= \dfrac{(x-3)(x+5)}{(x-2)(x+2)}$ ⟵ Simplifying

You need not carry out the multiplications in the numerator and denominator.

EXERCISE SET 9.2

Find the reciprocal.

1. $\dfrac{4}{x}$

2. $\dfrac{a+3}{a-1}$

3. $x^2 - y^2$

4. $\dfrac{1}{a+b}$

5. $\dfrac{x^2 + 2x - 5}{x^2 - 4x + 7}$

6. $\dfrac{x^2 - 3xy + y^2}{x^2 + 7xy - y^2}$

Divide and simplify.

7. $\dfrac{2}{5} \div \dfrac{4}{3}$

8. $\dfrac{5}{6} \div \dfrac{2}{3}$

9. $\dfrac{2}{x} \div \dfrac{8}{x}$

10. $\dfrac{x}{2} \div \dfrac{3}{x}$

11. $\dfrac{x^2}{y} \div \dfrac{x^3}{y^3}$

12. $\dfrac{a}{b^2} \div \dfrac{a^2}{b^3}$

13. $\dfrac{a+2}{a-3} \div \dfrac{a-1}{a+3}$

14. $\dfrac{y+2}{4} \div \dfrac{y}{2}$

15. $\dfrac{x^2 - 1}{x} \div \dfrac{x+1}{x-1}$

16. $\dfrac{4y - 8}{y + 2} \div \dfrac{y - 2}{y^2 - 4}$

17. $\dfrac{x+1}{6} \div \dfrac{x+1}{3}$

18. $\dfrac{a}{a-b} \div \dfrac{b}{a-b}$

19. $\dfrac{5x - 5}{16} \div \dfrac{x - 1}{6}$

20. $\dfrac{-4 + 2x}{8} \div \dfrac{x - 2}{2}$

21. $\dfrac{-6 + 3x}{5} \div \dfrac{4x - 8}{25}$

22. $\dfrac{-12 + 4x}{4} \div \dfrac{-6 + 2x}{6}$

23. $\dfrac{a+2}{a-1} \div \dfrac{3a+6}{a-5}$

24. $\dfrac{t-3}{t+2} \div \dfrac{4t - 12}{t+1}$

25. $\dfrac{x^2 - 4}{x} \div \dfrac{x - 2}{x + 2}$

26. $\dfrac{x+y}{x-y} \div \dfrac{x^2 + y}{x^2 - y^2}$

27. $\dfrac{x^2 - 9}{4x + 12} \div \dfrac{x - 3}{6}$

28. $\dfrac{x - b}{2x} \div \dfrac{x^2 - b^2}{5x^2}$

29. $\dfrac{c^2 + 3c}{c^2 + 2c - 3} \div \dfrac{c}{c + 1}$

30. $\dfrac{x - 5}{2x} \div \dfrac{x^2 - 25}{4x^2}$

31. $\dfrac{2y^2 - 7y + 3}{2y^2 + 3y - 2} \div \dfrac{6y^2 - 5y + 1}{3y^2 + 5y - 2}$

32. $\dfrac{x^2 - x - 20}{x^2 + 7x + 12} \div \dfrac{x^2 - 10x + 25}{x^2 + 6x + 9}$

33. $\dfrac{c^2 + 10c + 21}{c^2 - 2c - 15} \div (c^2 + 2c - 35)$

34. $\dfrac{1 - z}{1 + 2z - z^2} \div (1 - z)$

35. $\dfrac{(t+5)^3}{(t-5)^3} \div \dfrac{(t+5)^2}{(t-5)^2}$

36. $\dfrac{(y-3)^3}{(y+3)^3} \div \dfrac{(y-3)^2}{(y+3)^2}$

Problem-solving practice

37. Sixteen more than the square of a number is 8 times the number. Find the number.

38. A small chair company has 62 assembled chairs in stock. The staff can put together 2 chairs an hour. How many chairs will they have in stock after working 8 hours?

Simplify.

39. $\dfrac{2a^2 - 5ab}{c - 3d} \div (4a^2 - 25b^2)$

40. $x - 2a \div \dfrac{a^2 x^2 - 4a^4}{a^2 x + 2a^3}$

41. $\dfrac{3a^2 - 5ab - 12b^2}{3ab + 4b^2} \div (3b^2 - ab)$

42. $\dfrac{3x^2 - 2xy - y^2}{x^2 - y^2} \div 3x^2 + 4xy + y^2$

43. $xy \cdot \dfrac{y^2 - 4xy}{y - x} \div \dfrac{16x^2 y^2 - y^4}{4x^2 - 3xy - y^2}$

44. $\dfrac{z^2 - 8z + 16}{z^2 + 8z + 16} \div \dfrac{(z-4)^5}{(z+4)^5}$

45. $\dfrac{x^2 - x + xy - y}{x^2 + 6x - 7} \div \dfrac{x^2 + 2xy + y^2}{4x + 4y}$

46. $\dfrac{3x + 3y + 3}{9x} \div \dfrac{x^2 + 2xy + y^2 - 1}{x^4 + x^2}$

47. $\left(\dfrac{y^2 + 5y + 6}{y^2} \cdot \dfrac{3y^3 + 6y^2}{y^2 - y - 12}\right) \div \dfrac{y^2 - y}{y^2 - 2y - 8}$

48. $\dfrac{a^4 - 81b^4}{a^2 c - 6abc + 9b^2 c} \cdot \dfrac{a + 3b}{a^2 + 9b^2} \div \dfrac{a^2 + 6ab + 9b^2}{(a - 3b)^2}$

49. Show that

$$\dfrac{\dfrac{1}{a}}{b} \quad \text{and} \quad \dfrac{b}{a}$$

are equivalent expressions for the reciprocal of *a/b*.

9.3 ADDITION AND SUBTRACTION

We add and subtract fractional expressions as we do rational numbers. When denominators are the same, we add or subtract the numerators and keep the same denominator. When denominators are not the same, we multiply by 1 to obtain equivalent expressions with the same denominator.

Addition When Denominators Are the Same

We add fractional expressions as we add rational numbers using fractional notation.

> When denominators are the same, we add the numerators and keep the denominator.

EXAMPLES Add.

1. $\dfrac{x}{x+1} + \dfrac{2}{x+1} = \dfrac{x+2}{x+1}$

2. $\dfrac{2x^2 + 3x - 7}{2x + 1} + \dfrac{x^2 + x - 8}{2x + 1} = \dfrac{(2x^2 + 3x - 7) + (x^2 + x - 8)}{2x + 1}$

   $= \dfrac{3x^2 + 4x - 15}{2x + 1}$

Addition When Denominators Are Additive Inverses

When one denominator is the additive inverse of the other, we can first multiply one expression by $-1/-1$.

EXAMPLES Add.

3. $\dfrac{x}{2} + \dfrac{3}{-2} = \dfrac{x}{2} + \dfrac{-1}{-1} \cdot \dfrac{3}{-2}$ ⟵ Multiplying by $\dfrac{-1}{-1}$

   $= \dfrac{x}{2} + \dfrac{(-1)3}{(-1)(-2)}$

   $= \dfrac{x}{2} + \dfrac{-3}{2}$ ⟵ The denominators are now the same.

   $= \dfrac{x + (-3)}{2} = \dfrac{x - 3}{2}$

4. $\dfrac{3x + 4}{x - 2} + \dfrac{x - 7}{2 - x} = \dfrac{3x + 4}{x - 2} + \dfrac{-1}{-1} \cdot \dfrac{x - 7}{2 - x}$ ⟵ We could have chosen to multiply this expression by $-1/-1$.

   $= \dfrac{3x + 4}{x - 2} + \dfrac{-1(x - 7)}{-1(2 - x)}$ ⟵ Note: $-1(2 - x) = -2 + x = x - 2$

   $= \dfrac{3x + 4}{x - 2} + \dfrac{7 - x}{x - 2}$

   $= \dfrac{(3x + 4) + (7 - x)}{x - 2}$

   $= \dfrac{2x + 11}{x - 2}$

Subtraction When Denominators Are the Same

We subtract fractional expressions as we do rational numbers using fractional notation.

FRACTIONAL EXPRESSIONS AND EQUATIONS

> **When denominators are the same, we subtract the numerator and keep the denominator.**

EXAMPLE 5 Subtract: $\dfrac{3x}{x+2} - \dfrac{x-2}{x+2}$.

$$\dfrac{3x}{x+2} - \dfrac{x-2}{x+2} = \dfrac{3x-(x-2)}{x+2}$$

The parentheses are important to make sure that you subtract the entire numerator.

$$= \dfrac{3x - x + 2}{x+2}$$

$$= \dfrac{2x+2}{x+2}$$

Subtraction When Denominators Are Additive Inverses

When one denominator is the additive inverse of the other, we can first multiply one expression by $-1/-1$.

EXAMPLE 6 Subtract: $\dfrac{x}{5} - \dfrac{3x-4}{-5}$.

$$\dfrac{x}{5} - \dfrac{3x-4}{-5} = \dfrac{x}{5} - \dfrac{-1}{-1} \cdot \dfrac{3x-4}{-5}$$

$$= \dfrac{x}{5} - \dfrac{(-1)(3x-4)}{(-1)(-5)}$$

$$= \dfrac{x}{5} - \dfrac{4-3x}{5}$$

$$= \dfrac{x - (4-3x)}{5}$$

Remember the parentheses!

$$= \dfrac{x - 4 + 3x}{5}$$

$$= \dfrac{4x - 4}{5}$$

EXAMPLE 7 Subtract: $\dfrac{5y}{y-5} - \dfrac{2y-3}{5-y}$.

$$\frac{5y}{y-5} - \frac{2y-3}{5-y} = \frac{5y}{y-5} - \frac{-1}{-1} \cdot \frac{2y-3}{5-y}$$

$$= \frac{5y}{y-5} - \frac{(-1)(2y-3)}{(-1)(5-y)} = \frac{5y}{y-5} - \frac{3-2y}{y-5}$$

$$= \frac{5y - (3-2y)}{y-5} \quad \text{Remember the parentheses!}$$

$$= \frac{5y - 3 + 2y}{y-5}$$

$$= \frac{7y-3}{y-5}$$

EXERCISE SET 9.3

Add. Simplify, if possible.

1. $\dfrac{5}{12} + \dfrac{7}{12}$

2. $\dfrac{3}{14} + \dfrac{5}{14}$

3. $\dfrac{1}{3+x} + \dfrac{5}{3+x}$

4. $\dfrac{4x+1}{6x+5} + \dfrac{3x-7}{5+6x}$

5. $\dfrac{x^2+7x}{x^2-5x} + \dfrac{x^2-4x}{x^2-5x}$

6. $\dfrac{a}{x+y} + \dfrac{b}{y+x}$

7. $\dfrac{7}{8} + \dfrac{5}{-8}$

8. $\dfrac{11}{6} + \dfrac{5}{-6}$

9. $\dfrac{3}{t} + \dfrac{4}{-t}$

10. $\dfrac{5}{-a} + \dfrac{8}{a}$

11. $\dfrac{2x+7}{x-6} + \dfrac{3x}{6-x}$

12. $\dfrac{3x-2}{4x-3} + \dfrac{2x-5}{3-4x}$

13. $\dfrac{y^2}{y-3} + \dfrac{9}{3-y}$

14. $\dfrac{t^2}{t-2} + \dfrac{4}{2-t}$

15. $\dfrac{b-7}{b^2-16} + \dfrac{7-b}{16-b^2}$

16. $\dfrac{a-3}{a^2-25} + \dfrac{a-3}{25-a^2}$

17. $\dfrac{z}{(y+z)(y-z)} + \dfrac{y}{(z+y)(z-y)}$

18. $\dfrac{a^2}{a-b} + \dfrac{b^2}{b-a}$

[*Hint:* Multiply by $-1/-1$. Note that $(z+y)(z-y)(-1) = (z+y)(y-z)$.]

19. $\dfrac{x+3}{x-5} + \dfrac{2x-1}{5-x} + \dfrac{2(3x-1)}{x-5}$

20. $\dfrac{3(x-2)}{2x-3} + \dfrac{5(2x+1)}{2x-3} + \dfrac{3(x-1)}{3-2x}$

21. $\dfrac{2(4x+1)}{5x-7} + \dfrac{3(x-2)}{7-5x} + \dfrac{-10x-1}{5x-7}$

22. $\dfrac{5(x-2)}{3x-4} + \dfrac{2(x-3)}{4-3x} + \dfrac{3(5x+1)}{4-3x}$

23. $\dfrac{x+1}{(x+3)(x-3)} + \dfrac{4(x-3)}{(x-3)(x+3)} + \dfrac{(x-1)(x-3)}{(3-x)(x+3)}$

24. $\dfrac{2(x+5)}{(2x-3)(x-1)} + \dfrac{3x+4}{(2x-3)(1-x)} + \dfrac{x-5}{(3-2x)(x-1)}$

Subtract. Simplify, if possible.

25. $\dfrac{7}{8} - \dfrac{3}{8}$

26. $\dfrac{5}{y} - \dfrac{7}{y}$

27. $\dfrac{x}{x-1} - \dfrac{1}{x-1}$

28. $\dfrac{x^2}{x+4} - \dfrac{16}{x+4}$

29. $\dfrac{x+1}{x^2-2x+1} - \dfrac{5-3x}{x^2-2x+1}$

30. $\dfrac{2x-3}{x^2+3x-4} - \dfrac{x-7}{x^2+3x-4}$

31. $\dfrac{11}{6} - \dfrac{5}{-6}$

32. $\dfrac{7}{8} - \dfrac{5}{-8}$

33. $\dfrac{5}{a} - \dfrac{8}{-a}$

34. $\dfrac{3}{t} - \dfrac{4}{-t}$

35. $\dfrac{x}{4} - \dfrac{3x-5}{-4}$

36. $\dfrac{2}{x-1} - \dfrac{2}{1-x}$

37. $\dfrac{3-x}{x-7} - \dfrac{2x-5}{7-x}$

38. $\dfrac{t^2}{t-2} - \dfrac{4}{2-t}$

39. $\dfrac{x-8}{x^2-16} - \dfrac{x-8}{16-x^2}$

40. $\dfrac{x-2}{x^2-25} - \dfrac{6-x}{25-x^2}$

41. $\dfrac{4-x}{x-9} - \dfrac{3x-8}{9-x}$

42. $\dfrac{3-x}{x-7} - \dfrac{2x-5}{7-x}$

43. $\dfrac{2(x-1)}{2x-3} - \dfrac{3(x+2)}{2x-3} - \dfrac{x-1}{3-2x}$

44. $\dfrac{3(x-2)}{2x-3} - \dfrac{5(2x+1)}{2x-3} - \dfrac{3(x-1)}{3-2x}$

Perform the indicated operations and simplify.

45. $\dfrac{3(2x+5)}{x-1} - \dfrac{3(2x-3)}{1-x} + \dfrac{6x-1}{x-1}$

46. $\dfrac{2x-y}{x-y} + \dfrac{x-2y}{y-x} - \dfrac{3x-3y}{x-y}$

47. $\dfrac{x-y}{x^2-y^2} + \dfrac{x+y}{x^2-y^2} - \dfrac{2x}{x^2-y^2}$

48. $\dfrac{x+y}{2(x-y)} - \dfrac{2x-2y}{2(x-y)} + \dfrac{x-3y}{2(y-x)}$

49. $\dfrac{10}{2y-1} - \dfrac{6}{1-2y} + \dfrac{y}{2y-1} + \dfrac{y-4}{1-2y}$

50. $\dfrac{(x+3)(2x-1)}{(2x-3)(x-3)} - \dfrac{(x-3)(x+1)}{(3-x)(3-2x)} + \dfrac{(2x+1)(x+3)}{(3-2x)(x-3)}$

Problem-solving practice

51. The customs duty on phonograph records is 5% of the wholesale price. You buy 350 British records for a wholesale price of $2800. How much customs duty will you pay?

52. Some freight was shipped a total of 970 km, first by railroad and then by truck. The train averaged 100 km/h, the truck averaged 80 km/h, and the time for the delivery was $11\frac{1}{2}$ hr. How far did the freight ride on the train?

Simplify.

53. $\dfrac{x}{(x-y)(y-z)} - \dfrac{x}{(y-x)(z-y)}$

54. $\dfrac{x}{x-y} + \dfrac{y}{y-x} + \dfrac{x+y}{x-y} + \dfrac{x-y}{y-x}$

55. $\dfrac{b-c}{a-(b-c)} - \dfrac{b-a}{(b-a)-c}$

56. $\dfrac{x+y+1}{y-(x+1)} + \dfrac{x+y-1}{x-(y-1)} - \dfrac{x-y-1}{1-(y-x)}$

57. $\dfrac{x^2}{3x^2-5x-2} - \dfrac{2x}{3x+1} \cdot \dfrac{1}{x-2}$

58. $\dfrac{3}{x+4} \cdot \dfrac{2x+11}{x-3} - \dfrac{-1}{4+x} \cdot \dfrac{6x+3}{3-x}$

9.4 LEAST COMMON MULTIPLES

We find least common multiples of polynomials in a manner similar to that for finding least common multiples of natural numbers. It might be helpful to review Section 1.3. We learn to find least common multiples of denominators so that we can consider more complicated additions and subtractions of fractional expressions.

Finding Least Common Multiples

To add when denominators are different we first find a common denominator. For example, to add $\frac{5}{12}$ and $\frac{7}{30}$, we first look for a common multiple of both 12 and 30. Any multiple will do, but we prefer the smallest such number, the *least common multiple* (LCM). To find the LCM, we factor:

$$12 = 2 \cdot 2 \cdot 3,$$
$$30 = 2 \cdot 3 \cdot 5.$$

The LCM is the number that has 2 as a factor twice, 3 as a factor once, and 5 as a factor once:

$$\text{LCM} = 2 \cdot 2 \cdot 3 \cdot 5, \quad \text{or } 60.$$

> To find the LCM, we use each factor the greatest number of times that it appears in any one factorization.

EXAMPLE 1 Find the LCM of 24 and 36.

$$\left.\begin{array}{l}24 = 2 \cdot 2 \cdot 2 \cdot 3 \\ 36 = 2 \cdot 2 \cdot 3 \cdot 3\end{array}\right\} \quad \text{LCM} = 2 \cdot 2 \cdot 2 \cdot 3 \cdot 3, \quad \text{or } 72.$$

Adding Using the LCM

Let us finish adding $\frac{5}{12}$ and $\frac{7}{30}$:

$$\frac{5}{12} + \frac{7}{30} = \frac{5}{2 \cdot 2 \cdot 3} + \frac{7}{2 \cdot 3 \cdot 5}.$$

The LCM is $2 \cdot 2 \cdot 3 \cdot 5$. To get the LCM in the first denominator we need a 5. To get the LCM in the second denominator we need another 2. We get these numbers by multiplying by 1:

$$\frac{5}{12} + \frac{7}{30} = \frac{5}{2 \cdot 2 \cdot 3} \cdot \frac{5}{5} + \frac{7}{2 \cdot 3 \cdot 5} \cdot \frac{2}{2} \quad \text{Multiplying by 1}$$

$$= \frac{25}{2 \cdot 2 \cdot 3 \cdot 5} + \frac{14}{2 \cdot 3 \cdot 5 \cdot 2} \quad \text{The denominators are now the LCM.}$$

$$= \frac{39}{2 \cdot 2 \cdot 3 \cdot 5} \quad \text{Adding the numerators and keeping the LCM}$$

$$= \frac{13}{20}. \quad \text{Simplifying}$$

EXAMPLE 2 Add: $\frac{5}{12} + \frac{11}{18}$.

$$\left.\begin{array}{l}12 = 2 \cdot 2 \cdot 3 \\ 18 = 2 \cdot 3 \cdot 3\end{array}\right\} \quad \text{LCM} = 2 \cdot 2 \cdot 3 \cdot 3, \quad \text{or } 36.$$

$$\frac{5}{12} + \frac{11}{18} = \frac{5}{2 \cdot 2 \cdot 3} \cdot \frac{3}{3} + \frac{11}{2 \cdot 3 \cdot 3} \cdot \frac{2}{2} = \frac{37}{2 \cdot 2 \cdot 3 \cdot 3} = \frac{37}{36}$$

LCMs of Algebraic Expressions

To find the LCM of two or more algebraic expressions, we factor them. Then we use each factor the greatest number of times it occurs in any one expression.

EXAMPLE 3 Find the LCM of $12x$, $16y$, and $8xyz$.

$$\left.\begin{array}{l}12x = 2 \cdot 2 \cdot 3 \cdot x \\ 16y = 2 \cdot 2 \cdot 2 \cdot 2 \cdot y \\ 8xyz = 2 \cdot 2 \cdot 2 \cdot x \cdot y \cdot z\end{array}\right\} \quad \begin{aligned}\text{LCM} &= 2 \cdot 2 \cdot 2 \cdot 2 \cdot 3 \cdot x \cdot y \cdot z \\ &= 48xyz\end{aligned}$$

9.4 LEAST COMMON MULTIPLES

EXAMPLE 4 Find the LCM of $x^2 + 5x - 6$ and $x^2 - 1$.

$$\left.\begin{array}{r} x^2 + 5x - 6 = (x + 6)(x - 1) \\ x^2 - 1 = (x + 1)(x - 1) \end{array}\right\} \quad \text{LCM} = (x + 6)(x - 1)(x + 1)$$

EXAMPLE 5 Find the LCM of $x^2 + 4$, $x + 1$, and 5.

These expressions are not factorable, so the LCM is their product:

$$5(x^2 + 4)(x + 1).$$

The additive inverse of an LCM is also an LCM. For example, if $(x + 2)(x - 5)$ is an LCM, then $-(x + 2)(x - 5)$ is an LCM. We can name the latter

$$(x + 2)(-1)(x - 5), \quad \text{or} \quad (x + 2)(5 - x).$$

When finding an LCM, if factors that are additive inverses occur, we do not use them both. For example, if $(x - 5)$ occurs in one factorization and $(5 - x)$ occurs in another, since these are additive inverses, we do not use them both.

EXAMPLE 6 Find the LCM of $x^2 - 25$ and $10 - 2x$.

$$\left.\begin{array}{r} x^2 - 25 = (x + 5)(x - 5) \\ 10 - 2x = 2(5 - x) \end{array}\right\} \quad \begin{array}{l} \text{LCM} = 2 \cdot (x + 5)(x - 5) \leftarrow \\ \text{or } 2(x + 5)(5 - x) \leftarrow \end{array}$$

We can use $x - 5$ or $5 - x$ but not both.

EXAMPLE 7 Find the LCM of $x^2 - 4y^2$, $x^2 - 4xy + 4y^2$, and $x - 2y$.

$$\left.\begin{array}{r} x^2 - 4y^2 = (x - 2y)(x + 2y) \\ x^2 - 4xy + 4y^2 = (x - 2y)(x - 2y) \\ x - 2y = x - 2y \end{array}\right\} \quad \begin{array}{l} \text{LCM} = (x + 2y)(x - 2y)(x - 2y) \\ = (x + 2y)(x - 2y)^2 \end{array}$$

EXERCISE SET 9.4

Find the LCM.

1. 12, 27
2. 10, 15
3. 8, 9
4. 12, 15
5. 6, 9, 21
6. 8, 36, 40
7. 24, 36, 40
8. 3, 4, 5
9. 28, 42, 60

Add, first finding the LCM of the denominators. Simplify, if possible.

10. $\frac{7}{24} + \frac{11}{18}$
11. $\frac{7}{60} + \frac{6}{75}$
12. $\frac{1}{6} + \frac{3}{40} + \frac{2}{75}$
13. $\frac{5}{24} + \frac{3}{20} + \frac{7}{30}$
14. $\frac{2}{15} + \frac{5}{9} + \frac{3}{20}$
15. $\frac{1}{20} + \frac{1}{30} + \frac{2}{45}$

Find the LCM.

16. $6x^2, 12x^3$
17. $2a^2b, 8ab^2$
18. $2x^2, 6xy, 18y^2$
19. c^2d, cd^2, c^3d
20. $2(y-3), 6(3-y)$
21. $4(x-1), 8(1-x)$
22. $t, t+2, t-2$
23. $x, x+3, x-3$
24. x^2-4, x^2+5x+6
25. x^2+3x+2, x^2-4
26. t^3+4t^2+4t, t^2-4t
27. y^3-y^2, y^4-y^2
28. $a+1, (a-1)^2, a^2-1$
29. $x^2-y^2, 2x+2y, x^2+2xy+y^2$
30. m^2-5m+6, m^2-4m+4
31. $2x^2+5x+2, 2x^2-x-1$
32. $2+3x, 9x^2-4, 2-3x$
33. $3-2x, 4x^2-9, 3+2x$
34. $10v^2+30v, -5v^2-35v-60$
35. $12a^2+24a, -4a^2-20a-24$
36. $9x^3-9x^2-18x, 6x^5-24x^4+24x^3$
37. x^5-4x^3, x^3+4x^2+4x
38. $x^5+4x^4+4x^3, 3x^2-12, 2x+4$
39. $x^5+2x^4+x^3, 2x^3-2x, 5x-5$

○

Find the LCM.

40. 72, 90, 96
41. $8x^2-8, 6x^2-12x+6, 10-10x$

42. Two joggers leave the starting point of a circular course at the same time. One jogger completes one round in 6 minutes and the second jogger in 8 minutes. Assuming they continue to run at the same pace, after how many minutes will they meet again at the starting place?

43. If the LCM of two expressions is the same as one of the expressions, what is their relationship?

9.5 ADDITION WITH DIFFERENT DENOMINATORS

Now that we know how to find LCMs, we can add fractional expressions with different denominators. First, we find the LCM of the denominators (the least common denominator). Next, we multiply by 1 to make denominators the same. Then we add.

EXAMPLE 1 Add: $\dfrac{5x^2}{8} + \dfrac{7x}{12}$.

First, we find the LCM of the denominators:

$8 = 2 \cdot 2 \cdot 2$
$12 = 2 \cdot 2 \cdot 3$ } LCM $= 2 \cdot 2 \cdot 2 \cdot 3$

We multiply by 1 to get the LCM in each expression, and then add and simplify:

9.5 ADDITION WITH DIFFERENT DENOMINATORS

$$\frac{5x^2}{8} + \frac{7x}{12} = \frac{5x^2}{2 \cdot 2 \cdot 2} + \frac{7x}{2 \cdot 2 \cdot 3}$$

$$= \frac{5x^2}{2 \cdot 2 \cdot 2} \cdot \frac{3}{3} + \frac{7x}{2 \cdot 2 \cdot 3} \cdot \frac{2}{2} \quad \text{Multiplying by 1 to get the same denominators}$$

$$= \frac{15x^2 + 14x}{24}.$$

EXAMPLE 2 Add: $\dfrac{3}{x+1} + \dfrac{5}{x-1}$.

The denominators do not factor, so the LCM is their product. We multiply by 1 to get the LCM in each expression:

$$\frac{3}{x+1} \cdot \frac{x-1}{x-1} + \frac{5}{x-1} \cdot \frac{x+1}{x+1} = \frac{3(x-1) + 5(x+1)}{(x-1)(x+1)}$$

$$= \frac{3x - 3 + 5x + 5}{(x-1)(x+1)}$$

$$= \frac{8x + 2}{(x-1)(x+1)}.$$

The numerator and denominator have no common factor (other than 1), so we cannot simplify.

EXAMPLE 3 Add: $\dfrac{5}{x^2 + x} + \dfrac{4}{2x + 2}$.

First, we find the LCM of the denominators:

$$\left.\begin{array}{l} x^2 + x = x(x+1) \\ 2x + 2 = 2(x+1) \end{array}\right\} \quad \text{LCM} = 2 \cdot x(x+1)$$

We multiply by 1 to get the LCM in each expression, and then add and simplify.

$$\frac{5}{x(x+1)} \cdot \frac{2}{2} + \frac{4}{2(x+1)} \cdot \frac{x}{x} = \frac{10}{2x(x+1)} + \frac{4x}{2x(x+1)} \quad \text{Multiplying by 1}$$

$$= \frac{10 + 4x}{2x(x+1)} \quad \text{Adding}$$

$$= \frac{2(5 + 2x)}{2x(x+1)} \quad \text{Factoring the numerator}$$

$$= \frac{2}{2} \cdot \frac{5 + 2x}{x(x+1)} \quad \text{Factoring the fractional expression}$$

$$= \frac{5 + 2x}{x(x+1)} \quad \text{Simplifying}$$

EXAMPLE 4 Add: $\dfrac{x+4}{x-2} + \dfrac{x-7}{x+5}$.

First, we find the LCM of the denominators. It is just the product:

$$\text{LCM} = (x-2)(x+5).$$

We multiply by 1 to get the LCM in each expression, and then add and simplify:

$$\dfrac{x+4}{x-2} \cdot \dfrac{x+5}{x+5} + \dfrac{x-7}{x+5} \cdot \dfrac{x-2}{x-2} = \dfrac{(x+4)(x+5)}{(x-2)(x+5)} + \dfrac{(x-7)(x-2)}{(x-2)(x+5)}$$

$$= \dfrac{x^2+9x+20}{(x-2)(x+5)} + \dfrac{x^2-9x+14}{(x-2)(x+5)}$$

$$= \dfrac{x^2+9x+20+x^2-9x+14}{(x-2)(x+5)}$$

$$= \dfrac{2x^2+34}{(x-2)(x+5)}.$$

EXAMPLE 5 Add: $\dfrac{x}{x^2+11x+30} + \dfrac{-5}{x^2+9x+20}$.

$$\dfrac{x}{x^2+11x+30} + \dfrac{-5}{x^2+9x+20}$$

$$= \dfrac{x}{(x+5)(x+6)} + \dfrac{-5}{(x+5)(x+4)} \quad \text{Factoring denominators in order to find the LCM. The LCM is } (x+4)(x+5)(x+6).$$

$$= \dfrac{x}{(x+5)(x+6)} \cdot \dfrac{x+4}{x+4} + \dfrac{-5}{(x+5)(x+4)} \cdot \dfrac{x+6}{x+6} \quad \text{Multiplying by 1}$$

$$= \dfrac{x(x+4)}{(x+5)(x+6)(x+4)} + \dfrac{-5(x+6)}{(x+5)(x+4)(x+6)}$$

$$= \dfrac{x(x+4)+(-5)(x+6)}{(x+4)(x+5)(x+6)} \quad \text{Adding}$$

$$= \dfrac{x^2+4x-5x-30}{(x+4)(x+5)(x+6)}$$

$$= \dfrac{x^2-x-30}{(x+4)(x+5)(x+6)} = \dfrac{(x-6)(x+5)}{(x+4)(x+5)(x+6)}$$

$$= \dfrac{x+5}{x+5} \cdot \dfrac{x-6}{(x+4)(x+6)} = \dfrac{x-6}{(x+4)(x+6)}$$

← Always simplify at the end if possible.

EXERCISE SET 9.5

Add.

1. $\dfrac{2}{x} + \dfrac{5}{x^2}$

2. $\dfrac{4}{x} + \dfrac{8}{x^2}$

3. $\dfrac{5}{6r} + \dfrac{7}{8r}$

4. $\dfrac{2}{9t} + \dfrac{11}{6t}$

5. $\dfrac{x+y}{xy^2} + \dfrac{3x+y}{x^2y}$

6. $\dfrac{2c-d}{c^2d} + \dfrac{c+d}{cd^2}$

7. $\dfrac{3}{x-2} + \dfrac{3}{x+2}$

8. $\dfrac{2}{x-1} + \dfrac{2}{x+1}$

9. $\dfrac{3}{x+1} + \dfrac{2}{3x}$

10. $\dfrac{2}{x+5} + \dfrac{3}{4x}$

11. $\dfrac{2x}{x^2-16} + \dfrac{x}{x-4}$

12. $\dfrac{4x}{x^2-25} + \dfrac{x}{x+5}$

13. $\dfrac{5}{z+4} + \dfrac{3}{3z+12}$

14. $\dfrac{t}{t-3} + \dfrac{5}{4t-12}$

15. $\dfrac{3}{x-1} + \dfrac{2}{(x-1)^2}$

16. $\dfrac{2}{x+3} + \dfrac{4}{(x+3)^2}$

17. $\dfrac{4a}{5a-10} + \dfrac{3a}{10a-20}$

18. $\dfrac{3a}{4a-20} + \dfrac{9a}{6a-30}$

19. $\dfrac{x+4}{x} + \dfrac{x}{x+4}$

20. $\dfrac{x}{x-5} + \dfrac{x-5}{x}$

21. $\dfrac{x}{x^2+2x+1} + \dfrac{1}{x^2+5x+4}$

22. $\dfrac{7}{a^2+a-2} + \dfrac{5}{a^2-4a+3}$

23. $\dfrac{x+3}{x-5} + \dfrac{x-5}{x+3}$

24. $\dfrac{3x}{2y-3} + \dfrac{2x}{3y-2}$

25. $\dfrac{a}{a^2-1} + \dfrac{2a}{a^2-a}$

26. $\dfrac{3x+2}{3x+6} + \dfrac{x-2}{x^2-4}$

27. $\dfrac{6}{x-y} + \dfrac{4x}{y^2-x^2}$

28. $\dfrac{a-2}{3-a} + \dfrac{4-a^2}{a^2-9}$

29. $\dfrac{10}{x^2+x-6} + \dfrac{3x}{x^2-4x+4}$

30. $\dfrac{2}{z^2-z-6} + \dfrac{3}{z^2-9}$

○ ───

Find the perimeter and area of each figure.

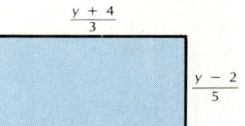

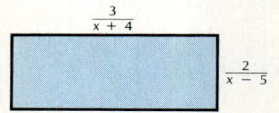

31. Rectangle with sides $\dfrac{y+4}{3}$ and $\dfrac{y-2}{5}$

32. Rectangle with sides $\dfrac{3}{x+4}$ and $\dfrac{2}{x-5}$

Add. Simplify, if possible.

33. $\dfrac{5}{z+2} + \dfrac{4z}{z^2-4} + 2$

34. $\dfrac{-2}{y^2-9} + \dfrac{4y}{(y-3)^2} + \dfrac{6}{3-y}$

35. $\dfrac{3z^2}{z^4-4} + \dfrac{5z^2-3}{2z^4+z^2-6}$

36. Find an expression equivalent to

$$\dfrac{a-3b}{a-b}$$

that is a sum of two fractional expressions. Answers can vary.

9.6 SUBTRACTION WITH DIFFERENT DENOMINATORS

Subtraction of fractional expressions is like addition, except that we subtract numerators.

Subtraction

To subtract fractional expressions, we first find the LCM of the denominators (the least common denominator). Next, we multiply by 1 to make denominators the same. Then we subtract.

EXAMPLE 1 Subtract: $\dfrac{x+2}{x-4} - \dfrac{x+1}{x+4}$.

$\text{LCM} = (x-4)(x+4)$

$\dfrac{x+2}{x-4} \cdot \dfrac{x+4}{x+4} - \dfrac{x+1}{x+4} \cdot \dfrac{x-4}{x-4} = \dfrac{(x+2)(x+4)}{(x-4)(x+4)} - \dfrac{(x+1)(x-4)}{(x-4)(x+4)}$

$= \dfrac{x^2 + 6x + 8}{(x-4)(x+4)} - \dfrac{x^2 - 3x - 4}{(x-4)(x+4)}$

$= \dfrac{x^2 + 6x + 8 - (x^2 - 3x - 4)}{(x-4)(x+4)}$ Subtracting numerators

$= \dfrac{x^2 + 6x + 8 - x^2 + 3x + 4}{(x-4)(x+4)}$ Don't forget parentheses.

$= \dfrac{9x + 12}{(x-4)(x+4)}$

Simplifying Combined Additions and Subtractions

EXAMPLE 2 Simplify: $\dfrac{1}{x} - \dfrac{1}{x^2} + \dfrac{2}{x+1}$.

$\text{LCM} = x^2(x+1)$

$\dfrac{1}{x} \cdot \dfrac{x(x+1)}{x(x+1)} - \dfrac{1}{x^2} \cdot \dfrac{x+1}{x+1} + \dfrac{2}{x+1} \cdot \dfrac{x^2}{x^2} = \dfrac{x(x+1)}{x^2(x+1)} - \dfrac{x+1}{x^2(x+1)} + \dfrac{2x^2}{x^2(x+1)}$

$= \dfrac{x(x+1) - (x+1) + 2x^2}{x^2(x+1)}$

$= \dfrac{x^2 + x - x - 1 + 2x^2}{x^2(x+1)}$

$= \dfrac{3x^2 - 1}{x^2(x+1)}$

EXERCISE SET 9.6

Subtract. Simplify, if possible.

1. $\dfrac{x-2}{6} - \dfrac{x+1}{3}$

2. $\dfrac{a+2}{2} - \dfrac{a-4}{4}$

3. $\dfrac{4z-9}{3z} - \dfrac{3z-8}{4z}$

4. $\dfrac{x-1}{4x} - \dfrac{2x+3}{x}$

5. $\dfrac{4x+2t}{3xt^2} - \dfrac{5x-3t}{x^2t}$

6. $\dfrac{5x+3y}{2x^2y} - \dfrac{3x+4y}{xy^2}$

7. $\dfrac{5}{x+5} - \dfrac{3}{x-5}$

8. $\dfrac{2z}{z-1} - \dfrac{3z}{z+1}$

9. $\dfrac{3}{2t^2-2t} - \dfrac{5}{2t-2}$

10. $\dfrac{8}{x^2-4} - \dfrac{3}{x+2}$

11. $\dfrac{2s}{t^2-s^2} - \dfrac{s}{t-s}$

12. $\dfrac{3}{12+x-x^2} - \dfrac{2}{x^2-9}$

13. $\dfrac{y-5}{y} - \dfrac{3y-1}{4y}$

14. $\dfrac{3x-2}{4x} - \dfrac{3x+1}{6x}$

15. $\dfrac{a}{x+a} - \dfrac{a}{x-a}$

16. $\dfrac{t}{y-t} - \dfrac{y}{y+t}$

17. $\dfrac{8x}{x^2-16} - \dfrac{5}{x+4}$

18. $\dfrac{5x}{x^2-9} - \dfrac{4}{x+3}$

19. $\dfrac{t^2}{2t^2-2t} - \dfrac{1}{2t-2}$

20. $\dfrac{4}{5b^2-5b} - \dfrac{3}{5b-5}$

21. $\dfrac{x}{x^2+5x+6} - \dfrac{2}{x^2+3x+2}$

22. $\dfrac{x}{x^2+11x+30} - \dfrac{5}{x^2+9x+20}$

Simplify.

23. $\dfrac{4y}{y^2-1} - \dfrac{2}{y} - \dfrac{2}{y+1}$

24. $\dfrac{x+6}{4-x^2} - \dfrac{x+3}{x+2} + \dfrac{x-3}{2-x}$

25. $\dfrac{2z}{1-2z} + \dfrac{3z}{2z+1} - \dfrac{3}{4z^2-1}$

26. $\dfrac{1}{x+y} + \dfrac{1}{x-y} - \dfrac{2x}{x^2-y^2}$

27. $\dfrac{5}{3-2x} + \dfrac{3}{2x-3} - \dfrac{x-3}{2x^2-x-3}$

28. $\dfrac{2r}{r^2-s^2} + \dfrac{1}{r+s} - \dfrac{1}{r-s}$

29. $\dfrac{3}{2c-1} - \dfrac{1}{c+2} - \dfrac{5}{2c^2+3c-2}$

30. $\dfrac{3y-1}{2y^2+y-3} - \dfrac{2-y}{y-1}$

31. $\dfrac{1}{x+y} - \dfrac{1}{x-y} + \dfrac{2x}{x^2-y^2}$

32. $\dfrac{1}{a-b} - \dfrac{1}{a+b} + \dfrac{2b}{a^2-b^2}$

Subtract. Simplify, if possible.

33. $\dfrac{1}{2xy-6x+ay-3a} - \dfrac{ay+xy}{(a^2-4x^2)(y^2-6y+9)}$

34. $\dfrac{x}{x^4-y^4} - \dfrac{1}{x^2+2xy+y^2}$

35. Find an expression equivalent to

$$\frac{5x^2 + 2xy}{x^2 + y^2}$$

that is a difference of two fractional expressions. Answers can vary.

9.7 COMPLEX FRACTIONAL EXPRESSIONS

A *complex fractional expression* is a fractional expression that has one or more fractional expressions somewhere in its numerator or denominator. Here are some examples:

$$\frac{1 + \frac{2}{x}}{3}, \quad \frac{\frac{x+y}{2}}{\frac{2x}{x+1}}, \quad \frac{\frac{1}{3} + \frac{1}{5}}{\frac{2}{x} - \frac{x}{y}}.$$

To simplify a complex fractional expression, we first add or subtract, if necessary, to get a single fractional expression in both numerator and denominator. Then we divide, by multiplying by the reciprocal of the denominator.

EXAMPLE 1 Simplify.

$$\frac{\frac{1}{5} + \frac{2}{5}}{\frac{7}{3}} = \frac{\frac{3}{5}}{\frac{7}{3}} \qquad \text{Adding in the numerator}$$

$$= \frac{3}{5} \cdot \frac{3}{7} \qquad \text{Multiplying by the reciprocal of the denominator}$$

$$= \frac{9}{35}$$

EXAMPLE 2 Simplify.

$$\frac{1 + \frac{2}{x}}{\frac{3}{4}} = \frac{1 \cdot \frac{x}{x} + \frac{2}{x}}{\frac{3}{4}} \qquad \text{Multiplying by } \frac{x}{x} \text{ to get a common denominator}$$

$$= \frac{\frac{x+2}{x}}{\frac{3}{4}} \qquad \text{Adding in the numerator}$$

$$= \frac{x+2}{x} \cdot \frac{4}{3} \qquad \text{Multiplying by the reciprocal of the denominator}$$

$$= \frac{4(x+2)}{3x}$$

9.7 COMPLEX FRACTIONAL EXPRESSIONS

EXAMPLE 3 Simplify.

$$\frac{\frac{3}{x} + \frac{1}{2x}}{\frac{1}{3x} - \frac{3}{4x}} = \left.\begin{array}{c}\dfrac{\dfrac{3}{x}\cdot\dfrac{2}{2} + \dfrac{1}{2x}}{\dfrac{1}{3x}\cdot\dfrac{4}{4} - \dfrac{3}{4x}\cdot\dfrac{3}{3}}\end{array}\right\} \begin{array}{l}\leftarrow \text{Finding the LCM, } 2x, \text{ and} \\ \text{multiplying by 1 in the numerator} \\ \leftarrow \text{Finding the LCM, } 12x, \text{ and} \\ \text{multiplying by 1 in the denominator}\end{array}$$

$$= \frac{\dfrac{6}{2x} + \dfrac{1}{2x}}{\dfrac{4}{12x} - \dfrac{9}{12x}} = \frac{\dfrac{7}{2x}}{\dfrac{-5}{12x}} \quad \begin{array}{l}\leftarrow \text{Adding in the numerator and} \\ \text{subtracting in the denominator}\end{array}$$

$$= \frac{7}{2x} \cdot \frac{12x}{-5} \qquad \text{Multiplying by the reciprocal of the denominator}$$

$$= \frac{2x}{2x} \cdot \frac{7 \cdot 6}{-5} \qquad \text{Factoring}$$

$$= \frac{42}{-5} = -\frac{42}{5} \qquad \text{Simplifying}$$

EXAMPLE 4 Simplify.

$$\frac{1 - \dfrac{1}{x}}{1 - \dfrac{1}{x^2}} = \left.\begin{array}{c}\dfrac{\dfrac{x}{x} - \dfrac{1}{x}}{\dfrac{x^2}{x^2} - \dfrac{1}{x^2}}\end{array}\right\} \begin{array}{l}\leftarrow \text{Finding the LCM, } x, \text{ and multiplying by 1} \\ \text{in the numerator} \\ \leftarrow \text{Finding the LCM, } x^2, \text{ and multiplying by 1} \\ \text{in the denominator}\end{array}$$

$$= \frac{\dfrac{x-1}{x}}{\dfrac{x^2-1}{x^2}} \quad \begin{array}{l}\leftarrow \text{Subtracting in the numerator and} \\ \text{subtracting in the denominator}\end{array}$$

$$= \frac{x-1}{x} \cdot \frac{x^2}{x^2-1} \qquad \text{Multiplying by the reciprocal of the divisor}$$

$$= \frac{(x-1)x^2}{x(x^2-1)} = \frac{x(x-1)}{x(x-1)} \cdot \frac{x}{x+1} \qquad \text{Factoring}$$

$$= \frac{x}{x+1} \qquad \text{Simplifying}$$

EXERCISE SET 9.7

Simplify.

1. $\dfrac{1+\dfrac{9}{16}}{1-\dfrac{3}{4}}$

2. $\dfrac{9-\dfrac{1}{4}}{3+\dfrac{1}{2}}$

3. $\dfrac{1-\dfrac{3}{5}}{1+\dfrac{1}{5}}$

4. $\dfrac{\dfrac{5}{27}-5}{\dfrac{1}{3}+1}$

5. $\dfrac{\dfrac{1}{x}+3}{\dfrac{1}{x}-5}$

6. $\dfrac{\dfrac{3}{s}+s}{\dfrac{s}{3}+s}$

7. $\dfrac{\dfrac{1}{2}+\dfrac{3}{4}}{\dfrac{5}{8}-\dfrac{5}{6}}$

8. $\dfrac{\dfrac{2}{3}-\dfrac{5}{6}}{\dfrac{3}{4}+\dfrac{7}{8}}$

9. $\dfrac{\dfrac{2}{y}+\dfrac{1}{2y}}{y+\dfrac{y}{2}}$

10. $\dfrac{4-\dfrac{1}{x^2}}{2-\dfrac{1}{x}}$

11. $\dfrac{8+\dfrac{8}{d}}{1+\dfrac{1}{d}}$

12. $\dfrac{2-\dfrac{3}{b}}{2-\dfrac{b}{3}}$

13. $\dfrac{\dfrac{1}{5}-\dfrac{1}{a}}{\dfrac{5-a}{5}}$

14. $\dfrac{2-\dfrac{1}{x}}{\dfrac{2}{x}}$

15. $\dfrac{\dfrac{x}{x-y}}{\dfrac{x^2}{x^2-y^2}}$

16. $\dfrac{\dfrac{x}{y}-\dfrac{y}{x}}{\dfrac{1}{y}+\dfrac{1}{x}}$

17. $\dfrac{x-3+\dfrac{2}{x}}{x-4+\dfrac{3}{x}}$

18. $\dfrac{1+\dfrac{a}{b-a}}{\dfrac{a}{a+b}-1}$

Problem-solving practice

19. A cast iron pipe has an outside diameter of 10 cm. Its wall thickness is 0.95 cm. Find the inside diameter.

20. The length of a rectangle is 3 yd greater than the width. The area of the rectangle is 10 yd². Find the perimeter.

21. Find the reciprocal of $\dfrac{2}{x-1}-\dfrac{1}{3x-2}$.

Simplify.

22. $\dfrac{\dfrac{a}{b}+\dfrac{c}{d}}{\dfrac{b}{a}+\dfrac{d}{c}}$

23. $\dfrac{\dfrac{a}{b}-\dfrac{c}{d}}{\dfrac{b}{a}-\dfrac{d}{c}}$

24. $\left[\dfrac{\dfrac{x+1}{x-1}+1}{\dfrac{x+1}{x-1}-1}\right]^5$

25. $1+\dfrac{1}{1+\dfrac{1}{1+\dfrac{1}{1+\dfrac{1}{x}}}}$

26. $\dfrac{\dfrac{z}{1-\dfrac{z}{2+2z}}-2z}{\dfrac{2z}{5z-2}-3}$

9.8 DIVISION OF POLYNOMIALS

In this section we will consider division of polynomials. You will see that such division is similar to what is done in arithmetic.

Divisor a Monomial

We first consider division by a monomial.

EXAMPLE 1 Divide: $x^3 + 10x^2 + 8x$ by $2x$.

We write a fractional expression to show division:

$$\frac{x^3 + 10x^2 + 8x}{2x}.$$

This is equivalent to

$$\frac{x^3}{2x} + \frac{10x^2}{2x} + \frac{8x}{2x}. \quad \text{To see this, add and get the original expression.}$$

Next, we do the separate divisions:

$$\frac{x^3}{2x} + \frac{10x^2}{2x} + \frac{8x}{2x} = \frac{1}{2}x^2 + 5x + 4.$$

EXAMPLE 2 Divide and check: $(5y^2 - 2y + 3) \div 2$.

$$\frac{5y^2 - 2y + 3}{2} = \frac{5y^2}{2} - \frac{2y}{2} + \frac{3}{2} = \frac{5}{2}y^2 - y + \frac{3}{2}$$

Check: $\frac{5}{2}y^2 - y + \frac{3}{2}$
$\underline{\,2}$
$5y^2 - 2y + 3$ We multiply.
                    The answer checks.

Try to write only the answer.

- - -

To divide by a monomial, we can divide each term by the monomial.

Divisor Not a Monomial

When the divisor is not a monomial, we use long division very much as we do in arithmetic.

EXAMPLE 3   Divide $x^2 + 5x + 6$ by $x + 2$.

$$\begin{array}{r} x \\ x+2\overline{\smash{)}x^2+5x+6} \\ \underline{x^2+2x} \\ 3x \end{array}$$

— Divide the first term by the first term, to get x. Ignore the term 2.
— Multiply x by the divisor.
— Subtract.

We now "bring down" the next term of the dividend, 6.

$$\begin{array}{r} x+3 \\ x+2\overline{\smash{)}x^2+5x+6} \\ \underline{x^2+2x} \\ 3x+6 \\ \underline{3x+6} \\ 0 \end{array}$$

— Divide the first term by the first term to get 3.

— Multiply 3 by the divisor.
— Subtract.

The quotient is $x + 3$.

To check, multiply the quotient by the divisor and add the remainder, if any, to see if you get the dividend:

$$(x+2)(x+3) = x^2 + 5x + 6.$$ The division checks.

EXAMPLE 4   Divide and check: $(x^2 + 2x - 12) \div (x - 3)$.

$$\begin{array}{r} x+5 \\ x-3\overline{\smash{)}x^2+2x-12} \\ \underline{x^2-3x} \\ 5x-12 \\ \underline{5x-15} \\ 3 \end{array}$$

Check:
$(x-3)(x+5) + 3 = x^2 + 2x - 15 + 3$
$ = x^2 + 2x - 12$

Quotient

Remainder

The quotient is $x + 5$ and the remainder is 3. We can write this $x + 5$, R 3. The answer can also be given in this way:

$$x + 5 + \frac{3}{x - 3}.$$

EXAMPLE 5 Divide: $(x^3 + 1) \div (x + 1)$.

$$\begin{array}{r} x^2 - x + 1 \\ x + 1 \overline{\smash{)}\, x^3 + 1} \\ \underline{x^3 + x^2} \\ -x^2 \\ \underline{-x^2 - x} \\ x + 1 \\ \underline{x + 1} \end{array}$$

Leave space for missing terms.

The answer is $x^2 - x + 1$.
You need not write a 0 remainder.

EXAMPLE 6 Divide: $(x^4 - 3x^2 + 1) \div (x - 4)$.

$$\begin{array}{r} x^3 + 4x^2 + 13x + 52 \\ x - 4 \overline{\smash{)}\, x^4 - 3x^2 + 1} \\ \underline{x^4 - 4x^3} \\ 4x^3 - 3x^2 \\ \underline{4x^3 - 16x^2} \\ 13x^2 \\ \underline{13x^2 - 52x} \\ 52x + 1 \\ \underline{52x - 208} \\ 209 \end{array}$$

The answer can be expressed as $x^3 + 4x^2 + 13x + 52$, R 209, or

$$x^3 + 4x^2 + 13x + 52 + \frac{209}{x - 4}.$$

EXERCISE SET 9.8

Divide and check.

1. $\dfrac{24x^4 - 4x^3 + x^2 - 16}{8}$

2. $\dfrac{12a^4 - 3a^2 + a - 6}{6}$

3. $\dfrac{u - 2u^2 - u^5}{u}$

4. $\dfrac{50x^5 - 7x^4 + x^2}{x}$

5. $(15t^3 + 24t^2 - 6t) \div 3t$
6. $(25t^3 + 15t^2 - 30t) \div 5t$
7. $(20x^6 - 20x^4 - 5x^2) \div (-5x^2)$
8. $(24x^6 + 32x^5 - 8x^2) \div (-8x^2)$
9. $(24x^5 - 40x^4 + 6x^3) \div (4x^3)$
10. $(18x^6 - 27x^5 - 3x^3) \div (9x^3)$
11. $\dfrac{9r^2s^2 + 3r^2s - 6rs^2}{-3rs}$
12. $\dfrac{4x^4y - 8x^6y^2 + 12x^8y^6}{4x^4y}$

Divide.

13. $(x^2 + 4x + 4) \div (x + 2)$
14. $(x^2 - 6x + 9) \div (x - 3)$
15. $(x^2 - 10x - 25) \div (x - 5)$
16. $(x^2 + 8x - 16) \div (x + 4)$
17. $(x^2 + 4x - 14) \div (x + 6)$
18. $(x^2 + 5x - 9) \div (x - 2)$
19. $\dfrac{x^2 - 9}{x + 3}$
20. $\dfrac{x^2 - 25}{x + 5}$
21. $\dfrac{x^5 + 1}{x + 1}$
22. $\dfrac{x^5 - 1}{x - 1}$
23. $\dfrac{8x^3 - 22x^2 - 5x + 12}{4x + 3}$
24. $\dfrac{2x^3 - 9x^2 + 11x - 3}{2x - 3}$
25. $(x^6 - 13x^3 + 42) \div (x^3 - 7)$
26. $(x^6 + 5x^3 - 24) \div (x^3 - 3)$
27. $(x^4 - 16) \div (x - 2)$
28. $(x^4 - 81) \div (x - 3)$
29. $(t^3 - t^2 + t - 1) \div (t - 1)$
30. $(t^3 - t^2 + t - 1) \div (t + 1)$

Problem-solving practice

31. The perimeter of a rectangle is 640 ft. The length is 15 ft greater than the width. Find the area of the rectangle.

32. When the base and the height of a triangle are increased by 2 in., the area is increased from 70 in^2 to 96 in^2. Find the sum of the base and the height of the original triangle.

Divide.

33. $(x^4 + 9x^2 + 20) \div (x^2 + 4)$
34. $(y^4 + a^2) \div (y + a)$
35. $(5a^3 + 8a^2 - 23a - 1) \div (5a^2 - 7a - 2)$
36. $(15y^3 - 30y + 7 - 19y^2) \div (3y^2 - 2 - 5y)$
37. $(6x^5 - 13x^3 + 5x + 3 - 4x^2 + 3x^4) \div (3x^3 - 2x - 1)$
38. $(5x^7 - 3x^4 + 2x^2 - 10x + 2) \div (x^2 - x + 1)$
39. $(a^6 - b^6) \div (a - b)$
40. $(x^5 + y^5) \div (x + y)$
41. Divide $6a^{3h} + 13a^{2h} - 4a^h - 15$ by $2a^h + 3$.

If the remainder is 0 when one polynomial is divided by another, the divisor is a factor of the dividend. Find the value(s) of c for which $x - 1$ is a factor of each polynomial.

42. $x^2 + 4x + c$
43. $2x^2 + 3cx - 8$
44. $c^2x^2 - 2cx + 1$

9.9 SOLVING FRACTIONAL EQUATIONS

A *fractional equation* is an equation containing one or more fractional expressions. Here are some examples:

$$\frac{2}{3} + \frac{5}{6} = \frac{x}{9}, \qquad x + \frac{6}{x} = -5, \qquad \frac{x^2}{x-1} = \frac{1}{x-1}.$$

In this section we learn a method for solving fractional equations.

> To solve a fractional equation, multiply on both sides by the LCM of all the denominators. This is called *clearing of fractions*.

We have used clearing of fractions in Section 3.3 and Section 6.7.

EXAMPLE 1 Solve: $\frac{2}{3} + \frac{5}{6} = \frac{x}{9}$.

The LCM of all denominators is 18, or $2 \cdot 3 \cdot 3$. We multiply on both sides by $2 \cdot 3 \cdot 3$:

$$2 \cdot 3 \cdot 3 \left(\frac{2}{3} + \frac{5}{6} \right) = 2 \cdot 3 \cdot 3 \cdot \frac{x}{9} \qquad \text{Multiplying on both sides by the LCM}$$

$$2 \cdot 3 \cdot 3 \cdot \frac{2}{3} + 2 \cdot 3 \cdot 3 \cdot \frac{5}{6} = 2 \cdot 3 \cdot 3 \cdot \frac{x}{9} \qquad \text{Multiplying to remove parentheses}$$

$$2 \cdot 3 \cdot 2 + 3 \cdot 5 = 2 \cdot x \qquad \text{Simplifying}$$

$$12 + 15 = 2x$$

$$27 = 2x$$

$$\frac{27}{2} = x.$$

Note that we did not use the LCM to add or subtract fractional expressions. We used it in such a way that all the denominators disappeared and the resulting equation was easier to solve.

> When clearing an equation of fractions, be sure to multiply *all* terms in the equation by the LCM.

EXAMPLE 2 Solve: $\frac{1}{x} = \frac{1}{4-x}$.

The LCM is $x(4 - x)$. We multiply on both sides by $x(4 - x)$:

$$x(4 - x) \cdot \frac{1}{x} = x(4 - x) \cdot \frac{1}{4 - x} \qquad \text{Multiplying on both sides by the LCM}$$

$$4 - x = x \qquad \text{Simplifying}$$
$$4 = 2x$$
$$x = 2.$$

Check:
$$\frac{\dfrac{1}{x} = \dfrac{1}{4 - x}}{\begin{array}{c|c} \dfrac{1}{2} & \dfrac{1}{4 - 2} \\ & \dfrac{1}{2} \end{array}}$$

This checks, so 2 is the solution.

The next examples show how important it is to multiply *all* terms in an equation by the LCM.

EXAMPLE 3 Solve: $\dfrac{x}{6} - \dfrac{x}{8} = \dfrac{1}{12}.$

The LCM is 24. We multiply on both sides by 24:

$$24\left(\frac{x}{6} - \frac{x}{8}\right) = 24 \cdot \frac{1}{12} \qquad \text{Multiplying on both sides by the LCM}$$

$$24 \cdot \frac{x}{6} - 24 \cdot \frac{x}{8} = 24 \cdot \frac{1}{12} \qquad \text{Multiplying to remove parentheses}$$

⎯⎯⎯⎯⎯⎯ Be sure to multiply *every* term by the LCM.

$$4x - 3x = 2 \qquad \text{Simplifying}$$
$$x = 2.$$

Check:
$$\frac{\dfrac{x}{6} - \dfrac{x}{8} = \dfrac{1}{12}}{\begin{array}{c|c} \dfrac{2}{6} - \dfrac{2}{8} & \dfrac{1}{12} \\ \dfrac{1}{3} - \dfrac{1}{4} & \\ \dfrac{4}{12} - \dfrac{3}{12} & \\ \dfrac{1}{12} & \end{array}}$$

This checks, so the solution is 2.

EXAMPLE 4 Solve: $\dfrac{2}{3x} + \dfrac{1}{x} = 10.$

9.9 SOLVING FRACTIONAL EQUATIONS

The LCM is $3x$. We multiply on both sides by $3x$:

$$3x\left(\frac{2}{3x} + \frac{1}{x}\right) = 3x \cdot 10 \qquad \text{Multiplying on both sides by the LCM}$$

$$3x \cdot \frac{2}{3x} + 3x \cdot \frac{1}{x} = 3x \cdot 10 \qquad \text{Multiplying to remove parentheses}$$

$$2 + 3 = 30x \qquad \text{Simplifying}$$

$$5 = 30x$$

$$\frac{5}{30} = x$$

$$\frac{1}{6} = x.$$

Check:
$$\frac{2}{3x} + \frac{1}{x} = 10$$

$$\begin{array}{c|c} \frac{2}{3(\frac{1}{6})} + \frac{1}{\frac{1}{6}} & 10 \\ \frac{2}{\frac{1}{2}} + 6 & \\ 2 \cdot 2 + 6 & \\ 4 + 6 & \\ 10 & \end{array}$$

This checks, so the solution is $\frac{1}{6}$.

EXAMPLE 5 Solve: $x + \dfrac{6}{x} = -5$.

The LCM is x. We multiply by x:

$$x\left(x + \frac{6}{x}\right) = -5x \qquad \text{Multiplying on both sides by } x$$

$$x^2 + x \cdot \frac{6}{x} = -5x \qquad \text{Note that } \textit{each term} \text{ on the left is now multiplied by } x.$$

$$x^2 + 6 = -5x \qquad \text{Simplifying}$$

$$x^2 + 5x + 6 = 0 \qquad \text{Adding } 5x \text{ to get a 0 on one side}$$

$$(x + 3)(x + 2) = 0 \qquad \text{Factoring}$$

$$x + 3 = 0 \quad \text{or} \quad x + 2 = 0 \qquad \text{Principle of zero products}$$

$$x = -3 \quad \text{or} \quad x = -2.$$

FRACTIONAL EXPRESSIONS AND EQUATIONS

Check:

$$\begin{array}{c|c} x + \dfrac{6}{x} = -5 \\ \hline -3 + \dfrac{6}{-3} & -5 \\ -5 & \end{array} \qquad \begin{array}{c|c} x + \dfrac{6}{x} = -5 \\ \hline -2 + \dfrac{6}{-2} & -5 \\ -5 & \end{array}$$

Both of these check, so there are two solutions, -3 and -2.

> It is important *always* to check when solving fractional equations.

EXAMPLE 6 Solve: $\dfrac{x^2}{x-1} = \dfrac{1}{x-1}$.

The LCM is $x - 1$. We multiply by $x - 1$:

$$(x-1) \cdot \frac{x^2}{x-1} = (x-1) \cdot \frac{1}{x-1} \qquad \text{Multiplying on both sides by } x-1$$

$$x^2 = 1 \qquad \text{Simplifying}$$
$$x^2 - 1 = 0 \qquad \text{Adding } -1 \text{ to get a 0 on one side}$$
$$(x-1)(x+1) = 0 \qquad \text{Factoring}$$
$$x - 1 = 0 \quad \text{or} \quad x + 1 = 0 \qquad \text{Principle of zero products}$$
$$x = 1 \quad \text{or} \quad x = -1.$$

Possible solutions are 1 and -1.

Check:

$$\begin{array}{c|c} \dfrac{x^2}{x-1} = \dfrac{1}{x-1} \\ \hline \dfrac{1^2}{1-1} & \dfrac{1}{1-1} \\ \dfrac{1}{0} & \dfrac{1}{0} \end{array} \qquad \begin{array}{c|c} \dfrac{x^2}{x-1} = \dfrac{1}{x-1} \\ \hline \dfrac{(-1)^2}{-1-1} & \dfrac{1}{-1-1} \\ -\dfrac{1}{2} & -\dfrac{1}{2} \end{array}$$

This number -1 is a solution, but 1 is not because it makes a denominator zero.

EXAMPLE 7 Solve: $\dfrac{3}{x-5} + \dfrac{1}{x+5} = \dfrac{2}{x^2-25}$.

9.9 SOLVING FRACTIONAL EQUATIONS

The LCM is $(x-5)(x+5)$. We multiply by $(x-5)(x+5)$:

$$(x-5)(x+5)\left[\frac{3}{x-5}+\frac{1}{x+5}\right]=(x-5)(x+5)\left[\frac{2}{x^2-25}\right]$$ Multiplying on both sides by the LCM

$$(x-5)(x+5)\cdot\frac{3}{x-5}+(x-5)(x+5)\cdot\frac{1}{x+5}=(x-5)(x+5)\cdot\frac{2}{x^2-25}$$

$$3(x+5)+(x-5)=2 \quad \text{Simplifying}$$
$$3x+15+x-5=2 \quad \text{Removing parentheses}$$
$$4x+10=2$$
$$4x=-8$$
$$x=-2.$$

This checks, so the solution is -2.

EXERCISE SET 9.9

Solve.

1. $\dfrac{3}{8}+\dfrac{4}{5}=\dfrac{x}{20}$

2. $\dfrac{3}{5}+\dfrac{2}{3}=\dfrac{x}{9}$

3. $\dfrac{2}{3}-\dfrac{5}{6}=\dfrac{1}{x}$

4. $\dfrac{1}{8}-\dfrac{3}{5}=\dfrac{1}{x}$

5. $\dfrac{1}{6}+\dfrac{1}{8}=\dfrac{1}{t}$

6. $\dfrac{1}{8}+\dfrac{1}{10}=\dfrac{1}{t}$

7. $x+\dfrac{4}{x}=-5$

8. $x+\dfrac{3}{x}=-4$

9. $\dfrac{x}{4}-\dfrac{4}{x}=0$

10. $\dfrac{x}{5}-\dfrac{5}{x}=0$

11. $\dfrac{5}{x}=\dfrac{6}{x}-\dfrac{1}{3}$

12. $\dfrac{4}{x}=\dfrac{5}{x}-\dfrac{1}{2}$

13. $\dfrac{5}{3x}+\dfrac{3}{x}=1$

14. $\dfrac{3}{4x}+\dfrac{5}{x}=1$

15. $\dfrac{x-7}{x+2}=\dfrac{1}{4}$

16. $\dfrac{a-2}{a+3}=\dfrac{3}{8}$

17. $\dfrac{2}{x+1}=\dfrac{1}{x-2}$

18. $\dfrac{5}{x-1}=\dfrac{3}{x+2}$

19. $\dfrac{x}{6}-\dfrac{x}{10}=\dfrac{1}{6}$

20. $\dfrac{x}{8}-\dfrac{x}{12}=\dfrac{1}{8}$

21. $\dfrac{x+1}{3}-\dfrac{x-1}{2}=1$

22. $\dfrac{x+2}{5}-\dfrac{x-2}{4}=1$

23. $\dfrac{a-3}{3a+2}=\dfrac{1}{5}$

24. $\dfrac{x-1}{2x+5}=\dfrac{1}{4}$

25. $\dfrac{x-1}{x-5}=\dfrac{4}{x-5}$

26. $\dfrac{x-7}{x-9}=\dfrac{2}{x-9}$

27. $\dfrac{2}{x+3}=\dfrac{5}{x}$

28. $\dfrac{3}{x+4} = \dfrac{4}{x}$

29. $\dfrac{x-2}{x-3} = \dfrac{x-1}{x+1}$

30. $\dfrac{2b-3}{3b+2} = \dfrac{2b+1}{3b-2}$

31. $\dfrac{1}{x+3} + \dfrac{1}{x-3} = \dfrac{1}{x^2-9}$

32. $\dfrac{4}{x-3} + \dfrac{2x}{x^2-9} = \dfrac{1}{x+3}$

33. $\dfrac{x}{x+4} - \dfrac{4}{x-4} = \dfrac{x^2+16}{x^2-16}$

34. $\dfrac{5}{y-3} - \dfrac{30}{y^2-9} = 1$

Solve.

35. $\dfrac{4}{y-2} - \dfrac{2y-3}{y^2-4} = \dfrac{5}{y+2}$

36. $\dfrac{x}{x^2+3x-4} + \dfrac{x+1}{x^2+6x+8} = \dfrac{2x}{x^2+x-2}$

37. $\dfrac{y}{y+0.2} - 1.2 = \dfrac{y-0.2}{y+0.2}$

38. $\dfrac{x^2}{x^2-4} = \dfrac{x}{x+2} - \dfrac{2x}{2-x}$

39. $4a - 3 = \dfrac{a+13}{a+1}$

40. $\dfrac{3x-9}{x-3} = \dfrac{5x-4}{2}$

41. $\dfrac{y^2-4}{y+3} = 2 - \dfrac{y-2}{y+3}$

42. $\dfrac{3a-5}{a^2+4a+3} + \dfrac{2a+2}{a+3} = \dfrac{a-3}{a+1}$

43. Solve and check:

$$\dfrac{n}{n-\frac{4}{9}} - \dfrac{n}{n+\frac{4}{9}} = \dfrac{1}{n}.$$

44. Suppose that

$$x = \dfrac{ab}{a+b} \quad \text{and} \quad y = \dfrac{ab}{a-b}.$$

Show that

$$\dfrac{y^2-x^2}{y^2+x^2} = \dfrac{2ab}{a^2+b^2}.$$

9.10 PROBLEM SOLVING

We now use our five-step process for problem solving and the new skills we have learned for solving fractional equations. This expands our problem-solving abilities.

EXAMPLE 1 The reciprocal of 2 less than a certain number is twice the reciprocal of the number itself. What is the number?

1. **Familiarize** The problem is stated explicitly enough that we can go right to the translation.

2. **Translate** Let x represent the number. Then 2 less than the number is $x - 2$, and the reciprocal of the number is $1/x$.

9.10 PROBLEM SOLVING

$$\underbrace{\begin{pmatrix} \text{The reciprocal of 2} \\ \text{less than the number} \end{pmatrix}}_{\dfrac{1}{x-2}} \text{ is } \underbrace{\begin{pmatrix} \text{Twice the reciprocal} \\ \text{of the number} \end{pmatrix}}_{2 \cdot \dfrac{1}{x}} \quad \text{Translating}$$

3. **Carry out** We solve the equation:

$$\frac{1}{x-2} = \frac{2}{x} \qquad \text{The LCM is } x(x-2).$$

$$x(x-2) \cdot \frac{1}{x-2} = x(x-2) \cdot \frac{2}{x} \qquad \text{Multiplying by the LCM}$$

$$x = 2(x-2) \qquad \text{Simplifying}$$

$$x = 2x - 4$$

$$x = 4.$$

4. **Check** Go to the original problem. The number to be checked is 4. Two less than 4 is 2. The reciprocal of 2 is $\frac{1}{2}$. The reciprocal of the number itself is $\frac{1}{4}$. Now $\frac{1}{2}$ is twice $\frac{1}{4}$, so the conditions are satisfied.

5. **State** The number is 4.

EXAMPLE 2 One car travels 20 km/h faster than another. While one of them goes 240 km, the other goes 160 km. Find their speeds.

1. **Familiarize** First make a drawing. We really do not know the directions in which the cars are traveling, but it does not matter.

Slow car 160 km
          r km/h

Fast car 240 km
          r + 20 km/h

Let r represent the speed of the slow car. Then $r + 20$ is the speed of the fast car. The cars travel the same length of time, so we can just use t for time. We can organize the information in a chart.

	Distance	Speed	Time
Slow car	160	r	t
Fast car	240	$r + 20$	t

We have the notions of distance, speed, and time in this problem. Are they related? Recall that we may need to find a formula that relates the parts of a problem. Indeed, you may need to look up such a formula. Actually we have considered two such formulas in Chapter 6. They are $r = d/t$ and $d = rt$.

2. **Translate** If we solve either the formula $r = d/t$ or $d = rt$ for t, we get $t = d/r$. Then, from the rows of the table, we get two different expressions for t. They are:

$$t = \frac{160}{r}$$

and

$$t = \frac{240}{r + 20}.$$

Since these times are the same, we get the following equation:

$$\frac{160}{r} = \frac{240}{r + 20}.$$

3. **Carry out** We solve the equation:

$$\text{LCM} = r(r + 20)$$

$$\frac{160 \cdot r(r + 20)}{r} = \frac{240 \cdot r(r + 20)}{r + 20} \quad \text{Multiplying on both sides by the LCM, } r(r + 20)$$

$160(r + 20) = 240r$ Simplifying
$160r + 3200 = 240r$ Removing parentheses
$3200 = 80r$ Adding $-160r$

$$\frac{3200}{80} = r \quad \text{Multiplying by } \tfrac{1}{80}$$

$40 = r$
$60 = r + 20.$

4. **Check** We check the speed of 40 km/h for the slow car and 60 km/h for the fast car. The fast car does travel 20 km/h faster than the slow car. The fast car will travel farther than the slow car. If the fast car goes 240 km at 60 km/h, the time it has traveled is $\frac{240}{60}$, or 4 hr. If the slow car goes 160 km at 40 km/h, the time it travels is $\frac{160}{40}$, or 4 hr. Since the times are the same, the speeds check.

5. **State** The slow car has a speed of 40 km/h and the fast car has a speed of 60 km/h.

9.10 PROBLEM SOLVING

EXAMPLE 3 The head of a secretarial pool examines work records and finds that it takes Helen 4 hr to type a certain report. It takes Willie 6 hr to type the same report. How long would it take them, working together, to type the same report?

1. **Familiarize** We familiarize ourselves with the problem by considering two incorrect ways of translating the problem to mathematical language.

   a) A common incorrect way to translate the problem is just to add the two times:

   $$4 \text{ hr} + 6 \text{ hr} = 10 \text{ hr}.$$

   Now think about this. Helen can do the job alone in 4 hr. If Helen and Willie work together, whatever time it takes them should be *less* than 4 hr. Thus we reject 10 hr as a solution, but we do have a partial check on any answer we get. The answer should be less than 4 hr.

   b) Suppose the two people split up the typing job in such a way that Helen does half the typing and Willie does the other half. Then

   $$\text{Helen types } \tfrac{1}{2} \text{ the report in } \tfrac{1}{2}(4 \text{ hr}), \quad \text{or } 2 \text{ hr,}$$

   and

   $$\text{Willie types } \tfrac{1}{2} \text{ the report in } \tfrac{1}{2}(6 \text{ hr}), \quad \text{or } 3 \text{ hr.}$$

   But, time is wasted since Helen would get done 1 hr early. In effect, they have not worked together to get the job done as fast as possible. If Helen helps Willie after completing her half, the entire job could be done in a time somewhere between 2 hr and 3 hr.

We proceed to a translation by considering how much of the job is finished in 1 hr, 2 hr, 3 hr, and so on. It takes Helen 4 hr to do the typing job alone. Then, in 1 hr, she can do $\tfrac{1}{4}$ of the job. It takes Willie 6 hr to do the job alone.

Then, in 1 hr, he can do $\frac{1}{6}$ of the job. Working together, they can do

$$\tfrac{1}{4} + \tfrac{1}{6}, \quad \text{or } \tfrac{5}{12} \text{ of the job.}$$

In 2 hr, Helen can do $2(\frac{1}{4})$ of the job and Willie can do $2(\frac{1}{6})$ of the job. Working together they can do

$$2(\tfrac{1}{4}) + 2(\tfrac{1}{6}), \quad \text{or } \tfrac{5}{6} \text{ of the job.}$$

Continuing this reasoning we can form a table like the following one.

	Fraction of the job completed		
Time	Helen	Willie	Together
1 hr	$\frac{1}{4}$	$\frac{1}{6}$	$\frac{1}{4} + \frac{1}{6}$, or $\frac{5}{12}$
2 hr	$2(\frac{1}{4})$	$2(\frac{1}{6})$	$2(\frac{1}{4}) + 2(\frac{1}{6})$, or $\frac{5}{6}$
3 hr	$3(\frac{1}{4})$	$3(\frac{1}{6})$	$3(\frac{1}{4}) + 3(\frac{1}{6})$, or $1\frac{1}{4}$
t hr	$t(\frac{1}{4})$	$t(\frac{1}{6})$	$t(\frac{1}{4}) + t(\frac{1}{6})$, or 1

From the table we see that if they worked 3 hr, the fraction of the job that they get done is $1\frac{1}{4}$, which is more of the job than needs to be done. We also see that the answer is somewhere between 2 hr and 3 hr. What we want is a number t such that the fraction of the job that gets completed is 1; that is, the job is just completed—not more ($1\frac{1}{4}$) and not less ($\frac{5}{6}$).

2. **Translate** From the table we see that the time we want is some number t for which

$$t\left(\frac{1}{4}\right) + t\left(\frac{1}{6}\right) = 1.$$

3. **Carry out** We solve the equation:

$$t\left(\frac{1}{4}\right) + t\left(\frac{1}{6}\right) = 1$$

$$12\left(\frac{t}{4} + \frac{t}{6}\right) = 12 \cdot 1 \qquad \text{The LCM is } 2 \cdot 2 \cdot 3, \text{ or } 12.$$

$$12 \cdot \frac{t}{4} + 12 \cdot \frac{t}{6} = 12$$

$$3t + 2t = 12$$

$$5t = 12$$

$$t = \frac{12}{5}, \text{ or } 2\frac{2}{5} \text{ hr.}$$

9.10 PROBLEM SOLVING

4. **Check** The check can be done by repeating the computations:

$$\frac{12}{5}\left(\frac{1}{4}\right) + \frac{12}{5}\left(\frac{1}{6}\right) = \frac{3}{5} + \frac{2}{5} = \frac{5}{5} = 1.$$

We also have another check in what we learned from our familiarization. The answer, $2\frac{2}{5}$ hr, is between 2 hr and 3 hr (see the table), and it is less than 4 hr, the time it takes Helen working alone.

5. **State** It takes $2\frac{2}{5}$ hr for them to do the job working together.

We now consider problems involving proportions. A proportion involves ratios. A *ratio* of two quantities is their quotient. For example, in the rectangle, the ratio of width to length is

$$\frac{2 \text{ cm}}{3 \text{ cm}}, \text{ or } \frac{2}{3}.$$

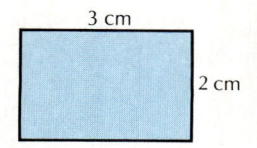

An older way to write this is 2:3, read "2 to 3." Percent can be considered a ratio. For example, 37% is the ratio of 37 to 100, 37/100. The ratio of two different kinds of measure is called a *rate*.

EXAMPLE 4 Betty Cuthbert of Australia set a world record in the 60-m dash of 7.2 sec. What was her rate, or *speed*, in m/sec?

$$\frac{60 \text{ m}}{7.2 \text{ sec}}, \text{ or } 8.3 \frac{\text{m}}{\text{sec}} \qquad \text{(Rounded to the nearest tenth)}$$

In applied problems a single ratio is often expressed in two ways. For example, it takes 9 gal of gas to drive 120 mi, and we wish to find how much will be required to go 550 mi. We can set up ratios:

$$\frac{9 \text{ gal}}{120 \text{ mi}} \qquad \frac{x \text{ gal}}{550 \text{ mi}}.$$

If we assume that the car uses gas at the same rate throughout the trip, the ratios are the same.

$$\begin{array}{c} \text{Gas} \longrightarrow \\ \text{Miles} \longrightarrow \end{array} \frac{9}{120} = \frac{x}{550} \begin{array}{c} \longleftarrow \text{Gas} \\ \longleftarrow \text{Miles} \end{array}$$

To solve, we multiply by 550 to get x alone on one side:

$$550 \cdot \frac{9}{120} = 550 \cdot \frac{x}{550}$$

$$\frac{550 \cdot 9}{120} = x$$

$$41.25 = x.$$

Thus 41.25 gal will be required. (Note that we could have multiplied by the LCM of 120 and 550, which is 6600, but in this case, that would have been more complicated.)

> An equality of ratios, $A/B = C/D$, is called a *proportion*. The numbers named in a true proportion are said to be *proportional*.

EXAMPLE 5 A pitcher gave up 71 earned runs in 285 innings in a recent year. At this rate, how many runs did the pitcher give up every 9 innings?

1. **Familiarize** The pitcher gives up 71 earned runs in 285 innings, and we wish to find how many runs the pitcher gives up every 9 innings. We can set up ratios:

$$\frac{A}{9} \qquad \frac{71}{285}$$

9.10 PROBLEM SOLVING

2. **Translate** If we assume that the pitcher gives up runs at the same rate every 9 innings, the ratios are the same, and we have an equation:

$$\text{Earned runs each 9 innings} \longrightarrow \frac{A}{9} = \frac{71}{285} \longleftarrow \text{Earned runs} \atop \longleftarrow \text{Innings pitched}$$

3. **Carry out** We solve the equation:

$$9 \cdot \frac{A}{9} = 9 \cdot \frac{71}{285} \quad \text{Multiplying by 9 to get } A \text{ alone}$$

$$A = \frac{9 \cdot 71}{285}$$

$$A = 2.24 \quad \text{(Rounded to the nearest hundredth)}$$

4. **Check** We leave the check to the student.

5. **State** The pitcher gives up 2.24 earned runs every 9 innings pitched. The variable A in the proportion stands for *earned run average*.

EXAMPLE 6 (*Estimating wildlife populations*). To determine the number of fish in a lake, a conservationist catches 225 fish, tags them, and throws them back into the lake. Later, 108 fish are caught. Fifteen of them are found to be tagged. Estimate how many fish are in the lake.

1. **Familiarize** The ratio of fish tagged to the total number of fish in the lake F is $225/F$. Of the 108 fish caught later, 15 fish were tagged. The ratio of fish tagged to fish caught is $15/108$.

2. **Translate** Assuming the two ratios are the same, we can translate to a proportion.

$$\text{Fish tagged originally} \longrightarrow \frac{225}{F} = \frac{15}{108} \longleftarrow \text{Tagged fish caught later} \atop \longleftarrow \text{Fish caught later}$$

3. **Carry out** We solve the proportion. This time we multiply by the LCM, which is $108F$:

$$108F \cdot \frac{225}{F} = 108F \cdot \frac{15}{108} \quad \text{Multiplying by } 108F$$

$$108 \cdot 225 = F \cdot 15$$

$$\frac{108 \cdot 225}{15} = F \quad \text{Multiplying by } \tfrac{1}{15}$$

$$1620 = F.$$

4. **Check** We leave the check to the student.

5. **State** We estimate that there are about 1620 fish in the lake.

EXERCISE SET 9.10

Problem solving

1. The reciprocal of 4 plus the reciprocal of 5 is the reciprocal of what number?

2. The reciprocal of 3 plus the reciprocal of 8 is the reciprocal of what number?

3. One number is 5 more than another. The quotient of the larger divided by the smaller is $\frac{4}{3}$. Find the numbers.

4. One number is 4 more than another. The quotient of the larger divided by the smaller is $\frac{5}{2}$. Find the numbers.

5. One car travels 40 km/h faster than another. While one travels 150 km, the other goes 350 km. Find their speeds.

6. One car travels 30 km/h faster than another. While one goes 250 km, the other goes 400 km. Find their speeds.

7. A person traveled 120 mi in one direction. The return trip was accomplished at double the speed, and took 3 hr less time. Find the speed going.

8. After making a trip of 126 mi, a person found that the trip would have taken 1 hr less time by increasing the speed by 8 mph. What was the actual speed?

9. The speed of a freight train is 14 km/h slower than the speed of a passenger train. The freight train travels 330 km in the same time that it takes the passenger train to travel 400 km. Find the speed of each train.

10. The speed of a freight train is 15 km/h slower than the speed of a passenger train. The freight train travels 390 km in the same time that it takes the passenger train to travel 480 km. Find the speed of each train.

11. It takes painter A 4 hr to paint a certain area of a house. It takes painter B 5 hr to do the same job. How long would it take them, working together, to do the painting job?

12. By checking work records, a carpenter finds that worker A can build a certain type of garage in 12 hr. Worker B can do the same job in 16 hr. How long would it take if they worked together?

13. By checking work records, a plumber finds that worker A can do a certain job in 12 hr. Worker B can do the same job in 9 hr. How long would it take if they worked together?

14. A tank can be filled in 18 hr by pipe A alone and in 24 hr by pipe B alone. How long would it take to fill the tank if both pipes were working?

Find the ratio of the following.

15. 54 days, 6 days

16. 800 mi, 50 gal

17. A black racer snake travels 4.6 km in 2 hr. What is the speed in km/h?

18. Light travels 558,000 mi in 3 sec. What is the speed in mi/sec?

19. The coffee beans from 14 trees are required to produce 7.7 kg of coffee (this is the average that each person in the United States drinks each year). How many trees are required to produce 320 kg of coffee?

20. Last season a minor league baseball player got 240 hits in 600 times at bat. This season his ratio of hits to number of times at bat is the same. He batted 500 times. How many hits has he had?

21. A student traveled 234 km in 14 days. At this same ratio, how far would the student travel in 42 days?

22. In a bread recipe, the ratio of milk to flour is $\frac{4}{3}$. If 5 cups of milk are used, how many cups of flour are used?

23. 10 cm³ of a normal specimen of human blood contains 1.2 g of hemoglobin. How many grams would 16 cm³ of the same blood contain?

24. The winner of an election for class president won by a vote of 3 to 2, with 324 votes. How many votes did the loser get?

25. To determine the number of trout in a lake, a conservationist catches 112 trout, tags them, and throws them back into the lake. Later, 82 trout are caught; 32 of them are tagged. How many trout are in the lake?

26. To determine the number of deer in a game preserve, a conservationist catches 318 deer, tags them, and lets them loose. Later, 168 deer are caught; 56 of them are tagged. How many deer are in the preserve?

27. The ratio of the weight of an object on the moon to the weight of an object on earth is 0.16 to 1.

a) How much would a 12-ton rocket weigh on the moon?
b) How much would a 90-kilogram astronaut weigh on the moon?

28. The ratio of the weight of an object on Mars to the weight of an object on earth is 0.4 to 1.

a) How much would a 12-ton rocket weigh on Mars?
b) How much would a 90-kilogram astronaut weigh on Mars?

29. Simplest fractional notation for a rational number is $\frac{9}{17}$. Find an equal ratio where the sum of the numerator and denominator is 104.

30. A baseball team has 12 more games to play. They have won 25 out of the 36 games they have played. How many more games must they win in order to finish with a 0.750 record?

○ ─────────────

31. The denominator of a fraction is 1 more than the numerator. If 2 is subtracted from both the numerator and denominator, the resulting fraction is $\frac{1}{2}$. Find the original fraction.

32. Ann and Betty work together and complete a job in 4 hr. It would take Betty 6 hr longer, working alone, to do the job than it would Ann. How long would it take each of them to do the job working alone?

33. The speed of a boat in still water is 10 mph. It travels 24 mi upstream and 24 mi downstream in a total time of 5 hr. What is the speed of the current?

34. Express 100 as the sum of two numbers for which the ratio of one number, increased by 5, to the other number, decreased by 5, is 4.

35. In a proportion

$$\frac{A}{B} = \frac{C}{D},$$

the numbers A and D are often called extremes, whereas the numbers B and C are called the means. Write four true proportions. Compare the product of the means with the product of the extremes.

36. Compare

$$\frac{A + B}{B} = \frac{C + D}{D}$$

with the proportion

$$\frac{A}{B} = \frac{C}{D}.$$

37. Rosina, Ng, and Oscar can complete a certain job in 3 days. Rosina can do the job in 8 days and Ng can do it in 10 days. How many days will it take Oscar to complete the job?

38. How soon after 5 o'clock will the hands on a clock first be together?

39. To reach an appointment 50 mi away Dr. Wright allowed 1 hr. After driving 30 mi she realized that her speed would have to be increased 15 mph for the remainder of the trip. What was her speed for the first 30 mi?

40. Together, Michelle, Sal, and Kristen can do a job in 1 hr and 20 min. To do the job alone, Michelle needs twice the time that Sal needs and 2 hr more than Kristen. How long would it take each to complete the job working alone?

9.11 FORMULAS

We have seen how the use of formulas is important in many applications of mathematics. We use our expanded equation-solving skills to solve formulas.

EXAMPLE 1 (*Gravitational force*). The gravitational force f between planets of mass M and m, at a distance d from each other, is given by

$$f = \frac{kMm}{d^2},$$

where k is a constant. Solve for m.

$$fd^2 = kMm \qquad \text{Multiplying by } d^2$$

$$\frac{fd^2}{kM} = m \qquad \text{Multiplying by } \frac{1}{kM}$$

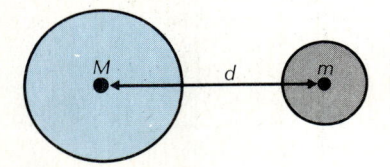

EXAMPLE 2 (*The area of a trapezoid*). The area A of a trapezoid is half the product of the height h and the sum of the lengths b_1 and b_2 of the parallel sides:

$$A = \tfrac{1}{2}(b_1 + b_2)h.$$

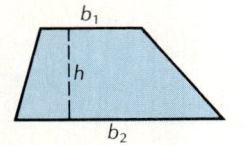

Solve for b_2.

We consider b_1 and b_2 to be different variables (or constants). The numbers 1 and 2 are called *subscripts*.

$$2A = (b_1 + b_2)h \qquad \text{Multiplying by 2}$$

$$2A = b_1 h + b_2 h$$

$$2A - b_1 h = b_2 h \qquad \text{Adding } -b_1 h$$

$$\frac{2A - b_1 h}{h} = b_2 \qquad \text{Multiplying by } \frac{1}{h}$$

Each of the following is also a correct answer:

$$\frac{2A}{h} - b_1 = b_2 \quad \text{and} \quad \frac{1}{h}(2A - b_1 h) = b_2.$$

In both of the examples, the letter for which we solved was on the right side of the equation. Ordinarily we put the letter for which we solve on the left. This is a matter of choice, since all equations are reversible.

EXAMPLE 3 Solve for P:

$$\frac{A}{P} = 1 + r.$$

This is an interest formula.

The LCM is P. We multiply by this.

$$P \cdot \frac{A}{P} = P(1 + r) \qquad \text{Multiplying by } P$$

$$A = P(1 + r) \qquad \text{Simplifying the left side}$$

$$\frac{A}{1 + r} = \frac{P(1 + r)}{1 + r} \qquad \text{Multiplying by } \frac{1}{1 + r}$$

$$\frac{A}{1 + r} = P \qquad \text{Simplifying}$$

EXAMPLE 4 If one person can do a job in a hr and another can do the same job in b hr, then working together they can do the same job in t hr, where a, b, and t are related by

$$\frac{t}{a} + \frac{t}{b} = 1.$$

Solve for t.

The LCM is ab. We multiply by this.

$$ab \cdot \left(\frac{t}{a} + \frac{t}{b}\right) = ab \cdot 1 \qquad \text{Multiplying by } ab$$

$$ab \cdot \frac{t}{a} + ab \cdot \frac{t}{b} = ab$$

$$bt + at = ab \qquad \text{Simplifying}$$

$$(b + a)t = ab \qquad \text{Factoring out } t$$

$$t = \frac{ab}{b + a} \qquad \text{Multiplying by } \frac{1}{b + a}$$

The answer can be used to solve problems such as Example 3 in Section 9.10:

$$t = \frac{4 \cdot 6}{6 + 4} = \frac{24}{10} = 2\frac{2}{5}.$$

EXERCISE SET 9.11

Solve for the letter indicated.

1. $S = 2\pi rh$; r
2. (*An interest formula*). $A = P(1 + rt)$; t
3. (*The area of a triangle*). $A = \frac{1}{2}bh$; b
4. $s = \frac{1}{2}gt^2$; g
5. $S = 180(n - 2)$; n
6. $S = \frac{n}{2}(a + l)$; a
7. $V = \frac{1}{3}k(B + b + 4M)$; b
8. $A = P + Prt$; P (*Hint:* Factor the right-hand side.)
9. $S(r - 1) = rl - a$; r
10. $T = mg - mf$; m (*Hint:* Factor the right-hand side.)
11. $A = \frac{1}{2}h(b_1 + b_2)$; h
12. (*The area of a right circular cylinder*). $S = 2\pi r(r + h)$; h
13. $r = \frac{v^2 pL}{a}$; a
14. $L = \frac{Mt - g}{t}$; M
15. $A = \frac{1}{2}h(b_1 + b_2)$; b_1
16. $l = a + (n - 1)d$; n
17. $A = \frac{\pi r^2 E}{180}$; E
18. $R = \frac{WL - x}{L}$; W
19. $V = -h(B + c + 4M)$; M
20. $W = I^2 R$; R
21. $y = \frac{v^2 pL}{a}$; L
22. $V = \frac{1}{3}bh$; b
23. $r = \frac{v^2 pL}{a}$; p
24. $P = 2(l + w)$; l
25. $\frac{a}{c} = n + bn$; n (*Hint:* Factor the right-hand side.)
26. $C = \frac{Ka - b}{a}$; K
27. (*A temperature conversion formula*). $C = \frac{5}{9}(F - 32)$; F
28. (*The volume of a sphere*). $V = \frac{4}{3}\pi r^3$; π
29. $f = \frac{gm - t}{m}$; g
30. $S = \frac{rl - a}{r - l}$; a
31. $\frac{1}{p} + \frac{1}{q} = \frac{1}{f}$; p
32. $\frac{1}{a} + \frac{1}{b} = \frac{1}{t}$; b
33. $\frac{A}{P} = 1 + r$; A
34. $\frac{2A}{h} = a + b$; h

35. (*An electricity formula*).
$\dfrac{1}{R} = \dfrac{1}{r_1} + \dfrac{1}{r_2}$; R

36. $\dfrac{1}{R} = \dfrac{1}{r_1} + \dfrac{1}{r_2}$; r_1

37. $\dfrac{A}{B} = \dfrac{C}{D}$; D

38. $\dfrac{A}{B} = \dfrac{C}{D}$; C

39. $h_1 = q\left(1 + \dfrac{h_2}{p}\right)$; h_2

40. $S = \dfrac{a - ar^n}{1 - r}$; a

41. In $V = \tfrac{4}{3}\pi r^3$, what is the effect on V when r is doubled?

42. The formula $C = \tfrac{5}{9}(F - 32)$ is used to convert Fahrenheit temperatures to Celsius temperatures. At what temperature are the Fahrenheit and Celsius readings the same?

43. For what value(s) of A will
$$\dfrac{3B}{A} = \dfrac{4C - 1}{A + 3}$$
be undefined?

44. In $N = a/c$, what is the effect on N when c increases? Assume that a, c, and N are positive.

Solve for the letter indicated.

45. $u = -F\left(E - \dfrac{P}{T}\right)$; T

46. $1 = a + (n - 1)d$; n

9.12 NEGATIVE EXPONENTS AND SCIENTIFIC NOTATION

When we considered properties of exponents in Section 4.1, we often got an answer like

$$\dfrac{1}{x^3}.$$

We write an expression, x^{-3}, equivalent to this fractional expression. It uses a negative exponent. We make the following agreement, which is a definition.

> If n is any positive integer,
>
> $$b^{-n} \text{ is given the meaning } \dfrac{1}{b^n}.$$
>
> In other words, b^n and b^{-n} are reciprocals.

EXAMPLE 1 Explain the meaning of 3^{-4} without using negative exponents.

$$3^{-4} \text{ means } \frac{1}{3^4}, \text{ or } \frac{1}{3 \cdot 3 \cdot 3 \cdot 3}, \text{ or } \frac{1}{81}.$$

Note that 3^{-4} is *not* a negative number.

EXAMPLE 2 Rename $\frac{1}{5^2}$ using a negative exponent.

$$\frac{1}{5^2} = 5^{-2}$$

EXAMPLE 3 Rename 4^{-3} using a positive exponent.

$$4^{-3} = \frac{1}{4^3}$$

Multiplying Using Exponents

We know that the property

$$a^m \cdot a^n = a^{m+n}$$

holds for any whole numbers m and n. Does it hold for any integers? Let us consider an example where both exponents are negative:

$$a^{-3} \cdot a^{-2} \text{ means } \frac{1}{a \cdot a \cdot a} \cdot \frac{1}{a \cdot a}.$$

This is equal to

$$\frac{1}{a \cdot a \cdot a \cdot a \cdot a}, \text{ or } \frac{1}{a^5}, \text{ or } a^{-5}.$$

Adding the exponents gives the correct result. This is true for all integers m and n.

For any integers m and n, and any nonzero number a,

$$a^m \cdot a^n = a^{m+n}.$$

In multiplication with exponential notation, we can add exponents if the bases are the same.

EXAMPLES Multiply and simplify.

4. $8^4 \cdot 8^3 = 8^{4+3}$ Adding exponents
   $= 8^7$
5. $7^{-3} \cdot 7^6 = 7^{-3+6}$
   $= 7^3$
6. $x \cdot x^8 = x^{1+8}$
   $= x^9$
7. $x^4 \cdot x^{-3} = x^{4+(-3)}$
   $= x^1$
   $= x$

Dividing Using Exponents

The other properties of exponents that we considered in Section 4.1 also hold for all integers, including negatives. We first consider division.

> For any integers m and n, and any nonzero number a,
> $$\frac{a^m}{a^n} = a^{m-n}.$$
> In division with exponential notation, we can subtract exponents if the bases are the same.

EXAMPLES Divide and simplify.

8. $\dfrac{5^4}{5^{-2}} = 5^{4-(-2)}$ Subtracting exponents
   $= 5^6$
9. $\dfrac{x}{x^7} = x^{1-7}$
   $= x^{-6}$
10. $\dfrac{b^{-4}}{b^{-5}} = b^{-4-(-5)}$
    $= b^1$
    $= b$

In Examples 8–10, it may help to think as follows: After writing the base, write the top exponent. Then write a subtraction sign. Then write the bottom exponent.

Then do the subtraction. For example,

$$\frac{x^{-3}}{x^{-5}} = x^{-3-(-5)}$$

Writing the base and the top exponent — Writing a subtraction sign — Writing the bottom exponent

Raising Powers to Powers

We now consider raising powers to powers.

For any integers m and n, and any nonzero number a,

$$(a^m)^n = a^{mn}.$$

To raise a power to a power, we can multiply the exponents.

EXAMPLES Simplify.

11. $(3^5)^4 = 3^{5 \cdot 4}$ Multiply exponents
    $= 3^{20}$

12. $(y^{-5})^7 = y^{-5 \cdot 7}$
    $= y^{-35}$

13. $(x^4)^{-2} = x^{4(-2)}$
    $= x^{-8}$

14. $(a^{-4})^{-6} = a^{(-4)(-6)}$
    $= a^{24}$

When several factors are in parentheses, raise each to the given power:

$$(a^m b^n)^t = (a^m)^t (b^n)^t = a^{mt} b^{nt}.$$

EXAMPLES Simplify.

15. $(5x^2 y^{-2})^3 = 5^3 (x^2)^3 (y^{-2})^3 = 125 x^6 y^{-6}$ Be careful to raise *every* factor in parentheses to the power.

16. $(3x^3 y^{-5} z^2)^4 = 3^4 (x^3)^4 (y^{-5})^4 (z^2)^4 = 81 x^{12} y^{-20} z^8$

Scientific Notation

The following are examples of scientific notation, which is useful when calculations involve very large or very small numbers:

$$6.4 \times 10^{23}, \quad 4.6 \times 10^{-4}, \quad 10^{15}.$$

9.12 NEGATIVE EXPONENTS AND SCIENTIFIC NOTATION

> *Scientific notation* for a number consists of exponential notation for a power of 10 and, if needed, decimal notation for a number between 1 and 10, and a multiplication sign:
>
> $N \times 10^n$ or 10^n.

We can convert to scientific notation by multiplying by 1, choosing a name like $10^b \cdot 10^{-b}$ for the number 1.

EXAMPLE 17 Light travels about 9,460,000,000,000 kilometers in one year. Write scientific notation for the number.

We want to move the decimal point 12 places, between the 9 and the 4, so we choose $10^{-12} \times 10^{12}$ as a name for 1. Then we multiply.

$9{,}460{,}000{,}000{,}000 \times 10^{-12} \times 10^{12}$ Multiplying by 1

$= 9.46 \times 10^{12}$ The 10^{-12} moved the decimal point 12 places to the left and we have scientific notation.

EXAMPLE 18 Write scientific notation for 0.0000000000156.

We want to move the decimal point 11 places. We choose $10^{11} \times 10^{-11}$ as a name for 1, and then multiply.

$0.0000000000156 \times 10^{11} \times 10^{-11}$ Multiplying by 1

$= 1.56 \times 10^{-11}$ The 10^{11} moved the decimal point 11 places to the right and we have scientific notation.

You should try to make conversions to scientific notation mentally as much as possible.

EXAMPLES Convert mentally to scientific notation.

19. $78{,}000 = 7.8 \times 10^4$ 7.8,000.

                                                            4 places

Large number, so the exponent is positive.

20. $0.0000057 = 5.7 \times 10^{-6}$ 0.000005.7

                                                                6 places

Small number, so the exponent is negative.

Multiplying and Dividing Using Scientific Notation

Multiplying. Consider the product

$$400 \cdot 2000 = 800{,}000.$$

In scientific notation this would be
$$(4 \times 10^2) \cdot (2 \times 10^3) = 8 \times 10^5.$$

Note that we could find this product by multiplying $4 \cdot 2$, to get 8, and $10^2 \cdot 10^3$ to get 10^5 (we do this by adding the exponents).

EXAMPLE 21 Multiply: $(1.8 \times 10^6) \cdot (2.3 \times 10^{-4})$.

a) Multiply 1.8 and 2.3:
$$1.8 \times 2.3 = 4.14.$$

b) Multiply 10^6 and 10^{-4}:
$$10^6 \cdot 10^{-4} = 10^{6+(-4)} \quad \text{Adding exponents}$$
$$= 10^2.$$

c) The answer is
$$4.14 \times 10^2.$$

EXAMPLE 22 Multiply: $(3.1 \times 10^5) \cdot (4.5 \times 10^{-3})$.

a) Multiply 3.1 and 4.5:
$$3.1 \times 4.5 = 13.95.$$

b) Multiply 10^5 and 10^{-3}:
$$10^5 \cdot 10^{-3} = 10^{5+(-3)} = 10^2.$$

c) The answer at this stage is
$$13.95 \times 10^2,$$

but this is *not* scientific notation, since 13.95 is not a number between 1 and 10. To find scientific notation we convert 13.95 to scientific notation and simplify:
$$13.95 \times 10^2 = (1.395 \times 10^1) \times 10^2$$
$$= 1.395 \times 10^3. \quad \text{Adding exponents}$$

Division. Consider the quotient
$$800{,}000 \div 400 = 2000.$$

In scientific notation this is
$$(8 \times 10^5) \div (4 \times 10^2) = 2 \times 10^3.$$

Note that we could find this product by dividing $8 \div 4$, to get 2, and $10^5 \div 10^2$, to

9.12 NEGATIVE EXPONENTS AND SCIENTIFIC NOTATION

get 10^3 (we do this by subtracting the exponents).

EXAMPLE 23 Divide: $(3.41 \times 10^5) \div (1.1 \times 10^{-3})$.

a) Divide 3.41 by 1.1:

$$3.41 \div 1.1 = 3.1.$$

b) Divide 10^5 by 10^{-3}:

$$10^5 \div 10^{-3} = 10^{5-(-3)} \quad \text{Subtracting exponents}$$
$$= 10^8.$$

c) The answer is

$$3.1 \times 10^8.$$

EXAMPLE 24 Divide: $(6.4 \times 10^{-7}) \div (8.0 \times 10^6)$.

a) Divide 6.4 by 8.0:

$$6.4 \div 8.0 = 0.8.$$

b) Divide 10^{-7} by 10^6:

$$10^{-7} \div 10^6 = 10^{-7-6} \quad \text{Subtracting exponents}$$
$$= 10^{-13}.$$

c) The answer at this stage is

$$0.8 \times 10^{-13},$$

but this is *not* scientific notation, since 0.8 is not a number between 1 and 10. To find scientific notation we convert 0.8 to scientific notation and simplify:

$$0.8 \times 10^{-13} = (8.0 \times 10^{-1}) \times 10^{-13}$$
$$= 8.0 \times 10^{-14}. \quad \text{Adding exponents}$$

EXERCISE SET 9.12

Explain the meaning of each without using negative exponents.

1. 3^{-2}
2. 2^{-3}
3. 10^{-4}
4. 5^{-6}

Rename, using negative exponents.

5. $\dfrac{1}{4^3}$
6. $\dfrac{1}{5^2}$
7. $\dfrac{1}{x^3}$
8. $\dfrac{1}{y^2}$

9. $\dfrac{1}{a^4}$ 10. $\dfrac{1}{t^5}$ 11. $\dfrac{1}{p^n}$ 12. $\dfrac{1}{m^n}$

Rename, using positive exponents.

13. 7^{-3} 14. 5^{-2} 15. a^{-3} 16. x^{-2}
17. y^{-4} 18. t^{-7} 19. z^{-n} 20. h^{-m}

Multiply and simplify.

21. $2^4 \cdot 2^3$ 22. $3^5 \cdot 3^2$ 23. $3^{-5} \cdot 3^8$ 24. $5^{-8} \cdot 5^9$
25. $x^{-2} \cdot x$ 26. $x \cdot x$ 27. $x^4 \cdot x^3$ 28. $x^9 \cdot x^4$
29. $x^{-7} \cdot x^{-6}$ 30. $y^{-5} \cdot y^{-8}$ 31. $t^8 \cdot t^{-8}$ 32. $m^{10} \cdot m^{-10}$

Divide and simplify.

33. $\dfrac{7^5}{7^2}$ 34. $\dfrac{4^7}{4^3}$ 35. $\dfrac{x}{x^{-1}}$ 36. $\dfrac{x^6}{x}$

37. $\dfrac{x^7}{x^{-2}}$ 38. $\dfrac{t^8}{t^{-3}}$ 39. $\dfrac{z^{-6}}{z^{-2}}$ 40. $\dfrac{y^{-7}}{y^{-3}}$

41. $\dfrac{x^{-5}}{x^{-8}}$ 42. $\dfrac{y^{-4}}{y^{-9}}$ 43. $\dfrac{m^{-9}}{m^{-9}}$ 44. $\dfrac{x^{-8}}{x^{-8}}$

Simplify.

45. $(2^3)^2$ 46. $(3^4)^3$ 47. $(5^2)^{-3}$ 48. $(9^3)^{-4}$
49. $(x^{-3})^{-4}$ 50. $(a^{-5})^{-6}$ 51. $(x^4 y^5)^{-3}$ 52. $(t^5 x^3)^{-4}$
53. $(x^{-6} y^{-2})^{-4}$ 54. $(x^{-2} y^{-7})^{-5}$ 55. $(3 x^3 y^{-8} z^{-3})^2$ 56. $(2 a^2 y^{-4} z^{-5})^3$

Convert to scientific notation.

57. 78,000,000,000
58. 3,700,000,000,000
59. 907,000,000,000,000,000
60. 168,000,000,000,000
61. 0.00000374
62. 0.000000000275
63. 0.000000018
64. 0.00000000002
65. 10,000,000
66. 100,000,000,000
67. 0.000000001
68. 0.0000001

Convert to decimal notation.

69. 7.84×10^8 70. 1.35×10^7 71. 8.764×10^{-10} 72. 9.043×10^{-3}
73. 10^8 74. 10^4 75. 10^{-4} 76. 10^{-7}

Multiply or divide and write scientific notation for the result.

77. $(3 \times 10^4)(2 \times 10^5)$
78. $(1.9 \times 10^8)(3.4 \times 10^{-3})$
79. $(5.2 \times 10^5)(6.5 \times 10^{-2})$
80. $(7.1 \times 10^{-7})(8.6 \times 10^{-5})$
81. $(9.9 \times 10^{-6})(8.23 \times 10^{-8})$
82. $(1.123 \times 10^4) \times 10^{-9}$
83. $\dfrac{8.5 \times 10^8}{3.4 \times 10^{-5}}$
84. $\dfrac{5.6 \times 10^{-2}}{2.5 \times 10^5}$
85. $(3.0 \times 10^6) \div (6.0 \times 10^9)$
86. $(1.5 \times 10^{-3}) \div (1.6 \times 10^{-6})$
87. $\dfrac{7.5 \times 10^{-9}}{2.5 \times 10^{12}}$
88. $\dfrac{4.0 \times 10^{-3}}{8.0 \times 10^{20}}$

Carry out the indicated operations. Write scientific notation for the result.

89. $\dfrac{(5.2 \times 10^6)(6.1 \times 10^{-11})}{1.28 \times 10^{-3}}$

90. $\dfrac{3.9 \times 10^{15}}{(8.0 \times 10^{-12})(3.2 \times 10^{19})}$

91. Perform the operations indicated. Express the result in scientific notation.

$$\{2.1 \times 10^6[(2.5 \times 10^{-3}) \div (5.0 \times 10^{-5})]\} \div (3.0 \times 10^{17})$$

92. Find the reciprocal and express in scientific notation.

a) 6.25×10^{-3} b) 4.0×10^{10}

SUMMARY AND REVIEW: CHAPTER 9

The following contains a summary of what you should be able to do after completing this chapter. The review exercises are for practice. Answers are at the back of the book. If you miss an exercise, restudy the section indicated alongside the answer.

You should be able to:

Simplify a fractional expression.

Simplify.

1. $\dfrac{4x^2 - 8x}{4x^2 + 4x}$

2. $\dfrac{14x^2 - x - 3}{2x^2 - 7x + 3}$

3. $\dfrac{(y-5)^2}{y^2 - 25}$

Multiply and divide fractional expressions and simplify.

Multiply or divide and simplify.

4. $\dfrac{a^2 - 36}{10a} \cdot \dfrac{2a}{a+6}$

5. $\dfrac{6t - 6}{2t^2 + t - 1} \cdot \dfrac{t^2 - 1}{t^2 - 2t + 1}$

6. $\dfrac{10 - 5t}{3} \div \dfrac{t - 2}{12t}$

7. $\dfrac{4x^4}{x^2 - 1} \div \dfrac{2x^3}{x^2 - 2x + 1}$

Find the least common multiple (LCM) of an algebraic expression by factoring.

Find the LCM.

8. $3x^2, 10xy, 15y^2$

9. $(a - 2), 4(2 - a)$

10. $y^2 - y - 2, y^2 - 4$

Add and subtract fractional expressions and simplify.

Add or subtract and simplify.

11. $\dfrac{x+8}{x+7} + \dfrac{10 - 4x}{x+7}$

12. $\dfrac{3}{3x - 9} + \dfrac{x - 2}{3 - x}$

13. $\dfrac{6x - 3}{x^2 - x - 12} - \dfrac{2x - 15}{x^2 - x - 12}$

14. $\dfrac{3x - 1}{2x} - \dfrac{x - 3}{x}$

15. $\dfrac{x + 3}{x - 2} - \dfrac{x}{2 - x}$

16. $\dfrac{2a}{a + 1} + \dfrac{4a}{a^2 - 1}$

17. $\dfrac{d^2}{d-c} + \dfrac{c^2}{c-d}$

18. $\dfrac{1}{x^2 - 25} - \dfrac{x-5}{x^2 - 4x - 5}$

19. $\dfrac{3x}{x+2} - \dfrac{x}{x-2} + \dfrac{8}{x^2 - 4}$

Simplify a complex fractional expression.

Simplify.

20. $\dfrac{\dfrac{1}{z} + 1}{\dfrac{1}{z^2} - 1}$

21. $\dfrac{\dfrac{c}{d} - \dfrac{d}{c}}{\dfrac{1}{c} + \dfrac{1}{d}}$

Divide a polynomial by a monomial or by a divisor that is not a monomial and check the result.

Divide.

22. $(10x^3 - x^2 + 6x) \div 2x$

23. $(6x^3 - 5x^2 - 13x + 13) \div (2x + 3)$

Solve a fractional equation.

Solve.

24. $\dfrac{3}{y} - \dfrac{1}{4} = \dfrac{1}{y}$

25. $\dfrac{15}{x} - \dfrac{15}{x+2} = 2$

Solve problems involving fractional equations using the five-step problem-solving process.

26. In checking records, a contractor finds that crew A can pave a certain length of highway in 9 hr. Crew B can do the same job in 12 hr. How long would it take if they worked together?

27. A lab is testing two high-speed trains. One train travels 40 km/h faster than the other. While one train travels 70 km, the other travels 60 km. Find their speeds.

28. The reciprocal of one more than a number is twice the reciprocal of the number itself. What is the number?

29. A sample of 250 batteries contained 8 defective batteries. How many defective batteries would you expect among 5000 batteries?

Solve a formula for a given letter when fractional expressions occur in the formula.

Solve for the letter indicated.

30. $\dfrac{1}{r} + \dfrac{1}{s} = \dfrac{1}{t}$; for s

31. $F = \dfrac{9C + 160}{5}$; for C

Rename a number with or without negative exponents.

32. Rename $\dfrac{1}{y^4}$ using negative exponents.

33. Rename 5^{-3} using positive exponents.

Use exponents in multiplying, dividing, and raising a power to a power.

Simplify.

34. $x^{-6} \cdot x^4$

35. $\dfrac{t^{-2}}{t^{-11}}$

36. $7^{-5} \cdot 7^{-5}$

37. $\dfrac{4^{-7}}{4^8}$

38. $(8^3)^3$

39. $(3a^{-6})^4$

40. $(x^{-2}yz^7)^{-5}$

Convert between scientific and ordinary decimal notation and multiply and divide using scientific notation.

Convert to scientific notation.

41. 0.0000278

42. $3{,}900{,}000{,}000$

Convert to decimal notation.

43. 5×10^{-8}

44. 1.28×10^4

Multiply or divide and write scientific notation for the result.

45. $(3.8 \times 10^4)(5.5 \times 10^{-1})$

46. $\dfrac{1.25 \times 10^{-8}}{2.5 \times 10^{-4}}$

---

47. Determine the replacements in the following expression that are not sensible.

$$\dfrac{x-5}{x^3 - 8x^2 + 15x}$$

Simplify.

48. $\dfrac{2a^2 + 5a - 3}{a^2} \cdot \dfrac{5a^3 + 30a^2}{2a^2 + 7a - 4} \div \dfrac{a^2 + 6a}{a^2 + 7a + 12}$

49. $\dfrac{12a}{(a-b)(b-c)} - \dfrac{2a}{(b-a)(c-b)}$

50. Divide: $(y^6 + b^6) \div (y + b)$.

TEST: CHAPTER 9

1. Simplify:

$$\dfrac{6x^2 + 17x + 7}{2x^2 + 7x + 3}.$$

2. Multiply and simplify:

$$\dfrac{a^2 - 25}{6a} \cdot \dfrac{3a}{a - 5}.$$

3. Divide and simplify:

$$\dfrac{25x^2 - 1}{9x^2 - 6x} \div \dfrac{5x^2 + 9x - 2}{3x^2 + x - 2}.$$

4. Find the LCM:

$$y^2 - 9, \quad y^2 + 10y + 21, \quad y^2 + 4y - 21.$$

Add or subtract. Simplify, if possible.

5. $\dfrac{16 + x}{x^3} + \dfrac{7 - 4x}{x^3}$

6. $\dfrac{5 - t}{t^2 + 1} - \dfrac{t - 3}{t^2 + 1}$

7. $\dfrac{x - 4}{x - 3} + \dfrac{x - 1}{3 - x}$

8. $\dfrac{x - 4}{x - 3} - \dfrac{x - 1}{3 - x}$

9. $\dfrac{5}{t - 1} + \dfrac{3}{t}$

10. $\dfrac{1}{x^2 - 16} - \dfrac{x + 4}{x^2 - 3x - 4}$

11. $\dfrac{1}{x - 1} + \dfrac{4}{x^2 - 1} - \dfrac{2}{x^2 - 2x + 1}$

Divide.

12. $(12x^4 + 9x^3 - 15x^2) \div 3x^2$

13. $(6x^3 - 8x^2 - 14x + 13) \div (3x + 2)$

14. Simplify: $\dfrac{9 - \dfrac{1}{y^2}}{3 - \dfrac{1}{y}}$.

Solve.

15. $\dfrac{7}{y} - \dfrac{1}{3} = \dfrac{1}{4}$

16. $\dfrac{15}{x} - \dfrac{15}{x-2} = -2$

17. The reciprocal of 3 less than a number is 4 times the reciprocal of the number itself. What is the number?

18. A sample of 125 spark plugs contained 4 defective spark plugs. How many defective spark plugs would you expect among 500 spark plugs?

19. One car travels 20 km/h faster than another. While one goes 225 km, the other goes 325 km. Find their speeds.

20. Solve $L = \dfrac{Mt - g}{t}$ for t.

21. Multiply and simplify $z^{-3} \cdot z^{12}$.

22. Divide and simplify $\dfrac{r^{-4}}{r^{-9}}$.

23. Simplify $(2xy^{-7})^4$.

24. Convert 0.0000000000000317 to scientific notation.

25. Convert 7×10^{13} to decimal notation.

26. Divide. Write scientific notation for the answer.
$$(1.242 \times 10^{11}) \div (5.4 \times 10^{15})$$

27. Team A and team B work together and complete a job in $2\tfrac{6}{7}$ hr. It would take team B 6 hr longer, working alone, to do the job than it would team A. How long would it take each of them to do the job working alone?

28. Simplify: $1 + \dfrac{1}{1 + \dfrac{1}{1 + \dfrac{1}{a}}}$.

CUMULATIVE REVIEW: CHAPTERS 1–9

1. Evaluate $xy + (xy)^2 + x + y$ for $x = 5$ and $y = -2$.

Compute and simplify.

2. $-1448.37 \div 4.62$

3. $\tfrac{11}{9} - \tfrac{4}{18} + \tfrac{1}{6} - \tfrac{2}{3} + \tfrac{1}{2}$

Remove parentheses and simplify.

4. $-7(x - 1) + 2x$

5. $8[4 - (6x - 5)]$

6. $x - [x - (x - 1)]$

Solve.

7. $5(a + 3) = 8a + 18$

8. $-\frac{1}{3} = x - \frac{4}{7}$

9. $-3(m + 2) - 4 = -12 - (-2 + m)$

Collect like terms.

10. $10a^2 + 6a - 8a^2 + 3a - 2a^2 - 8a$

11. $(8m + 6n) - (12n + 7m) + 2(4m - 11n)$

Multiply.

12. $(2x - 7)(x - 12)$

13. $(1 - 5x)^2$

14. $(5x - 4)(5x + 4)$

15. $(2x^2 + 3)(x^4 - x^2 - 8)$

16. $(n - 4k)^2$

17. $(2x + 10y)(x - 13y)$

Factor.

18. $36x + 24y + 12z$

19. $-21x - 28w$

20. $x^3y^3 + x^2y^2 - 4xy$

21. $a^8 + a^7 - a^6$

22. $4 - 9m^4$

23. $-a^2 + 100$

24. $9x^2 - 9b^2$

25. $x^2 - 4x - 12$

26. $s^2 - 16s + 15$

27. $t^3 - 5t^2 + 6t$

28. $7x^2 - 6x - 1$

29. $20y^2 + 19y + 3$

30. $6y^2 + 9y - 15$

31. $a^5 + a^4 - 5a^2 - 5a$

32. $5a^2 - 28am - 12m^2$

33. $x^8 - y^8$

Graph on a plane.

34. $y = -2x + 4$

35. $x = -4$

36. $5x - 2y = 10$

Solve these systems.

37. $6x + 3y = -6,$
$-2x + 5y = 14$

38. $2x + 3y = -3,$
$y = 2x - 9$

Solve.

39. $8x - 2 \geq 7x + 5$

40. $-4x \geq 24$

41. $-3x < 30 + 2x$

42. $5x + 3 \geq 6(x - 4) + 7$

Graph on a number line.

43. $5x - 1 < 24$

44. $|x| \geq 3$

Graph on a plane.

45. $y \leq 5x$

46. $2y - 3x > -6$

Find each of the following.

47. $\{1, 4, 7, 11, 15, 19, 23, 27, 31\} \cap \{3, 6, 9, 12, 15, 18, 21, 24, 27, 30\}$

48. $\{0, 1, 3\} \cup \{-1, 0, 1\}$

Find the indicated output for the function $g(x) = 2x^2 - x + 3$.

49. $g(0)$

50. $g(2)$

51. $g(-1)$

Find an equation of variation where y varies directly as x.

52. $y = 132$, when $x = 12$

53. $y = 100$, when $x = \frac{1}{5}$

Find an equation of variation where y varies inversely as x.

54. $y = 1$, when $x = 15$

55. $y = \frac{1}{700}$, when $x = 35$

56. Simplify:

$$\frac{2x^2 + 9x - 35}{x^2 - 49}.$$

57. Divide and simplify:

$$\frac{9x^4}{x^2 - 25} \div \frac{3x}{x^2 - 10x + 25}.$$

58. Subtract:

$$\frac{x^2}{x - 3} - \frac{9}{3 - x}.$$

59. Simplify:

$$\frac{\frac{2}{3x} + 4}{3 - \frac{2}{x}}.$$

60. Divide: $(3x^3 - 15x^2 - 4x + 5) \div (x - 5)$.

Solve.

61. $\frac{4}{3x} = \frac{2}{x} - \frac{1}{3}$

62. $\frac{12}{x} = \frac{48}{x + 9}$

Solve for the letter indicated.

63. $\frac{E}{e} = \frac{R + M}{R}$; for M

64. $\frac{1}{T} = \frac{1}{R} + \frac{1}{S}$; for S

Simplify.

65. $\frac{x^{-3}}{x^9}$

66. $6^{-5} \cdot 6^{-8}$

67. $(2a^{-3})^4$

Convert to scientific notation.

68. 0.0000328

69. $830,000,000$

Problem solving

70. During the library marathon, Phil read twice as many books as Howard, but only half as many as Larry. If the three boys read 21 books, how many did Howard read?

71. Two-thirds of what number is 32?

72. After a 20% reduction, an item is on sale for $144. What was the marked price (the price before reduction)?

73. The current in a stream moves at a speed of 2 km/h. A boat travels 23 km upstream and 54 km downstream in a total time of 3 hr. What is the speed of the boat in still water?

74. The length of a rectangle is 3 cm greater than its width. The area is 108 cm². Find the length and width.

75. Evaluate $a^3b^3 - a^2b + ab^2 - 4a^2b^2$ when $a = 0.1$ and $b = -0.2$.

76. Solve:
$$\tfrac{2}{5}x - \tfrac{5}{9}x + \tfrac{1}{3} > 5 - 9x.$$

77. Find the area of a triangle whose base is
$$\frac{4}{x^2 - 4}$$
and whose height is
$$\frac{x - 2}{2}.$$
Simplify the answer.

78. Find the reciprocal of
$$\frac{1}{y - 4} - \frac{3}{2y + 1}.$$

RADICAL EXPRESSIONS AND EQUATIONS

10

How can we use the length of the skid marks of a car to estimate its speed before the brakes were applied? A radical expression can be used to find the answer.

When we multiply the number 3 by itself we get $3 \cdot 3$, or 9. We say that 3 is a square root of 9. We also know that -3 is a square root of 9 because $(-3)(-3) = 9$. This chapter introduces square roots and certain algebraic expressions that involve them, called *radical expressions*.

Certain rational numbers such as 2 do not have square roots that are rational numbers. The rational numbers and numbers that we call *irrational* make up a set known as the *real numbers*, which we introduce in this chapter.

The remainder of the chapter consists of studying manipulations of radical expressions in addition, subtraction, multiplication, division, and simplifying. Finally, we consider another equation-solving principle and apply it to problem solving.

10.1 SQUARE ROOTS AND REAL NUMBERS

In this section we study square roots, irrational numbers, and real numbers.

Square Roots

When we raise a number to the second power, we have squared the number. Sometimes we may need to find the number that was squared. We call this process finding a square root of a number.

> The number c is a *square root* of a if $c^2 = a$.

Every positive number has two square roots. For example, the square roots of 25 are 5 and -5 because $5^2 = 25$ and $(-5)^2 = 25$. The positive square root is also called the *principal square root*. The symbol $\sqrt{}$ is called a *radical* symbol. The radical symbol refers only to the principal root. Thus $\sqrt{25} = 5$. To name the negative square root of a number, we use $-\sqrt{}$. The number 0 has only one square root, 0.

EXAMPLES

1. Find the square roots of 81.

   9 and -9 are square roots.

2. Find $\sqrt{225}$.

   $\sqrt{225} = 15,$ taking the principal root.

3. Find $-\sqrt{64}$.

   $\sqrt{64} = 8,$ so $-\sqrt{64} = -8.$

Irrational Numbers

Recall that all rational numbers can be named by fractional notation a/b, where a and b are integers and $b \neq 0$. Rational numbers can be named in other ways, such as with decimal notation, but they can all be named with fractional notation. Suppose we try to find a rational number for $\sqrt{2}$. We look for a number a/b for which $(a/b)^2 = 2$.

We can find rational numbers whose squares are quite close to 2:

$$\left(\frac{14}{10}\right)^2 = (1.4)^2 = 1.96,$$

$$\left(\frac{141}{100}\right)^2 = (1.41)^2 = 1.9881,$$

$$\left(\frac{1414}{1000}\right)^2 = (1.414)^2 = 1.999396,$$

$$\left(\frac{14142}{10000}\right)^2 = (1.4142)^2 = 1.99996164,$$

$$\left(\frac{141421}{100000}\right)^2 = (1.41421)^2 = 1.99998992.$$

Actually, we can never find one whose square is exactly 2. That can be proved, but we will not do so here. Since $\sqrt{2}$ is not a rational number, we call it an *irrational number*.

> An *irrational number* is a number that cannot be named by fractional notation a/b, where a and b are integers and $b \neq 0$.

The square roots of most whole numbers are irrational. Only the perfect squares 1, 4, 9, 16, 25, 36, 49, 64, 81, 100, etc., have rational square roots.

EXAMPLES Identify the rational numbers and the irrational numbers.

4. $\sqrt{3}$ $\sqrt{3}$ is irrational, since 3 is not a perfect square.
5. $\sqrt{25}$ $\sqrt{25}$ is rational, since 25 is a perfect square.
6. $\sqrt{35}$ $\sqrt{35}$ is irrational, since 35 is not a perfect square.

Real Numbers

The rational numbers are very close together on the number line. Yet no matter how close together two rational numbers are, we can find many rational numbers between them. By averaging, we can find the number *halfway* between. This process can be repeated indefinitely. For instance, the number halfway between $\frac{1}{32}$ and $\frac{2}{32}$ is $\frac{3}{64}$, the average of $\frac{1}{32}$ and $\frac{2}{32}$. In turn, $\frac{5}{128}$ is halfway between $\frac{1}{32}$ and $\frac{3}{64}$. It looks as if the rational numbers fill up the number line, but they do not. There are many points on the line for which there are no rational numbers. These points correspond to irrational numbers.

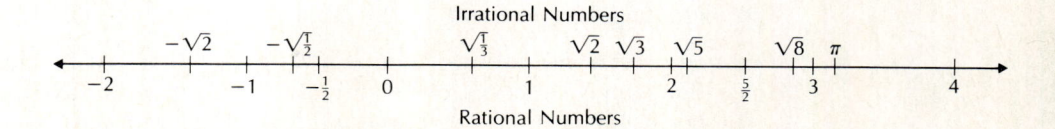

> The *real numbers* consist of the rational numbers and the irrational numbers. There is a real number for each point on a number line.

$$\text{Real Numbers}$$
$$\swarrow \qquad \searrow$$
$$\text{Rational Numbers} \qquad \text{Irrational Numbers}$$

We know that decimal notation for a rational number either ends or repeats a group of digits. For example,

$\frac{1}{4} = 0.25$ The decimal ends.

$\frac{1}{3} = 0.3333\ldots$ The 3 repeats.

$\frac{5}{11} = 0.45\overline{45}$ The bar indicates that "45" repeats.

Decimal notation for an irrational number never ends and does not repeat. The number π is an example:

$$\pi = 3.1415926535\ldots$$

Decimal notation for π never ends and never repeats. The numbers 3.1416, 3.14, or $\frac{22}{7}$ are only rational number approximations for π. Decimal notation for $\frac{22}{7}$ is 3.142857142857.... It repeats.

Here are some other examples of irrational numbers:

2.818118111811118111118...   No group of digits repeats.

0.0350355035550355550...   No group of digits repeats.

> Decimal notation for an irrational number is nonrepeating and nonending.

EXAMPLES Identify the rational numbers and the irrational numbers.

7. $\frac{18}{37}$ Rational, since it is the ratio of two integers.

8. 2.565656...   (Numeral repeats). Rational, since the digits "56" repeat.

9. 4.020020002...   (Numeral does not repeat). Irrational, since the decimal notation neither ends nor repeats.

Approximating Square Roots

We often need to use rational numbers to approximate square roots that are irrational. Such approximations can be found using a table such as Table 2 at the back of the book. They can also be found on a calculator with a square root key [$\sqrt{}$].

EXAMPLE 10 Use your calculator or Table 2 to approximate $\sqrt{10}$. Round to three decimal places.

You will need to consult the instruction manual to find square roots on your calculator. Calculators vary in their methods of operation.

$$\sqrt{10} \approx 3.162277660 \quad \text{Using a calculator with a 10-digit readout}$$

Different calculators give different numbers of digits in their readouts. This may cause some variance in answers. We round to the third decimal place. Then

$$\sqrt{10} \approx 3.162. \quad \text{This can also be found in Table 2.}$$

The symbol $\approx$ means "is approximately equal to."

It would be helpful before beginning the exercise set or continuing this chapter to list the squares of numbers from 0 to 20, and to memorize the list.

Problem Solving

We now consider an application involving a formula with a square root. Since it just involves evaluating an expression, we need not use the entire problem-solving process.

EXAMPLE 11 (*An application: Parking lot arrival spaces*). A parking lot has attendants to park cars, and it uses spaces for cars to be left before they are taken to permanent parking stalls. The number N of such spaces needed is approximated by the formula

$$N = 2.5\sqrt{A},$$

where A is the average number of arrivals in peak hours. Find the number of spaces needed when the average number of arrivals in peak hours is 49 and 77.

We substitute 49 into the formula:
$$N = 2.5\sqrt{49} = 2.5(7) = 17.5 \approx 18.$$

Note that we round up to 18 spaces because 17.5 spaces would give us a half space, which we could not use. To ensure that we have enough, we need 18. We substitute 77 into the formula. We use a calculator or Table 2 to find an approximation:
$$N = 2.5\sqrt{77} \approx 2.5(8.775) = 21.938 \approx 22.$$

When the average number of arrivals is 49, about 18 spaces are needed. When the average number of arrivals is 77, about 22 spaces are needed.

EXERCISE SET 10.1

Find the square roots of each number.

1. 1
2. 4
3. 16
4. 9
5. 100
6. 121
7. 169
8. 144

Simplify.

9. $\sqrt{4}$
10. $\sqrt{1}$
11. $-\sqrt{9}$
12. $-\sqrt{25}$
13. $-\sqrt{64}$
14. $-\sqrt{81}$
15. $-\sqrt{225}$
16. $\sqrt{400}$
17. $\sqrt{361}$
18. $\sqrt{441}$

Identify each square root as rational or irrational.

19. $\sqrt{2}$
20. $\sqrt{6}$
21. $\sqrt{8}$
22. $\sqrt{10}$
23. $\sqrt{49}$
24. $\sqrt{100}$
25. $\sqrt{98}$
26. $\sqrt{75}$
27. $-\sqrt{4}$
28. $-\sqrt{1}$
29. $-\sqrt{12}$
30. $-\sqrt{14}$

Identify each number as rational or irrational.

31. 4.23
32. 0.03
33. 23
34. -19
35. $-\frac{2}{3}$
36. 0
37. $\frac{2.3}{0.01}$
38. 0.424242... (numeral repeats)
39. 0.156156156... (numeral repeats)
40. 4.282282228... (numeral does not repeat)
41. 7.767767776... (numeral does not repeat)
42. $-34.69191191119...$ (numeral does not repeat)
43. $-63.030030003...$ (numeral does not repeat)

Use your calculator or Table 2 to approximate these square roots. Round to three decimal places.

44. $\sqrt{5}$
45. $\sqrt{6}$
46. $\sqrt{17}$
47. $\sqrt{19}$
48. $\sqrt{93}$
49. $\sqrt{43}$
50. $\sqrt{40}$

Problem solving

Use the formula $N = 2.5\sqrt{A}$ of Example 11.

51. Find the number of spaces needed when the average number of arrivals is 25 and 89.

52. Find the number of spaces needed when the average number of arrivals is 62 and 100.

53. Simplify $\sqrt{\sqrt{16}}$.

54. Simplify $\sqrt{3^2 + 4^2}$.

55. Between what two consecutive integers is $-\sqrt{33}$?

Use a calculator to approximate these square roots. Round to three decimal places.

56. $\sqrt{12.8}$ 57. $\sqrt{198}$ 58. $\sqrt{930}$ 59. $\sqrt{4932}$ 60. $\sqrt{1043.89}$

61. What number is halfway between x and y?

62. Find a number that is the square of an integer and the cube of a different integer.

63. Find the number one-third of the way from $2\frac{3}{4}$ to $4\frac{5}{8}$.

64. Find the number one-fifth of the way from -1 to $-\frac{4}{3}$.

65. What number is one-fourth of the way between z and w if $z < w$?

10.2 RADICAL EXPRESSIONS

In this section we consider radical notation for square roots and the ways in which we can manipulate such notation to get equivalent expressions. Such manipulation can be important in problem solving.

Radicands

When an expression is written under a radical, we have a *radical expression*. Here are some examples:

$$\sqrt{14}, \quad \sqrt{x}, \quad \sqrt{x^2 + 4}, \quad \sqrt{\frac{x^2 - 5}{2}}.$$

The expression written under the radical is called the *radicand*.

EXAMPLES Identify the radicand in each expression.

1. $\sqrt{x}$ The radicand is x.
2. $\sqrt{y^2 - 5}$ The radicand is $y^2 - 5$.

Meaningless Expressions

The square of any number is always positive. For example, $8^2 = 64$ and $(-11)^2 = 121$. There are no real numbers that can be squared to get negative numbers.

• •
Radical expressions with negative radicands have no meaning in the real-number system.

Thus the following expressions do not represent real numbers:
$$\sqrt{-100}, \quad \sqrt{-49}, \quad -\sqrt{-3}.$$

Later in your study of mathematics, you may encounter a number system in which negative numbers have square roots.

EXAMPLE 3 Determine whether 6 is a sensible replacement in the expression $\sqrt{1-y}$.

If we replace y by 6, we get $\sqrt{1-6} = \sqrt{-5}$, which has no meaning because the radicand is negative.

EXAMPLES Determine the sensible replacements in each expression.

4. $\sqrt{x}$ Any number greater than or equal to 0 is sensible.
5. $\sqrt{x+2}$ We solve the inequality $x + 2 \geq 0$. Any number greater than or equal to -2 is sensible.
6. $\sqrt{x^2}$ Squares of numbers are never negative. All replacements are sensible.
7. $\sqrt{x^2 + 1}$ Since x^2 is never negative, $x^2 + 1$ is never negative. All replacements are sensible.

Perfect-Square Radicands

We now consider radical expressions with a perfect-square radicand, $\sqrt{x^2}$. If x represents a nonnegative number, $\sqrt{x^2}$ simplifies to x. If x represents a negative number, $\sqrt{x^2}$ simplifies to $-x$ (the additive inverse of x). That is because $\sqrt{a}$ denotes the *principal* square root of a.

Suppose $x = 3$. Then $\sqrt{x^2} = \sqrt{3^2}$, which is $\sqrt{9}$, or 3.
Suppose $x = -3$. Then $\sqrt{x^2} = \sqrt{(-3)^2}$, which is $\sqrt{9}$ or 3, the absolute value of -3.

In either case we have $\sqrt{x^2} = |x|$.

• •

Any radical expression $\sqrt{A^2}$ can be simplified to $|A|$.

EXAMPLES Simplify.

8. $\sqrt{(3x)^2} = |3x|$ Absolute-value notation is necessary.
9. $\sqrt{a^2 b^2} = \sqrt{(ab)^2} = |ab|$
10. $\sqrt{x^2 + 2x + 1} = \sqrt{(x+1)^2} = |x+1|$

We can sometimes simplify absolute-value notation. In Example 8, $|3x|$ simplifies to $|3| \cdot |x|$, or $3|x|$. In Example 9, we can change $|ab|$ to $|a| \cdot |b|$ if we wish. The absolute value of a product is always the product of the absolute values.

10.2 RADICAL EXPRESSIONS

For any real numbers a and b, $|a \cdot b| = |a| \cdot |b|$.

EXERCISE SET 10.2

Identify the radicand in each expression.

1. $\sqrt{a-4}$
2. $\sqrt{t+3}$
3. $5\sqrt{t^2+1}$
4. $8\sqrt{x^2+5}$
5. $x^2y\sqrt{\dfrac{3}{x+2}}$
6. $ab^2\sqrt{\dfrac{a}{a-b}}$

Which of these expressions are meaningless? Write "yes" or "no."

7. $\sqrt{-16}$
8. $\sqrt{-81}$
9. $-\sqrt{81}$
10. $-\sqrt{64}$

Determine the sensible replacements in each expression.

11. $\sqrt{5x}$
12. $\sqrt{3y}$
13. $\sqrt{t-5}$
14. $\sqrt{y-8}$
15. $\sqrt{y+8}$
16. $\sqrt{x+6}$
17. $\sqrt{x+20}$
18. $\sqrt{m-18}$
19. $\sqrt{2y-7}$
20. $\sqrt{3x+8}$
21. $\sqrt{t^2+5}$
22. $\sqrt{y^2+1}$

Simplify.

23. $\sqrt{t^2}$
24. $\sqrt{x^2}$
25. $\sqrt{9x^2}$
26. $\sqrt{4a^2}$
27. $\sqrt{(-7)^2}$
28. $\sqrt{(-5)^2}$
29. $\sqrt{(-4d)^2}$
30. $\sqrt{(-3b)^2}$
31. $\sqrt{(x+3)^2}$
32. $\sqrt{(x-7)^2}$
33. $\sqrt{a^2-10a+25}$
34. $\sqrt{x^2+2x+1}$

Problem-solving practice

35. The amount F that a family spends on food varies directly as its income I. A family making $19,600 a year will spend $5096 on food. At this rate, how much would a family making $20,500 spend on food?

36. A collection of dimes and nickels is worth $15.25. There are 157 coins in all. How many of each kind are there?

Solve.

37. $\sqrt{x^2} = 6$
38. $\sqrt{y^2} = -7$
39. $-\sqrt{x^2} = -3$
40. $t^2 = 49$

Simplify.

41. $\sqrt{(3a)^2}$
42. $(\sqrt{3a})^2$
43. $\sqrt{\dfrac{144x^8}{36y^6}}$
44. $\sqrt{\dfrac{y^{12}}{8100}}$
45. $\sqrt{\dfrac{169}{m^{16}}}$
46. $\sqrt{\dfrac{p^2}{3600}}$

Determine the sensible replacements in each expression.

47. $\sqrt{m(m+3)}$
48. $\sqrt{x^2(x-3)}$
49. $\sqrt{(x+3)(x-2)}$
50. $\sqrt{x^2+7x+12}$
51. $\sqrt{x^2-4}$
52. $\sqrt{4x^2-1}$

10.3 MULTIPLYING AND FACTORING

In this section, we look at multiplying and factoring with radical notation, making sure, as usual, that we obtain equivalent expressions.

Multiplying Radical Expressions

To see how to multiply with radical notation, look at the following examples.

EXAMPLE 1 Simplify.

a) $\sqrt{9} \cdot \sqrt{4} = 3 \cdot 2 = 6$ This is a product of square roots.
b) $\sqrt{9 \cdot 4} = \sqrt{36} = 6$ This is the square root of a product.

EXAMPLE 2 Simplify.

a) $\sqrt{4} \cdot \sqrt{25} = 2 \cdot 5 = 10$
b) $\sqrt{4 \cdot 25} = \sqrt{100} = 10$

EXAMPLE 3 $\sqrt{-9}\sqrt{-4}$ Meaningless (negative radicands)

We can multiply radical expressions by multiplying the radicands, provided that the radicands are not negative.

> For any nonnegative radicands A and B, $\sqrt{A} \cdot \sqrt{B} = \sqrt{A \cdot B}$. (The product of square roots, provided they exist, is the square root of the product of the radicands.)

EXAMPLES Multiply. Assume that all radicands are nonnegative.

4. $\sqrt{5}\sqrt{7} = \sqrt{5 \cdot 7} = \sqrt{35}$
5. $\sqrt{8}\sqrt{8} = \sqrt{8 \cdot 8} = \sqrt{64} = 8$
6. $\sqrt{\frac{2}{3}}\sqrt{\frac{4}{5}} = \sqrt{\frac{2}{3} \cdot \frac{4}{5}} = \sqrt{\frac{8}{15}}$
7. $\sqrt{2x}\sqrt{3x-1} = \sqrt{2x(3x-1)} = \sqrt{6x^2 - 2x}$

To prove the property $\sqrt{A} \cdot \sqrt{B} = \sqrt{AB}$, we consider a product $\sqrt{A}\sqrt{B}$, where A and B are not negative. We square this product, to show that we get AB; thus the product is the square root of AB (or $\sqrt{AB}$):

$$(\sqrt{A}\sqrt{B})^2 = (\sqrt{A}\sqrt{B})(\sqrt{A}\sqrt{B}) = (\sqrt{A}\sqrt{A})(\sqrt{B}\sqrt{B}) = AB.$$

Factoring and Simplifying

We know that for nonnegative radicands,
$$\sqrt{A}\sqrt{B} = \sqrt{AB}.$$

To factor radical expressions we can use this equation. It may help to think of it in reverse.
$$\sqrt{AB} = \sqrt{A}\sqrt{B}$$

In some cases we can simplify after factoring. A radical expression is simplified when its radicand has no factors that are perfect squares.

EXAMPLES Simplify by factoring.

8. $\sqrt{18} = \sqrt{9 \cdot 2}$ Factoring the radicand (9 is a perfect square)
   $= \sqrt{9} \cdot \sqrt{2}$ Factoring the radical expression
   $= 3\sqrt{2}$ $3\sqrt{2}$ means $3 \cdot \sqrt{2}$
   The radicand has no factors that are perfect squares.

9. $\sqrt{25x} = \sqrt{25} \cdot \sqrt{x} = 5\sqrt{x}$

10. $\sqrt{x^2 - 6x + 9} = |x - 3|$

11. $\sqrt{36x^2} = \sqrt{36}\sqrt{x^2} = 6|x|$

Approximating Square Roots

If we are using a table to approximate square roots, some numbers can be too large to find in the table. For example, Table 2 goes only to 100. We may still be able to find approximate square roots for other numbers. We do this by first factoring out the largest perfect square, if there is one. If there is none, we use any factorization we can find that will give smaller factors.

EXAMPLES Approximate these square roots.

12. $\sqrt{160} = \sqrt{16 \cdot 10}$ Factoring the radicand (make one factor a perfect square, if you can)
    $= \sqrt{16}\sqrt{10}$ Factoring the radical expression
    $= 4\sqrt{10}$
    $\approx 4(3.162)$ From Table 2, $\sqrt{10} \approx 3.162$.
    ≈ 12.648

13. $\sqrt{341} = \sqrt{11 \cdot 31}$ Factoring (there is no perfect-square factor)
    $= \sqrt{11}\sqrt{31}$
    $\approx 3.317 \times 5.568$ From Table 2
    ≈ 18.469 Rounded to three decimal places

EXERCISE SET 10.3

Multiply.

1. $\sqrt{2}\sqrt{3}$
2. $\sqrt{3}\sqrt{5}$
3. $\sqrt{4}\sqrt{3}$
4. $\sqrt{2}\sqrt{9}$
5. $\sqrt{\frac{2}{5}}\sqrt{\frac{3}{4}}$
6. $\sqrt{\frac{3}{8}}\sqrt{\frac{1}{5}}$
7. $\sqrt{17}\sqrt{17}$
8. $\sqrt{18}\sqrt{18}$
9. $\sqrt{25}\sqrt{3}$
10. $\sqrt{36}\sqrt{2}$
11. $\sqrt{2}\sqrt{x}$
12. $\sqrt{3}\sqrt{a}$
13. $\sqrt{0.24}\sqrt{3}$
14. $\sqrt{2}\sqrt{0.56}$
15. $\sqrt{x}\sqrt{t}$
16. $\sqrt{a}\sqrt{y}$
17. $\sqrt{x}\sqrt{x-3}$
18. $\sqrt{x}\sqrt{x+1}$
19. $\sqrt{5}\sqrt{2x-1}$
20. $\sqrt{3}\sqrt{4x+2}$
21. $\sqrt{x+2}\sqrt{x+1}$
22. $\sqrt{x-3}\sqrt{x+4}$
23. $\sqrt{x-3}\sqrt{2x+4}$
24. $\sqrt{2x+5}\sqrt{x-4}$
25. $\sqrt{x+4}\sqrt{x-4}$
26. $\sqrt{x-2}\sqrt{x+2}$
27. $\sqrt{x+y}\sqrt{x-y}$
28. $\sqrt{a-b}\sqrt{a+b}$
29. $\sqrt{-3}\sqrt{2x}$
30. $\sqrt{-5}\sqrt{-4x}$

Simplify by factoring.

31. $\sqrt{12}$
32. $\sqrt{8}$
33. $\sqrt{75}$
34. $\sqrt{50}$
35. $\sqrt{20}$
36. $\sqrt{45}$
37. $\sqrt{200}$
38. $\sqrt{300}$
39. $\sqrt{3x}$
40. $\sqrt{5y}$
41. $\sqrt{9x}$
42. $\sqrt{4y}$
43. $\sqrt{16a}$
44. $\sqrt{49b}$
45. $\sqrt{64y^2}$
46. $\sqrt{9x^2}$
47. $\sqrt{13x^2}$
48. $\sqrt{29t^2}$
49. $\sqrt{8t^2}$
50. $\sqrt{125a^2}$

Approximate these square roots using Table 2. Round to three decimal places.

51. $\sqrt{125}$
52. $\sqrt{180}$
53. $\sqrt{360}$
54. $\sqrt{105}$
55. $\sqrt{300}$
56. $\sqrt{143}$
57. $\sqrt{122}$
58. $\sqrt{2000}$

Problem-solving practice

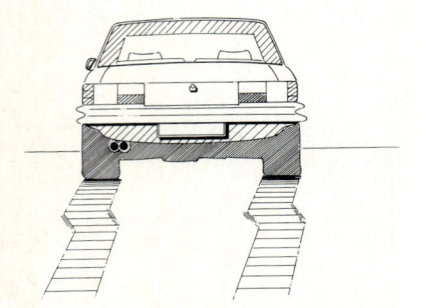

(*Speed of a skidding car*). How do police determine the speed of a car after an accident? The formula

$$r = 2\sqrt{5L}$$

can be used to approximate the speed r, in miles per hour, of a car that has left a skid mark of length L, in feet.

59. What was the speed of a car that left skid marks of 20 ft? of 150 ft?

60. What was the speed of a car that left skid marks of 30 ft? of 70 ft?

Factor.

61. $\sqrt{3x-3}$
62. $\sqrt{x^2-x-2}$
63. $\sqrt{x^2-4}$
64. $\sqrt{2x^2-5x-12}$
65. $\sqrt{x^3-2x^2}$
66. $\sqrt{a^2-b^2}$

Simplify.

67. $\sqrt{0.01}$
68. $\sqrt{0.25}$
69. $\sqrt{x^4}$
70. $\sqrt{9a^6}$

71. Find $\sqrt{49}$, $\sqrt{490}$, $\sqrt{4900}$, $\sqrt{49{,}000}$ and $\sqrt{490{,}000}$. What pattern do you see?

Use the proper symbol ($>$, $<$, or $=$) between each pair of values to make a true sentence. Assume that x is positive.

72. $15 \quad 4\sqrt{14}$
73. $15\sqrt{2} \quad \sqrt{450}$
74. $16 \quad \sqrt{15}\sqrt{17}$
75. $3\sqrt{11} \quad 7\sqrt{2}$
76. $5\sqrt{7} \quad 4\sqrt{11}$
77. $8 \quad \sqrt{15}+\sqrt{17}$
78. $3\sqrt{x} \quad 2\sqrt{2.5x}$
79. $4\sqrt{x} \quad 5\sqrt{0.64x}$
80. $90\sqrt{100x} \quad 100\sqrt{90x}$

Simplify.

81. $\sqrt{z^8 w^{10}}$
82. $\sqrt{x^{-2}}$

10.4 SIMPLIFYING RADICAL EXPRESSIONS

We now consider simplifying square roots of powers and of products.

In many formulas and problems involving radical notation, variables do not represent negative numbers. Thus absolute-value notation is not necessary. From now on we shall assume that all radicands are nonnegative.

To simplify radical expressions, we usually try to factor out as many perfect-square factors as possible. Compare the following:

$$\sqrt{50} = \sqrt{10 \cdot 5} = \sqrt{10}\sqrt{5};$$
$$\sqrt{50} = \sqrt{25 \cdot 2} = \sqrt{25}\sqrt{2} = 5\sqrt{2}.$$

In the second case, we factored out the perfect square 25. If you do not recognize perfect squares, try factoring the radicand into its prime factors. For example,

$$\sqrt{50} = \sqrt{2 \cdot \underline{5 \cdot 5}} = 5\sqrt{2}. \quad \text{— Perfect square}$$

Radical expressions, such as $5\sqrt{2}$, where the radicand has no perfect-square factors, are considered to be in simplest form.

EXAMPLES Simplify by factoring. Assume that all expressions under radicals represent nonnegative numbers.

1. $\sqrt{48t} = \sqrt{16 \cdot 3t}$ Identifying perfect-square factors and factoring the radicand
   $= \sqrt{16}\sqrt{3t}$ Factoring into a product of radicals
   $= 4\sqrt{3t}$ Taking the square root

2. $\sqrt{20t^2} = \sqrt{4 \cdot t^2 \cdot 5}$ Identifying perfect-square factors and factoring the radicand
   $= \sqrt{4}\sqrt{t^2}\sqrt{5}$ Factoring into a product of several radicals
   $= 2t\sqrt{5}$ Taking square roots (absolute-value notation is not necessary because expressions are not negative)

3. $\sqrt{3x^2 + 6x + 3} = \sqrt{3(x^2 + 2x + 1)}$ Factoring the radicand
   $= \sqrt{3}\sqrt{x^2 + 2x + 1}$ Factoring into a product of radicals
   $= \sqrt{3}\sqrt{(x+1)^2}$
   $= \sqrt{3}(x+1)$ Taking the square root (absolute-value notation is not necessary because expressions are not negative)

Simplifying Square Roots of Powers

To take the square root of an even power such as x^8, we note that $x^8 = (x^4)^2 = x^4 \cdot x^4$. Then

$$\sqrt{x^8} = \sqrt{(x^4)^2} = \sqrt{x^4 \cdot x^4} = x^4.$$

We can find the answer by taking half the exponent. That is,

$$\sqrt{x^8} = x^4 \quad \tfrac{1}{2}(8) = 4$$

Again, absolute-value notation is not necessary because expressions are not negative.

EXAMPLES Simplify by factoring. Assume that all expressions under radicals represent nonnegative numbers.

4. $\sqrt{x^6} = x^3$ $\tfrac{1}{2}(6) = 3$ (absolute-value notation is not necessary because expressions are not negative)

5. $\sqrt{x^{10}} = x^5$

6. $\sqrt{t^{22}} = t^{11}$

When odd powers occur, express the power in terms of the next lower even power. Then simplify the even power.

EXAMPLE 7 Simplify by factoring. Assume that all expressions under radicals represent nonnegative numbers.

$$\sqrt{x^9} = \sqrt{x^8 \cdot x}$$
$$= \sqrt{x^8}\sqrt{x}$$
$$= x^4\sqrt{x}$$

Multiplying and Simplifying

Sometimes we can simplify after multiplying. We factor the new radicand and look for perfect-square factors.

EXAMPLE 8 Multiply and then simplify by factoring.

$$\sqrt{2}\sqrt{14} = \sqrt{2 \cdot 14} \quad \text{Multiplying}$$
$$= \sqrt{2 \cdot 2 \cdot 7} \quad \text{Looking for perfect-square factors and factoring}$$
$$\text{Perfect square}$$
$$= \sqrt{2 \cdot 2}\sqrt{7}$$
$$= 2\sqrt{7}$$

EXAMPLE 9 Multiply and then simplify by factoring. Assume that all expressions under radicals represent nonnegative numbers.

$$\sqrt{3x^2}\sqrt{9x^3} = \sqrt{3 \cdot 9x^5} \quad \text{Multiplying}$$
$$= \sqrt{3 \cdot 9 \cdot x^4 \cdot x} \quad \text{Looking for perfect-square factors and factoring}$$
$$= \sqrt{9 \cdot x^4 \cdot 3 \cdot x}$$
$$\text{Perfect squares}$$
$$= \sqrt{9}\sqrt{x^4}\sqrt{3x}$$
$$= 3x^2\sqrt{3x}$$

EXERCISE SET 10.4

Simplify by factoring. Assume that all expressions under radicals represent nonnegative numbers.

1. $\sqrt{180}$
2. $\sqrt{448}$
3. $\sqrt{48x}$
4. $\sqrt{40m}$
5. $\sqrt{288y}$
6. $\sqrt{363p}$
7. $\sqrt{20x^2}$
8. $\sqrt{28x^2}$
9. $\sqrt{8x^2 + 8x + 2}$
10. $\sqrt{27x^2 - 36x + 12}$
11. $\sqrt{36y + 12y^2 + y^3}$
12. $\sqrt{x - 2x^2 + x^3}$
13. $\sqrt{x^6}$
14. $\sqrt{x^{10}}$
15. $\sqrt{x^{12}}$
16. $\sqrt{x^{16}}$
17. $\sqrt{x^5}$
18. $\sqrt{x^3}$
19. $\sqrt{t^{19}}$
20. $\sqrt{p^{17}}$
21. $\sqrt{(y-2)^8}$
22. $\sqrt{(x+3)^6}$
23. $\sqrt{4(x+5)^{10}}$
24. $\sqrt{16(a-7)^4}$
25. $\sqrt{36m^3}$
26. $\sqrt{250y^3}$
27. $\sqrt{8a^5}$
28. $\sqrt{12b^7}$
29. $\sqrt{448x^6y^3}$
30. $\sqrt{243x^5y^4}$

Multiply and then simplify by factoring. Assume that all expressions under radicals represent nonnegative numbers.

31. $\sqrt{3}\sqrt{18}$
32. $\sqrt{5}\sqrt{10}$
33. $\sqrt{15}\sqrt{6}$
34. $\sqrt{3}\sqrt{27}$
35. $\sqrt{18}\sqrt{14x}$
36. $\sqrt{12}\sqrt{18x}$

37. $\sqrt{3x}\sqrt{12y}$
38. $\sqrt{7x}\sqrt{21y}$
39. $\sqrt{10}\sqrt{10}$
40. $\sqrt{11}\sqrt{11x}$
41. $\sqrt{5b}\sqrt{15b}$
42. $\sqrt{6a}\sqrt{18a}$
43. $\sqrt{2t}\sqrt{2t}$
44. $\sqrt{3a}\sqrt{3a}$
45. $\sqrt{ab}\sqrt{ac}$
46. $\sqrt{xy}\sqrt{xz}$
47. $\sqrt{2x^2y}\sqrt{4xy^2}$
48. $\sqrt{15mn^2}\sqrt{5m^2n}$
49. $\sqrt{18x^2y^3}\sqrt{6xy^4}$
50. $\sqrt{12x^3y^2}\sqrt{8xy}$
51. $\sqrt{50ab}\sqrt{10a^2b^4}$

○ ─────────────────────────

Multiply and then simplify by factoring. Assume that all expressions under radicals represent nonnegative numbers.

52. $(\sqrt{2y})(\sqrt{3})(\sqrt{8y})$
53. $\sqrt{a}(\sqrt{a^3}-5)$
54. $\sqrt{27(x+1)}\sqrt{12y(x+1)^2}$
55. $\sqrt{18(x-2)}\sqrt{20(x-2)^3}$
56. $\sqrt{x}\sqrt{2x}\sqrt{10x^5}$
57. $\sqrt{2^{109}}\sqrt{x^{306}}\sqrt{x^{11}}$

Simplify.

58. $\sqrt{x^{8n}}$
59. $\sqrt{0.04x^{4n}}$
60. We know that $\sqrt{A}\cdot\sqrt{B}=\sqrt{AB}$. Determine whether it is true that $\sqrt{A}+\sqrt{B}=\sqrt{A+B}$.
61. Determine whether it is true that $\sqrt{A}-\sqrt{B}=\sqrt{A-B}$.
62. Simplify $\sqrt{y^n}$, where n is an even whole number ≥ 2.
63. Simplify $\sqrt{y^n}$, where n is an odd whole number ≥ 3.
64. Multiply $(x^2+\sqrt{2}xy+y^2)$ by $(x^2-\sqrt{2}xy+y^2)$. Use your result to factor x^8+y^8.

The number c is called the *cube root* of a if $c^3=a$. We denote the cube root of a as $\sqrt[3]{a}$. For example, $\sqrt[3]{125}=5$ and $\sqrt[3]{-125}=-5$. Note that each real number has exactly one cube root. In particular, negative numbers have cube roots. Find each of the following cube roots.

65. $\sqrt[3]{8}$
66. $\sqrt[3]{-8}$
67. $\sqrt[3]{-27}$
68. $\sqrt[3]{64}$
69. $\sqrt[3]{x^3}$
70. $\sqrt[3]{t^6}$

10.5 SIMPLIFYING SQUARE ROOTS OF QUOTIENTS

In this section we proceed as we did in the preceding section, except that we shall look at division instead of multiplication.

Simplifying

Consider the expressions

$$\sqrt{\frac{25}{16}} \quad \text{and} \quad \frac{\sqrt{25}}{\sqrt{16}}.$$

10.5 SIMPLIFYING SQUARE ROOTS OF QUOTIENTS

Let us evaluate them separately.

a) $\sqrt{\dfrac{25}{16}} = \dfrac{5}{4}$ because $\dfrac{5}{4} \cdot \dfrac{5}{4} = \dfrac{25}{16}$

b) $\dfrac{\sqrt{25}}{\sqrt{16}} = \dfrac{5}{4}$ since $\sqrt{25} = 5$ and $\sqrt{16} = 4$

We see that both expressions represent the same number. This suggests that we can take the square root of a quotient by taking the square root of the numerator and denominator separately. That is true.

> For any nonnegative number A and any positive number B,
> $$\sqrt{\dfrac{A}{B}} = \dfrac{\sqrt{A}}{\sqrt{B}}.$$
> **(We can take the square root of the numerator and denominator separately.)**

EXAMPLES Simplify by taking the square roots of the numerator and denominator. Assume that all expressions under radicals represent positive numbers.

1. $\sqrt{\dfrac{25}{9}} = \dfrac{\sqrt{25}}{\sqrt{9}} = \dfrac{5}{3}$ Taking the square root of the numerator and denominator

2. $\sqrt{\dfrac{1}{16}} = \dfrac{\sqrt{1}}{\sqrt{16}} = \dfrac{1}{4}$ Taking the square root of the numerator and denominator

3. $\sqrt{\dfrac{49}{t^2}} = \dfrac{\sqrt{49}}{\sqrt{t^2}} = \dfrac{7}{t}$

We are assuming that no expression represents 0 or a negative number. Thus we need not be concerned about zero denominators or absolute-value signs.

Sometimes a fractional expression can be simplified to one that has a perfect-square numerator and denominator.

EXAMPLE 4 Simplify: $\sqrt{\dfrac{18}{50}}$.

$$\sqrt{\dfrac{18}{50}} = \sqrt{\dfrac{9 \cdot 2}{25 \cdot 2}} = \sqrt{\dfrac{9}{25} \cdot \dfrac{2}{2}} = \sqrt{\dfrac{9}{25} \cdot 1} = \sqrt{\dfrac{9}{25}} = \dfrac{3}{5}$$

EXAMPLE 5 Simplify: $\sqrt{\dfrac{2560}{2890}}$.

$$\sqrt{\dfrac{2560}{2890}} = \sqrt{\dfrac{256 \cdot 10}{289 \cdot 10}} = \sqrt{\dfrac{256}{289} \cdot \dfrac{10}{10}} = \sqrt{\dfrac{256}{289} \cdot 1} = \sqrt{\dfrac{256}{289}} = \dfrac{16}{17}$$

Rationalizing Denominators

When neither the numerator nor the denominator is a perfect square, we can simplify to an expression that has a whole-number radicand. We then have $a\sqrt{b}$, where b is a whole number with no perfect-square factors.

EXAMPLE 6 Simplify: $\sqrt{\dfrac{2}{3}}$.

We multiply by 1, choosing 3/3 for 1. This makes the denominator a perfect square.

$$\sqrt{\dfrac{2}{3}} = \sqrt{\dfrac{2}{3} \cdot \dfrac{3}{3}} \quad \text{\color{blue}Multiplying by 1}$$

$$= \sqrt{\dfrac{6}{9}}$$

$$= \dfrac{\sqrt{6}}{\sqrt{9}}$$

$$= \dfrac{\sqrt{6}}{3}$$

We can always multiply by 1 to make a denominator a perfect square. Then we can take the square root of the denominator.

EXAMPLE 7 Rationalize the denominator: $\sqrt{\dfrac{5}{12}}$.

$$\sqrt{\dfrac{5}{12}} = \sqrt{\dfrac{5}{12} \cdot \dfrac{3}{3}} = \sqrt{\dfrac{15}{36}} = \dfrac{\sqrt{15}}{\sqrt{36}} = \dfrac{\sqrt{15}}{6}$$

Approximating Square Roots of Fractions

We can use a calculator or Table 2 to approximate square roots of fractions. There are various ways to do it.

10.5 SIMPLIFYING SQUARE ROOTS OF QUOTIENTS

EXAMPLE 8 Approximate $\sqrt{\dfrac{3}{5}}$ to three decimal places.

Method 1. Suppose we are using a calculator. Then we divide and approximate the square root:

$$\sqrt{\dfrac{3}{5}} = \sqrt{0.6} \approx 0.774596669 \approx 0.775.$$

Method 2. Suppose we are using a table such as Table 2. We first rationalize the denominator. Next, we use a calculator or Table 2 to approximate the square root in the numerator. Then we divide:

$$\sqrt{\dfrac{3}{5}} = \sqrt{\dfrac{3}{5} \cdot \dfrac{5}{5}} = \sqrt{\dfrac{15}{25}} = \dfrac{\sqrt{15}}{\sqrt{25}} = \dfrac{\sqrt{15}}{5} \approx \dfrac{3.873}{5} \approx 0.775. \quad \text{Rounding to three decimal places}$$

EXERCISE SET 10.5

Simplify. Assume that all expressions under radicals represent positive numbers.

1. $\sqrt{\dfrac{9}{49}}$
2. $\sqrt{\dfrac{16}{25}}$
3. $\sqrt{\dfrac{1}{36}}$
4. $\sqrt{\dfrac{1}{4}}$
5. $-\sqrt{\dfrac{16}{81}}$
6. $-\sqrt{\dfrac{25}{49}}$
7. $\sqrt{\dfrac{64}{289}}$
8. $\sqrt{\dfrac{81}{361}}$
9. $\sqrt{\dfrac{1690}{1960}}$
10. $\sqrt{\dfrac{1440}{6250}}$
11. $\sqrt{\dfrac{36}{a^2}}$
12. $\sqrt{\dfrac{25}{x^2}}$
13. $\sqrt{\dfrac{9a^2}{625}}$
14. $\sqrt{\dfrac{x^2 y^2}{256}}$

Rationalize the denominator. Assume that all expressions under radicals represent positive numbers.

15. $\sqrt{\dfrac{2}{5}}$
16. $\sqrt{\dfrac{2}{7}}$
17. $\sqrt{\dfrac{3}{8}}$
18. $\sqrt{\dfrac{7}{8}}$
19. $\sqrt{\dfrac{7}{12}}$
20. $\sqrt{\dfrac{1}{12}}$
21. $\sqrt{\dfrac{1}{18}}$
22. $\sqrt{\dfrac{5}{18}}$
23. $\sqrt{\dfrac{1}{2}}$
24. $\sqrt{\dfrac{1}{3}}$
25. $\sqrt{\dfrac{8}{3}}$
26. $\sqrt{\dfrac{12}{5}}$
27. $\sqrt{\dfrac{3}{x}}$
28. $\sqrt{\dfrac{2}{x}}$
29. $\sqrt{\dfrac{x}{y}}$
30. $\sqrt{\dfrac{a}{b}}$
31. $\sqrt{\dfrac{x^2}{18}}$
32. $\sqrt{\dfrac{x^2}{20}}$

Approximate to three decimal places.

33. $\sqrt{\dfrac{1}{3}}$
34. $\sqrt{\dfrac{3}{2}}$
35. $\sqrt{\dfrac{7}{8}}$
36. $\sqrt{\dfrac{3}{8}}$
37. $\sqrt{\dfrac{1}{12}}$
38. $\sqrt{\dfrac{5}{12}}$

39. $\sqrt{\dfrac{1}{2}}$ 40. $\sqrt{\dfrac{1}{7}}$ 41. $\sqrt{\dfrac{7}{20}}$ 42. $\sqrt{\dfrac{3}{20}}$ 43. $\sqrt{\dfrac{13}{18}}$ 44. $\sqrt{\dfrac{11}{18}}$

Problem-solving practice

The period T of a pendulum is the time it takes to move from one side to the other and back. A formula for the period is

$$T = 2\pi \sqrt{\dfrac{L}{32}},$$

where T is in seconds and L is in feet. Use 3.14 for π.

45. Find the periods of pendulums of lengths 2 ft, 8 ft, 64 ft, and 100 ft.

46. Find the period of a pendulum of length $\tfrac{2}{3}$ in.

47. The pendulum of a grandfather clock is $32/\pi^2$ ft long. How long does it take to swing from one side to the other?

Rationalize the denominator.

48. $\sqrt{\dfrac{5}{1600}}$ 49. $\sqrt{\dfrac{3}{1000}}$ 50. $\sqrt{\dfrac{1}{5x^3}}$ 51. $\sqrt{\dfrac{3x^2y}{a^2x^5}}$

52. $\sqrt{\dfrac{3a}{b}}$ 53. $\sqrt{\dfrac{1}{5zw^2}}$ 54. $\sqrt{0.007}$ 55. $\sqrt{0.012}$

Simplify. Assume that all expressions under radicals represent positive numbers.

56. $\sqrt{\dfrac{1}{x^2} - \dfrac{2}{xy} + \dfrac{1}{y^2}}$ 57. $\sqrt{2 - \dfrac{4}{z^2} + \dfrac{2}{z^4}}$

Rationalize the denominator. Assume that all expressions under radicals represent positive numbers.

58. $\sqrt{\dfrac{1}{3x^5}}$ 59. $\sqrt{\dfrac{3}{8x}}$ 60. $\dfrac{3}{\sqrt{6z}}$ 61. $\sqrt{\dfrac{3y}{2+y^2}}$

10.6 DIVISION

In this section we find quotients of square roots, and we continue to study rationalizing denominators.

Quotients of Square Roots

We know that for positive radicands,

$$\sqrt{\frac{A}{B}} = \frac{\sqrt{A}}{\sqrt{B}}.$$

To find quotients we can use this equation. It may help to think of it in reverse.

• •

> For any nonnegative radicand A and any positive number B,
> $$\frac{\sqrt{A}}{\sqrt{B}} = \sqrt{\frac{A}{B}}.$$

(The quotient of square roots, provided they exist, is the square root of the quotient of the radicands.)

EXAMPLES Divide and simplify. Assume that all expressions under radicals represent positive numbers.

1. $\dfrac{\sqrt{27}}{\sqrt{3}} = \sqrt{\dfrac{27}{3}} = \sqrt{9} = 3$

2. $\dfrac{\sqrt{7}}{\sqrt{14}} = \sqrt{\dfrac{7}{14}} = \sqrt{\dfrac{1}{2}} = \sqrt{\dfrac{1}{2} \cdot \dfrac{2}{2}} = \sqrt{\dfrac{2}{4}} = \dfrac{\sqrt{2}}{\sqrt{4}} = \dfrac{\sqrt{2}}{2}$

3. $\dfrac{\sqrt{30a^3}}{\sqrt{6a^2}} = \sqrt{\dfrac{30a^3}{6a^2}} = \sqrt{5a}$

Rationalizing Denominators

Expressions with radicals are considered simpler if there are no radicals in the denominators. We can simplify by multiplying by 1, but this time we do it a bit differently. We choose notation like $\sqrt{a}/\sqrt{a}$.

EXAMPLE 4 Rationalize the denominator.

$$\frac{\sqrt{2}}{\sqrt{3}} = \frac{\sqrt{2}}{\sqrt{3}} \cdot \frac{\sqrt{3}}{\sqrt{3}} = \frac{\sqrt{2} \cdot \sqrt{3}}{\sqrt{3} \cdot \sqrt{3}} = \frac{\sqrt{6}}{3}, \quad \text{or} \quad \frac{1}{3}\sqrt{6}$$

EXAMPLES Rationalize the denominator. Assume that all expressions under radicals represent positive numbers.

5. $\dfrac{\sqrt{49a^5}}{\sqrt{12}} = \dfrac{\sqrt{49a^5}}{\sqrt{12}} \cdot \dfrac{\sqrt{3}}{\sqrt{3}} = \dfrac{\sqrt{49a^5}\sqrt{3}}{\sqrt{12}\sqrt{3}} = \dfrac{\sqrt{49a^5}\sqrt{3}}{\sqrt{36}}$

$= \dfrac{\sqrt{49a^4 \cdot 3a}}{\sqrt{36}} = \dfrac{7a^2\sqrt{3a}}{6}$

6. $\dfrac{\sqrt{5}}{\sqrt{x}} = \dfrac{\sqrt{5}}{\sqrt{x}} \cdot \dfrac{\sqrt{x}}{\sqrt{x}}$ Multiplying by 1

$= \dfrac{\sqrt{5}\sqrt{x}}{\sqrt{x}\sqrt{x}}$

$= \dfrac{\sqrt{5x}}{x}$

EXERCISE SET 10.6

Divide and simplify. Assume that all expressions under radicals represent positive numbers.

1. $\dfrac{\sqrt{18}}{\sqrt{2}}$ 2. $\dfrac{\sqrt{20}}{\sqrt{5}}$ 3. $\dfrac{\sqrt{60}}{\sqrt{15}}$ 4. $\dfrac{\sqrt{108}}{\sqrt{3}}$

5. $\dfrac{\sqrt{75}}{\sqrt{15}}$ 6. $\dfrac{\sqrt{18}}{\sqrt{3}}$ 7. $\dfrac{\sqrt{3}}{\sqrt{75}}$ 8. $\dfrac{\sqrt{3}}{\sqrt{48}}$

9. $\dfrac{\sqrt{12}}{\sqrt{75}}$ 10. $\dfrac{\sqrt{18}}{\sqrt{32}}$ 11. $\dfrac{\sqrt{8x}}{\sqrt{2x}}$ 12. $\dfrac{\sqrt{18b}}{\sqrt{2b}}$

13. $\dfrac{\sqrt{63y^3}}{\sqrt{7y}}$ 14. $\dfrac{\sqrt{48x^3}}{\sqrt{3x}}$ 15. $\dfrac{\sqrt{15x^5}}{\sqrt{3x}}$ 16. $\dfrac{\sqrt{30a^5}}{\sqrt{5a}}$

17. $\dfrac{\sqrt{3x}}{\sqrt{\dfrac{3x}{4}}}$ 18. $\dfrac{\sqrt{5x}}{\sqrt{\dfrac{5x}{9}}}$ 19. $\dfrac{\sqrt{\dfrac{5}{6}}}{\sqrt{\dfrac{2}{3}}}$ 20. $\dfrac{\sqrt{\dfrac{3}{7}}}{\sqrt{\dfrac{3}{14}}}$

Rationalize the denominator. Assume that all variables under radicals represent positive numbers.

21. $\dfrac{\sqrt{7}}{\sqrt{3}}$ 22. $\dfrac{\sqrt{2}}{\sqrt{5}}$ 23. $\dfrac{\sqrt{9}}{\sqrt{8}}$ 24. $\dfrac{\sqrt{4}}{\sqrt{27}}$

25. $\dfrac{\sqrt{2}}{\sqrt{5}}$ 26. $\dfrac{\sqrt{3}}{\sqrt{2}}$ 27. $\dfrac{2}{\sqrt{2}}$ 28. $\dfrac{3}{\sqrt{3}}$

29. $\dfrac{\sqrt{5}}{\sqrt{11}}$ 30. $\dfrac{\sqrt{7}}{\sqrt{27}}$ 31. $\dfrac{\sqrt{7}}{\sqrt{12}}$ 32. $\dfrac{\sqrt{5}}{\sqrt{18}}$

33. $\dfrac{\sqrt{48}}{\sqrt{32}}$ 34. $\dfrac{\sqrt{56}}{\sqrt{40}}$ 35. $\dfrac{\sqrt{450}}{\sqrt{18}}$ 36. $\dfrac{\sqrt{224}}{\sqrt{14}}$

37. $\dfrac{\sqrt{3}}{\sqrt{x}}$ 38. $\dfrac{\sqrt{2}}{\sqrt{y}}$ 39. $\dfrac{4y}{\sqrt{3}}$ 40. $\dfrac{8x}{\sqrt{5}}$

41. $\dfrac{\sqrt{a^3}}{\sqrt{8}}$ 42. $\dfrac{\sqrt{x^3}}{\sqrt{27}}$ 43. $\dfrac{\sqrt{56}}{\sqrt{12x}}$ 44. $\dfrac{\sqrt{45}}{\sqrt{8a}}$

45. $\dfrac{\sqrt{27c}}{\sqrt{32c^3}}$ 46. $\dfrac{\sqrt{7x^3}}{\sqrt{12x}}$ 47. $\dfrac{\sqrt{y^5}}{\sqrt{xy^2}}$ 48. $\dfrac{\sqrt{x^3}}{\sqrt{xy}}$

49. $\dfrac{\sqrt{16a^4b^6}}{\sqrt{128a^6b^6}}$ 50. $\dfrac{\sqrt{45mn^2}}{\sqrt{32m}}$

Problem-solving practice

51. The weight J of an object on Jupiter varies directly as its weight E on earth. An object that weighs 225 kg on earth has a weight of 594 kg on Jupiter. What is the weight on Jupiter of an object that has a weight of 115 kg on earth?

52. The perimeter of a rectangle is 76 cm. The width is 17 cm less than the length. Find the area of the rectangle.

Perform the indicated operations and simplify. Assume that all expressions under radicals represent positive numbers. Rationalize denominators where appropriate.

53. $\dfrac{\sqrt{2}}{3\sqrt{3}}$ 54. $\dfrac{3\sqrt{6}}{6\sqrt{2}}$ 55. $\dfrac{5\sqrt{2}}{3\sqrt{5}}$ 56. $\dfrac{3\sqrt{15}}{5\sqrt{32}}$

57. $\dfrac{4\sqrt{\dfrac{6}{7}}}{\sqrt{\dfrac{12}{63}}}$ 58. $\dfrac{\sqrt{\dfrac{2}{3}}}{\sqrt{\dfrac{3}{2}}}$ 59. $\dfrac{\sqrt{\dfrac{3x}{2}}}{\sqrt{\dfrac{x^3}{5}}}$

10.7 ADDITION AND SUBTRACTION

We can add any two real numbers. The sum of 5 and $\sqrt{2}$ can be expressed as

$$5 + \sqrt{2}.$$

We cannot simplify this unless we use rational approximations. When we have *like radicals*, however, a sum can be simplified using the distributive laws and collecting like terms. *Like radicals* have the same radicands.

EXAMPLE 1 Add $3\sqrt{5}$ and $4\sqrt{5}$ and simplify by collecting like radical terms, if possible.

$$3\sqrt{5} + 4\sqrt{5} = (3 + 4)\sqrt{5} \quad \text{Using the distributive law to factor out } \sqrt{5}$$
$$= 7\sqrt{5}$$

To simplify like this, the radicands must be the same. Sometimes we can make them the same.

EXAMPLES Add or subtract. Simplify, if possible, by collecting like radical terms.

2. $\sqrt{2} - \sqrt{8} = \sqrt{2} - \sqrt{4 \cdot 2}$ Factoring 8
$= \sqrt{2} - \sqrt{4}\sqrt{2}$
$= \sqrt{2} - 2\sqrt{2}$
$= 1\sqrt{2} - 2\sqrt{2}$
$= (1 - 2)\sqrt{2}$ Using the distributive law to factor out the common factor $\sqrt{2}$
$= -1 \cdot \sqrt{2}$
$= -\sqrt{2}$

3. $\sqrt{x^3 - x^2} + \sqrt{4x - 4} = \sqrt{x^2(x-1)} + \sqrt{4(x-1)}$ Factoring radicands
$= \sqrt{x^2}\sqrt{x-1} + \sqrt{4}\sqrt{x-1}$
$= x\sqrt{x-1} + 2\sqrt{x-1}$ Assuming that all expressions under radicals are nonnegative
$= (x + 2)\sqrt{x-1}$ Using the distributive law to factor out the common factor $\sqrt{x-1}$

A sum of square roots is not the square root of a sum. For example,

$$\sqrt{9} + \sqrt{16} = 3 + 4$$
$$= 7$$

but

$$\sqrt{9 + 16} = \sqrt{25}$$
$$= 5.$$

In general,

$$\sqrt{a} + \sqrt{b} \neq \sqrt{a + b}.$$

Sometimes rationalizing denominators enables us to factor and then combine expressions.

EXAMPLE 4 Add: $\sqrt{3} + \sqrt{\frac{1}{3}}$.

$$\sqrt{3} + \sqrt{\frac{1}{3}} = \sqrt{3} + \sqrt{\frac{1}{3} \cdot \frac{3}{3}} \qquad \text{Multiplying by 1}$$

$$= \sqrt{3} + \sqrt{\frac{3}{9}}$$

$$= \sqrt{3} + \frac{\sqrt{3}}{\sqrt{9}}$$

$$= \sqrt{3} + \frac{\sqrt{3}}{3}$$

$$= 1 \cdot \sqrt{3} + \frac{1}{3}\sqrt{3}$$

$$= \left(1 + \frac{1}{3}\right)\sqrt{3} \qquad \text{Factoring and simplifying}$$

$$= \frac{4}{3}\sqrt{3}$$

EXERCISE SET 10.7

Add or subtract. Simplify by collecting like radical terms, if possible. Assume that all expressions under radicals represent nonnegative numbers.

1. $3\sqrt{2} + 4\sqrt{2}$
2. $8\sqrt{3} + 3\sqrt{3}$
3. $7\sqrt{5} - 3\sqrt{5}$
4. $8\sqrt{2} - 5\sqrt{2}$
5. $6\sqrt{x} + 7\sqrt{x}$
6. $9\sqrt{y} + 3\sqrt{y}$
7. $9\sqrt{x} - 11\sqrt{x}$
8. $6\sqrt{a} - 14\sqrt{a}$
9. $5\sqrt{8} + 15\sqrt{2}$
10. $3\sqrt{12} + 2\sqrt{3}$
11. $\sqrt{27} - 2\sqrt{3}$
12. $7\sqrt{50} - 3\sqrt{2}$
13. $\sqrt{45} - \sqrt{20}$
14. $\sqrt{27} - \sqrt{12}$
15. $\sqrt{72} + \sqrt{98}$
16. $\sqrt{45} + \sqrt{80}$
17. $2\sqrt{12} + \sqrt{27} - \sqrt{48}$
18. $9\sqrt{8} - \sqrt{72} + \sqrt{98}$
19. $3\sqrt{18} - 2\sqrt{32} - 5\sqrt{50}$
20. $\sqrt{18} - 3\sqrt{8} + \sqrt{50}$
21. $2\sqrt{27} - 3\sqrt{48} + 2\sqrt{18}$
22. $3\sqrt{48} - 2\sqrt{27} - 2\sqrt{18}$

23. $\sqrt{4x} + \sqrt{81x^3}$
24. $\sqrt{12x^2} + \sqrt{27}$
25. $\sqrt{27} - \sqrt{12x^2}$
26. $\sqrt{81x^3} - \sqrt{4x}$
27. $\sqrt{8x+8} + \sqrt{2x+2}$
28. $\sqrt{12x+12} + \sqrt{3x+3}$
29. $\sqrt{x^5 - x^2} + \sqrt{9x^3 - 9}$
30. $\sqrt{16x - 16} + \sqrt{25x^3 - 25x^2}$
31. $3x\sqrt{y^3 x} - x\sqrt{yx^3} + y\sqrt{y^3 x}$
32. $4a\sqrt{a^2 b} + a\sqrt{a^2 b^3} - 5\sqrt{b^3}$
33. $\sqrt{8(a+b)^3} - \sqrt{32(a+b)^3}$
34. $\sqrt{x^2 y + 6xy + 9y} + \sqrt{y^3}$
35. $\sqrt{3} - \sqrt{\frac{1}{3}}$
36. $\sqrt{2} - \sqrt{\frac{1}{2}}$
37. $5\sqrt{2} + 3\sqrt{\frac{1}{2}}$
38. $4\sqrt{3} + 2\sqrt{\frac{1}{3}}$
39. $\sqrt{\frac{2}{3}} - \sqrt{\frac{1}{6}}$
40. $\sqrt{\frac{1}{2}} - \sqrt{\frac{1}{8}}$
41. $\sqrt{\frac{1}{12}} - \sqrt{\frac{1}{27}}$
42. $\sqrt{\frac{5}{3}} - \sqrt{\frac{6}{5}}$

43. Three students were asked to simplify $\sqrt{10} + \sqrt{50}$. Their answers were $\sqrt{10}(1 + \sqrt{5})$, $\sqrt{10} + 5\sqrt{2}$, and $\sqrt{2}(5 + \sqrt{5})$.

a) Which, if any, is incorrect?

b) Which is in simplest form?

Add or subtract.

44. $\sqrt{125} - \sqrt{45} + 2\sqrt{5}$
45. $3\sqrt{\frac{1}{2}} + \frac{5}{2}\sqrt{18} + \sqrt{98}$
46. $\frac{3}{5}\sqrt{24} + \frac{2}{5}\sqrt{150} - \sqrt{96}$
47. $\frac{1}{3}\sqrt{27} + \sqrt{8} + \sqrt{300} - \sqrt{18} - \sqrt{162}$
48. $\sqrt{ab^6} + b\sqrt{a^3} + a\sqrt{a}$
49. $x\sqrt{2y} - \sqrt{8x^2 y} + \frac{x}{3}\sqrt{18y}$
50. $7x\sqrt{12xy^2} - 9y\sqrt{27x^3} + 5\sqrt{300x^3 y^2}$
51. $\sqrt{x} + \sqrt{\frac{1}{x}}$
52. $5\sqrt{\frac{3}{10}} + 2\sqrt{\frac{5}{6}} - 6\sqrt{\frac{15}{32}}$
53. $2\sqrt{\frac{2a}{b}} - 4\sqrt{\frac{b}{2a^3}} + 5\sqrt{\frac{1}{8}a^3 b}$
54. $\sqrt{1+x^2} + \frac{1}{\sqrt{1+x^2}}$

55. Can you find any pairs of numbers a and b, for which

$$\sqrt{a} + \sqrt{b} = \sqrt{a+b}?$$

If so, name them.

Multiply. Simplify by collecting like radical terms, if possible. Assume that all expressions under radicals represent nonnegative numbers.

56. $(\sqrt{5} + 7)(\sqrt{5} - 7)$
57. $(1 + \sqrt{5})(1 - \sqrt{5})$
58. $(\sqrt{6} - \sqrt{3})(\sqrt{6} + \sqrt{3})$
59. $(\sqrt{2} + \sqrt{6})(\sqrt{2} - \sqrt{6})$
60. $(3\sqrt{5} - 2)(\sqrt{5} + 1)$
61. $(\sqrt{5} - 2\sqrt{2})(\sqrt{10} - 1)$
62. $(\sqrt{x} - \sqrt{y})(\sqrt{x} + 5\sqrt{y})$
63. $(b^2 - b\sqrt{z} + z)(b + \sqrt{z})$

64. Evaluate for $a = 1$, $b = 3$, $c = 2$, $d = 4$.

a) $\sqrt{a^2 + c^2}$, $\sqrt{a^2} + \sqrt{c^2}$
b) $\sqrt{b^2 + c^2}$, $\sqrt{b^2} + \sqrt{c^2}$
c) $\sqrt{a^2 + d^2}$, $\sqrt{a^2} + \sqrt{d^2}$
d) $\sqrt{b^2 + d^2}$, $\sqrt{b^2} + \sqrt{d^2}$
e) $\sqrt{a^2 + b^2}$, $\sqrt{a^2} + \sqrt{b^2}$
f) $\sqrt{c^2 + d^2}$, $\sqrt{c^2} + \sqrt{d^2}$

65. Can you find any numbers that make the following true?

$$\sqrt{x^2 + y^2} = \sqrt{x^2} + \sqrt{y^2}$$

Observe that $(\sqrt{a} + \sqrt{b})(\sqrt{a} - \sqrt{b}) = (\sqrt{a})^2 - (\sqrt{b})^2 = a - b$. Use this to rationalize denominators in Exercises 66–69.

66. $\dfrac{2}{\sqrt{3} - \sqrt{5}}$

67. $\dfrac{1 - \sqrt{7}}{3 + \sqrt{7}}$

68. $\dfrac{\sqrt{3} - \sqrt{2}}{2\sqrt{3} + \sqrt{2}}$

69. $\dfrac{2 - \sqrt{x + y}}{3 + \sqrt{x + y}}$

10.8 RIGHT TRIANGLES AND PROBLEM SOLVING

We now learn a special property regarding right triangles. Then we use it with our skills in manipulating radical expressions to solve problems.

Right Triangles

In a right triangle, the longest side is called the *hypotenuse*. It is also the side opposite the right angle. The other two sides are called *legs*. We generally use the letters *a* and *b* for the lengths of the legs and *c* for the length of the hypotenuse. They are related as follows.

> **THE PYTHAGOREAN PROPERTY OF RIGHT TRIANGLES**
>
> **In any right triangle, if *a* and *b* are the lengths of the legs and *c* is the length of the hypotenuse, then**
>
> $$a^2 + b^2 = c^2.$$

The equation $a^2 + b^2 = c^2$ is called the Pythagorean equation.

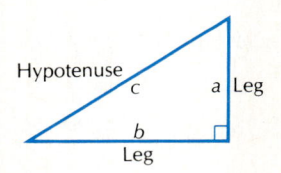

If we know the lengths of any two sides of a triangle, we can find the length of the third side.

EXAMPLE 1 Find the length of the hypotenuse of this right triangle. Give an exact answer and an approximation to three decimal places.

$$4^2 + 5^2 = c^2 \qquad \text{Substituting in the Pythagorean equation}$$
$$16 + 25 = c^2$$
$$41 = c^2$$
$$c = \sqrt{41}$$
$$c \approx 6.403 \qquad \text{Use your calculator.}$$

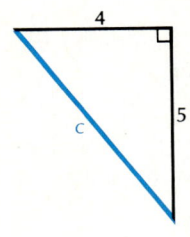

EXAMPLE 2 Find the length of the leg of this right triangle. Give an exact answer and an approximation to three decimal places.

$$10^2 + b^2 = 12^2 \qquad \text{Substituting in the Pythagorean equation}$$
$$100 + b^2 = 144$$
$$b^2 = 144 - 100 = 44$$
$$b = \sqrt{44}$$
$$b \approx 6.633 \qquad \text{Use your calculator.}$$

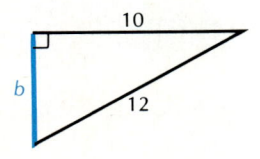

EXAMPLE 3 Find the length of the leg of this right triangle. Give an exact answer and an approximation to three decimal places.

$$1^2 + b^2 = (\sqrt{7})^2 \qquad \text{Substituting in the Pythagorean equation}$$
$$1 + b^2 = 7$$
$$b^2 = 7 - 1 = 6$$
$$b = \sqrt{6}$$
$$b \approx 2.449$$

10.8 RIGHT TRIANGLES AND PROBLEM SOLVING   421

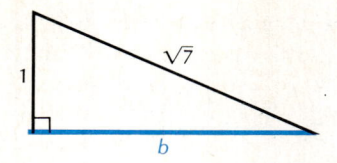

EXAMPLE 4 Find the length of the leg of this right triangle. Give an exact answer and an approximation to three decimal places.

$$a^2 + 10^2 = 15^2$$
$$a^2 + 100 = 225$$
$$a^2 = 225 - 100$$
$$a^2 = 125$$
$$a = \sqrt{125}$$
$$a \approx 11.180$$

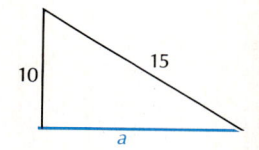

Problem Solving

The Pythagorean property can be used for problem solving. We continue to use our five-step process.

EXAMPLE 5 A 12-ft ladder is leaning against a building. The bottom of the ladder is 7 ft from the building. How high is the top of the ladder? Give an exact answer and an approximation to three decimal places.

1. **Familiarize** We first make a drawing. In it we see a right triangle. We label the unknown height h.

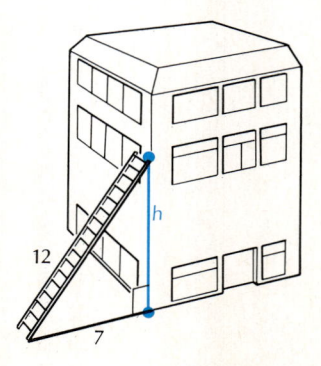

2. **Translate** Recall that in problem solving we may have to look up a formula, or we may use a formula that we know. In this case we use the Pythagorean property. We substitute 7 for a, h for b, and 12 for c in the Pythagorean equation:

$$7^2 + h^2 = 12^2.$$

This gives us a translation of the problem.

3. **Carry out** We solve the equation:

$$7^2 + h^2 = 144$$
$$49 + h^2 = 144$$
$$h^2 = 144 - 49$$
$$h^2 = 95$$

Exact answer: $\quad h = \sqrt{95}$

Approximation: $\quad h \approx 9.747$ ft.

4. **Check** We check by substituting 7, $\sqrt{95}$, and 12 into the Pythagorean equation:

$$\begin{array}{c|c} a^2 + b^2 = c^2 \\ \hline 7^2 + (\sqrt{95})^2 & 12^2 \\ 49 + 95 & 144 \\ 144 & \end{array}$$

We could have used the approximation, 9.747, but the resulting check would not have been exact.

5. **State** The top of the ladder is $\sqrt{95}$, or about 9.747 ft from the ground.

EXAMPLE 6 A slow-pitch softball diamond is a square 65 ft on a side. How far is it from home to second base? Give an exact answer and an approximation to three decimal places.

1. **Familiarize** We first make a drawing. We note that the first and second base lines, together with a line from home to second, form a right triangle. We label the unknown distance d.

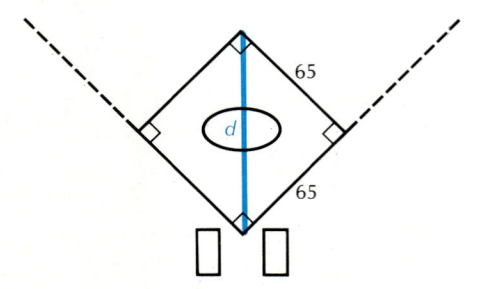

2. **Translate** We substitute 65 for a, 65 for b, and d for c in the Pythagorean

equation:
$$65^2 + 65^2 = d^2.$$

This gives us a translation of the problem.

3. **Carry out** We solve the equation:
$$65^2 + 65^2 = d^2$$
$$4225 + 4225 = d^2$$
$$8450 = d^2$$

Exact answer: $\sqrt{8450} = d$

Approximation: $91.924 \approx d.$

If you use Table 2 to find an approximation, you will need to simplify before finding an approximation in the table:
$$d = \sqrt{8450} = \sqrt{25 \cdot 169 \cdot 2} = \sqrt{25}\sqrt{169}\sqrt{2} \approx 5(13)(1.414) = 91.910.$$

Note that we get a variance in the last two decimal places.

4. **Check** We check by substituting 65, 65, and $\sqrt{8450}$ into the Pythagorean equation:

$$\begin{array}{c|c} a^2 + b^2 = c^2 & \\ \hline 65^2 + 65^2 & (\sqrt{8450})^2 \\ 4225 + 4225 & 8450 \\ 8450 & \end{array}$$

5. **State** The distance from home to second is about 91.924 ft.

EXERCISE SET 10.8

Find the length of the third side of each right triangle. Give an exact answer and an approximation to three decimal places.

1.

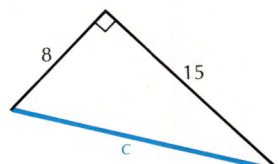

   Triangle with sides 8, 15, and c

2.

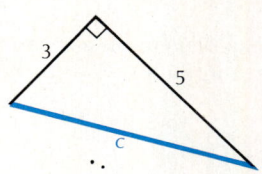

   Triangle with sides 3, 5, and c

3.

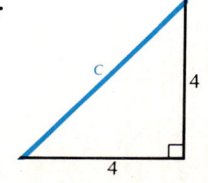

   Triangle with legs 4 and 4, hypotenuse c

4.

   Triangle with sides 7, 7, and c

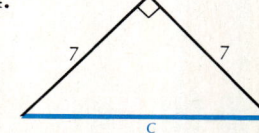

5.

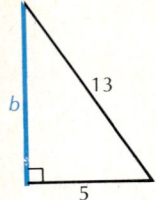

Triangle with legs 5 and b, hypotenuse 13.

6.

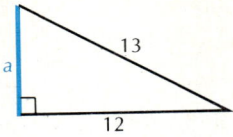

Triangle with legs a and 12, hypotenuse 13.

7.

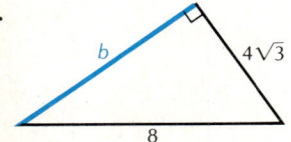

Right triangle with legs 8 and $4\sqrt{3}$, hypotenuse b.

8.

Right triangle with legs 6 and $\sqrt{5}$, hypotenuse b.

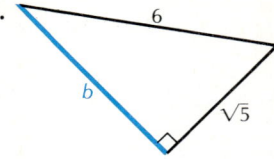

In a right triangle, find the length of the side not given. Give an exact answer and an approximation to three decimal places.

9. $a = 10, b = 24$

10. $a = 5, b = 12$

11. $a = 9, c = 15$

12. $a = 18, c = 30$

13. $b = 1, c = \sqrt{5}$

14. $b = 1, c = \sqrt{2}$

15. $a = 1, c = \sqrt{3}$

16. $a = \sqrt{3}, b = \sqrt{5}$

17. $c = 10, b = 5\sqrt{3}$

18. $a = 5, b = 5$

Problem solving

Don't forget to make drawings. Give an exact answer and an approximation to three decimal places.

19. A 10-m ladder is leaning against a building. The bottom of the ladder is 5 m from the building. How high is the top of the ladder?

20. Find the length of a diagonal of a square whose sides are 3 cm long.

21. How long is a guy wire reaching from the top of a 12-ft pole to a point 8 ft from the pole?

22. How long must a wire be to reach from the top of a 13-m telephone pole to a point on the ground 9 m from the foot of the pole?

23. A little league baseball diamond is a square 60 ft on a side. How far is it from home to second base?

24. A baseball diamond is a square 90 ft on a side. How far is it from first to third base?

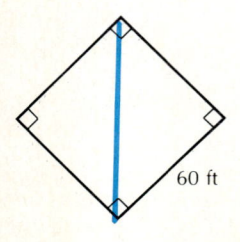

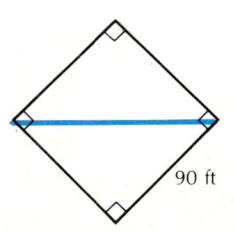

10.8 RIGHT TRIANGLES AND PROBLEM SOLVING

An *equilateral* triangle is shown at the right.

25. Find an expression for its height h in terms of a.

26. Find an expression for its area A in terms of a.

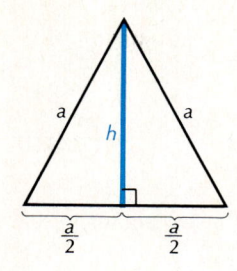

27. The length and width of a rectangle are given by consecutive integers. The area of the rectangle is 90 cm². Find the length of the diagonal of the rectangle.

28. Two cars leave a service station at the same time. One car travels east at a speed of 50 mph, and the other travels south at a speed of 60 mph. After one-half hour, how far apart are they?

29. Figure *ABCD* is a square. Find *AC*.

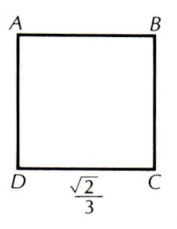

30. Suppose an outfielder catches the ball on the third-base line on a baseball diamond about 40 ft behind third base. About how far would the outfielder have to throw the ball to first base? (Be sure to make a drawing.)

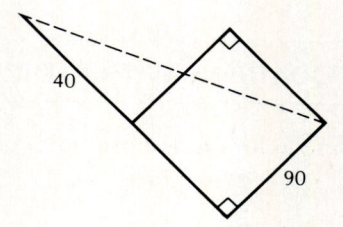

31. The diagonal of a square has a length of $8\sqrt{2}$ ft. Find the length of a side of the square.

32. Find the length of the diagonal of a rectangle whose length is 12 in. and whose width is 7 in.

33. A right triangle has sides whose lengths are consecutive integers. Find the lengths of the sides.

34. Find the length of the diagonal of a cube with side s.

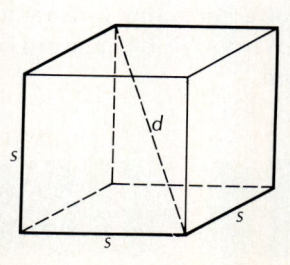

Find x.

35.

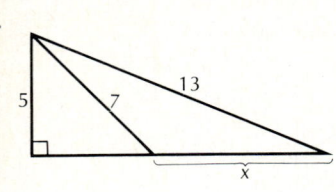

36.

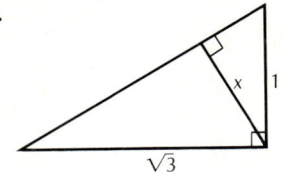

37.

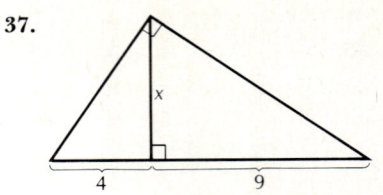

38. It is known that 2 miles of fencing will enclose a square plot of land whose area is 160 acres. How large a square, in acres, will 4 miles of fencing enclose?

39. The area of square PQRS is 100 ft², and A, B, C, and D are midpoints of the sides on which they lie. Find the area of square ABCD.

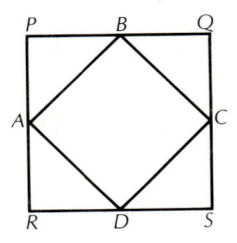

10.9 EQUATIONS WITH RADICALS

The following are examples of equations with radicals:

$$\sqrt{2x} - 4 = 7,$$
$$\sqrt{x+1} = \sqrt{2x-5}.$$

We now learn to solve such equations and to use them in problem solving.

Solving Equations with Radicals

To solve equations with radicals, we first convert them to equations without radicals. We do this by squaring both sides of the equation. The following new principle is used.

THE PRINCIPLE OF SQUARING

If an equation $a = b$ is true, then the equation $a^2 = b^2$ is true.

10.9 EQUATIONS WITH RADICALS

EXAMPLE 1 Solve: $\sqrt{2x} - 4 = 7$.

$$\sqrt{2x} - 4 = 7$$
$$\sqrt{2x} = 11 \qquad \text{Adding 4 to get the radical alone on one side}$$
$$(\sqrt{2x})^2 = 11^2 \qquad \text{Squaring both sides}$$
$$2x = 121$$
$$x = \frac{121}{2}$$

Check:
$$\begin{array}{c|c} \sqrt{2x} - 4 = 7 \\ \hline \sqrt{2 \cdot \frac{121}{2}} - 4 & 7 \\ \sqrt{121} - 4 & \\ 11 - 4 & \\ 7 & \end{array}$$

The solution is $\frac{121}{2}$.

It is important to check when using the principle of squaring. This principle may not produce equivalent equations. When we square both sides of an equation, the new equation may have solutions that the first one does not. For example, the equation

$$x = 1$$

has just *one* solution, the number 1. When we square both sides we get

$$x^2 = 1,$$

which has *two* solutions, 1 and -1. Thus the equations $x = 1$ and $x^2 = 1$ do not have the same solutions. They are not equivalent.

EXAMPLE 2 Solve: $\sqrt{x + 1} = \sqrt{2x - 5}$.

$$(\sqrt{x + 1})^2 = (\sqrt{2x - 5})^2 \qquad \text{Squaring both sides}$$
$$x + 1 = 2x - 5$$
$$x = 6$$

Since 6 checks, it is the solution.

Problem Solving

Radical equations can be used in problem solving. Let's look at the following example.

How far can you see from a given height? There is a formula for this. At a height of h meters you can see V kilometers to the horizon. These numbers are related as follows:

$$V = 3.5\sqrt{h}.$$

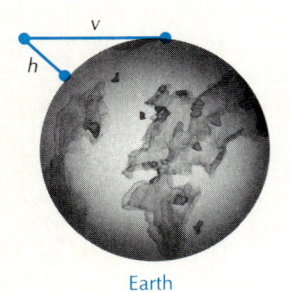

Earth

EXAMPLE 3 How far to the horizon can you see through an airplane window at a height, or altitude, of 9000 m?

We substitute 9000 for h in equation (1) and find an approximation.

Method 1. We use a calculator and approximate $\sqrt{9000}$ directly:
$$V = 3.5\sqrt{9000} \approx 3.5(94.868) = 332.038.$$

Method 2. We simplify and then approximate:
$$V = 3.5\sqrt{9000} = 3.5\sqrt{900 \cdot 10} = 3.5 \times 30 \times \sqrt{10}$$
$$V \approx 3.5 \times 30 \times 3.162 \approx 332.010 \text{ km}.$$

You can see about 332 km at 9000 m.

EXAMPLE 4 A person can see 50.4 km to the horizon from the top of a cliff. How high is the cliff?

1. **Familiarize** We first make a drawing.

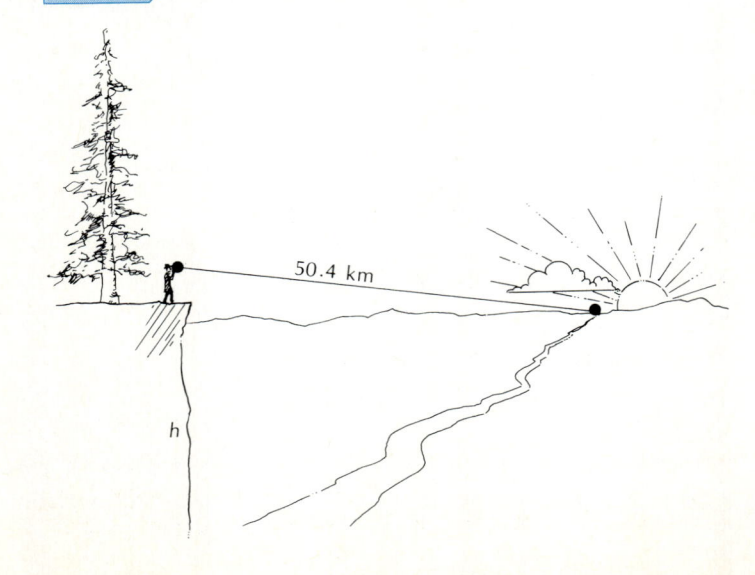

2. **Translate** We substitute 50.4 for V in the equation $V = 3.5\sqrt{h}$:

$$50.4 = 3.5\sqrt{h}.$$

3. **Carry out** We solve the equation for h:

$$50.4 = 3.5\sqrt{h}$$

$$\frac{50.4}{3.5} = \sqrt{h}$$

$$14.4 = \sqrt{h}$$

$$(14.4)^2 = (\sqrt{h})^2$$

$$207.36 \text{ m} = h.$$

4. **Check** We leave the check to the student.

5. **State** The cliff is about 207 m high.

EXERCISE SET 10.9

Solve.

1. $\sqrt{x} = 5$
2. $\sqrt{x} = 7$
3. $\sqrt{x} = 6.2$
4. $\sqrt{x} = 4.3$
5. $\sqrt{x+3} = 20$
6. $\sqrt{x+4} = 11$
7. $\sqrt{2x+4} = 25$
8. $\sqrt{2x+1} = 13$
9. $3 + \sqrt{x-1} = 5$
10. $4 + \sqrt{y-3} = 11$
11. $6 - 2\sqrt{3n} = 0$
12. $8 - 4\sqrt{5n} = 0$
13. $\sqrt{5x-7} = \sqrt{x+10}$
14. $\sqrt{4x-5} = \sqrt{x+9}$
15. $\sqrt{x} = -7$
16. $\sqrt{x} = -5$
17. $\sqrt{2y+6} = \sqrt{2y-5}$
18. $2\sqrt{3x-2} = \sqrt{2x-3}$

Problem solving

Use $V = 3.5\sqrt{h}$ for Exercises 19–22.

19. How far can you see to the horizon through an airplane window at a height of 9800 m?

20. How far can a sailor see to the horizon from the top of a 24-m mast?

21. A person can see 371 km to the horizon from an airplane window. How high is the airplane?

22. A sailor can see 99.4 km to the horizon from the top of a mast. How high is the mast?

The formula $r = 2\sqrt{5L}$ can be used to approximate the speed r, in mph, of a car that has left a skid mark of length L, in ft.

23. How far will a car skid at 50 mph? at 70 mph?

24. How far will a car skid at 60 mph? at 100 mph?

25. Find a number such that twice its square root is 14.

26. Find a number such that the additive inverse of 3 times its square root is -33.

27. Find a number such that the square root of 4 more than 5 times the number is 8.

The formula $T = 2\pi\sqrt{L/32}$ can be used to find the period T, in sec, of a pendulum of length L, in ft.

28. What is the length of a pendulum that has a period of 1.6 sec? Use 3.14 for π.

29. What is the length of a pendulum that has a period of 3 sec? Use 3.14 for π.

○

Solve.

30. $\sqrt{5x^2 + 5} = 5$

31. $\sqrt{x} = -x$

32. $x - 1 = \sqrt{x + 5}$

33. $\sqrt{y^2 + 6} + y - 3 = 0$

34. $\sqrt{x - 5} + \sqrt{x} = 5$
(Use the principle of squaring twice.)

35. $\sqrt{3x + 1} = 1 + \sqrt{x + 4}$

36. $4 + \sqrt{10 - x} = 6 + \sqrt{4 - x}$

37. $x = (x - 2)\sqrt{x}$

38. Solve $A = \sqrt{1 + a^2/b^2}$ for b.

The formula $t = \sqrt{2s/g}$ gives the time in seconds for an object, initially at rest, to fall s feet. Use this formula in Exercises 39–41.

39. Solve the formula for s.

40. If $g = 32.2$, find the distance an object falls in the first 5 seconds.

41. Find the distance an object falls in the first 10 seconds. Let $g = 32.2$.

42. A mountain climber stops at some point in the climb and views the horizon. The mountain climber uses the formula $V = 3.5\sqrt{h}$ to determine the distance to the horizon. After climbing another 100 m, the climber again computes the distance to the horizon, and finds that it is 20 km farther than before. At what height was the climber when the first computation was made? (*Hint:* Use a system of equations.)

SUMMARY AND REVIEW: CHAPTER 10

The following contains a summary of what you should be able to do after completing this chapter. The review exercises are for practice. Answers are at the back of the book. If you miss an exercise, restudy the section indicated alongside the answer.

You should be able to:

Find the square roots of a number and simplify radical expressions with a perfect-square radicand.

Find the square roots of each number.

1. 64   **2.** 25   **3.** 196   **4.** 400

Simplify.

5. $\sqrt{36}$ **6.** $-\sqrt{81}$ **7.** $\sqrt{49}$ **8.** $-\sqrt{169}$

Identify a given square root or other real number as rational or irrational.

Identify each number as rational or irrational.

9. $\sqrt{3}$ **10.** $\sqrt{36}$ **11.** $-\sqrt{12}$ **12.** $-\sqrt{4}$ **13.** $-\frac{4}{5}$ **14.** 5.56

15. 0.272727... (numeral repeats)   **16.** 0.313313331... (numeral does not repeat)

Use a calculator or Table 2 to approximate an expression involving a square root to three decimal places.

Approximate these square roots to three decimal places.

17. $\sqrt{3}$
18. $\sqrt{99}$
19. $\sqrt{108}$
20. $\sqrt{320}$
21. $\sqrt{\frac{1}{8}}$
22. $\sqrt{\frac{11}{20}}$

Identify the radicands of a radical expression.

Identify the radicand in each expression.

23. $\sqrt{x^2 + 4}$
24. $\sqrt{5ab^3}$

Identify a meaningless radical expression, and determine the sensible replacements in a radical expression.

Which of these expressions are meaningless? Write "yes" or "no."

25. $\sqrt{-22}$
26. $-\sqrt{49}$
27. $\sqrt{-36}$
28. $\sqrt{-100}$

Determine the sensible replacements in each expression.

29. $\sqrt{x + 7}$
30. $\sqrt{2y - 20}$

Simplify a radical expression with a perfect-square radicand.

Simplify. *Do not* assume expressions in radicands represent nonnegative numbers.

31. $\sqrt{m^2}$
32. $\sqrt{49t^2}$
33. $\sqrt{p^2}$
34. $\sqrt{(x-4)^2}$

Note: For the remainder of this review, *assume that all expressions in radicands represent nonnegative numbers.*

Multiply with radical notation and, where possible, simplify the result.

Multiply. Remember, radicands are nonnegative.

35. $\sqrt{3}\sqrt{7}$
36. $\sqrt{a}\sqrt{t}$
37. $\sqrt{x-3}\sqrt{x+3}$
38. $\sqrt{2x}\sqrt{3y}$

Simplify a radical expression by factoring.

Simplify by factoring.

39. $-\sqrt{48}$
40. $\sqrt{32t^2}$
41. $\sqrt{x^2 + 16x + 64}$
42. $\sqrt{t^2 - 49}$

Simplify a radical expression where the radicand is a power.

Simplify.

43. $\sqrt{x^8}$
44. $\sqrt{m^{15}}$

Multiply and simplify a radical expression.

Multiply and simplify.

45. $\sqrt{6}\sqrt{10}$
46. $\sqrt{5x}\sqrt{8x}$
47. $\sqrt{5x}\sqrt{10xy^2}$
48. $\sqrt{20a^3b}\sqrt{5a^2b^2}$

Simplify a radical expression with a fractional radicand.

Simplify.

49. $\sqrt{\frac{25}{64}}$
50. $\sqrt{\frac{20}{45}}$
51. $\sqrt{\frac{49}{t^2}}$

Rationalize the denominator of a radical expression and approximate the square roots of a fraction.

Rationalize the denominator.

52. $\sqrt{\frac{1}{2}}$
53. $\sqrt{\frac{1}{8}}$
54. $\sqrt{\frac{5}{y}}$
55. $\frac{2}{\sqrt{3}}$

Divide with radical notation and, where possible, simplify the result.

Divide and simplify.

56. $\dfrac{\sqrt{27}}{\sqrt{45}}$

57. $\dfrac{\sqrt{45x^2y}}{\sqrt{54y}}$

Add or subtract with radical notation, using the distributive law to simplify.

Add or subtract.

58. $10\sqrt{5} + 3\sqrt{5}$

59. $\sqrt{80} - \sqrt{45}$

60. $3\sqrt{2} - 5\sqrt{\tfrac{1}{2}}$

Given the lengths of any two sides of a right triangle, find the length of the third side, and solve problems involving right triangles.

In a right triangle, find the length of the side not given.

61. $a = 15$, $c = 25$

62. $a = 1$, $b = \sqrt{2}$

63. Find the length of the diagonal of a square whose sides are 7 m long.

Solve an equation with a radical.

Solve.

64. $\sqrt{x - 3} = 7$

65. $\sqrt{5x + 3} = \sqrt{2x - 1}$

Solve a problem involving the solution of a radical equation.

Solve.

66. The formula $r = 2\sqrt{5L}$ can be used to approximate the speed r, in mph, of a car that has left a skid mark of length L, in ft. How far will a car skid at 90 mph?

○ ──────────────────────────────────

67. Simplify: $\sqrt{\sqrt{\sqrt{256}}}$.

68. Solve: $\sqrt{x^2} = -10$.

69. Use square roots to factor $x^2 - 5$.

70. Solve $A = \sqrt{a^2 + b^2}$ for b.

TEST: CHAPTER 10

1. Find the square roots of 81.

Simplify.

2. $\sqrt{64}$

3. $-\sqrt{25}$

Identify each number as rational or irrational.

4. $-\sqrt{10}$

5. $\sqrt{16}$

6. $\tfrac{1}{4}$

7. $0.4324432444\ldots$ (numeral does not repeat)

8. $0.136136136\ldots$ (numeral repeats)

Approximate these expressions involving square roots to three decimal places.

9. $\sqrt{116}$

10. $\sqrt{87}$

11. $\dfrac{3}{\sqrt{3}}$

12. Identify the radicand in $\sqrt{4 - y^3}$.

TEST: CHAPTER 10

Which of these expressions are meaningless? Write "yes" or "no."

13. $\sqrt{24}$

14. $\sqrt{-23}$

15. Determine the sensible replacements in $\sqrt{8-x}$.

Simplify. *Do not* assume that expressions in radicands represent nonnegative numbers.

16. $\sqrt{a^2}$

17. $\sqrt{36y^2}$

Note: For the remainder of this test, *assume that all expressions in radicands represent nonnegative numbers.*

Multiply. Remember, radicands are nonnegative.

18. $\sqrt{5}\sqrt{6}$

19. $\sqrt{x-8}\sqrt{x+8}$

Simplify by factoring.

20. $\sqrt{27}$

21. $\sqrt{25x-25}$

22. $\sqrt{t^5}$

Multiply and simplify.

23. $\sqrt{5}\sqrt{10}$

24. $\sqrt{3ab}\sqrt{6ab^3}$

Simplify.

25. $\sqrt{\dfrac{27}{12}}$

26. $\sqrt{\dfrac{144}{a^2}}$

Rationalize the denominator.

27. $\sqrt{\dfrac{2}{5}}$

28. $\sqrt{\dfrac{2x}{y}}$

Divide and simplify.

29. $\dfrac{\sqrt{27}}{\sqrt{32}}$

30. $\dfrac{\sqrt{35x}}{\sqrt{80xy^2}}$

Add or subtract.

31. $3\sqrt{18} - 5\sqrt{18}$

32. $\sqrt{5} + \sqrt{\dfrac{1}{5}}$

33. In a right triangle, $a = 8$ and $b = 4$. Find c.

34. Solve: $\sqrt{3x+2} = 14$.

35. A person can see 247.49 km to the horizon from an airplane window. How high is the airplane? Use the formula $V = 3.5\sqrt{h}$.

○ ───

36. Solve: $\sqrt{1-x} + 1 = \sqrt{6-x}$.

37. Simplify: $\sqrt{y^{16n}}$.

QUADRATIC EQUATIONS

11

How long would it take an object to drop from the top of the Sears Tower in Chicago? A quadratic equation can be used to solve such a problem.

A quadratic equation contains a polynomial of second degree. In this chapter we first learn to solve quadratic equations by factoring. We actually began this study in Chapter 5. Certain quadratic equations are difficult to solve by factoring. For this reason, we learn to use the quadratic formula, which is a "recipe" for finding solutions of quadratic equations.

We apply our skills for solving quadratic equations to problem solving. Then we graph quadratic equations.

11.1 INTRODUCTION TO QUADRATIC EQUATIONS

The following are *quadratic equations*. They contain polynomials of second degree.

$$x^2 + 7x - 5 = 0, \quad 3t^2 - \tfrac{1}{2}t = 9, \quad 5y^2 = -6y, \quad 3m^2 = 0$$

In this section we learn about the standard form of a quadratic equation. We also learn to solve quadratic equations of the type $ax^2 + c = 0$ and to use them for problem solving.

Standard Form

The quadratic equation

$$4x^2 + 7x - 5 = 0$$

is said to be in *standard form*. The quadratic equation

$$4x^2 = 5 - 7x$$

is equivalent to the preceding equation, but it is not in standard form.

> An equation of the type $ax^2 + bx + c = 0$, where a, b, and c are real-number constants and $a > 0$, is called the *standard form of a quadratic equation*.

To write standard form for a quadratic equation, we find an equivalent equation that is in standard form.

EXAMPLES Write standard form and determine a, b, and c.

1. $4x^2 + 7x - 5 = 0$ The equation is already in standard form.
   $a = 4; b = 7; c = -5$

2. $3x^2 - 0.5x = 9$
   $3x^2 - 0.5x - 9 = 0$ Adding -9. This is standard form.
   $a = 3; b = -0.5; c = -9$

3. $-4y^2 = 5y$
   $-4y^2 - 5y = 0$ Adding $-5y$
       Not positive!
   $4y^2 + 5y = 0$ Multiplying by -1. This is standard form.
   $a = 4; b = 5; c = 0$

Solving Equations of the Type $ax^2 + c = 0$

When b is 0, we solve for x^2 and take the principal square root.

EXAMPLE 4 Solve: $5x^2 = 15$.

$$5x^2 = 15$$
$$x^2 = 3 \qquad \text{Solving for } x^2\text{; multiplying by } \tfrac{1}{5}$$
$$|x| = \sqrt{3}$$

Since $|x|$ is either x or $-x$,

$$x = \sqrt{3} \quad \text{or} \quad -x = \sqrt{3}$$
$$x = \sqrt{3} \quad \text{or} \quad x = -\sqrt{3}.$$

Check: For $\sqrt{3}$:

$$\begin{array}{c|c} 5x^2 = 15 \\ \hline 5(\sqrt{3})^2 & 15 \\ 5 \cdot 3 & \\ 15 & \end{array}$$

For $-\sqrt{3}$:

$$\begin{array}{c|c} 5x^2 = 15 \\ \hline 5(-\sqrt{3})^2 & 15 \\ 5 \cdot 3 & \\ 15 & \end{array}$$

The solutions are $\sqrt{3}$ and $-\sqrt{3}$.

EXAMPLE 5 Solve: $\tfrac{1}{3}x^2 = 0$.

$$\tfrac{1}{3}x^2 = 0$$
$$x^2 = 0 \qquad \text{Multiplying by 3}$$
$$|x| = \sqrt{0} \qquad \text{Taking the principal square root}$$
$$|x| = 0$$

The only number with absolute value 0 is 0. It checks, so it is the solution.

EXAMPLE 6 Solve: $-3x^2 + 7 = 0$.

$$-3x^2 + 7 = 0$$
$$-3x^2 = -7 \qquad \text{Adding } -7$$
$$x^2 = \frac{-7}{-3} \qquad \text{Multiplying by } -\frac{1}{3}$$
$$x^2 = \frac{7}{3}$$
$$|x| = \sqrt{\frac{7}{3}} \qquad \text{Taking the principal square root}$$
$$x = \sqrt{\frac{7}{3}} \quad \text{or} \quad x = -\sqrt{\frac{7}{3}}$$
$$x = \sqrt{\frac{7}{3} \cdot \frac{3}{3}} \quad \text{or} \quad x = -\sqrt{\frac{7}{3} \cdot \frac{3}{3}} \qquad \text{Rationalizing the denominators}$$
$$x = \frac{\sqrt{21}}{3} \quad \text{or} \quad x = -\frac{\sqrt{21}}{3}$$

Check:

$$\begin{array}{c|c} -3x^2 + 7 = 0 & -3x^2 + 7 = 0 \\ \hline -3\left(\dfrac{\sqrt{21}}{3}\right)^2 + 7 \; \bigg| \; 0 & -3\left(-\dfrac{\sqrt{21}}{3}\right)^2 + 7 \; \bigg| \; 0 \\ -3 \cdot \dfrac{21}{9} + 7 & -3 \cdot \dfrac{21}{9} + 7 \\ -7 + 7 & -7 + 7 \\ 0 & 0 \end{array}$$

The solutions are $\dfrac{\sqrt{21}}{3}$ and $-\dfrac{\sqrt{21}}{3}$.

Problem Solving

Quadratic equations can be used for problem solving.

EXAMPLE 7 The Sears Tower in Chicago is 1451 ft tall. How long would it take an object to fall from the top?

1. **Familiarize** If we did not know anything about this problem, we might consider looking up a formula in a mathematics or physics book. Actually, we studied such a formula earlier. A formula that fits this situation is

$$s = 16t^2,$$

where s is the distance, in feet, traveled by a body falling freely from rest in t seconds. This formula is actually an approximation in that it does not account for air resistance. In this problem we know the distance s to be 1451. We want the time t.

11.1 INTRODUCTION TO QUADRATIC EQUATIONS

$s = 16t^2$

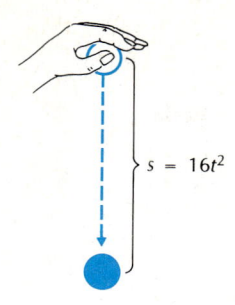

2. **Translate** We know the distance of 1451 and need to solve for t. We substitute 1451 for s:

$$1451 = 16t^2.$$

This gives us a translation.

3. **Carry out** We solve the equation:

$$1451 = 16t^2$$

$$\frac{1451}{16} = t^2 \qquad \text{Solving for } t^2$$

$$90.6875 = t^2 \qquad \text{Dividing}$$

$$\sqrt{90.6875} = |t| \qquad \text{Taking the principal square root}$$

$$9.5 \approx |t| \qquad \text{Using a calculator to find the square root and rounding to the nearest tenth}$$

$$9.5 \approx t \quad \text{or} \quad -9.5 \approx t.$$

4. **Check** The number -9.5 cannot be a solution because time cannot be negative in this situation. We substitute 9.5 in the original equation:

$$s = 16(9.5)^2 = 16(90.25) = 1444.$$

This is close. Remember we approximated a solution. Thus we have a check.

5. **State** It takes about 9.5 sec for the object to fall.

EXERCISE SET 11.1

Write standard form and determine a, b, and c.

1. $x^2 - 3x + 2 = 0$
2. $x^2 - 8x - 5 = 0$
3. $2x^2 = 3$
4. $5x^2 = 9$
5. $7x^2 = 4x - 3$
6. $9x^2 = x + 5$
7. $5 = -2x^2 + 3x$
8. $2x = x^2 - 5$
9. $2x - 1 = 3x^2 + 7$

QUADRATIC EQUATIONS

Solve.

10. $x^2 = 121$
11. $x^2 = 10$
12. $5x^2 = 35$
13. $3x^2 = 30$
14. $5x^2 = 3$
15. $2x^2 = 5$
16. $4x^2 - 25 = 0$
17. $9x^2 - 4 = 0$
18. $3x^2 - 49 = 0$
19. $5x^2 - 16 = 0$
20. $4y^2 - 3 = 9$
21. $49y^2 - 16 = 0$
22. $25y^2 - 36 = 0$
23. $5x^2 - 100 = 0$
24. $100x^2 - 5 = 0$

Problem solving

Use the formula $s = 16t^2$.

25. A body falls 1000 ft. How many seconds does this take?

26. A body falls 2496 ft. How many seconds does this take?

27. The world record for free-fall, to the ground without a parachute, by a woman is 175 ft and is held by Kitty O'Neill. Approximately how long did the fall take?

28. The world record for free-fall, to the ground without a parachute, by a man is 311 ft and is held by Dar Robinson. Approximately how long did the fall take?

Stuntwoman Kitty O'Neill, leaping 127 feet from the top of the Valley Hilton Hotel in Los Angeles during the filming of a television series. She established a new world record for a high fall for women.

Solve.

29. 🧮 $4.82x^2 = 12{,}000$
30. $\dfrac{x}{4} = \dfrac{9}{x}$
31. $1 = \dfrac{1}{3}x^2$

32. $\dfrac{x}{9} = \dfrac{36}{4x}$
33. $\dfrac{4}{m^2 - 7} = 1$

Solve for x.

34. $\dfrac{1}{4}x^2 + \dfrac{1}{6} = \dfrac{2}{3}x^2$
35. $\dfrac{4}{x^2 - 3} = \dfrac{6}{x^2}$
36. $3ax^2 - 9b = 3b^2$
37. $x^2 + 9a^2 = 9 + ax^2$

11.2 SOLVING BY FACTORING

Sometimes we can use factoring and the principle of zero products to solve quadratic equations.

Equations of the Type $ax^2 + bx = 0$

When c is 0 (and $b \neq 0$), we can factor and use the principle of zero products.

EXAMPLE 1 Solve: $7x^2 + 2x = 0$.

$$7x^2 + 2x = 0$$
$$x(7x + 2) = 0 \qquad \text{Factoring}$$
$$x = 0 \quad \text{or} \quad 7x + 2 = 0 \qquad \text{Principle of zero products}$$
$$x = 0 \quad \text{or} \quad 7x = -2$$
$$x = 0 \quad \text{or} \quad x = -\tfrac{2}{7}$$

Check:

$$\begin{array}{c|c} 7x^2 + 2x = 0 & 7x^2 + 2x = 0 \\ \hline 7 \cdot 0^2 + 2 \cdot 0 \ \big|\ 0 & 7(-\tfrac{2}{7})^2 + 2(-\tfrac{2}{7}) \ \big|\ 0 \\ 0 & 7(\tfrac{4}{49}) - \tfrac{4}{7} \\ & \tfrac{4}{7} - \tfrac{4}{7} \\ & 0 \end{array}$$

The solutions are 0 and $-\tfrac{2}{7}$.

When we use the principle of zero products, we need not check except to detect errors in solving.

EXAMPLE 2 Solve: $20x^2 - 15x = 0$.

$$20x^2 - 15x = 0$$
$$5x(4x - 3) = 0 \qquad \text{Factoring}$$
$$5x = 0 \quad \text{or} \quad 4x - 3 = 0 \qquad \text{Principle of zero products}$$
$$x = 0 \quad \text{or} \quad 4x = 3$$
$$x = 0 \quad \text{or} \quad x = \tfrac{3}{4}$$

The solutions are 0 and $\tfrac{3}{4}$.

A quadratic equation of this type will always have 0 as one solution and a non-zero number as the other solution.

Equations of the Type $ax^2 + bx + c = 0$

When neither b nor c is 0, we can sometimes solve by factoring.

EXAMPLE 3 Solve: $5x^2 - 8x + 3 = 0$.

$$5x^2 - 8x + 3 = 0$$
$$(5x - 3)(x - 1) = 0 \quad \text{Factoring}$$
$$5x - 3 = 0 \quad \text{or} \quad x - 1 = 0$$
$$5x = 3 \quad \text{or} \quad x = 1$$
$$x = \tfrac{3}{5} \quad \text{or} \quad x = 1$$

The solutions are $\tfrac{3}{5}$ and 1.

EXAMPLE 4 Solve: $(y - 3)(y - 2) = 6(y - 3)$.

We write standard form and then try to factor:

$$y^2 - 5y + 6 = 6y - 18 \quad \text{Multiplying}$$
$$y^2 - 11y + 24 = 0 \quad \text{Standard form}$$
$$(y - 8)(y - 3) = 0$$
$$y - 8 = 0 \quad \text{or} \quad y - 3 = 0$$
$$y = 8 \quad \text{or} \quad y = 3.$$

The solutions are 8 and 3.

EXAMPLE 5 The number of diagonals d of a polygon of n sides is given by the formula

$$d = \frac{n^2 - 3n}{2}.$$

If a polygon has 27 diagonals, how many sides does it have?

1. **Familiarize** We can make a drawing to become familiar with the problem. We draw an octagon (8 sides). We count the diagonals and see that there are 20. Let us check this in the formula. We evaluate the formula for $n = 8$:

$$d = \frac{8^2 - 3(8)}{2} = \frac{64 - 24}{2} = \frac{40}{2} = 20.$$

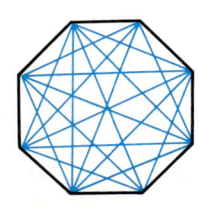

11.2 SOLVING BY FACTORING

2. **Translate** We know that the number of diagonals is 27. We substitute 27 for d:

$$27 = \frac{n^2 - 3n}{2}.$$

This gives us a translation.

3. **Carry out** We solve the equation for n. We first reverse the equation for convenience.

$$\frac{n^2 - 3n}{2} = 27$$

$$n^2 - 3n = 54 \quad \text{Multiplying by 2 to clear of fractions}$$

$$n^2 - 3n - 54 = 0$$

$$(n - 9)(n + 6) = 0$$

$$n - 9 = 0 \quad \text{or} \quad n + 6 = 0$$

$$n = 9 \quad \text{or} \quad n = -6$$

4. **Check** Since the number of sides cannot be negative, -6 cannot be a solution. We leave it to the student to show that 9 checks by substitution.

5. **State** The polygon has 9 sides (it is a nonagon).

EXERCISE SET 11.2

Solve.

1. $x^2 + 7x = 0$
2. $x^2 + 5x = 0$
3. $3x^2 + 6x = 0$
4. $4x^2 + 8x = 0$
5. $5x^2 = 2x$
6. $7x = 3x^2$
7. $4x^2 + 4x = 0$
8. $2x^2 - 2x = 0$
9. $0 = 10x^2 - 30x$
10. $0 = 10x^2 - 50x$
11. $11x = 55x^2$
12. $33x^2 = -11x$
13. $14t^2 = 3t$
14. $8m = 17m^2$
15. $5y^2 - 3y^2 = 72y + 9y$
16. $x^2 - 16x + 48 = 0$
17. $x^2 + 8x - 48 = 0$
18. $x^2 + 6 + 7x = 0$
19. $5 + 6x + x^2 = 0$
20. $t^2 + 4t = 21$
21. $18 = 7p + p^2$
22. $m^2 + 14 = 9m$
23. $-15 = -8y + y^2$
24. $x^2 + 10x + 25 = 0$
25. $x^2 + 6x + 9 = 0$
26. $x^2 + 1 = 2x$
27. $r^2 = 8r - 16$
28. $2x^2 - 13x + 15 = 0$
29. $6x^2 + x - 2 = 0$
30. $3a^2 = 10a + 8$
31. $15b - 9b^2 = 4$
32. $3x^2 - 7x = 20$
33. $6x^2 - 4x = 10$
34. $2t^2 + 12t = -10$
35. $12w^2 - 5w = 2$
36. $6z^2 + z - 1 = 0$
37. $t(t - 5) = 14$
38. $3y^2 + 8y = 12y + 15$
39. $t(9 + t) = 4(2t + 5)$
40. $(2x - 3)(x + 1) = 4(2x - 3)$
41. $16(p - 1) = p(p + 8)$
42. $(t - 1)(t + 3) = t - 1$
43. $(x - 2)(x + 2) = x + 2$
44. $m(3m + 1) = 2$

Problem solving

Use $d = \dfrac{n^2 - 3n}{2}$ for Exercises 45–48.

45. A hexagon is a figure with 6 sides. How many diagonals does a hexagon have?

46. A decagon is a figure with 10 sides. How many diagonals does a decagon have?

47. A polygon has 14 diagonals. How many sides does it have?

48. A polygon has 9 diagonals. How many sides does it have?

Solve.

49. $4m^2 - (m+1)^2 = 0$

50. $x^2 + \sqrt{3}\,x = 0$

51. $\sqrt{5}\,x^2 - x = 0$

52. $\sqrt{7}\,x^2 + \sqrt{3}\,x = 0$

53. $\dfrac{5}{y+4} - \dfrac{3}{y-2} = 4$

54. $\dfrac{2z+11}{2z+8} = \dfrac{3z-1}{z-1}$

55. Solve for x: $ax^2 + bx = 0$.

56. Solve: $0.0025x^2 + 70{,}400x = 0$.

Solve.

57. $y^4 - 4y^2 + 4 = 0$ (*Hint:* Let $x = y^2$. Then $x^2 = y^4$. Write a quadratic equation in x and solve. Remember to solve for y after finding x.)

58. $z - 10\sqrt{z} + 9 = 0$ (Let $x = \sqrt{z}$.)

11.3 COMPLETING THE SQUARE

The equation $(x-5)^2 = 9$ can be solved by taking the square root on both sides. We will see that other equations can be made to look like this one.

Solving Equations of the Type $(x + k)^2 = d$

In equations of the type $(x+k)^2 = d$, we have the square of a binomial equal to a constant.

EXAMPLE 1   Solve: $(x-5)^2 = 9$.

$$(x-5)^2 = 9$$
$$|x-5| = \sqrt{9} \quad \text{Taking the principal square root}$$
$$|x-5| = 3$$
$$x - 5 = 3 \quad \text{or} \quad x - 5 = -3$$
$$x = 8 \quad \text{or} \quad x = 2$$

The solutions are 8 and 2.

EXAMPLE 2 Solve: $(x + 2)^2 = 7$.

$$(x + 2)^2 = 7$$
$$|x + 2| = \sqrt{7}$$
$$x + 2 = \sqrt{7} \quad \text{or} \quad x + 2 = -\sqrt{7}$$
$$x = -2 + \sqrt{7} \quad \text{or} \quad x = -2 - \sqrt{7}$$

The solutions are $-2 + \sqrt{7}$ and $-2 - \sqrt{7}$, or simply $-2 \pm \sqrt{7}$ (read "-2 plus or minus $\sqrt{7}$").

Completing the Square

The following is the square of a binomial:

$$x^2 + 10x + 25.$$

An equivalent expression is $(x + 5)^2$. We could find the 25 from $10x$, by taking half the coefficient of x and squaring it.

We can make $x^2 + 10x$ the square of a binomial by adding the proper number to it. This is called *completing the square*.

$$x^2 + 10x$$
$$\downarrow$$
$$\frac{10}{2} = 5 \qquad \text{Taking half the } x\text{-coefficient}$$
$$\downarrow$$
$$5^2 = 25 \qquad \text{Squaring}$$
$$\downarrow$$
$$x^2 + 10x + 25 \qquad \text{Adding}$$

The trinomial $x^2 + 10x + 25$ is the square of $x + 5$.

EXAMPLES Complete the square.

3. $x^2 - 12x$

$$\left(\frac{-12}{2}\right)^2 = (-6)^2 = 36 \qquad \text{Taking half the } x\text{-coefficient and squaring}$$
$$\downarrow$$
$$x^2 - 12x + 36$$

The trinomial $x^2 - 12x + 36$ is the square of $x - 6$.

4. $x^2 - 5x$

$\left(\dfrac{-5}{2}\right)^2 = \dfrac{25}{4}$

$x^2 - 5x + \dfrac{25}{4}$

The trinomial $x^2 - 5x + \dfrac{25}{4}$ is the square of $x - \dfrac{5}{2}$.

Problem Solving: Compound Interest

We studied compound interest in Section 4.1. Recall that if you put money in a savings account, the bank will pay you interest. At the end of a year, the bank will start paying you interest on both the original amount and the interest. This is called *compounding interest annually*.

> If an amount of money P is invested at interest rate r, compounded annually, then in t years it will grow to the amount A given by
> $$A = P(1 + r)^t.$$

We can use quadratic equations to solve certain interest problems. Before we do that let us review the use of the preceding formula.

EXAMPLE 5 $1000 invested at 16% for 2 years compounded annually will grow to what amount?

$A = P(1 + r)^t$
$A = 1000(1 + 0.16)^2$ Substituting into the formula
$A = 1000(1.16)^2$
$A = 1000(1.3456)$
$A = 1345.60$ Computing

The amount is $1345.60.

EXAMPLE 6 $2560 is invested at interest rate r compounded annually. In 2 years it grows to $2890. What is the interest rate?

1. **Familiarize** We know that $2560 is originally invested. Thus P is $2560. That amount grows to $2890 in 2 years. Thus A is $2890 and t is 2.

2. **Translate** We substitute 2560 for P, 2890 for A, and 2 for t in the formula, and obtain a translation:

$A = P(1 + r)^t$
$2890 = 2560(1 + r)^2$.

3. **Carry out** We solve the equation:

$$2890 = 2560(1 + r)^2$$

$$\frac{2890}{2560} = (1 + r)^2$$

$$\frac{289}{256} = (1 + r)^2$$

$$\sqrt{\frac{289}{256}} = |1 + r| \qquad \text{Taking the principal square root}$$

$$\frac{17}{16} = |1 + r|.$$

We then have

$$\frac{17}{16} = 1 + r \quad \text{or} \quad -\frac{17}{16} = 1 + r$$

$$-\frac{16}{16} + \frac{17}{16} = r \quad \text{or} \quad -\frac{16}{16} - \frac{17}{16} = r$$

$$\frac{1}{16} = r \quad \text{or} \quad -\frac{33}{16} = r.$$

4. **Check** Since the interest rate cannot be negative,

$$\frac{1}{16} = r$$

$$0.0625 = r$$

or $6.25\% = r$. This checks in the formula.

5. **State** The interest rate must be 6.25% in order for $2560 to grow to $2890 in 2 years.

EXERCISE SET 11.3

Solve.
1. $(x - 2)^2 = 49$
2. $(x + 1)^2 = 6$
3. $(x + 3)^2 = 21$
4. $(x - 3)^2 = 6$
5. $(x + 13)^2 = 8$
6. $(x - 13)^2 = 64$
7. $(x - 7)^2 = 12$
8. $(x + 1)^2 = 14$
9. $(x + 9)^2 = 34$
10. $(t + 2)^2 = 25$
11. $(x + \frac{3}{2})^2 = \frac{7}{2}$
12. $(y - \frac{3}{4})^2 = \frac{17}{16}$

Complete the square.

13. $x^2 - 2x$
14. $x^2 - 4x$
15. $x^2 + 18x$
16. $x^2 + 22x$
17. $x^2 - x$
18. $x^2 + x$
19. $t^2 + 5t$
20. $y^2 - 9y$
21. $x^2 - \frac{3}{2}x$
22. $x^2 + \frac{4}{3}x$
23. $m^2 + \frac{9}{2}m$
24. $r^2 - \frac{2}{5}r$

Problem solving

Use $A = P(1 + r)^t$ for Exercises 25–32. What is the interest rate?

25. $1000 grows to $1210 in 2 years
26. $1000 grows to $1440 in 2 years
27. $2560 grows to $3610 in 2 years
28. $4000 grows to $4410 in 2 years
29. $6250 grows to $7290 in 2 years
30. $6250 grows to $6760 in 2 years
31. $2500 grows to $3600 in 2 years
32. $1600 grows to $2500 in 2 years

Factor the left side of the equation. Then solve.

33. $x^2 + 2x + 1 = 81$
34. $x^2 - 2x + 1 = 16$
35. $x^2 + 4x + 4 = 29$
36. $y^2 + 16y + 64 = 15$
37. $t^2 + 3t + \frac{9}{4} = \frac{49}{4}$
38. $m^2 - \frac{3}{2}m + \frac{9}{16} = \frac{17}{16}$
39. $9x^2 - 24x + 16 = 2$
40. $0.81x^2 + 0.36x + 0.04 = 5.76$
41. $64y^2 + 48y + 9 = 100$
42. In order for $2000 to double itself in 2 years, what would the interest rate have to be?
43. ▦ $1000 is invested at interest rate r. In 2 years it grows to $1267.88. What is the interest rate?
44. ▦ $4000 is invested at interest rate r. In 2 years it grows to $5267.03. What is the interest rate?
45. ▦ In two years you want to have $3000. How much do you need to invest now if you can get an interest rate of 15.75% compounded annually?

Solve (for x or y).

46. $\dfrac{x - 1}{9} = \dfrac{1}{x - 1}$
47. $(y - b)^2 = 4b^2$
48. $2(3x + 1)^2 = 8$
49. $5(5x - 2)^2 - 7 = 13$

Complete the square.

50. $x^2 - ax$
51. $x^2 + (2b - 4)x$
52. $ax^2 + bx$

11.4 SOLVING BY COMPLETING THE SQUARE

We have seen that a quadratic equation $(x + k)^2 = d$ can be solved by taking the principal square root on both sides. An equation such as $x^2 + 6x + 8 = 0$ can be put in this form by completing the square. Then we can solve as before.

11.4 SOLVING BY COMPLETING THE SQUARE

EXAMPLE 1 Solve: $x^2 + 6x + 8 = 0$.

$$x^2 + 6x + 8 = 0$$
$$x^2 + 6x = -8 \quad \text{Adding } -8$$

We take half of 6 and square it, to get 9. Then we add 9 on *both* sides of the equation. This makes the left side the square of a binomial. We have *completed the square*.

$$x^2 + 6x + 9 = -8 + 9$$
$$(x + 3)^2 = 1$$
$$|x + 3| = \sqrt{1} \quad \text{Taking the principal square root}$$
$$|x + 3| = 1$$
$$x + 3 = 1 \quad \text{or} \quad x + 3 = -1$$
$$x = -2 \quad \text{or} \quad x = -4$$

The solutions are -2 and -4.

This method of solving is called *completing the square*.

EXAMPLE 2 Solve $x^2 - 4x - 7 = 0$ by completing the square.

$$x^2 - 4x - 7 = 0$$
$$x^2 - 4x = 7 \quad \text{Adding 7}$$
$$x^2 - 4x + 4 = 7 + 4 \quad \text{Adding 4: } (\tfrac{-4}{2})^2 = (-2)^2 = 4$$
$$(x - 2)^2 = 11$$
$$|x - 2| = \sqrt{11}$$
$$x - 2 = \sqrt{11} \quad \text{or} \quad x - 2 = -\sqrt{11}$$
$$x = 2 + \sqrt{11} \quad \text{or} \quad x = 2 - \sqrt{11}$$

The solutions are $2 \pm \sqrt{11}$.

EXAMPLE 3 Solve $x^2 + 3x - 10 = 0$ by completing the square.

$$x^2 + 3x - 10 = 0$$
$$x^2 + 3x = 10$$
$$x^2 + 3x + \frac{9}{4} = 10 + \frac{9}{4} \quad \text{Adding } \frac{9}{4}: \left(\frac{3}{2}\right)^2 = \frac{9}{4}$$
$$\left(x + \frac{3}{2}\right)^2 = \frac{40}{4} + \frac{9}{4} = \frac{49}{4}$$
$$\left|x + \frac{3}{2}\right| = \sqrt{\frac{49}{4}} = \frac{7}{2}$$

We then have

$$x + \frac{3}{2} = \frac{7}{2} \quad \text{or} \quad x + \frac{3}{2} = -\frac{7}{2}$$

$$x = \frac{4}{2} \quad \text{or} \quad x = -\frac{10}{2}$$

$$x = 2 \quad \text{or} \quad x = -5.$$

The solutions are 2 and -5.

When the coefficient of x^2 is not 1, we can make it 1.

EXAMPLE 4 Solve $2x^2 - 3x - 1 = 0$ by completing the square.

$$2x^2 - 3x - 1 = 0$$

$$\frac{1}{2}(2x^2 - 3x - 1) = \frac{1}{2} \cdot 0 \qquad \text{Multiplying by } \frac{1}{2} \text{ to make the } x^2\text{-coefficient 1}$$

$$x^2 - \frac{3}{2}x - \frac{1}{2} = 0$$

$$x^2 - \frac{3}{2}x = \frac{1}{2}$$

$$x^2 - \frac{3}{2}x + \frac{9}{16} = \frac{1}{2} + \frac{9}{16} \qquad \text{Adding } \frac{9}{16}: \left[\frac{1}{2}\left(-\frac{3}{2}\right)\right]^2 = \left[-\frac{3}{4}\right]^2 = \frac{9}{16}$$

$$\left(x - \frac{3}{4}\right)^2 = \frac{8}{16} + \frac{9}{16}$$

$$\left(x - \frac{3}{4}\right)^2 = \frac{17}{16}$$

$$\left|x - \frac{3}{4}\right| = \sqrt{\frac{17}{16}}$$

$$\left|x - \frac{3}{4}\right| = \frac{\sqrt{17}}{4}$$

$$x - \frac{3}{4} = \frac{\sqrt{17}}{4} \quad \text{or} \quad x - \frac{3}{4} = -\frac{\sqrt{17}}{4}$$

$$x = \frac{3}{4} + \frac{\sqrt{17}}{4} \quad \text{or} \quad x = \frac{3}{4} - \frac{\sqrt{17}}{4}$$

The solutions are $\frac{3 \pm \sqrt{17}}{4}$.

EXERCISE SET 11.4

Solve by completing the square. Show your work.

1. $x^2 - 6x - 16 = 0$
2. $x^2 + 8x + 15 = 0$
3. $x^2 + 22x + 21 = 0$
4. $x^2 + 14x - 15 = 0$
5. $x^2 - 2x - 5 = 0$
6. $x^2 - 4x - 11 = 0$
7. $x^2 - 22x + 102 = 0$
8. $x^2 - 18x + 74 = 0$
9. $x^2 + 10x - 4 = 0$
10. $x^2 - 10x - 4 = 0$
11. $x^2 - 7x - 2 = 0$
12. $x^2 + 7x - 2 = 0$
13. $x^2 + 3x - 28 = 0$
14. $x^2 - 3x - 28 = 0$
15. $x^2 + \frac{3}{2}x - \frac{1}{2} = 0$
16. $x^2 - \frac{3}{2}x - 2 = 0$
17. $2x^2 + 3x - 17 = 0$
18. $2x^2 - 3x - 1 = 0$
19. $3x^2 + 4x - 1 = 0$
20. $3x^2 - 4x - 3 = 0$
21. $2x^2 - 9x - 5 = 0$
22. $2x^2 - 5x - 12 = 0$
23. $4x^2 + 12x - 7 = 0$
24. $6x^2 + 11x - 10 = 0$

Problem-solving practice

25. There were 12,000 people at a rock concert. A ticket cost $7.00 at the door and $6.50 if it was bought in advance. Total receipts were $81,165. How many people bought their tickets in advance?

26. It is known that 100 g of a certain kind of milk contains 3.5 g of protein. How many grams of protein are contained in 450 g of milk?

○ ─────────

Find b such that each trinomial is a square.

27. $x^2 + bx + 36$
28. $x^2 + bx + 55$
29. $x^2 + bx + 128$
30. $4x^2 + bx + 16$
31. $x^2 + bx + c$
32. $ax^2 + bx + c$

Solve for x by completing the square.

33. $x^2 - ax - 6a^2 = 0$
34. $x^2 + 4bx + 2b = 0$
35. $x^2 - x - c^2 - c = 0$
36. $3x^2 - bx + 1 = 0$
37. $ax^2 + 4x + 3 = 0$
38. $4x^2 + 4x + c = 0$
39. $kx^2 + mx + n = 0$
40. $b^2x^2 - 2bx + c^2 = 0$

11.5 THE QUADRATIC FORMULA

There are at least two reasons for learning to complete the square. One is to enhance your ability to graph certain second-degree equations, which you will encounter later in mathematics. The other is to prove a general formula that can be used to solve quadratic equations.

Solving by Using the Quadratic Formula

Each time you solve by completing the square, you do about the same thing. In situations like this in mathematics, when we do about the same kind of computation many times, we look for a formula so we can speed up our work. Consider any

quadratic equation in standard form:
$$ax^2 + bx + c = 0, \quad a > 0.$$
Let's solve by completing the square.

$$x^2 + \frac{b}{a}x + \frac{c}{a} = 0 \qquad \text{Multiplying by } \frac{1}{a}$$

$$x^2 + \frac{b}{a}x = -\frac{c}{a} \qquad \text{Adding } -\frac{c}{a}$$

Half of $\frac{b}{a}$ is $\frac{b}{2a}$. The square is $\frac{b^2}{4a^2}$. We add $\frac{b^2}{4a^2}$ on both sides.

$$x^2 + \frac{b}{a}x + \frac{b^2}{4a^2} = -\frac{c}{a} + \frac{b^2}{4a^2}$$

$$\left(x + \frac{b}{2a}\right)^2 = -\frac{4ac}{4a^2} + \frac{b^2}{4a^2}$$

$$\left(x + \frac{b}{2a}\right)^2 = \frac{b^2 - 4ac}{4a^2}$$

$$\left|x + \frac{b}{2a}\right| = \sqrt{\frac{b^2 - 4ac}{4a^2}} \qquad \text{Taking the principal square root}$$

$$x + \frac{b}{2a} = \sqrt{\frac{b^2 - 4ac}{4a^2}} \quad \text{or} \quad x + \frac{b}{2a} = -\sqrt{\frac{b^2 - 4ac}{4a^2}}$$

Since $a > 0$, $|a| = a$. Then

$$x + \frac{b}{2a} = \frac{\sqrt{b^2 - 4ac}}{2a} \quad \text{or} \quad x + \frac{b}{2a} = -\frac{\sqrt{b^2 - 4ac}}{2a}.$$

Thus,
$$x + \frac{b}{2a} = \pm\frac{\sqrt{b^2 - 4ac}}{2a},$$
so
$$x = -\frac{b}{2a} + \frac{\sqrt{b^2 - 4ac}}{2a} \quad \text{or} \quad x = -\frac{b}{2a} - \frac{\sqrt{b^2 - 4ac}}{2a}.$$

The solutions are given by the following.

• • • • • • • • • • • • • • • • • •

THE QUADRATIC FORMULA

$$x = \frac{-b \pm \sqrt{b^2 - 4ac}}{2a}$$

11.5 THE QUADRATIC FORMULA

EXAMPLE 1 Solve $5x^2 - 8x = -3$ using the quadratic formula.

First find standard form and determine a, b, and c:

$$5x^2 - 8x + 3 = 0$$
$$a = 5, \quad b = -8, \quad c = 3.$$

Then use the quadratic formula:

$$x = \frac{-b \pm \sqrt{b^2 - 4ac}}{2a}$$

$$x = \frac{-(-8) \pm \sqrt{(-8)^2 - 4 \cdot 5 \cdot 3}}{2 \cdot 5} \quad \text{Substituting}$$

$$x = \frac{8 \pm \sqrt{64 - 60}}{10}$$

$$x = \frac{8 \pm \sqrt{4}}{10}$$

$$x = \frac{8 \pm 2}{10}$$

$$x = \frac{8 + 2}{10} \quad \text{or} \quad x = \frac{8 - 2}{10}$$

$$x = \frac{10}{10} \quad \text{or} \quad x = \frac{6}{10}$$

$$x = 1 \quad \text{or} \quad x = \frac{3}{5}.$$

The solutions are 1 and $\frac{3}{5}$.

It turns out that the equation in Example 1 could have been solved by factoring, which would actually have been easier.

To solve a quadratic equation:

1. Try factoring.
2. If it is not possible to factor or if factoring seems difficult, use the quadratic formula.

The solutions of a quadratic equation can always be found using the quadratic formula. They cannot *always* be found by factoring. When $b^2 - 4ac \geq 0$, the equation has solutions. When $b^2 - 4ac < 0$, the equation has no real-number solutions. The expression $b^2 - 4ac$ is called the *discriminant*.

When using the quadratic formula, it is wise to compute the discriminant first. If it is negative, there are no real-number solutions.

EXAMPLE 2 Solve $3x^2 = 7 - 2x$ using the quadratic formula.

Find standard form and determine a, b, and c:
$$3x^2 + 2x - 7 = 0$$
$$a = 3, \quad b = 2, \quad c = -7.$$

We compute the discriminant:
$$b^2 - 4ac = 2^2 - 4 \cdot 3 \cdot (-7) = 4 + 84 = 88.$$

This is positive, so there are solutions. They are given by

$$x = \frac{-2 \pm \sqrt{88}}{6} \qquad \text{Substituting into the quadratic formula}$$

$$x = \frac{-2 \pm \sqrt{4 \cdot 22}}{6} = \frac{-2 \pm 2\sqrt{22}}{6}$$

$$x = \frac{2(-1 \pm \sqrt{22})}{2 \cdot 3} \qquad \text{Factoring out 2 in the numerator and denominator}$$

$$x = \frac{-1 \pm \sqrt{22}}{3}.$$

The solutions are $\dfrac{-1 + \sqrt{22}}{3}$ and $\dfrac{-1 - \sqrt{22}}{3}$.

Approximating Solutions

A calculator or Table 2 can be used to approximate solutions.

EXAMPLE 3 Use a calculator or Table 2 to approximate to the nearest tenth the solutions to the equation in Example 2.

Using a calculator or Table 2, we see that $\sqrt{22} \approx 4.690$:

$$\frac{-1 + \sqrt{22}}{3} \approx \frac{-1 + 4.690}{3} \qquad \frac{-1 - \sqrt{22}}{3} \approx \frac{-1 - 4.690}{3}$$

$$\approx \frac{3.69}{3} \qquad\qquad\qquad \approx \frac{-5.69}{3}$$

$$\approx 1.2 \quad \text{to the nearest tenth;} \qquad \approx -1.9 \quad \text{to the nearest tenth.}$$

EXERCISE SET 11.5

Solve. Try factoring first. If factoring is not possible or is difficult, use the quadratic formula.

1. $x^2 - 4x = 21$
2. $x^2 + 7x = 18$
3. $x^2 = 6x - 9$
4. $x^2 = 8x - 16$
5. $3y^2 - 2y - 8 = 0$
6. $3y^2 - 7y + 4 = 0$
7. $4x^2 + 12x = 7$
8. $4x^2 + 4x = 15$
9. $x^2 - 9 = 0$
10. $x^2 - 4 = 0$
11. $x^2 - 2x - 2 = 0$
12. $x^2 - 4x - 7 = 0$
13. $y^2 - 10y + 22 = 0$
14. $y^2 + 6y - 1 = 0$
15. $x^2 + 4x + 4 = 7$
16. $x^2 - 2x + 1 = 5$
17. $3x^2 + 8x + 2 = 0$
18. $3x^2 - 4x - 2 = 0$
19. $2x^2 - 5x = 1$
20. $2x^2 + 2x = 3$
21. $4y^2 - 4y - 1 = 0$
22. $4y^2 + 4y - 1 = 0$
23. $3x^2 + 5x = 0$
24. $5x^2 - 2x = 0$
25. $2t^2 + 6t + 5 = 0$
26. $4y^2 + 3y + 2 = 0$
27. $4x^2 = 100$
28. $5t^2 = 80$
29. $3x^2 = 5x + 4$
30. $2x^2 + 3x = 1$
31. $2y^2 - 6y = 10$
32. $5m^2 = 3 + 11m$
33. $3p^2 + 2p = 3$

Solve using the quadratic formula. Use a calculator or Table 2 to approximate the solutions to the nearest tenth.

34. $x^2 - 4x - 7 = 0$
35. $x^2 + 2x - 2 = 0$
36. $y^2 - 6y - 1 = 0$
37. $y^2 + 10y + 22 = 0$
38. $4x^2 + 4x = 1$
39. $4x^2 = 4x + 1$
40. $3x^2 + 4x - 2 = 0$
41. $3x^2 - 8x + 2 = 0$
42. $2y^2 + 2y - 3 = 0$

○ ─────────────

Solve for x or y.

43. $5x + x(x - 7) = 0$
44. $x(3x + 7) - 3x = 0$
45. $3 - x(x - 3) = 4$
46. $x(5x - 7) = 1$
47. $(y + 4)(y + 3) = 15$
48. $(y + 5)(y - 1) = 27$
49. $x^2 + (x + 2)^2 = 7$
50. $x^2 + (x + 1)^2 = 5$
51. $(x + 2)^2 + (x + 1)^2 = 0$
52. $(x + 3)^2 + (x + 1)^2 = 0$
53. $ax^2 + 2x = 3$
54. $2bx^2 - 5x + 3b = 0$
55. $4x^2 - 4cx + c^2 - 3d^2 = 0$
56. $0.8x^2 + 0.16x - 0.09 = 0$
57. $bdx^2 + bcx - ac = adx$
58. $\frac{1}{2}x^2 + bx + (b - \frac{1}{2}) = 0$

59. a) In $ax^2 + bx + c = 0$, $b^2 > 4ac$. Will the equation have real-number solutions? Does it make any difference whether b is positive, negative, or zero?
    b) In $ax^2 + bx + c = 0$, $ac < 0$. Will the equation have real-number solutions? Does it make any difference whether b is positive, negative, or zero?
    c) In $ax^2 + bx + c = 0$, a and c are both positive. Under what conditions will the equation have real-number solutions?

60. Use the two roots given by the quadratic formula to find a formula for the sum of the solutions of any quadratic equation. What is the product of the solutions? Without solving, find the sum and product of the solutions of $2x^2 + 5x - 3 = 0$.

61. One solution of the equation $2x^2 + bx - 3 = 0$ is known to be -5. Use the results of the preceding exercise to find the other solution.

11.6 FRACTIONAL AND RADICAL EQUATIONS

In solving a fractional or radical equation, we may obtain a quadratic equation after a few steps. When that happens, we know how to finish solving.

Fractional Equations

Recall that to solve a fractional equation we multiply on both sides by the LCM of all the denominators.

EXAMPLE 1 Solve: $\dfrac{3}{x-1} + \dfrac{5}{x+1} = 2$.

The LCM is $(x-1)(x+1)$. We multiply by this:

$$(x-1)(x+1) \cdot \left(\dfrac{3}{x-1} + \dfrac{5}{x+1}\right) = 2 \cdot (x-1)(x+1)$$

We use the distributive law on the left.

$$(x-1)(x+1) \cdot \dfrac{3}{x-1} + (x-1)(x+1) \cdot \dfrac{5}{x+1} = 2(x-1)(x+1)$$

$$3(x+1) + 5(x-1) = 2(x-1)(x+1)$$
$$3x + 3 + 5x - 5 = 2(x^2 - 1)$$
$$8x - 2 = 2x^2 - 2$$
$$-2x^2 + 8x = 0$$
$$2x^2 - 8x = 0 \qquad \text{Multiplying by } -1$$
$$2x(x - 4) = 0 \qquad \text{Factoring}$$
$$2x = 0 \quad \text{or} \quad x - 4 = 0$$
$$x = 0 \quad \text{or} \quad x = 4$$

Both numbers check. The solutions are 0 and 4.

Radical Equations

We can solve some radical equations by first using the principle of squaring to find a quadratic equation. When we do this we must be sure to check.

EXAMPLE 2 Solve: $x - 5 = \sqrt{x + 7}$.

$$x - 5 = \sqrt{x + 7}$$
$$(x - 5)^2 = (\sqrt{x + 7})^2 \quad \text{Principle of squaring}$$
$$x^2 - 10x + 25 = x + 7$$
$$x^2 - 11x + 18 = 0$$
$$(x - 9)(x - 2) = 0$$
$$x - 9 = 0 \quad \text{or} \quad x - 2 = 0$$
$$x = 9 \quad \text{or} \quad x = 2$$

Check:

$x - 5 = \sqrt{x + 7}$		$x - 5 = \sqrt{x + 7}$	
$9 - 5$	$\sqrt{9 + 7}$	$2 - 5$	$\sqrt{2 + 7}$
4	4	-3	3

The number 9 checks, but 2 does not. Thus the solution is 9.

EXAMPLE 3 Solve: $\sqrt{27 - 3x} + 3 = x$.

$$\sqrt{27 - 3x} + 3 = x$$
$$\sqrt{27 - 3x} = x - 3 \quad \text{Adding } -3 \text{ to get the radical alone on one side}$$
$$(\sqrt{27 - 3x})^2 = (x - 3)^2 \quad \text{Principle of squaring}$$
$$27 - 3x = x^2 - 6x + 9$$
$$0 = x^2 - 3x - 18 \quad \text{We can have 0 on the left.}$$
$$0 = (x - 6)(x + 3) \quad \text{Factoring}$$
$$x - 6 = 0 \quad \text{or} \quad x + 3 = 0$$
$$x = 6 \quad \text{or} \quad x = -3$$

Check:

$\sqrt{27 - 3x} + 3 = x$		$\sqrt{27 - 3x} + 3 = x$	
$\sqrt{27 - 3 \cdot 6} + 3$	6	$\sqrt{27 - 3 \cdot (-3)} + 3$	-3
$\sqrt{9} + 3$		$\sqrt{27 + 9} + 3$	
$3 + 3$		$\sqrt{36} + 3$	
6		$6 + 3$	
		9	

There is only one solution, 6.

EXERCISE SET 11.6

Solve.

1. $\dfrac{8}{x+2} + \dfrac{8}{x-2} = 3$

2. $\dfrac{24}{x-2} + \dfrac{24}{x+2} = 5$

3. $\dfrac{1}{x} + \dfrac{1}{x+6} = \dfrac{1}{4}$

4. $\dfrac{1}{x} + \dfrac{1}{x+9} = \dfrac{1}{20}$

5. $1 + \dfrac{12}{x^2-4} = \dfrac{3}{x-2}$

6. $\dfrac{5}{t-3} - \dfrac{30}{t^2-9} = 1$

7. $\dfrac{r}{r-1} + \dfrac{2}{r^2-1} = \dfrac{8}{r+1}$

8. $\dfrac{x+2}{x^2-2} = \dfrac{2}{3-x}$

9. $\dfrac{4-x}{x-4} + \dfrac{x+3}{x-3} = 0$

10. $\dfrac{y+2}{y} = \dfrac{1}{y+2}$

11. $\dfrac{x^2}{x-4} - \dfrac{7}{x-4} = 0$

12. $\dfrac{x^2}{x+3} - \dfrac{5}{x+3} = 0$

13. $x + 2 = \dfrac{3}{x+2}$

14. $x - 3 = \dfrac{5}{x-3}$

15. $\dfrac{1}{x} + \dfrac{1}{x+6} = \dfrac{1}{5}$

16. $\dfrac{1}{x} + \dfrac{1}{x+1} = \dfrac{1}{3}$

17. $x - 7 = \sqrt{x-5}$

18. $\sqrt{x+7} = x - 5$

19. $\sqrt{x+18} = x - 2$

20. $x - 9 = \sqrt{x-3}$

21. $2\sqrt{x-1} = x - 1$

22. $x + 4 = 4\sqrt{x+1}$

23. $\sqrt{5x+21} = x + 3$

24. $\sqrt{27-3x} = x - 3$

25. $x = 1 + 6\sqrt{x-9}$

26. $\sqrt{2x-1} + 2 = x$

27. $\sqrt{x^2+6} - x + 3 = 0$

28. $\sqrt{x^2+5} - x + 2 = 0$

29. $\sqrt{(p+6)(p+1)} - 2 = p + 1$

30. $\sqrt{(4x+5)(x+4)} = 2x + 5$

Problem-solving practice

31. A student rode a bike 10 mi to the university from home. On the way back home, the bike had a flat tire. The student walked the rest of the way home. When riding, the speed of the student was 15 mph. When walking, the speed was 5 mph. The return trip took 1 hr and 36 min. How far did the student have to walk?

32. Dried apricots are 5% protein and dried prunes are 2% protein. How much of each type of fruit should be used to make a 100-g mixture that is 3% protein?

○ ─────────

Solve.

33. $\dfrac{4}{x} - \dfrac{5}{2x+3} = 2$

34. $\dfrac{7}{1+x} - 1 = \dfrac{5x}{x^2+3x+2}$

35. $\dfrac{x}{x+1} = 4 + \dfrac{1}{3x^2-3}$

36. $\dfrac{1}{1+x} - 1 = \dfrac{5x}{x^2+3x+2}$

37. $\sqrt{x+3} = \dfrac{8}{\sqrt{x-9}}$

38. $\dfrac{12}{\sqrt{5x+6}} = \sqrt{2x+5}$

39. $2\sqrt{x-1} - \sqrt{3x-5} = \sqrt{x-9}$

40. $\sqrt{y+1} - \sqrt{2y-5} = \sqrt{y-2}$

41. Using two methods, solve $x + 1 + 3\sqrt{x+1} - 4 = 0$.

a) First, use the principle of squaring. (Remember to check.)

b) Second, let $y = \sqrt{x+1}$. (Then $y^2 = x+1$.) Write a quadratic equation in y and solve. (Remember to solve for x after finding y.)

11.7 FORMULAS

We now solve formulas involving quadratic, radical, and fractional equations. Recall that in order to solve a formula for a given letter, we try to get the letter alone on one side.

EXAMPLE 1 Solve for h: $V = 3.5\sqrt{h}$ (the distance to the horizon).

$$V = 3.5\sqrt{h}$$
$$V^2 = (3.5)^2(\sqrt{h})^2 \quad \text{Squaring both sides}$$
$$V^2 = 12.25h$$
$$\dfrac{V^2}{12.25} = h \quad \text{Multiplying by } \dfrac{1}{12.25} \text{ to get } h \text{ alone}$$

EXAMPLE 2 Solve for g: $T = 2\pi\sqrt{\dfrac{L}{g}}$ (the period of a pendulum).

$$T = 2\pi\sqrt{\dfrac{L}{g}}$$
$$T^2 = (2\pi)^2\left(\sqrt{\dfrac{L}{g}}\right)^2 \quad \text{Squaring both sides}$$
$$T^2 = 4\pi^2\dfrac{L}{g} = \dfrac{4\pi^2 L}{g}$$
$$gT^2 = 4\pi^2 L \quad \text{Multiplying by } g \text{ to clear of fractions}$$
$$g = \dfrac{4\pi^2 L}{T^2} \quad \text{Multiplying by } \dfrac{1}{T^2} \text{ to get } g \text{ alone}$$

In most formulas the letters represent nonnegative numbers, so you don't need to use absolute values when taking square roots.

EXAMPLE 3 (*Torricelli's theorem*). The speed v of a liquid leaving a tank from an orifice is related to the height h of the top of the water above the orifice by the formula

$$h = \frac{v^2}{2g}.$$

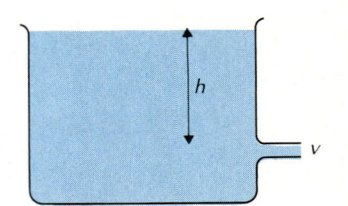

Solve for v:

$$h = \frac{v^2}{2g}$$

$2gh = v^2$ Multiplying by $2g$ to clear of fractions

$\sqrt{2gh} = v.$ Taking the square root

EXAMPLE 4 Solve for r: $A = P(1 + r)^2$ (a compound interest formula).

$$A = P(1 + r)^2$$

$\dfrac{A}{P} = (1 + r)^2$ Multiplying by $\dfrac{1}{P}$

$\sqrt{\dfrac{A}{P}} = 1 + r$ Taking the square root

$-1 + \sqrt{\dfrac{A}{P}} = r$ Adding -1 to get r alone

    Sometimes you need to use the quadratic formula to solve a formula for a certain letter.

EXAMPLE 5 Solve for n: $d = \dfrac{n^2 - 3n}{2}$ (the number of diagonals of a polygon).

$$d = \dfrac{n^2 - 3n}{2}$$

$n^2 - 3n = 2d$ Multiplying by 2 to clear of fractions

$n^2 - 3n - 2d = 0$ Finding standard form

$a = 1, \quad b = -3, \quad c = -2d$ All letters are considered constants except n.

$n = \dfrac{-b \pm \sqrt{b^2 - 4ac}}{2a}$

$n = \dfrac{-(-3) \pm \sqrt{(-3)^2 - 4 \cdot 1 \cdot (-2d)}}{2 \cdot 1}$ Substituting into the quadratic formula

$n = \dfrac{3 \pm \sqrt{9 + 8d}}{2}$

EXAMPLE 6 Solve for t: $S = gt + 16t^2$.

$$S = gt + 16t^2$$

$16t^2 + gt - S = 0$ Finding standard form

$a = 16, \quad b = g, \quad c = -S$

$t = \dfrac{-b \pm \sqrt{b^2 - 4ac}}{2a}$

$t = \dfrac{-g \pm \sqrt{g^2 - 4 \cdot 16 \cdot (-S)}}{2 \cdot 16}$ Substituting into the quadratic formula

$t = \dfrac{-g \pm \sqrt{g^2 + 64S}}{32}$

EXERCISE SET 11.7

Solve for the indicated letter.

1. $N = 2.5\sqrt{A}$; for A
2. $T = 2\pi\sqrt{\dfrac{L}{32}}$; for L
3. $Q = \sqrt{\dfrac{aT}{c}}$; for T
4. $v = \sqrt{\dfrac{2gE}{m}}$; for E
5. $E = mc^2$; for c
6. $S = 4\pi r^2$; for r
7. $Q = ad^2 - cd$; for d
8. $P = kA^2 + mA$; for A

9. $c^2 = a^2 + b^2$; for a

10. $c = \sqrt{a^2 + b^2}$; for b

11. $s = 16t^2$; for t

12. $V = \pi r^2 h$; for r

13. $A = \pi r^2 + 2\pi rh$; for r

14. $A = 2\pi r^2 + 2\pi rh$; for r

15. $A = \dfrac{\pi r^2 S}{360}$; for r

16. $H = \dfrac{D^2 N}{2.5}$; for D

17. $c = \sqrt{a^2 + b^2}$; for a

18. $c^2 = a^2 + b^2$; for b

19. $h = \dfrac{a}{2}\sqrt{3}$; for a

(the height of an equilateral triangle with sides of length a)

20. $d = s\sqrt{2}$; for s

(the hypotenuse of an isosceles right triangle with s the length of the two sides that have the same length)

21. The circumference C of a circle is given by $C = 2\pi r$.
   a) Solve $C = 2\pi r$ for r.
   b) The area is given by $A = \pi r^2$. Express the area in terms of the circumference C.

22. In reference to Exercise 21, express the circumference C in terms of the area A.

23. Solve $n = aT^2 - 4T + m$ for T.

24. Solve $y = ax^2 + bx + c$ for x.

25. Solve $3ax^2 - x - 3ax + 1 = 0$ for x.

26. Solve $6y^2 - 8xy = 9y - 3x - 2x^2$ for x in terms of y.

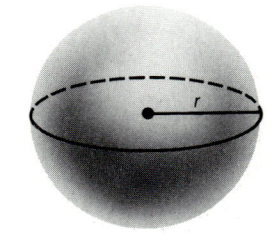

27. The volume V of a sphere with radius r is given by
$$V = \tfrac{4}{3}\pi r^3.$$
Solve for r.

11.8 PROBLEM SOLVING

We now use our new equation-solving procedures and the five-step process for problem solving.

EXAMPLE 1 A picture frame measures 20 cm by 14 cm. 160 square centimeters of picture shows. Find the width of the frame.

1. **Familiarize** We first make a drawing and label it with both known and unknown information. We have let x represent the width of the frame. The length of the frame is 20 cm and the width is 14 cm. The length of the picture is $20 - 2x$, and the width of the picture is $14 - 2x$.

11.8 PROBLEM SOLVING

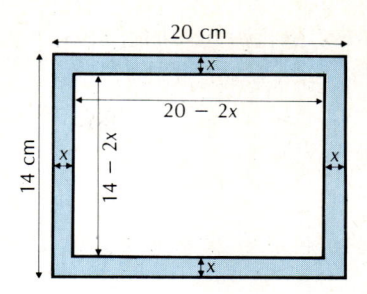

[Figure: Rectangle 20 cm by 14 cm with inner rectangle (20 − 2x) by (14 − 2x), border width x.]

2. **Translate** Recall that area is length × width. Thus we have two expressions for the area of the picture: $(20 - 2x)(14 - 2x)$ and 160. This gives us a translation:

$$(20 - 2x)(14 - 2x) = 160.$$

3. **Carry out** We solve the equation:

$$280 - 68x + 4x^2 = 160$$
$$4x^2 - 68x + 120 = 0$$
$$x^2 - 17x + 30 = 0 \quad \text{Multiplying by } \tfrac{1}{4}$$
$$(x - 15)(x - 2) = 0 \quad \text{Factoring}$$
$$x - 15 = 0 \quad \text{or} \quad x - 2 = 0 \quad \text{Principle of zero products}$$
$$x = 15 \quad \text{or} \quad x = 2$$

4. **Check** We check in the original problem. We know that 15 is not a solution because when $x = 15$, $20 - 2x = -10$, and the length of the picture cannot be negative. When $x = 2$, $20 - 2x = 16$. This is the length. When $x = 2$, $14 - 2x = 10$. This is the width. The area is 16×10, or 160. This checks.

5. **State** The width of the frame is 2 cm.

EXAMPLE 2 The hypotenuse of a right triangle is 6 m long. One leg is 1 m longer than the other. Find the lengths of the legs. Round to the nearest tenth.

1. **Familiarize** We first make a drawing and label it. We have let x represent the length of one leg. Then $x + 1$ represents the length of the other leg.

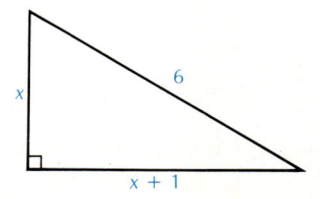

[Figure: Right triangle with legs x and $x + 1$, hypotenuse 6.]

2. **Translate** To translate we use the Pythagorean equation:

$$x^2 + (x+1)^2 = 6^2.$$

3. **Carry out** We solve the equation:

$$x^2 + (x+1)^2 = 6^2$$
$$x^2 + x^2 + 2x + 1 = 36$$
$$2x^2 + 2x - 35 = 0.$$

Since we cannot factor, we use the quadratic formula:

$$a = 2, \quad b = 2, \quad c = -35$$

$$x = \frac{-b \pm \sqrt{b^2 - 4ac}}{2a}$$

$$= \frac{-2 \pm \sqrt{2^2 - 4 \cdot 2(-35)}}{2 \cdot 2}$$

$$= \frac{-2 \pm \sqrt{4 + 280}}{4} = \frac{-2 \pm \sqrt{284}}{4}$$

$$= \frac{-2 \pm \sqrt{4 \cdot 71}}{4}$$

$$= \frac{-2 \pm 2 \cdot \sqrt{71}}{2 \cdot 2}$$

$$= \frac{-1 \pm \sqrt{71}}{2}.$$

Using a calculator or Table 2 we get an approximation: $\sqrt{71} \approx 8.426$:

$$\frac{-1 + \sqrt{71}}{2} \approx 3.7, \qquad \frac{-1 - \sqrt{71}}{2} \approx -4.7.$$

4. **Check** Since the length of a leg cannot be negative, -4.7 does not check. But 3.7 does check. If the smaller leg is 3.7, the other leg is 4.7. Then,

$$3.7^2 + 4.7^2 = 13.69 + 22.09 = 35.78$$

and using a calculator, we get $\sqrt{35.78} \approx 5.98 \approx 6$. Note that our check is not exact since we are using an approximation.

5. **State** One leg is about 3.7 m long and the other is about 4.7 m long.

11.8 PROBLEM SOLVING

EXAMPLE 3 The current in a stream moves at a speed of 2 km/h. A boat travels 24 km upstream and 24 km downstream in a total time of 5 hr. What is the speed of the boat in still water?

1. **Familiarize** First make a drawing. The distances are the same. Let r represent the speed of the boat in still water. Then when the boat is traveling upstream its speed is $r - 2$. When it is traveling downstream, its speed is $r + 2$. We let t_1 represent the time it takes the boat to go upstream, and t_2 the time it takes to go downstream. We summarize in a table.

	d	r	t
Upstream	24	$r - 2$	t_1
Downstream	24	$r + 2$	t_2

Upstream
$r - 2$
t_1 hours — 24 km

Downstream
$r + 2$
t_2 hours — 24 km

2. **Translate** Recall the basic formula for motion: $r = d/t$. From it we can obtain an equation for time: $t = d/r$. Total time consists of the time to go upstream, t_1, plus the time to go downstream, t_2. Using $t = d/r$ and the rows of the table, we have

$$t_1 = \frac{24}{r - 2} \quad \text{and} \quad t_2 = \frac{24}{r + 2}.$$

Since the total time is 5 hours, $t_1 + t_2 = 5$, and we have

$$\frac{24}{r - 2} + \frac{24}{r + 2} = 5.$$

3. **Carry out** We solve the equation. The LCM of the denominators is

$$(r-2)(r+2).$$

We multiply on both sides by the LCM:

$$(r-2)(r+2) \cdot \left[\frac{24}{r-2} + \frac{24}{r+2}\right] = (r-2)(r+2) \cdot 5 \quad \text{Multiplying by the LCM}$$

$$(r-2)(r+2) \cdot \frac{24}{r-2} + (r-2)(r+2) \cdot \frac{24}{r+2} = (r^2-4)5$$

$$24(r+2) + 24(r-2) = 5r^2 - 20$$
$$24r + 48 + 24r - 48 = 5r^2 - 20$$
$$-5r^2 + 48r + 20 = 0$$
$$5r^2 - 48r - 20 = 0 \quad \text{Multiplying by } -1$$
$$(5r+2)(r-10) = 0 \quad \text{Factoring}$$
$$5r+2 = 0 \quad \text{or} \quad r-10 = 0 \quad \text{Principle of zero products}$$
$$5r = -2 \quad \text{or} \quad r = 10$$
$$r = -\tfrac{2}{5} \quad \text{or} \quad r = 10.$$

4. **Check** Since speed cannot be negative, $-\tfrac{2}{5}$ cannot be a solution. But suppose the speed of the boat in still water is 10 km/h. The speed upstream is then $10 - 2$, or 8 km/h. The speed downstream is $10 + 2$, or 12 km/h. The time upstream, using $t = d/r$, is 24/8, or 3 hr. The time downstream is 24/12, or 2 hr. The total time is 5 hr. This checks.

5. **State** The speed of the boat in still water is 10 km/h.

EXERCISE SET 11.8

Problem solving

1. A picture frame measures 20 cm by 12 cm. There are 84 cm² of picture showing. Find the width of the frame.

2. A picture frame measures 18 cm by 14 cm. There are 192 cm² of picture showing. Find the width of the frame.

3. The hypotenuse of a right triangle is 25 ft long. One leg is 17 ft longer than the other. Find the lengths of the legs.

4. The hypotenuse of a right triangle is 26 yd long. One leg is 14 yd longer than the other. Find the lengths of the legs.

5. The length of a rectangle is 2 cm greater than the width. The area is 80 cm². Find the length and width.

6. The length of a rectangle is 3 m greater than the width. The area is 70 m². Find the length and width.

7. The width of a rectangle is 4 cm less than the length. The area is 320 cm². Find the length and width.

8. The width of a rectangle is 3 cm less than the length. The area is 340 cm². Find the length and width.

9. The length of a rectangle is twice the width. The area is 50 m². Find the length and width.

10. The length of a rectangle is twice the width. The area is 32 cm². Find the length and width.

Find approximate answers for Exercises 11–16. Round to the nearest tenth.

11. The hypotenuse of a right triangle is 8 m long. One leg is 2 m longer than the other. Find the lengths of the legs.

12. The hypotenuse of a right triangle is 5 cm long. One leg is 2 cm longer than the other. Find the lengths of the legs.

13. The length of a rectangle is 2 in. greater than the width. The area is 20 in². Find the length and width.

14. The length of a rectangle is 3 ft greater than the width. The area is 15 ft². Find the length and width.

15. The length of a rectangle is twice the width. The area is 10 m². Find the length and width.

16. The length of a rectangle is twice the width. The area is 20 cm². Find the length and width.

17. The current in a stream moves at a speed of 3 km/h. A boat travels 40 km upstream and 40 km downstream in a total time of 14 hr. What is the speed of the boat in still water?

18. The current in a stream moves at a speed of 3 km/h. A boat travels 45 km upstream and 45 km downstream in a total time of 8 hr. What is the speed of the boat in still water?

19. The current in a stream moves at a speed of 4 mph. A boat travels 4 mi upstream and 12 mi downstream in a total time of 2 hr. What is the speed of the boat in still water?

20. The current in a stream moves at a speed of 4 mph. A boat travels 5 mi upstream and 13 mi downstream in a total time of 2 hr. What is the speed of the boat in still water?

21. The speed of a boat in still water is 10 km/h. The boat travels 12 km upstream and 28 km downstream in a total time of 4 hr. What is the speed of the stream?

22. The speed of a boat in still water is 8 km/h. The boat travels 60 km upstream and 60 km downstream in a total time of 16 hr. What is the speed of the stream?

23. An airplane flies 738 mi against the wind and 1062 mi with the wind in a total time of 9 hr. The speed of the airplane in still air is 200 mph. What is the speed of the wind?

24. An airplane flies 520 km against the wind and 680 km with the wind in a total time of 4 hr. The speed of the airplane in still air is 300 km/h. What is the speed of the wind?

25. The speed of a boat in still water is 9 km/h. The boat travels 80 km upstream and 80 km downstream in a total time of 18 hr. What is the speed of the stream?

26. The speed of a boat in still water is 8 km/h. The boat travels 60 km upstream and 60 km downstream in a total time of 16 hr. What is the speed of the stream?

27. Find the area of a square for which the diagonal is one unit longer than the length of the sides.

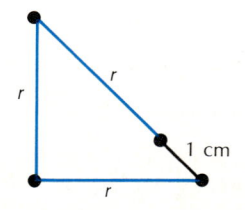

28. Two consecutive integers have squares that differ by 75. Find the integers.

29. Find r in this figure. Round to the nearest hundredth.

30. A 20-ft pole is struck by lightning, and, while not completely broken, falls over and touches the ground 10 ft from the bottom of the pole. How high up did the pole break?

31. What should the diameter d of a pizza be so that it has the same area as two 10-in. pizzas? Do you get more to eat with a 13-in. pizza or two 10-in. pizzas?

32. In this figure, the area of the shaded region is 24 cm². Find r if $R = 6$ cm. Round to the nearest hundredth.

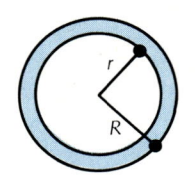

33. A ladder 10 ft long leans against a wall. The bottom of the ladder is 6 ft from the wall. How much would the lower end of the ladder have to be pulled away so that the top end would be pulled down the same amount?

34. Trains A and B leave the same city at right angles at the same time. Train B travels 5 mph faster than train A. After 2 hr they are 50 mi apart. Find the speed of each train.

35. Find the side of a square whose diagonal is 3 cm longer than a side.

36. Two pipes are connected to the same tank. When filling together they can fill the tank in 2 hr. The larger pipe, filling alone, can fill the tank in 3 hr less than the smaller one. How long would the smaller one take, filling alone, to fill the tank?

11.9 GRAPHS OF QUADRATIC EQUATIONS AND FUNCTIONS

In this section you will learn to graph equations of the form

$$y = ax^2 + bx + c, \quad a \neq 0.$$

The polynomial on the right is of second degree, or *quadratic*. If you studied functions in Chapter 8, you will recognize that the equation is that of a quadratic function

$$f(x) = ax^2 + bx + c, \quad a \neq 0.$$

Examples of the types of equations we are going to graph are

$$y = x^2 \quad \text{and} \quad y = x^2 + 2x - 3.$$

Graphing Quadratic Equations, $y = ax^2 + bx + c$

Graphs of quadratic equations, $y = ax^2 + bx + c$ (where $a \neq 0$), are always cup-shaped. They all have a *line of symmetry* like the dashed line shown in the figure. If you fold on this line, the two halves will match exactly. The curve goes on forever.

Line of symmetry

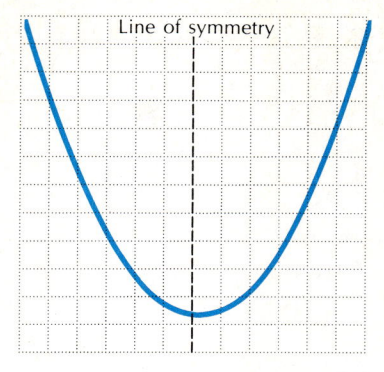

These curves are called *parabolas*. Some parabolas are thin and others are flat, but they all have the same general shape.

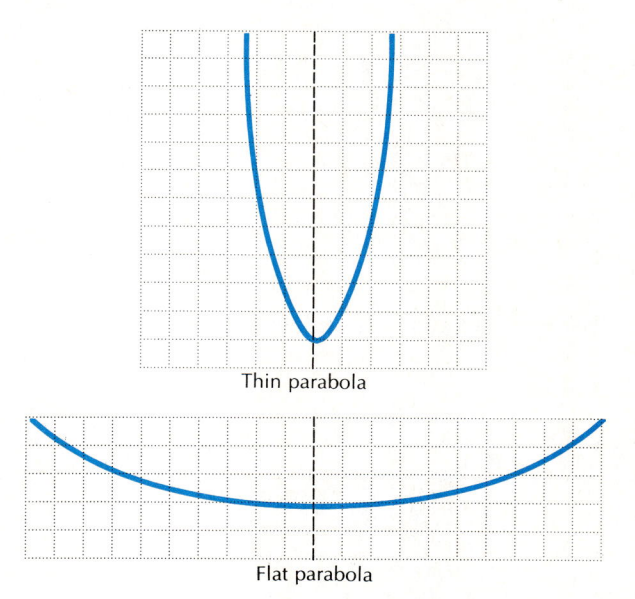

Thin parabola

Flat parabola

To graph a quadratic equation, we begin by choosing some numbers for x and computing the corresponding values of y.

EXAMPLE 1 Graph: $y = x^2$.

We plot the ordered pairs resulting from the computations shown in the table and connect the points with a smooth curve.

QUADRATIC EQUATIONS

x	y
−2	4
−1	1
0	0
1	1
2	4

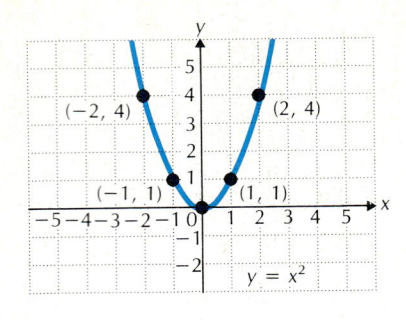

EXAMPLE 2 Graph: $y = x^2 + 2x - 3$.

x	y
1	0
0	−3
−1	−4
−2	−3
−3	0
−4	5
2	5

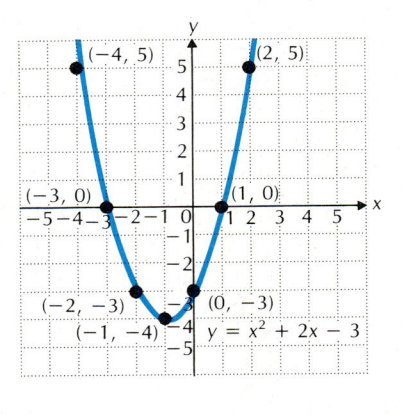

EXAMPLE 3 Graph: $y = -2x^2 + 3$.

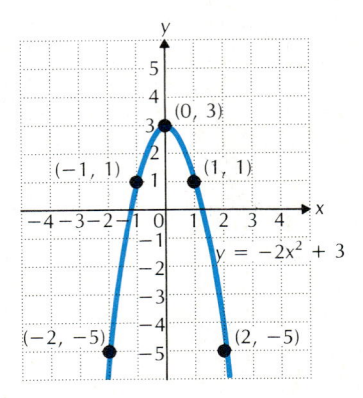

x	y
0	3
1	1
−1	1
2	−5
−2	−5

The graphs in Examples 1 and 2 open upward and the coefficients of x^2 are both 1, which is positive. The graph in Example 3 opens downward and the coefficient of x^2 is -2, which is negative.

11.9 GRAPHS OF QUADRATIC EQUATIONS AND FUNCTIONS

> Graphs of quadratic equations $y = ax^2 + bx + c$ are all parabolas. They are *smooth* cup-shaped symmetric curves, with no sharp points or kinks in them.
>
> The graph of $y = ax^2 + bx + c$ opens upward if $a > 0$. It opens downward if $a < 0$.
>
> In drawing parabolas, be sure to plot enough points to see the general shape of each graph.

If your graphs look like any of the following, they are incorrect.

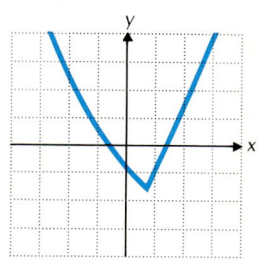

a) Sharp point is wrong.

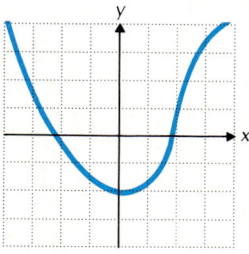

b) Outward nonsymmetric curve is wrong.

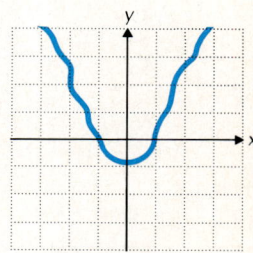

c) Kinks are wrong.

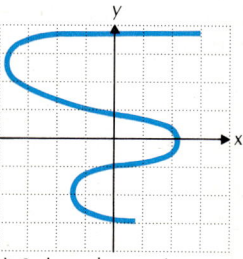

d) S-shaped curve is wrong.

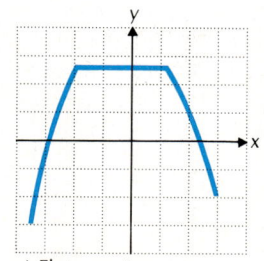

e) Flat nose is wrong.

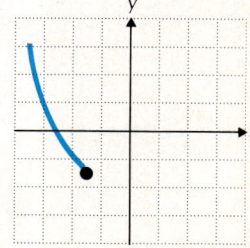

f) Half a cup-shaped curve is wrong.

Approximating Solutions of $0 = ax^2 + bx + c$

We can use graphing to approximate the solutions of quadratic equations, $0 = ax^2 + bx + c$. We graph the equation $y = ax^2 + bx + c$. If the graph crosses the x-axis, the points of crossing will give us solutions.

EXAMPLE 4 Approximate the solutions of

$$-2x^2 + 3 = 0$$

by graphing.

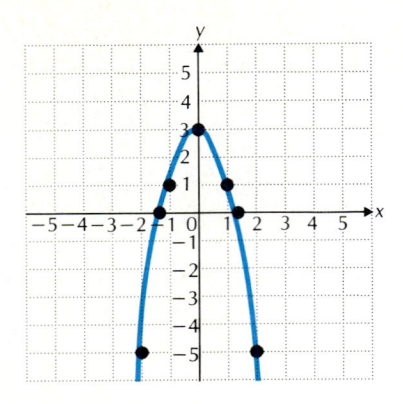

The graph was found in Example 3. The graph crosses the x-axis at about $(-1.2, 0)$ and $(1.2, 0)$. So the solutions are about -1.2 and 1.2.

EXERCISE SET 11.9

Without graphing, tell whether the graph of the equation opens upward or downward. Then graph the equation using graph paper.

1. $y = x^2 + 1$
2. $y = 2x^2$
3. $y = -1 \cdot x^2$
4. $y = x^2 - 1$
5. $y = -x^2 + 2x$
6. $y = x^2 + x - 6$
7. $y = 8 - x - x^2$
8. $y = x^2 + 2x + 1$
9. $y = x^2 - 2x + 1$
10. $y = -\frac{1}{2}x^2$
11. $y = -x^2 + 2x + 3$
12. $y = -x^2 - 2x + 3$
13. $y = -2x^2 - 4x + 1$
14. $y = 2x^2 + 4x - 1$
15. $y = \frac{1}{4}x^2$
16. $y = -0.1x^2$
17. $y = 3 - x^2$
18. $y = x^2 + 3$
19. $y = -x^2 + x - 1$
20. $y = x^2 + 2x$
21. $y = -2x^2$
22. $y = -x^2 - 1$
23. $y = x^2 - x - 6$
24. $y = 8 + x - x^2$

Approximate the solutions by graphing.

25. $x^2 - 5 = 0$
26. $x^2 - 3 = 0$
27. $x^2 + 2x = 0$
28. $x^2 - 2x = 0$
29. $8 - x - x^2 = 0$
30. $8 + x - x^2 = 0$
31. $x^2 + 10x + 25 = 0$
32. $x^2 - 8x + 16 = 0$
33. $-2x^2 - 4x + 1 = 0$
34. $2x^2 + 4x - 1 = 0$
35. $x^2 + 5 = 0$
36. $x^2 + 3 = 0$

37. Graph the equation $y = x^2 - x - 6$. Use your graph to approximate the solutions of the following equations.

a) $x^2 - x - 6 = 2$ (*Hint:* Graph $y = 2$ on the same set of axes as your graph of $y = x^2 - x - 6$.)
b) $x^2 - x - 6 = -3$

38. Graph $y = 2x^2 - 4x + 7$. Use the graph to approximate the solutions of $2x^2 - 4x + 7 = 0$. Solve $2x^2 - 4x + 7 = 0$ using the quadratic formula. What might you guess is true of any quadratic equation whose graph does not cross the x-axis?

39. The y-intercept of the linear equation $y = mx + b$ is $(0, b)$. What is the y-intercept of the quadratic equation $y = ax^2 + bx + c$?

40. The graph of a quadratic equation has either a high point (maximum) or a low point (minimum). Examine the graphs in Exercises 1–24. What characteristic of each equation tells whether the graph has a maximum or minimum?

41. Find an equation of the line that is the axis of symmetry of each graph in Exercises 1–10.

42. Using the same set of axes, graph $y = x^2$, $y = 2x^2$, $y = 3x^2$, $y = \frac{1}{2}x^2$, $y = -x^2$, $y = -2x^2$, $y = -3x^2$, and $y = -\frac{1}{2}x^2$. Describe the change in the graph of a quadratic equation $y = ax^2$ as $|a|$ increases.

43. What is the largest rectangular area that can be enclosed with 16 feet of fence?

$$2w + 2l = 16$$
$$w + l = 8$$
$$w = 8 - l$$

(*Hint:* Find $A = lw$ in terms of l. Graph the resulting quadratic equation and find its maximum.)

SUMMARY AND REVIEW: CHAPTER 11

The following contains a summary of what you should be able to do after completing this chapter. The review exercises are for practice. Answers are at the back of the book. If you miss an exercise, restudy the section indicated alongside the answer.

You should be able to:

Solve a quadratic equation by factoring or the quadratic formula.

Solve.

1. $8x^2 = 24$
2. $5x^2 - 8x + 3 = 0$
3. $x^2 - 2x - 10 = 0$
4. $3y^2 + 5y = 2$
5. $(x + 8)^2 = 13$
6. $9x^2 = 0$
7. $5t^2 - 7t = 0$
8. $9x^2 - 6x - 9 = 0$
9. $x^2 + 6x = 9$
10. $1 + 4x^2 = 8x$
11. $6 + 3y = y^2$
12. $3m = 4 + 5m^2$
13. $3x^2 = 4x$
14. $40 = 5y^2$

Solve a quadratic equation by completing the square.

Solve by completing the square. Show your work.

15. $3x^2 - 2x - 5 = 0$
16. $x^2 - 5x + 2 = 0$

Use a calculator or Table 2 to approximate the solutions of a quadratic equation.

Approximate the solutions to the nearest tenth.

17. $x^2 - 5x + 2 = 0$

18. $4y^2 + 8y + 1 = 0$

Solve certain fractional equations by first deriving a quadratic equation, solve certain radical equations by first using the principle of squaring to derive a quadratic equation, and solve a formula for a letter.

Solve.

19. $\dfrac{15}{x} - \dfrac{15}{x+2} = 2$

20. $x + \dfrac{1}{x} = 2$

21. $\sqrt{x+5} = x - 1$

22. $1 + x = \sqrt{1 + 5x}$

23. Solve for T: $V = \dfrac{1}{2}\sqrt{1 + \dfrac{T}{L}}$.

Without graphing, tell whether the graph of an equation of the type $y = ax^2 + bx + c$ opens upward or downward. Then graph the equation. Approximate the solutions of $0 = ax^2 + bx + c$ by graphing.

Without graphing, tell whether the graph of the equation opens upward or downward. Then graph the equation.

24. $y = 2 - x^2$

25. $y = x^2 - 4x - 2$

26. Approximate the solutions of $0 = x^2 - 4x - 2$ by graphing.

Solve problems using the five-step process and the solution of quadratic equations.

27. The hypotenuse of a right triangle is 5 m long. One leg is 3 m longer than the other. Find the lengths of the legs. Round to the nearest tenth.

28. $1000 is invested at interest rate r, compounded annually. In 2 years it grows to $1690. What is the interest rate?

29. The length of a rectangle is 3 m greater than the width. The area is 70 m². Find the length and width.

30. The current in a stream moves at a speed of 2 km/h. A boat travels 56 km upstream and 64 km downstream in a total time of 4 hr. What is the speed of the boat in still water?

31. Two consecutive integers have squares that differ by 63. Find the integers.

32. Find b such that the trinomial $x^2 + bx + 49$ is a square.

33. Solve $x - 4\sqrt{x} - 5 = 0$.

34. A square with sides of length s has the same area as a circle with radius of 5 in. Find s.

TEST: CHAPTER 11

Solve.

1. $7x^2 = 35$

2. $7x^2 + 8x = 0$

3. $48 = t^2 + 2t$

4. $3y^2 + 5y = 2$

5. $(x + 8)^2 = 13$

6. $x^2 = x + 3$

7. $m^2 - 3m = 7$

8. $10 = 4x + x^2$

9. $3x^2 - 7x + 1 = 0$

10. $x - \dfrac{2}{x} = 1$

11. $\dfrac{4}{x} - \dfrac{4}{x+2} = 1$

12. $\sqrt{6x + 13} = x + 3$

13. Solve by completing the square. Show your work.
$$x^2 - 4x - 10 = 0$$

14. Approximate the solutions to the nearest tenth.
$$x^2 - 4x - 10 = 0$$

15. Solve for n: $d = an^2 + bn$.

16. a) Without graphing, tell whether the graph of the equation $y = -x^2 + x + 5$ opens upward or downward.
    b) Graph the equation.

17. Approximate the solutions of $0 = -x^2 + x + 5$ by graphing.

Problem solving

18. $4000 is invested at interest rate r, compounded annually. In 2 years it grows to $6250. What is the interest rate?

19. The width of a rectangle is 4 m less than the length. The area is 16.25 m^2. Find the length and the width.

20. The current in a stream moves at a speed of 2 km/h. A boat travels 44 km upstream and 52 km downstream in a total of 4 hr. What is the speed of the boat in still water?

○ ─────────────

21. Find the side of a square whose diagonal is 5 ft longer than a side.

22. Solve this system for x. Use the substitution method.
$$x - y = 2,$$
$$xy = 4$$

CUMULATIVE REVIEW: CHAPTERS 1–11

1. What is the meaning of x^3?
2. Evaluate $(x - 3)^2 + 5$ for $x = 10$.
3. Use commutative and associative laws to write an expression equivalent to $6(xy)$.
4. Find the LCM of 15 and 48.
5. Find the absolute value: $|-7|$.

Compute and simplify.

6. $-6 + 12 + (-4) + 7$
7. $2.8 - (-12.2)$
8. $-\tfrac{3}{8} \div \tfrac{5}{2}$
9. $-9(7)$

10. Remove parentheses and simplify: $4m + 9 - (6m + 13)$.

Solve.

11. $3x = -24$

12. $3x + 7 = 2x - 5$

13. $3(y - 1) - 2(y + 2) = 0$

14. $x^2 - 8x + 15 = 0$
15. $y - x = 1,$
    $y = 3 - x$
16. $x + y = 17,$
    $x - y = 7$
17. $4x - 3y = 3,$
    $3x - 2y = 4$
18. $x^2 - x - 6 = 0$
19. $x^2 + 3x = 5$
20. $3 - x = \sqrt{x^2 - 3}$
21. $5 - 9x \leqslant 19 + 5x$
22. $-\frac{7}{8}x + 7 = \frac{3}{8}x - 3$
23. $0.6x - 1.8 = 1.2x$
24. $-3x > 24$
25. $23 - 19y - 3y \geqslant -12$
26. $3y^2 = 30$
27. $(x - 3)^2 = 6$
28. $\dfrac{6x - 2}{2x - 1} = \dfrac{9x}{3x + 1}$
29. $\dfrac{2x}{x + 1} = 2 - \dfrac{5}{2x}$
30. $\dfrac{2x}{x + 3} + \dfrac{6}{x} + 7 = \dfrac{18}{x^2 + 3x}$
31. $\sqrt{x + 9} = \sqrt{2x - 3}$

Solve each formula for the given letter.

32. $A = \dfrac{4b}{t},$ for b
33. $\dfrac{1}{t} = \dfrac{1}{m} - \dfrac{1}{n},$ for m
34. $r = \sqrt{\dfrac{A}{\pi}},$ for A
35. $y = ax^2 - bx,$ for x

Simplify.

36. $x^{-6} \cdot x^2$
37. $\dfrac{y^3}{y^{-4}}$
38. $(2y^6)^2$

39. Collect like terms and arrange in descending order.
$$2x - 3 + 5x^3 - 2x^3 + 7x^3 + x$$

Compute and simplify.

40. $(4x^3 + 3x^2 - 5) + (3x^3 - 5x^2 + 4x - 12)$
41. $(6x^2 - 4x + 1) - (-2x^2 + 7)$
42. $-2y^2(4y^2 - 3y + 1)$
43. $(2t - 3)(3t^2 - 4t + 2)$
44. $(t - \frac{1}{4})(t + \frac{1}{4})$
45. $(3m - 2)^2$
46. $(15x^2y^3 + 10xy^2 + 5) - (5xy^2 - x^2y^2 - 2)$
47. $(x^2 - 0.2y)(x^2 + 0.2y)$
48. $(3p + 4q^2)^2$
49. $\dfrac{4}{2x - 6} \cdot \dfrac{x - 3}{x + 3}$
50. $\dfrac{3a^4}{a^2 - 1} \div \dfrac{2a^3}{a^2 - 2a + 1}$
51. $\dfrac{3}{3x - 1} + \dfrac{4}{5x}$
52. $\dfrac{2}{x^2 - 16} - \dfrac{x - 3}{x^2 - 9x + 20}$

Factor.

53. $8x^2 - 4x$
54. $25x^2 - 4$
55. $6y^2 - 5y - 6$
56. $m^2 - 8m + 16$
57. $x^3 - 8x^2 - 5x + 40$
58. $3a^4 + 6a^2 - 72$
59. $16x^4 - 1$
60. $49a^2b^2 - 4$

61. $9x^2 + 30xy + 25y^2$
62. $2ac - 6ab - 3db + dc$
63. $15x^2 + 14xy - 8y^2$

Simplify.

64. $\dfrac{\dfrac{3}{x} + \dfrac{1}{2x}}{\dfrac{1}{3x} - \dfrac{3}{4x}}$
65. $\sqrt{49}$
66. $-\sqrt{625}$
67. $\sqrt{64x^2}$

Note: For the remainder of this review, *assume that all expressions in radicands represent nonnegative numbers.*

68. Multiply: $\sqrt{a+b}\sqrt{a+b}$.
69. Multiply and simplify: $\sqrt{32ab}\sqrt{6a^4b^2}$.

Simplify.

70. $\sqrt{150}$
71. $\sqrt{243x^3y^2}$
72. $\sqrt{\dfrac{100}{81}}$
73. $\sqrt{\dfrac{64}{x^2}}$
74. $4\sqrt{12} + 2\sqrt{48}$
75. Divide and simplify: $\dfrac{\sqrt{72}}{\sqrt{45}}$.
76. In a right triangle, $a = 9$ and $c = 41$. Find b.

Graph in the plane.

77. $y = \tfrac{1}{3}x - 2$
78. $2x + 3y = -6$
79. $y = -3$
80. $4x - 3y > 12$
81. $y = x^2 + 2x + 1$
82. Solve $9x^2 - 12x - 2 = 0$ by completing the square. Show your work.
83. Approximate the solutions of $4x^2 = 4x + 1$ to the nearest tenth.

Problem solving

84. What percent of 52 is 13?
85. 12 is what percent of 60?
86. The speed of a boat in still water is 8 km/h. It travels 60 km upstream and 60 km downstream in a total time of 16 hr. What is the speed of the stream?
87. The length of a rectangle is 7 m more than the width. The length of a diagonal is 13 m. Find the length.
88. Three-fifths of the automobiles entering the city each morning will be parked in city parking lots. There are 3654 such parking spaces. How many cars enter the city each morning?
89. A candy shop mixes nuts worth $1.10 per lb with another variety worth $0.80 per lb to make 42 lb of a mixture worth $0.90 per lb. How many lb of each kind of nuts should be used?
90. In checking records a contractor finds that crew A can pave a certain length of highway in 8 hr. Crew B can do the same job in 10 hr. How long would they take if they worked together?
91. A student's paycheck varies directly as the number of hours worked. The pay was $242.52 for 43 hr of work. What would the pay be for 80 hr of work? Explain the meaning of the variation constant.
92. For the function f described by
$$f(x) = 2x^2 + 7x - 4,$$
find $f(0)$, $f(-4)$, and $f(\tfrac{1}{2})$.
93. Find the intersection:
$$\{a, b, c, d, e\} \cap \{g, h, c, d, f\}.$$

94. Find the union:

$\{a, b, c, d, e\} \cup \{g, h, c, d, f\}$.

95. Solve: $|x| = 12$.
96. Simplify: $\sqrt[3]{8}$.
97. Find b such that the trinomial $x^2 - bx + 225$ is a square.
98. Find x.

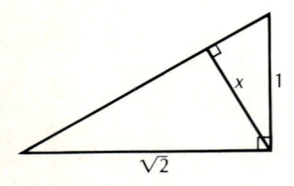

Determine whether each pair of expressions is equivalent.

99. $x^2 - 9$, $(x - 3)(x + 3)$
100. $\dfrac{x + 3}{3}$, x
101. $(x + 5)^2$, $x^2 + 25$
102. $\sqrt{x^2 + 16}$, $x + 4$
103. $\sqrt{x^2}$, $|x|$

TABLES

TABLE 1 FRACTIONAL AND DECIMAL EQUIVALENTS

Fractional notation	Decimal notation	Percent notation
$\frac{1}{10}$	0.1	10%
$\frac{1}{8}$	0.125	12.5% or $12\frac{1}{2}$%
$\frac{1}{6}$	$0.16\overline{6}$	$16.6\overline{6}$% or $16\frac{2}{3}$%
$\frac{1}{5}$	0.2	20%
$\frac{1}{4}$	0.25	25%
$\frac{3}{10}$	0.3	30%
$\frac{1}{3}$	$0.333\overline{3}$	$33.3\overline{3}$% or $33\frac{1}{3}$%
$\frac{3}{8}$	0.375	37.5% or $37\frac{1}{2}$%
$\frac{2}{5}$	0.4	40%
$\frac{1}{2}$	0.5	50%
$\frac{3}{5}$	0.6	60%
$\frac{5}{8}$	0.625	62.5% or $62\frac{1}{2}$%
$\frac{2}{3}$	$0.666\overline{6}$	$66.6\overline{6}$% or $66\frac{2}{3}$%
$\frac{7}{10}$	0.7	70%
$\frac{3}{4}$	0.75	75%
$\frac{4}{5}$	0.8	80%
$\frac{5}{6}$	$0.83\overline{3}$	$83.3\overline{3}$% or $83\frac{1}{3}$%
$\frac{7}{8}$	0.875	87.5% or $87\frac{1}{2}$%
$\frac{9}{10}$	0.9	90%
$\frac{1}{1}$	1	100%

TABLE 2 SQUARE ROOTS

N	$\sqrt{N}$	N	$\sqrt{N}$	N	$\sqrt{N}$	N	$\sqrt{N}$
2	1.414	27	5.196	52	7.211	77	8.775
3	1.732	28	5.292	53	7.280	78	8.832
4	2	29	5.385	54	7.348	79	8.888
5	2.236	30	5.477	55	7.416	80	8.944
6	2.449	31	5.568	56	7.483	81	9
7	2.646	32	5.657	57	7.550	82	9.055
8	2.828	33	5.745	58	7.616	83	9.110
9	3	34	5.831	59	7.681	84	9.165
10	3.162	35	5.916	60	7.746	85	9.220
11	3.317	36	6	61	7.810	86	9.274
12	3.464	37	6.083	62	7.874	87	9.327
13	3.606	38	6.164	63	7.937	88	9.381
14	3.742	39	6.245	64	8	89	9.434
15	3.873	40	6.325	65	8.062	90	9.487
16	4	41	6.403	66	8.124	91	9.539
17	4.123	42	6.481	67	8.185	92	9.592
18	4.243	43	6.557	68	8.246	93	9.644
19	4.359	44	6.633	69	8.307	94	9.695
20	4.472	45	6.708	70	8.367	95	9.747
21	4.583	46	6.782	71	8.426	96	9.798
22	4.690	47	6.856	72	8.485	97	9.849
23	4.796	48	6.928	73	8.544	98	9.899
24	4.899	49	7	74	8.602	99	9.950
25	5	50	7.071	75	8.660	100	10
26	5.099	51	7.141	76	8.718		

TABLE 3 GEOMETRIC FORMULAS

Plane Geometry:

Rectangle
Area: $A = lw$
Perimeter: $P = 2l + 2w$

Square
Area: $A = s^2$
Perimeter: $P = 4s$

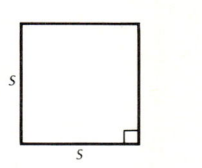

Triangle
Area: $A = \frac{1}{2}bh$

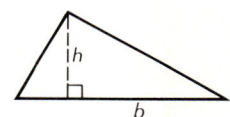

Sum of Angle Measures:
$A + B + C = 180°$

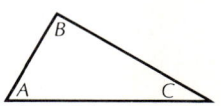

Right Triangle
Pythagorean Theorem (Equation):
$a^2 + b^2 = c^2$

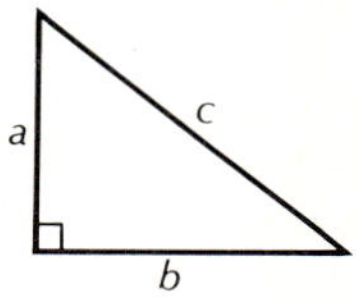

Parallelogram
Area: $A = bh$

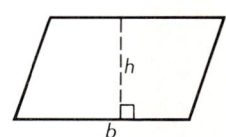

Trapezoid
Area: $A = \frac{1}{2}h(a + b)$

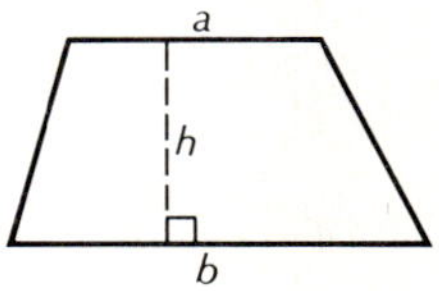

Circle
Area: $A = \pi r^2$
Circumference:
$C = \pi D = 2\pi r$
($\frac{22}{7}$ and 3.14 are different approximations for π)

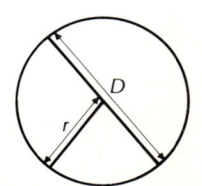

Solid Geometry:

Rectangular Solid
Volume: $V = lwh$

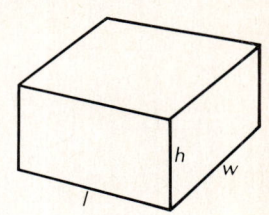

Cube
Volume: $V = s^3$

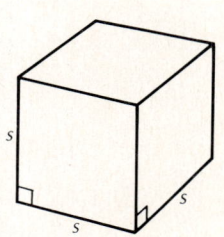

Right Circular Cylinder
Volume: $V = \pi r^2 h$
Lateral Surface Area:
$L = 2\pi rh$
Total Surface Area:
$S = 2\pi rh + 2\pi r^2$

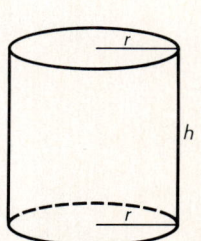

Right Circular Cone
Volume: $V = \frac{1}{3}\pi r^2 h$
Lateral Surface Area:
$L = \pi rs$
Total Surface Area:
$S = \pi r^2 + \pi rs$
Slant Height:
$s = \sqrt{r^2 + h^2}$

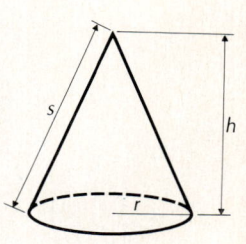

Sphere
Volume: $V = \frac{4}{3}\pi r^3$
Surface Area: $S = 4\pi r^2$

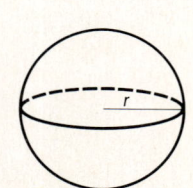

ANSWERS

CHAPTER 1

Exercise Set 1.1, pp. 4–5

These exercises are meant to be open-ended, and for the most part do not have absolute answers.

Exercise Set 1.2, pp. 8–9

1. 23, 28, 41   **3.** 112 ft² (square feet)   **5.** 42
7. 3   **9.** 1   **11.** 6   **13.** 2   **15.** $\frac{1}{2}$ **17.** 20
19. $b+6$ **21.** $c-9$ **23.** $q+6$ **25.** $a+b$
27. $y-x$ **29.** $w+x$ **31.** $n-m$ **33.** $r+s$
35. $2x$ **37.** $5t$ **39.** $3b$ **41.** $m+1$ **43.** $h-43$
45. $y+2x$ **47.** $2x-3$ **49.** $b-2$ **51.** 98% x
53. 6   **55.** 6   **57.** 9   **59.** d **61.** $2v$
63. $n+0.10n$

Exercise Set 1.3, pp. 14–16

1. $1 \times 21, 3 \times 7$ **3.** $1 \times 45, 3 \times 15, 5 \times 9, 5 \times 3 \times 3$
5. $1 \times 49, 7 \times 7$ **7.** $1 \times 28, 2 \times 14, 4 \times 7, 2 \times 2 \times 7$
9. $1 \times 76, 2 \times 38, 4 \times 19, 2 \times 2 \times 19$
11. $1 \times 56, 2 \times 28, 4 \times 14, 7 \times 8, 2 \times 2 \times 2 \times 7$
13. $1 \times 93, 3 \times 31$
15. $1 \times 144, 2 \times 72, 3 \times 48, 4 \times 36, 6 \times 24, 8 \times 18, 9 \times 16, 12 \times 12$ There are many others.
17. 2×7 **19.** 3×11 **21.** 3×3 **23.** 7×7
25. $2 \times 3 \times 3$ **27.** $2 \times 2 \times 2 \times 5$
29. $2 \times 3 \times 3 \times 5$ **31.** $2 \times 3 \times 5 \times 7$
33. 7×13 **35.** 7×17 **37.** 21, 42, 63
39. 24, 48, 72   **41.** 60, 120, 180   **43.** 60, 120, 180
45. 15   **47.** 120   **49.** 299   **51.** 420   **53.** 36
55. 360   **57.** 150   **59.** 120   **61.** 72   **63.** 315
65. 30   **67.** 72   **69.** 60   **71.** 36
73. (a) No, (b) No, (c) No, (d) Yes. The greatest number of times the 2 occurs in the factorizations of 8 and 12 is 3 times. The greatest number of times the 3 occurs is once. Thus the LCM must contain only three factors of 2 and one factor of 3.
75. 70,200   **77.** Every 60 yr   **79.** Every 420 yr

Exercise Set 1.4, pp. 23–24

1. $8+y$ **3.** nm **5.** $9+yx, yx+9, xy+9$
7. $ba+c, c+ba, c+ab$ **9.** $\frac{40}{48}$ **11.** $\frac{600}{700}$ **13.** $\frac{st}{20t}$
15. $\frac{2}{5}$ **17.** $\frac{7}{2}$ **19.** $\frac{1}{7}$ **21.** 8   **23.** y **25.** $\frac{1}{9y}$

27. $\frac{8}{3b}$ **29.** $\frac{p}{2}$ **31.** $\frac{9z}{19t}$ **33.** $\frac{1}{8}$ **35.** $\frac{51}{8}$ **37.** 1
39. $\frac{7}{6}$ **41.** $\frac{5}{6}$ **43.** $\frac{1}{2}$ **45.** $\frac{5}{18}$ **47.** $\frac{31}{60}$ **49.** $\frac{35}{18}$
51. $\frac{10}{3}$ **53.** $\frac{1}{2}$ **55.** $\frac{5}{36}$ **57.** 500 **59.** $\frac{3}{40}$
61. No **63.** Yes, commutative law of addition
65. Yes, commutative law of multiplication **67.** No
69. $\frac{2}{3}$ **71.** $\frac{3sb}{2}$ **73.** $\frac{r}{g}$
75. No. ($12 \div 4$ does not equal $4 \div 12$)

Exercise Set 1.5, pp. 26–27

1. $2 \cdot 2 \cdot 2 \cdot 2$ **3.** $(1.4) \cdot (1.4) \cdot (1.4) \cdot (1.4) \cdot (1.4)$
5. $n \cdot n \cdot n \cdot n \cdot n$ **7.** $7p \cdot 7p$
9. $19k \cdot 19k \cdot 19k \cdot 19k$ **11.** $10pq \cdot 10pq \cdot 10pq$
13. 10^6 **15.** x^7 **17.** $(3y)^4$ **19.** 27 **21.** 19
23. 256 **25.** 1 **27.** 37 **29.** $576m^2$ **31.** 10^5
33. 1 **35.** 111 **37.** 216; 24 **39.** 1728; 192
41. 400; 80 **43.** 3025; 605 **45.** x^3y^3 **47.** 0
49. 127
51. Answers may vary. For a 10-digit readout, 6^{13} and larger will be too large. **53.** Yes **55.** Yes
57. No

Exercise Set 1.6, pp. 31–32

1. 19 **3.** 86 **5.** 7 **7.** 5 **9.** 12 **11.** 324
13. 100 **15.** 512 **17.** 22 **19.** 1 **21.** 4
23. 1500 **25.** 76 **27.** 60 **29.** 125 **31.** 96
33. 500 **35.** $\frac{11}{5}$ **37.** 3 **39.** 32
41. $5 + (x + y)$ **43.** $6 \cdot (x \cdot y)$
45. $3 + (x + y), 3 + (y + x), (3 + y) + x$
47. $(6 + x) + y, 6 + (y + x), (6 + y) + x$
49. $a \cdot (b \cdot 5), (b \cdot a) \cdot 5, b \cdot (a \cdot 5)$
51. $(5 \cdot x) \cdot y, 5 \cdot (y \cdot x), (5 \cdot y) \cdot x$
53. Any number except 0
55. Any number except 2 **57.** $x^2 + 7$
59. $(x + 7)^2$ **61.** $\frac{x+3}{(x+3)^2}$ **63.** $2(x - 1)$

Exercise Set 1.7, pp. 35–36

1. $2b + 10$ **3.** $7 + 7t$ **5.** $3x + 3$ **7.** $4 + 4y$
9. $30x + 12$ **11.** $7x + 28 + 42y$ **13.** $2(x + 2)$
15. $5(6 + y)$ **17.** $7(2x + 3y)$ **19.** $5(x + 2 + 3y)$
21. $9(x + 3)$ **23.** $3(3x + y)$ **25.** $8(a + 2b + 8)$
27. $11(x + 4y + 11)$ **29.** $19a$ **31.** $11a$
33. $8x + 9z$ **35.** $7x + 15y^2$ **37.** $101a + 92$
39. $11a + 11b$ **41.** $14u^2 + 13t + 2$
43. $50 + 6t + 8y$ **45.** $1b$ or b **47.** $\frac{13}{4}y$ or $3\frac{1}{4}y$
49. $P(1 + rt)$ **51.** $\frac{3}{2}$

53. $6x + 6x^2 + 3x^3 = 3x(2 + 2x + x^2)$
55. $7x + 7xy + 7y = 7(x + xy + y)$

Exercise Set 1.8, pp. 41–42

1. $x + 60 = 112$ **3.** $42y = 2352$ **5.** $s + 35 = 64$
7. 15 **9.** 19 **11.** 9 **13.** 30 **15.** 20 **17.** 7
19. 17 **21.** 86 **23.** 377 **25.** 2818 **27.** 5.66
29. $\frac{5}{7}$ **31.** 6 **33.** $\frac{5}{4}$ **35.** 0.24 **37.** 3.1
39 8.5 **41.** 18 **43.** $\frac{5}{4}$ **45.** $\frac{3}{25}$ **47.** 8 **49.** 8
51. Answers may vary. $3x = 2$ **53.** 470,188 **55.** 5

Exercise Set 1.9, pp. 46–48

1. 52 **3.** 142 **5.** 319 **7.** 56 **9.** 72 **11.** 3
13. 64 **15.** \$67 **17.** \$0.26 **19.** 667.5
21. 6.64 hr **23.** 64.8°C **25.** 383,631.7 km
27. 59°F **29.** 25.75¢ **31.** 39.37
33. Salary now

Exercise Set 1.10, pp. 53–55

1. 0.76 **3.** 0.547 **5.** 1 **7.** 0.0061
9. $\frac{20}{100}$, or $\frac{1}{5}$ **11.** $\frac{786}{1000}$, or $\frac{393}{500}$ **13.** $\frac{25}{200}$, or $\frac{1}{8}$
15. $\frac{42}{100,000}$, or $\frac{21}{50,000}$ **17.** 454% **19.** 99.8%
21. 200% **23.** 7.2% **25.** $\frac{100}{8}$%, or 12.5%
27. 68% **29.** 17% **31.** 70% **33.** 60%
35. $\frac{200}{3}$%, or $66\frac{2}{3}$% **37.** $\frac{700}{4}$%, or 175% **39.** 75%
41. 25% **43.** 24% **45.** 150 **47.** 2.5 **49.** 546
51. 125% **53.** 0.8 **55.** 5% **57.** 86.4%
59. \$800 **61.** 36 cm³, 436 cm³ **63.** \$7200
65. 26% **67.** 90% **69.** 12 **71.** 345%, or 3.45
73. 2.5%, or 0.025 **75.** \$20 **77.** 40%, 70%, 95%
79. 20%

Summary and Review: Chapter 1, pp. 55–57

1. [1.2] 15 **2.** [1.2] 6 **3.** [1.2] 5 **4.** [1.2] 4
5. [1.5] 9 **6.** [1.5] 1 **7.** [1.6] 25 **8.** [1.6] 32
9. [1.6] 119 **10.** [1.6] 29 **11.** [1.6] 7
12. [1.2] $z - 8$ **13.** [1.2] $3x$
14. [1.10] 19%x, or $0.19x$ **15.** [1.2] $x - 1$
16. [1.5] $2m \cdot 2m \cdot 2m$ **17.** [1.5] $(6z)^2$ or $36z^2$
18. [1.4] $y + 4$ **19.** [1.4] ba
20. [1.4] $qp + 2$, or $2 + pq$ **21.** [1.6] $3 + (x + 1)$
22. [1.6] $(m \cdot 4) \cdot n$
23. [1.6] Answers may vary. $(n + m) + 1$, $n + (m + 1), (n + 1) + m$
24. [1.6] Answers may vary. $(4 \cdot y) \cdot x, x \cdot (y \cdot 4)$, $y \cdot (4 \cdot x)$ **25.** [1.4] $\frac{12}{30}$ **26.** [1.4] $\frac{2xz}{yz}$
27. [1.3] $2 \cdot 2 \cdot 23$ **28.** [1.3] $2 \cdot 2 \cdot 2 \cdot 5 \cdot 5 \cdot 7$

CHAPTERS 1–2   A-3

29. [1.3] 96   **30.** [1.3] 90   **31.** [1.4] $\frac{5}{12}$
32. [1.4] $\frac{5a}{9b}$ **33.** [1.4] $\frac{31}{36}$ **34.** [1.4] $\frac{1}{4}$
35. [1.4] $\frac{3}{5}$ **36.** [1.4] $\frac{72}{25}$ **37.** [1.7] $18x + 30y$
38. [1.7] $40x + 24y + 16$ **39.** [1.7] $6(3x + y)$
40. [1.7] $4(9x + 4 + y)$ **41.** [1.7] $40y + 10a$
42. [1.7] $60x + 12b$ **43.** [1.10] 0.047
44. [1.10] $\frac{3}{5}$ **45.** [1.10] 88.6%   **46.** [1.10] 62.5%
47. [1.10] 116%   **48.** [1.8] 39   **49.** [1.8] $\frac{1}{8}$
50. [1.8] 18   **51.** [1.8] 26   **52.** See Section 1.9
53. [1.9] 67   **54.** [1.9] 50   **55.** [1.9] 32
56. [1.10] 250   **57.** [1.10] 30   **58.** [1.10] $14,200
59. [1.10] 0.0000006%   **60.** [1.10] 40%
61. [1.10] 20%, 10%, and 10%   **62.** [1.5] 25,281
63. [1.8] 4   **64.** [1.10] (a) 250; (b) 226

CHAPTER 2

Exercise Set 2.1, pp. 63–64

1. 5, −12   **3.** −17, 12   **5.** −1286, 29,028
7. 750, −125   **9.** 20, −150, 300   **11.** >
13. <   **15.** <   **17.** <   **19.** >   **21.** <
23. 3   **25.** 10   **27.** 0   **29.** 24   **31.** 5
33. $-5 > x$ **35.** $120 > -20$
37. $-500{,}000 < 1{,}000{,}000$ **39.** $-2 > -10$
41. −23, −17, 0, 4   **43.** −7, 7   **45.** <
47. =   **49.** <
51. $-100, -5, 0, 1^7, |3|, \frac{14}{4}, 4, |-6|, 7^1$

Exercise Set 2.2, p. 67

1. Answers may vary. $-\frac{9}{7}, 0, 4\frac{1}{2}, -1.97, -491, 128,$
$\frac{3}{11}, -\frac{1}{7}, 0.000011, -26\frac{1}{3}$

3. number line with $\frac{10}{3}$ marked between 3 and 4

5. number line with -4.3 marked between -5 and -4

7. >   **9.** <   **11.** <
13. <   **15.** $\frac{2}{3}$ **17.** 0   **19.** 1   **21.** $-\frac{13}{5}, \frac{13}{5}$
23. Answers may vary. $-1\frac{3}{4}$ **25.** (a) 1; (b) $\frac{1}{9}$

Exercise Set 2.3, pp. 72–74

1. −7   **3.** −4   **5.** 0   **7.** −8   **9.** −7
11. −27   **13.** 0   **15.** −42   **17.** 0   **19.** 0
21. 3   **23.** −9   **25.** 7   **27.** 2   **29.** −26
31. −22   **33.** 32   **35.** 0   **37.** 45   **39.** −1.8
41. −8.1   **43.** $-\frac{1}{5}$ **45.** $-\frac{8}{7}$ **47.** $-\frac{3}{8}$ **49.** $-\frac{29}{35}$

51. $-\frac{11}{15}$ **53.** −6.3   **55.** −4.3   **57.** $\frac{7}{16}$ **59.** 39
61. 50   **63.** −1093   **65.** −24   **67.** 9
69. 26.9   **71.** −9   **73.** $\frac{14}{3}$ **75.** −0.101
77. −65   **79.** $\frac{5}{3}$ **81.** 1   **83.** −7   **85.** 14
87. 0   **89.** 8 yards   **91.** 999 mb
93. When x is positive   **95.** Positive   **97.** Negative
99. 0

Exercise Set 2.4, pp. 78–79

1. −4   **3.** −7   **5.** −6   **7.** 0   **9.** −4
11. −7   **13.** −6   **15.** 0   **17.** 0   **19.** 14
21. 11   **23.** −14   **25.** 5   **27.** −7   **29.** −1
31. 18   **33.** −5   **35.** −3   **37.** −21   **39.** 5
41. −8   **43.** 12   **45.** −23   **47.** −68
49. −73   **51.** 116   **53.** 0   **55.** $-\frac{1}{4}$ **57.** $\frac{1}{12}$
59. $-\frac{17}{12}$ **61.** $\frac{1}{8}$ **63.** 19.9   **65.** −9
67. −0.01   **69.** −193   **71.** 500   **73.** −2.8
75. −3.53   **77.** $-\frac{1}{2}$ **79.** $\frac{6}{7}$ **81.** $-\frac{41}{30}$
83. $-\frac{1}{156}$ **85.** 37   **87.** −62   **89.** −139
91. 6   **93.** $10x + 7$ **95.** $15x + 66$
97. −$330.54   **99.** $y + \$215.50$ **101.** 116 m
103. −309,882   **105.** 1   **107.** 2
109. False. $3 - 0 \neq 0 - 3$ **111.** True
113. True by definition

Exercise Set 2.5, pp. 81–82

1. −16   **3.** −42   **5.** −24   **7.** −72   **9.** 16
11. 42   **13.** −120   **15.** −238   **17.** 1200
19. 98   **21.** −72   **23.** −12.4   **25.** 24
27. 21.7   **29.** $-\frac{2}{5}$ **31.** $\frac{1}{12}$ **33.** −17.01
35. $-\frac{5}{12}$ **37.** 420   **39.** $\frac{2}{7}$ **41.** −60   **43.** 150
45. $-\frac{2}{45}$ **47.** 1911   **49.** 50.4   **51.** $\frac{10}{189}$
53. −960   **55.** 17.64   **57.** $-\frac{5}{784}$ **59.** 0
61. −720   **63.** −30,240   **65.** 72   **67.** −6
69. 1944   **71.** 16   **73.** −1   **75.** 13   **77.** −79
79. (a) Either m is negative or n is negative;
(b) Either m or n is zero;
(c) Both m and n are positive or both are negative

Exercise Set 2.6, pp. 86–88

1. −6   **3.** −13   **5.** −2   **7.** 4   **9.** −8   **11.** 2
13. −12   **15.** −8   **17.** Not possible   **19.** $-\frac{88}{9}$
21. $\frac{7}{15}$ **23.** $-\frac{13}{47}$ **25.** $\frac{1}{13}$ **27.** $\frac{1}{4.3}$, or $\frac{10}{43}$
29. −7.1, or $-\frac{71}{10}$ **31.** $\frac{q}{p}$ **33.** $4y$ **35.** $\frac{3b}{2a}$
37. $3(\frac{1}{19})$ **39.** $6(-\frac{1}{13})$ **41.** $13.9(-\frac{1}{1.5})$
43. $x \cdot y$ **45.** $(3x + 4)\frac{1}{5}$ **47.** $(5a - b)\left(\dfrac{1}{5a + b}\right)$

49. $-\frac{9}{8}$ **51.** $\frac{5}{3}$ **53.** $\frac{9}{14}$ **55.** $\frac{9}{64}$ **57.** -2
59. $\frac{11}{13}$ **61.** -16.2 **63.** $\frac{69}{25}$ **65.** -8 **67.** $-\frac{95}{77}$
69. 1 **71.** No, $\frac{4}{2} \ne \frac{2}{4}$ **73.** -1 and 1
75. You get the original number. The reciprocal of the reciprocal is the original number. **77.** Positive
79. Positive

Exercise Set 2.7, pp. 90–91

1. 7 **3.** 12 **5.** -12.71 **7.** $7x - 14$
9. $-7y + 14$ **11.** $45x + 54y - 72$
13. $-4x + 12y + 8z$ **15.** $-3.72x + 9.92y - 3.41$
17. $8(x-3)$ **19.** $4(8-y)$ **21.** $2(4x+5y-11)$
23. $a(x-7)$ **25.** $a(x-y-z)$ **27.** $4x, 3z$
29. $7x, 8y, -9z$ **31.** $12x, -13.2y, \frac{5}{8}z, -4.5$
33. $8x$ **35.** $5n$ **37.** $4x + 2y$ **39.** $7x + y$
41. $0.8x + 0.5y$ **43.** $\frac{3}{5}x + \frac{3}{5}y$ **45.** $\pi r(2+s)$
47. $18x - 23$ **49.** $2x$ **51.** $1.08P$ **53.** $8(x-y)$
55. $3b - 4a$ **57.** $2500(x+y)$

Exercise Set 2.8, pp. 96–98

1. $-2x - 7$ **3.** $-5x + 8$ **5.** $-4a + 3b - 7c$
7. $-6x + 8y - 5$ **9.** $-3x + 5y + 6$
11. $8x + 6y + 43$ **13.** $5x - 3$ **15.** $-3a + 9$
17. $5x - 6$ **19.** $-19x + 2y$ **21.** $9y - 25z$
23. 7 **25.** -40 **27.** 19 **29.** $12x + 30$
31. $3x + 30$ **33.** $9x - 18$ **35.** -16 **37.** -334
39. 14 **41.** 1880 **43.** 4682.688 **45.** 12
47. 8 **49.** 16 **51.** A^2+2*A*B+B^2
53. 2*(3−B)/C **55.** A/B−C/D
57. $6y - (-2x + 3a - c)$
59. $6m - (-3n + 5m - 4b)$ **61.** $-4z$ **63.** $x - 3$
65. False **67.** False **69.** False **71.** True
73. Yes. $(-a)(-b) = (-1)(a)(-1)(b) = (-1)^2(ab) = 1ab = ab$

Summary and Review: Chapter 2, pp. 98–99

1. [2.1] $-45, 72$ **2.** [2.1] 38 **3.** [2.1] 7
4. [2.2] $\frac{5}{2}$ **5.** [2.2] (number line: -2.5 between -3 and -2)
6. [2.2] (number line: $\frac{8}{9}$ between 0 and 1) **7.** [2.1] $<$ **8.** [2.1] $>$
9. [2.2] $>$ **10.** [2.2] $<$ **11.** [2.3] -3.8
12. [2.3] $\frac{3}{4}$ **13.** [2.3] 34 **14.** [2.3] 5
15. [2.6] $\frac{8}{3}$ **16.** [2.6] $-\frac{1}{7}$ **17.** [2.6] $\frac{2y}{x}$
18. [2.3] -3 **19.** [2.3] $-\frac{7}{12}$ **20.** [2.3] -4

21. [2.3] -5 **22.** [2.4] 4 **23.** [2.4] $-\frac{7}{5}$
24. [2.4] -7.9 **25.** [2.5] 54 **26.** [2.5] -9.18
27. [2.5] $-\frac{2}{7}$ **28.** [2.5] -210 **29.** [2.6] -7
30. [2.6] -3 **31.** [2.6] $\frac{3}{4}$ **32.** [2.3] 8-yd gain
33. [2.4] $-\$130$ **34.** [2.7] $15x - 35$
35. [2.7] $-8x + 10$ **36.** [2.7] $4x + 15$
37. [2.7] $-24 + 48x$ **38.** [2.7] $2(x-7)$
39. [2.7] $6(x-1)$ **40.** [2.7] $5(x+2)$
41. [2.7] $3(4-x)$ **42.** [2.7] $7a - 3b$
43. [2.7] $-2x + 5y$ **44.** [2.7] $5x - y$
45. [2.7] $-a + 8b$ **46.** [2.8] 19 **47.** [2.8] 8
48. [2.8] $6x + 3$ **49.** [2.8] $11y - 16$
50. [2.8] $-3a + 9$ **51.** [2.8] $-2b + 21$
52. [2.8] 6 **53.** [2.8] $12y - 34$
54. [2.8] $5x + 24$ **55.** [2.8] $-15x + 25$
56. [2.4] $-\frac{5}{8}$ **57.** [2.2] (a) $\frac{3}{11}$; (b) $\frac{10}{11}$
58. [2.2] No solution **59.** [2.8] -2.1

CHAPTER 3

Exercise Set 3.1, pp. 103–104

1. 4 **3.** -20 **5.** -14 **7.** -18 **9.** 15
11. -14 **13.** 2 **15.** 20 **17.** -6 **19.** $\frac{7}{3}$
21. $-\frac{7}{4}$ **23.** $\frac{41}{24}$ **25.** $-\frac{1}{20}$ **27.** 5.1 **29.** 12.4
31. -5 **33.** $\frac{11}{6}$ **35.** $-\frac{10}{21}$ **37.** 19°C
39. 342.246 **41.** -10 **43.** All real numbers
45. $b + 3$ **47.** $a + 4$ **49.** $-\frac{5}{17}$
51. Subtraction is addition of inverses.

Exercise Set 3.2, pp. 107–108

1. 6 **3.** 9 **5.** 12 **7.** -40 **9.** 1 **11.** -7
13. -6 **15.** 6 **17.** -63 **19.** 36 **21.** -21
23. $-\frac{3}{5}$ **25.** $\frac{3}{2}$ **27.** $\frac{9}{2}$ **29.** 7 **31.** -7 **33.** 8
35. 15.9 **37.** 670 mph **39.** -8655
41. No solution **43.** $\frac{b}{3a}$ **45.** $\frac{4b}{a}$
47. No solution **49.** 250

Exercise Set 3.3, pp. 112–113

1. 5 **3.** 8 **5.** 10 **7.** 14 **9.** -8 **11.** -8
13. -7 **15.** 15 **17.** 18 **19.** 6 **21.** 4 **23.** 6
25. 5 **27.** -3 **29.** 1 **31.** -20 **33.** 6
35. 7 **37.** 7 **39.** 2 **41.** 5 **43.** 2 **45.** 10
47. 4 **49.** 8 **51.** -1 **53.** $-\frac{4}{3}$ **55.** $\frac{2}{5}$
57. -2 **59.** -4 **61.** 0.8 **63.** 3 **65.** 28
67. 4 **69.** 3.6 billion **71.** 4.42
73. $x = -\frac{1}{3}, y = -5$ **75.** 2 **77.** $-\frac{7}{2}$ **79.** -2

Exercise Set 3.4, pp. 114–115

1. 6 **3.** 2 **5.** 6 **7.** 8 **9.** 4 **11.** 1
13. 17 **15.** −8 **17.** −$\frac{5}{3}$ **19.** −3 **21.** 2
23. 5 **25.** 4538 **27.** $\frac{837,353}{1929}$ **29.** −2
31. $a + 4$

Exercise Set 3.5, pp. 121–123

1. 19 **3.** −10 **5.** 40 **7.** 20 m, 40 m, 120 m
9. 37, 39 **11.** 56, 58 **13.** 35, 36, 37
15. 61, 63, 65
17. The length is 90 m; the width is 65 m.
19. The length is 49 m; the width is 27 m.
21. 22.5° **23.** $4400 **25.** $16 **27.** 450.5 mi
29. 28°, 84°, 68° **31.** 1984 **33.** 20 **35.** 19¢
37. 12 cm, 9 cm **39.** 30 **41.** 7.5%
43. 5 half dollars, 10 quarters, 20 dimes, 60 nickels

Exercise Set 3.6, pp. 127–129

1. (a) 68 ft^2; (b) $w = \frac{A}{l}$; (c) 8 cm
3. (a) $17\frac{3}{16}$ m^2; (b) $h = \frac{2A}{b}$; (c) 84 yd **5.** $b = \frac{A}{h}$
7. $r = \frac{d}{t}$ **9.** $P = \frac{I}{rt}$ **11.** $a = \frac{F}{m}$ **13.** $w = \frac{P - 2l}{2}$
15. $r^2 = \frac{A}{\pi}$ **17.** $m = \frac{E}{c^2}$ **19.** $b = 3A - a - c$
21. $t = \frac{3k}{v}$ **23.** $b = \frac{2A - ah}{h}$ **25.** $D^2 = \frac{2.5H}{N}$
27. $S = \frac{360A}{\pi r^2}$ **29.** $t = \frac{R - 3.85}{-0.0075}$, or $\frac{3.85 - R}{0.0075}$
31. Not necessarily **33.** A increases by $2h$ units
35. $c - ax$ **37.** $-\frac{b}{a}$ **39.** $\frac{t^2}{v}$ **41.** $\frac{g}{20} - 2n$
43. $\frac{y}{1-b}$ **45.** $\frac{1-de}{d}$, or $\frac{1}{d} - e$ **47.** $\frac{m - ax^2 - c}{x}$
49. 720° **51.** $\frac{A + 360}{180}$, or $\frac{A}{180} + 2$

Summary and Review: Chapter 3, pp. 130–131

1. [3.1] −22 **2.** [3.2] 7 **3.** [3.2] −192
4. [3.1] 1 **5.** [3.2] −$\frac{7}{3}$ **6.** [3.1] 25 **7.** [3.1] $\frac{1}{2}$
8. [3.2] −$\frac{15}{64}$ **9.** [3.1] 9.99 **10.** [3.3] −8
11. [3.3] −5 **12.** [3.3] −$\frac{1}{3}$ **13.** [3.3] 4
14. [3.3] 3 **15.** [3.3] 4 **16.** [3.3] 16

17. [3.4] 6 **18.** [3.4] −3 **19.** [3.4] 12
20. [3.4] 4 **21.** [3.6] $d = \frac{C}{\pi}$ **22.** [3.6] $B = \frac{3V}{h}$
23. [3.6] $a = 2A - b$ **24.** [3.5] $591
25. [3.5] 27 **26.** [3.5] 3 m, 5 m **27.** [3.5] 9
28. [3.5] 57, 59 **29.** [3.5] $w = 11$ cm, $l = 17$ cm
30. [3.5] $220 **31.** [3.5] $26,087
32. [3.5] 35°, 85°, 60°
33. [3.5] Amazon: 6437 km; Nile: 6671 km
34. [3.5] 140 **35.** [3.5] $14,150
36. [3.3] −23, 23 **37.** [3.2] −20, 20
38. [3.2] −24, 24 **39.** [3.6] $a = \frac{y - 3}{2 - b}$

Cumulative Review: Chapters 1–3, pp. 132–133

1. [1.2] $\frac{3}{2}$ **2.** [1.2] $\frac{15}{4}$ **3.** [1.6] 24 **4.** [1.6] 28
5. [1.6] 215 **6.** [1.6] 55 **7.** [1.2] $2w - 4$
8. [1.2] $3(x + y)$
9. [1.6] Answers may vary. $(q + r) + 6$, $6 + (q + r)$, $(r + 6) + q$
10. [1.3] $2 \cdot 2 \cdot 2 \cdot 3 \cdot 3 \cdot 3 \cdot 3$ **11.** [1.3] 120
12. [1.4] $\frac{5}{9}$ **13.** [1.4] $\frac{3}{2y}$ **14.** [1.10] 0.026
15. [1.10] $\frac{4}{5}$ **16.** [1.10] 190% **17.** [1.10] 87.5%
18. [2.1] > **19.** [2.2] > **20.** [2.2] >
21. [2.6] Additive inverse: −$\frac{2}{5}$; reciprocal: $\frac{5}{2}$
22. [2.3, 2.4] −9.2 **23.** [2.3, 2.4] 0 **24.** [2.5] $\frac{5}{6}$
25. [2.5] −105 **26.** [2.6] −9 **27.** [2.6] −0.3
28. [2.6] $\frac{32}{125}$ **29.** [1.7] $15x + 25y + 10z$
30. [2.7] $-12x - 8$ **31.** [2.7] $-12y + 24x$
32. [2.7] $5x + 5$ **33.** [1.7] $2(32 + 9x + 12y)$
34. [2.7] $8(2y - 7)$ **35.** [2.7] $-2(x + 4)$
36. [2.7] $5(a - 3b + 5)$ **37.** [1.7] $15b + 22y$
38. [1.7] $9y + 6z + 4$ **39.** [2.7] $-3a - 9d + 1$
40. [2.7] $-2.6x - 5.2y$ **41.** [2.8] $3x - 1$
42. [2.8] $-2x - y$ **43.** [2.8] $-7x + 6$
44. [2.8] $8x$ **45.** [2.8] $5x - 13$ **46.** [3.1] 4.5
47. [3.2] $\frac{4}{25}$ **48.** [3.1] 10.9 **49.** [3.1] $3\frac{5}{6}$
50. [3.2] −48 **51.** [3.2] 12 **52.** [3.2] −6.2
53. [3.3] −3 **54.** [3.3] −$\frac{12}{5}$ **55.** [3.3] 8
56. [3.4] 7 **57.** [3.3] −$\frac{4}{5}$ **58.** [3.3] −$\frac{10}{3}$
59. [3.6] $h = \frac{2A}{b + c}$ **60.** [1.10] 30%
61. [1.10] 50 **62.** [3.5] 22 **63.** [3.5] $45
64. [3.5] $1500 **65.** [3.5] 50 m, 53 m, 40 m
66. [1.5] $\frac{1}{64}$
67. [2.1] $-\left|-\frac{5}{2}\right|, -0.2^2, -0.2^3, |0|, (-0.2)^2, \left(-\frac{2}{5}\right)^2, \left|-\frac{1}{5}\right|$ **68.** [3.3] −5, 5 **69.** [3.4] All reals
70. [3.4] No solution **71.** [3.3, 3.4] 3

CHAPTER 4

Exercise Set 4.1, pp. 140–141

1. 2^7 3. 8^{14} 5. x^7 7. 9^{38} 9. $(3y)^{12}$
11. $(7y)^{17}$ 13. 7^3 15. 8^6 17. y^4 19. $\dfrac{1}{16^6}$
21. $\dfrac{1}{m^6}$ 23. $\dfrac{1}{(8x)^4}$ 25. 1 27. 1 29. 3^{12}
31. 6^{72} 33. x^{60} 35. n^{105} 37. $25a^{10}$
39. $81x^{20}y^{36}z^{12}$ 41. x^{200} 43. $2508.80
45. $22,318.40 47. 2^{13}
49. No, $(5y)^0 = 1$ and $5(y)^0 = 5 \cdot 1 = 5$
51. $\tfrac{1}{5}$ 53. 3^{11} 55. 1 57. 2
59. $3^{-2} = 3^{0-2} = \dfrac{3^0}{3^2} = \dfrac{1}{3^2}$
61. No, $(a^b)^c = a^{bc}$, $b \cdot c \neq b^c$ 63. $a = 81$
65. < 67. < 69. True 71. False 73. False

Exercise Set 4.2, pp. 145–147

1. -18 3. 19 5. -12 7. Approximately 449
9. 2 11. 4 13. 11 15. 68 ft 17. 1024 ft
19. 400 gal 21. $2, -3x, x^2$ 23. $6x^2$ and $-3x^2$
25. $2x^4$ and $-3x^4$, $5x$ and $-7x$ 27. $-3x$
29. $-8x$ 31. $11x^3 + 4$ 33. $x^3 - x$ 35. $4b^5$
37. $\tfrac{3}{4}x^5 - 2x - 42$ 39. x^4 41. $\tfrac{15}{16}x^3 - \tfrac{7}{6}x^2$
43. $3x^2 + 2x + 1$ 45. $3x^6$ 47. 50

Exercise Set 4.3, pp. 149–150

1. $x^5 + 6x^3 + 2x^2 + x + 1$
3. $15x^9 + 7x^8 + 5x^3 - x^2 + x$
5. $-5y^8 + y^7 + 9y^6 + 8y^3 - 7y^2$ 7. $x^6 + x^4$
9. $13x^3 - 9x + 8$ 11. $-5x^2 + 9x$
13. $12x^4 - 2x + \tfrac{1}{4}$ 15. 1, 0; 1 17. 2, 1, 0; 2
19. 3, 2, 1, 0; 3 21. 2, 1, 6, 4; 6 23. $-3, 6$
25. 5, 3, 3 27. $-7, 6, 3, 7$
29. $-5, 6, -3, 8, -2$ 31. x^2, x 33. x^3, x^2, x^0
35. None missing 37. Trinomial
39. None of these 41. Binomial 43. Monomial
45. $0.125 47. Any ax^5 plus 3 terms of degree <5
49. 10
51. $ax^3 + (a-3)x^2 + 3(a-3)x + (a+2)$,
$a + a - 3 + 3(a-3) + (a+2) = -4$, $a = 1$,
$x^3 - 2x^2 - 6x + 3$ is the polynomial

Exercise Set 4.4, pp. 152–155

1. $-x + 5$ 3. $x^2 - 5x - 1$
5. $3x^5 + 13x^2 + 6x - 3$
7. $-4x^4 + 6x^3 + 6x^2 + 2x + 4$ 9. $6 + 12x^2$
11. $5x^4 - 2x^3 - 7x^2 - 5x$
13. $9x^8 + 8x^7 - 3x^4 + 2x^2 - 2x + 5$
15. $-\tfrac{1}{2}x^4 + \tfrac{2}{3}x^3 + x^2$
17. $0.01x^5 + x^4 - 0.2x^3 + 0.2x + 0.06$
19. $-3x^4 + 3x^2 + 4x$
21. $3x^5 - 3x^4 - 3x^3 + x^2 + 3x$
23. $5x^3 - 9x^2 + 4x - 7$
25. $\tfrac{1}{4}x^4 - \tfrac{1}{4}x^3 + \tfrac{3}{2}x^2 + 6\tfrac{3}{4}x + \tfrac{1}{4}$
27. $-x^4 + 3x^3 + 2x + 1$ 29. $x^4 + 4x^2 + 12x - 1$
31. $x^5 - 6x^4 + 4x^3 - x^2 + 1$
33. $7x^4 - 2x^3 + 7x^2 + 4x + 9$
35. $-6 + 3x + x^2 + 3x^3 + x^4 + 3x^5$
37. $1.05x^4 + 0.36x^3 + 14.22x^2 + x + 0.97$
39. (a) $5x^2 + 4x$; (b) 57, 352 41. $14y + 17$
43. $9r + 99 + r^2 + 11r = r^2 + 20r + 99$
45. $(x+3)^2 = x^2 + 3x + 3x + 3^2 = x^2 + 6x + 9$
47. (a) Compare $(ax + b) + (cx + d)$ and
$(cx + d) + (ax + b)$.
$(ax + b) + (cx + d) = ax + b + cx + d =$
$(a + c)x + (b + d)$ (collecting like terms);
$(cx + d) + (ax + b) = cx + d + ax + b =$
$(c + a)x + (d + b) = (a + c)x + (b + d)$
(commutative property of addition of real numbers).
Since $(ax + b) + (cx + d) = (cx + d) + (ax + b)$,
addition of binomials is commutative. (b) Compare
$(ax^2 + bx + c) + (dx^2 + ex + f)$ and
$(dx^2 + ex + f) + (ax^2 + bx + c)$ as in (a).
49. 20, 25, 30

Exercise Set 4.5, pp. 158–160

1. $-(-5x), 5x$
3. $-(-x^2 + 10x - 2), x^2 - 10x + 2$
5. $-(12x^4 - 3x^3 + 3), -12x^4 + 3x^3 - 3$
7. $-3x + 7$ 9. $-4x^2 + 3x - 2$
11. $4x^4 + 6x^2 - \tfrac{3}{4}x + 8$ 13. $2x^2 + 14$
15. $-2x^5 - 6x^4 + x + 2$ 17. $9x^2 + 9x - 8$
19. $\tfrac{3}{4}x^3 - \tfrac{1}{2}x$ 21. $0.06x^3 - 0.05x^2 + 0.01x + 1$
23. $3x + 6$ 25. $4x^3 - 3x^2 + x + 1$
27. $11x^4 + 12x^3 - 9x^2 - 8x - 9$
29. $-4x^5 + 9x^4 + 6x^2 + 16x + 6$
31. $x^4 - x^3 + x^2 - x$ 33. $\pi r^2 - 9\pi$
35. $z^2 - 24(z - 3) - (24 \cdot 3) - 3(z - 24) =$
$z^2 - 27z + 72$ 37. $144 - 4x^2$ 39. $y - 9$
41. $11a^2 - 18a - 4$ 43. $-10y^2 - 2y - 10$
45. $-3y^4 - y^3 + 5y - 2$
47. $569.607x^3 - 15.168x$

Exercise Set 4.6, pp. 162–164

1. $42x^2$ 3. x^4 5. $-x^8$ 7. $6x^6$ 9. $28t^8$
11. $-0.02x^{10}$ 13. $\tfrac{1}{15}x^4$ 15. 0 17. $8x^2 - 12x$
19. $-10x^3 + 5x^2$ 21. $-4x^4 + 4x^3$

23. $4y^7 - 24y^6$ 25. $20x^{38} - 50x^{25} + 25x^{14}$
27. $-66y^{108} + 42y^{58} - 66y^{49} + 360y^{12} - 54y^8$
29. $x^2 + 7x + 10$ 31. $x^2 + 4x - 12$
33. $x^2 - 10x + 21$ 35. $x^2 - 36$
37. $18 + 12x + 2x^2$ 39. $9x^2 - 24x + 16$
41. $4y^2 - 1$ 43. $x^2 + \frac{17}{6}x + 2$ 45. $x^3 + x^2 + 4$
47. $12x^3 - 10x^2 - x + 1$
49. $3y^4 + 18y^3 - 18y - 3$ 51. $x^6 - 2x^5 + 2x^4 - x^3$
53. $-20x^5 + 25x^4 - 4x^3 - 5x^2 - 2$
55. $1 - 2x + 3x^2 - 2x^3 + x^4$
57. $4x^4 - 12x^3 - 5x^2 + 17x + 6$
59. $6a^4 - 19a^3 + 31a^2 - 26a + 8$
61. $-6x^4 + 4x^3 + 36x^2 - 20x + 2$
63. $3x^3 + 3x^7 + 3x^{11}$
65. $x^4 - 3x^3 + 3x^2 - 4x + 4$
67. $x^4 + 3x^3 - 5x^2 + 16$ 69. $a^2 - 2ab + b^2$
71. $25y^2 + 60y + 36$
73. $84t^2 + 32t - (6t^2 - 8t) = 78t^2 + 40t$
75. $x^3 - 5x^2 + 8x - 4$ 77. 8 ft × 16 ft
79. (a) $2x^2 - 18x + 28$; (b) 0
81. (a) $2x^2 - 112$; (b) $-6x$

Exercise Set 4.7, pp. 165–167

1. $4x^2 + 4x$ 3. $-3x^2 + 3x$ 5. $x^5 + x^2$
7. $6x^3 - 18x^2 + 3x$ 9. $x^3 + 3x + x^2 + 3$
11. $x^4 + x^3 + 2x + 2$ 13. $x^2 - x - 6$
15. $9x^2 + 15x + 6$ 17. $5x^2 + 4x - 12$
19. $9x^2 - 1$ 21. $4x^2 - 6x + 2$ 23. $x^2 - \frac{1}{16}$
25. $x^2 - 0.01$ 29. $2x^3 + 2x^2 + 6x + 6$
29. $-2x^2 - 11x + 6$ 31. $x^2 + 14x + 49$
33. $1 - x - 6x^2$ 35. $x^5 - x^2 + 3x^3 - 3$
37. $x^3 - x^2 - 2x + 2$ 39. $3x^6 - 6x^2 - 2x^4 + 4$
41. $6x^7 + 18x^5 + 4x^2 + 12$ 43. $8x^6 + 65x^3 + 8$
45. $4x^3 - 12x^2 + 3x - 9$ 47. $4x^6 + 4x^5 + x^4 + x^3$
49. $8y^3 + 72y^2 + 160y$
51. $-2x^4 + x^3 + 5x^2 - x - 2$ 53. -7
55. $-\frac{5}{2}$ 57. $w(w+1)(w+2) = w^3 + 3w^2 + 2w$
59. $h(h-1)(h-2) = h^3 - 3h^2 + 2h$
61. $F^2 - (F-17)(F-7)$ or
$17F + 7(F-17) = 24F - 119$
63. Each trip is $0.70 + (3\frac{6}{7})(7)(0.10) = 3.40$,
$3.40(11x + 6) = 37.40x + 20.40$.

Exercise Set 4.8, pp. 170–171

1. $x^2 - 16$ 3. $4x^2 - 1$ 5. $25m^2 - 4$
7. $4x^4 - 9$ 9. $9x^8 - 16$ 11. $x^{12} - x^4$
13. $x^8 - 9x^2$ 15. $x^{24} - 9$ 17. $4x^{16} - 9$
19. $x^2 + 4x + 4$ 21. $9x^4 + 6x^2 + 1$
23. $x^2 - x + \frac{1}{4}$ 25. $9 + 6x + x^2$

27. $x^4 + 2x^2 + 1$ 29. $4 - 12x^4 + 9x^8$
31. $25 + 60x^2 + 36x^4$ 33. $9 - 12x^3 + 4x^6$
35. $4x^3 + 24x^2 - 12x$ 37. $4x^4 - 2x^2 + \frac{1}{4}$
39. $-1 + 9p^2$ 41. $15t^5 - 3t^4 + 3t^3$
43. $36x^8 + 48x^4 + 16$ 45. $12x^3 + 8x^2 + 15x + 10$
47. $64 - 96x^4 + 36x^8$
49. $4567.0564x^2 + 435.891x + 10.400625$
51. $(100 - 7)(100 + 7) = 10{,}000 - 49 = 9951$
53. $a^2 + 10a + 24$ 55. $x^2 + 10x + 25$ 57. 6
59. $\frac{1}{17}$
61. (a) $a^2 + ab$; (b) $ab + b^2$;
    (c) $a^2 + ab - (b^2 + ab) = a^2 - b^2$;
    (d) $a(a-b) + b(a-b) = a^2 - b^2$
63. $100(x^2 + x) + 25$. Add the first digit to its square, multiply by 100, and add 25.

Summary and Review: Chapter 4, pp. 171–173

1. [4.1] 7^6 2. [4.1] y^{11} 3. [4.1] $(3x)^{14}$
4. [4.1] t^8 5. [4.1] 4^3 6. [4.1] $\frac{1}{a^3}$ 7. [4.1] 1
8. [4.1] x^{12} 9. [4.1] $9t^8$ 10. [4.1] $-8x^3y^6$
11. [4.2] -17 12. [4.2] 10 13. [4.2] $3x^2, 6x, \frac{1}{2}$
14. [4.2] $-4y^5, 7y^2, -3y, -2$ 15. [4.3] x^2, x^0
16. [4.3] 6, 17 17. [4.3] 4, 6, $-5, \frac{5}{3}$
18. [4.3] 3, 1, 0; 3 19. [4.3] 0, 4, 9, 6, 3; 9
20. [4.3] Binomial 21. [4.3] None of these
22. [4.3] Monomial 23. [4.2] $-x^2 + 9x$
24. [4.2] $-\frac{1}{4}x^3 + 4x^2 + 7$ 25. [4.2] $-3x^5 + 25$
26. [4.3] $-2x^2 - 3x + 2$
27. [4.3] $10x^4 - 7x^2 - x - \frac{1}{2}$
28. [4.4] $x^5 - 2x^4 + 6x^3 + 3x^2 - 9$
29. [4.4] $2x^5 - 6x^4 + 2x^3 - 2x^2 + 2$
30. [4.4] $\frac{3}{4}x^4 + \frac{1}{4}x^3 - \frac{1}{3}x^2 - \frac{7}{4}x + \frac{3}{8}$
31. [4.5] $2x^2 - 4x - 6$
32. [4.5] $x^5 - 3x^3 - 2x^2 + 8$
33. [4.5] $-x^5 + x^4 - 5x^3 - 2x^2 + 2x$
34. (a) [4.4] $4L + 8$; (b) [4.5] $L^2 + 4L$
35. [4.6] $-12x^3$ 36. [4.8] $49x^2 + 14x + 1$
37. [4.7] $x^2 + \frac{7}{6}x + \frac{1}{3}$
38. [4.7] $0.3x^2 + 0.65x - 8.45$
39. [4.6] $12x^3 - 23x^2 + 13x - 2$
40. [4.8] $x^2 - 18x + 81$
41. [4.6] $15x^7 - 40x^6 + 50x^5 + 10x^4$
42. [4.7] $x^2 - 3x - 28$
43. [4.7] $x^2 - 1.05x + 0.225$
44. [4.6] $x^7 + x^5 - 3x^4 + 3x^3 - 2x^2 + 5x - 3$
45. [4.8] $9x^4 - 12x^3 + 4x^2$
46. [4.7] $2x^4 - 11x^2 - 21$
47. [4.6] $4x^5 - 5x^4 - 8x^3 + 22x^2 - 15x$
48. [4.8] $9x^4 - 16$ 49. [4.8] $4 - x^2$
50. [4.7] $13x^2 - 172x + 39$

51. [4.6–4.8] (a) 3; (b) 2; (c) 1
52. [4.1, 4.2] $-28x^8$
53. [4.3] $8x^4 + 4x^3 + 5x - 2$
54. [4.7] $-4x^6 + 3x^4 - 20x^3 + x^2 - 16$
55. [4.8] $\frac{94}{13}$

CHAPTER 5

Exercise Set 5.1, pp. 178–179

1. Answers may vary. $(6x)(x^2)$, $(3x^2)(2x)$, $(2x^2)(3x)$
3. Answers may vary. $(-3x^2)(3x^3)$, $(-x)(9x^4)$, $(3x^2)(-3x^3)$
5. Answers may vary. $(6x)(4x^3)$, $(-3x^2)(-8x^2)$, $(2x^3)(12x)$ **7.** $x(x-4)$ **9.** $2x(x+3)$
11. $x^2(x+6)$ **13.** $8x^2(x^2-3)$
15. $2(x^2+x-4)$ **17.** $17x(x^4+2x^2+3)$
19. $x^2(6x^2-10x+3)$ **21.** $x^2(x^3+x^2+x-1)$
23. $2x^3(x^4-x^3-32x^2+2)$
25. $0.8x(2x^3-3x^2+4x+8)$
27. $\frac{1}{3}x^3(5x^3+4x^2+x+1)$ **29.** $(y+1)(y+4)$
31. $(x-1)(x-4)$ **33.** $(2x+3)(3x+2)$
35. $(x-4)(3x-4)$ **37.** $(5x+3)(7x-8)$
39. $(2x-3)(2x+3)$ **41.** $(2x^2+5)(x^2+3)$
43. $(x+3)(2x^2+1)$ **45.** $(2x-3)(4x^2+3)$
47. $(3x-4)(4x^2+1)$ **49.** $(x+8)(x^2-3)$
51. No   **53.** No   **55.** No   **57.** Yes   **59.** Yes
61. $(2x^2+3)(2x^3+3)$
63. $(x^4+1)(x^2+1)$
65. $x^2(x+1)^2 - (x^2+1)^2 = x^4 + 2x^3 + x^2 - (x^4 + 2x^2 + 1) = 2x^3 - 2x^2 + x^2 - 1 = 2x^2(x-1) + (x+1)(x-1) = (2x^2+x+1)(x-1)$

Exercise Set 5.2, pp. 181–182

1. Yes   **3.** No   **5.** No   **7.** Yes   **9.** No
11. $(x-6)(x+6)$ **13.** $(x-1)(x+1)$
15. $(5x-2)(5x+2)$ **17.** $(3a-4)(3a+4)$
19. $6(2x-3)(2x+3)$ **21.** $x(4-9x)(4+9x)$
23. $x^2(x^7-3)(x^7+3)$ **25.** $(5a^2-3)(5a^2+3)$
27. $(11a^4-10)(11a^4+10)$
29. $(10y^3-7)(10y^3+7)$
31. $(x^2+4)(x+2)(x-2)$
33. $5(x^2+4)(x+2)(x-2)$
35. $(x^4+1)(x^2+1)(x+1)(x-1)$
37. $(x^4+9)(x^2+3)(x^2-3)$ **39.** $(\frac{1}{5}+x)(\frac{1}{5}-x)$
41. $(2+\frac{1}{3}y)(2-\frac{1}{3}y)$ **43.** $(1+a^2)(1+a)(1-a)$
45. $100°$, $25°$, $55°$ **47.** $3x^3(x+2)(x-2)$
49. $2x(3x+\frac{2}{5})(3x-\frac{2}{5})$ **51.** $x(x+\frac{1}{1.3})(x-\frac{1}{1.3})$

53. $(0.8x+1.1)(0.8x-1.1)$ **55.** $x(x+6)$
57. $3(a-1)(3a+11)$
59. $(y^4+16)(y^2+4)(y+2)(y-2)$
61. $\left(x-\frac{1}{x}\right)\left(x+\frac{1}{x}\right)$, or $\frac{1}{x^2}(x^2+1)(x+1)(x-1)$
63. Irreducible   **65.** Prime, irreducible
67. Irreducible

Exercise Set 5.3, pp. 184–185

1. Yes   **3.** No   **5.** No   **7.** No   **9.** No
11. $(x-8)^2$ **13.** $(x+7)^2$ **15.** $(x+1)^2$
17. $(x-2)^2$ **19.** $(y+3)^2$ **21.** $2(x-10)^2$
23. $x(x+12)^2$ **25.** $3(2x+3)^2$ **27.** $(8-7x)^2$
29. $(a^2+7)^2$ **31.** $(y^3-8)^2$ **33.** $(3x^5+2)^2$
35. $(1-a^3)^2$ **37.** $(\frac{1}{3}a+\frac{1}{2})^2$
39. Length: 144.5 m; width: 125.5 m
41. $x(27x^2-13)$ **43.** Not possible
45. $2x(3x+1)^2$ **47.** $2(81x^2-41)$
49. Not possible   **51.** $(9+x^2)(3+x)(3-x)$
53. $(a+3)^2$ **55.** $(7x+4)^2$ **57.** $(a+1)^2$
59. No. $(x+3)^2(x-3)^2 = [(x+3)(x-3)]^2 = (x^2-9)^2 = x^4 - 18x^2 + 81$ **61.** $(x^n+5)^2$
63. $(y+3)^2 - (x+4)^2 = (y+x+7)(y-x-1)$
65. 16
67. $x^2+a^2x+a^2 = x^2+2ax+a^2$, $a^2x = 2ax$, $a^2 = 2a$. Then $a=0$ or $a=2$.

Exercise Set 5.4, pp. 188–189

1. $(x+5)(x+3)$ **3.** $(x+4)(x+3)$
5. $(x-3)^2$ **7.** $(x+7)(x+2)$ **9.** $(b+4)(b+1)$
11. $(x+\frac{1}{3})^2$ **13.** $(d-5)(d-2)$
15. $(y-10)(y-1)$ **17.** $(x+7)(x-6)$
19. $(x+2)(x-9)$ **21.** $(x+2)(x-8)$
23. $(y+5)(y-9)$ **25.** $(x+9)(x-11)$
27. $(c+8)(c-7)$ **29.** $(a+7)(a-5)$
31. $(x+10)^2$ **33.** $(x-25)(x+4)$
35. $(x-24)(x+3)$ **37.** $(x-16)(x-9)$
39. $(a+12)(a-11)$ **41.** $(x-15)(x-8)$
43. $(12+x)(9-x)$
45. $15, -15, 27, -27, 51, -51$
47. $(x+\frac{1}{4})(x-\frac{3}{4})$ **49.** $(x+5)(x-\frac{5}{7})$
51. $2x^2(4-\pi)$

Exercise Set 5.5, pp. 194–195

1. $(2x+1)(x-4)$ **3.** $(5x-9)(x+2)$
5. $(2x+7)(3x+1)$ **7.** $(3x+1)(x+1)$
9. $(2x+5)(2x-3)$ **11.** $(2x+1)(x-1)$

CHAPTER 5

13. $(3x+8)(3x-2)$ **15.** $(3x+1)(x-2)$
17. $(3x+4)(4x+5)$ **19.** $(7x-1)(2x+3)$
21. $(3x+4)(3x+2)$ **23.** $(7-3x)^2$
25. $(x+2)(24x-1)$ **27.** $(7x+4)(5x-11)$
29. $2(5-x)(2+x)$ **31.** $4(3x-2)(x+3)$
33. $6(5x-9)(x+1)$ **35.** $2(3x+5)(x-1)$
37. $(3x-1)(x-1)$ **39.** $4(3x+2)(x-3)$
41. $(2x+1)(x-1)$ **43.** $(3x-8)(3x+2)$
45. $5(3x+1)(x-2)$ **47.** $x(3x+4)(4x+5)$
49. $x^2(2x+3)(7x-1)$ **51.** $3x(8x-1)(7x-1)$
53. Lamps: 500 watts; air conditioner: 2000 watts; television: 50 watts **55.** Not factorable
57. $(5x^2-3)(3x^2-2)$ **59.** $3x(2x-3)(3x+1)$
61. Not factorable **63.** $(3x+7)(3x-7)(3x-7)$
65. $(-3x^m+4)(5x^m-2)$ **67.** $x(x^n-1)^2$

Exercise Set 5.6, pp. 197–198

1. $2(x+8)(x-8)$ **3.** $(a-5)^2$
5. $(2x-3)(x-4)$ **7.** $x(x+12)^2$
9. $(x-2)(x+2)(x+3)$ **11.** $6(2x+3)(2x-3)$
13. $4x(x-2)(5x+9)$ **15.** Not factorable
17. $x(x-3)(x^2+7)$ **19.** $x^3(x-7)^2$
21. $-2(x-2)(x+5)$ **23.** Not factorable
25. $4(x^2+4)(x+2)(x-2)$
27. $(y^4+1)(y^2+1)(y+1)(1-y)$
29. $x^3(x-3)(x-1)$ **31.** $(6a-\frac{5}{4})^2$
33. $(a+1)(a+1)(a-1)(a-1)$ **35.** $(3.5x-1)^2$
37. $5(x+1.8)(x+0.8)$ **39.** $(y-2)(y+3)(y-3)$
41. $(a+4)(a^2+1)$ **43.** $(x+3)(x-3)(x^2+2)$
45. $(x+2)(x-2)(x-1)$ **47.** $(y-1)^3$
49. $(a^2+9)(a+3)(a-3)$

Exercise Set 5.7, pp. 201–202

1. $-8, -6$ **3.** $3, -5$ **5.** $-12, 11$ **7.** $0, -5$
9. $0, 13$ **11.** $0, -10$ **13.** $-\frac{5}{2}, -4$ **15.** $\frac{1}{3}, -2$
17. $-\frac{1}{5}, 3$ **19.** $4, \frac{1}{4}$ **21.** $0, \frac{2}{3}$ **23.** $0, 18$
25. $\frac{1}{9}, \frac{1}{10}$ **27.** $2, 6$ **29.** $\frac{1}{3}, 20$ **31.** $0, \frac{2}{3}, \frac{1}{2}$
33. $-5, -1$ **35.** $2, -9$ **37.** $5, 3$ **39.** $0, 8$
41. $0, -19$ **43.** $4, -4$ **45.** $\frac{2}{3}, -\frac{2}{3}$ **47.** -3
49. 4 **51.** $0, \frac{6}{5}$ **53.** $-1, \frac{5}{3}$ **55.** $\frac{2}{3}, -\frac{1}{4}$
57. $7, -2$ **59.** $\frac{9}{8}, -\frac{9}{8}$ **61.** $-3, 1$ **63.** $\frac{4}{5}, \frac{3}{2}$
65. $4, -5$ **67.** $9, -3$ **69.** $\frac{1}{8}, -\frac{1}{8}$ **71.** $4, -4$
73. (a) $x^2 + 2x - 3 = 0$; (b) $x^2 - 2x - 3 = 0$;
(c) $x^2 - 4x + 4 = 0$; (d) $x^2 - 7x + 12 = 0$;
(e) $x^2 + x - 12 = 0$; (f) $x^2 - x - 12 = 0$;
(g) $x^2 + 7x + 12 = 0$; (h) $x^2 - x + \frac{1}{4} = 0$ or $4x^2 - 4x + 1 = 0$; (i) $x^2 - 25 = 0$;
(j) $x^3 - \frac{14}{40}x^2 + \frac{1}{40}x = 0$ or $40x^3 - 14x^2 + x = 0$
75. $-2000, -51.546392$

Exercise Set 5.8, pp. 206–208

1. $-\frac{3}{4}, 1$ **3.** $2, 4$ **5.** 13 and 14, -13 and -14
7. 12 and 14, -12 and -14
9. 15 and 17, -15 and -17
11. Length: 12 m, width: 8 m **13.** 5
15. Height: 4 cm; base: 14 cm **17.** 6 m
19. 5 and 7 **21.** 506 **23.** 12 **25.** 780 **27.** 20
29. 5 ft **31.** (a) 2 seconds; (b) 4.2 seconds
33. 7 m **35.** 5 in.

Exercise Set 5.9, pp. 214–217

1. -1 **3.** -7 **5.** $\$12{,}597.12$ **7.** 44.4624 in^2
9. Coefficients: $1, -2, 3, -5$; degrees: $4, 2, 2, 0; 4$
11. Coefficients: $17, -3, -7$; degrees: $5, 5, 0; 5$
13. $-a - 2b$ **15.** $3x^2y - 2xy^2 + x^2$
17. $8u^2v - 5uv^2$ **19.** $-8au + 10av$
21. $x^2 - 4xy + 3y^2$ **23.** $3r + 7$
25. $-x^2 - 8xy - y^2$ **27.** $2ab$
29. $-2a + 10b - 5c + 8d$ **31.** $6z^2 + 7zu - 3u^2$
33. $a^4b^2 - 7a^2b + 10$
35. $a^3 + a^3 - a^2y - ay + a + y - 1$ **37.** $a^6 - b^2c^2$
39. $y^6x + y^4x + y^4 + 2y^2 + 1$
41. $12x^2y^2 + 2xy - 2$ **43.** $12 - c^2d^2 - c^4d^4$
45. $m^3 + m^2n - mn^2 - n^3$
47. $x^9y^9 - x^6y^6 + x^5y^5 - x^2y^2$ **49.** $x^2 + 2xh + h^2$
51. $r^6t^4 - 8r^3t^2 + 16$ **53.** $p^8 + 2m^2n^2p^4 + m^4n^4$
55. $4a^6 - 2a^3b^3 + \frac{1}{4}b^6$ **57.** $3a^3 - 12a^2b + 12ab^2$
59. $4a^2 - b^2$ **61.** $c^4 - d^2$ **63.** $a^2b^2 - c^2d^4$
65. $x^2 + 2xy + y^2 - 9$ **67.** $x^2 - y^2 - 2yz - z^2$
69. $a^2 - b^2 - 2bc - c^2$ **71.** $12n^2(1+2n)$
73. $9xy(xy - 4)$ **75.** $2\pi r(h+r)$
77. $(a+b)(2x+1)$ **79.** $(x-1-y)(x+1)$
81. $(n+p)(n+2)$ **83.** $(2x+z)(x-2)$
85. $(x-y)^2$ **87.** $(3c+d)^2$ **89.** $(7m^2-8n)^2$
91. $(y^2+5z^2)^2$ **93.** $(\frac{1}{2}a + \frac{1}{3}b)^2$
95. $(a+b)(a-2b)$ **97.** $(m+20n)(m-18n)$
99. $(mn-8)(mn+4)$ **101.** $a^3(ab+5)(ab-2)$
103. $a^3(a-b)(a+5b)$ **105.** $(x^3-y)(x^3+2y)$
107. $(x-y)(x+y)$
109. $7(p^2+q^2)(p+q)(p-q)$
111. $(9a^2+b^2)(3a-b)(3a+b)$
113. $4xy - 4y^2$ **115.** $2xy + \pi x^2$
117. $(A+B)^3 = A^3 + 3A^2B + 3AB^2 + B^3$
119. $[(y+4)+x]^2$

Summary and Review: Chapter 5, pp. 217–218

1. [5.1] Answers may vary. $(-5x)(2x)$, $(-x)(10x)$, $(-2x)(5x)$

2. [5.1] Answers may vary. $(4x^2)(9x^3)$, $(18x)(2x^4)$, $(-6x^3)(-6x^2)$
3. [5.2] $5(1 + 2x^3)(1 - 2x^3)$ **4.** [5.1] $x(x - 3)$
5. [5.2] $(3x + 2)(3x - 2)$ **6.** [5.4] $(x + 6)(x - 2)$
7. [5.3] $(x + 7)^2$ **8.** [5.1] $3x(2x^2 + 4x + 1)$
9. [5.1] $(x^2 + 3)(x + 1)$
10. [5.5] $(3x - 1)(2x - 1)$
11. [5.2] $(x^2 + 9)(x + 3)(x - 3)$
12. [5.5] $3x(3x - 5)(x + 3)$
13. [5.2] $2(x + 5)(x - 5)$
14. [5.1] $(x^3 - 2)(x + 4)$
15. [5.2] $(4x^2 + 1)(2x + 1)(2x - 1)$
16. [5.1] $4x^4(2x^2 - 8x + 1)$ **17.** [5.3] $3(2x + 5)^2$
18. [5.2] Not factorable **19.** [5.4] $x(x - 6)(x + 5)$
20. [5.2] $(2x + 5)(2x - 5)$ **21.** [5.3] $(3x - 5)^2$
22. [5.4] $2(3x + 4)(x - 6)$ **23.** [5.3] $(x - 3)^2$
24. [5.5] $(2x + 1)(x - 4)$ **25.** [5.3] $2(3x - 1)^2$
26. [5.2] $3(x + 3)(x - 3)$ **27.** [5.4] $(x - 5)(x - 3)$
28. [5.3] $(5x - 2)^2$ **29.** [5.7] $1, -3$
30. [5.7] $-7, 5$ **31.** [5.7] $-4, 3$
32. [5.7] $\frac{2}{3}, 1$ **33.** [5.7] $\frac{3}{2}, -4$ **34.** [5.7] $8, -2$
35. [5.8] $3, -2$ **36.** [5.8] -18 and -16, 16 and 18
37. [5.8] -19 and -17, 17 and 19
38. [5.8] $\frac{5}{2}, -2$ **39.** [5.9] 49
40. [5.9] Coefficients: $1, -7, 9, -8$; degrees: $6, 2, 2, 0$; 6
41. [5.9] Coefficients: $1, -1, 1$; degrees: $16, 40, 23$; 40
42. [5.9] $9w - y - 5$
43. [5.9] $m^6 - 2m^2n + 2m^2n^2 + 8n^2m - 6m^3$
44. [5.9] $-9xy - 2y^2$
45. [5.9] $11x^3y^2 - 8x^2y - 6x - 6x^2 + 6$
46. [5.9] $p^3 - q^3$ **47.** [5.9] $9a^8 - 2a^4b^3 + \frac{1}{9}b^6$
48. [5.9] $(xy + 4)(xy - 3)$ **49.** [5.9] $3(2a + 7b)^2$
50. [5.9] $(m + t)(m + 5)$ **51.** [5.8] $2\frac{1}{2}$ cm
52. [5.8] $0, 2$ **53.** [5.8] $l = 12, w = 6$
54. [5.2] No real solution **55.** [5.7] $2, -3, \frac{5}{2}$

CHAPTER 6

Exercise Set 6.1, pp. 224–225

1. [Graph showing points $(2, 5)$, $(0, 4)$, $(-1, 3)$, $(-5, 0)$, $(5, 0)$, $(3, -2)$, $(-2, -4)$, $(0, -5)$]

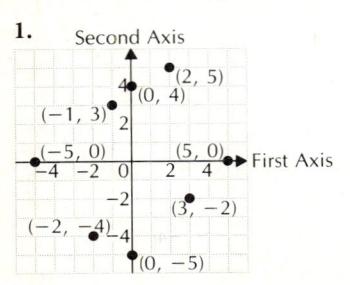

3. II **5.** IV **7.** III **9.** I
11. Negative; negative
13. A: $(3, 3)$; B: $(0, -4)$; C: $(-5, 0)$; D: $(-1, -1)$; E: $(2, 0)$
15. [Graph showing points $(2, 1)$, $(1, -1)$, $(0, -3)$, $(-1, -5)$]
17. 44.330 mi
19. I or IV
21. I or III
23. $(-1, -5)$
25. Answers may vary. **27.** 26
[Graph with descending series of points]

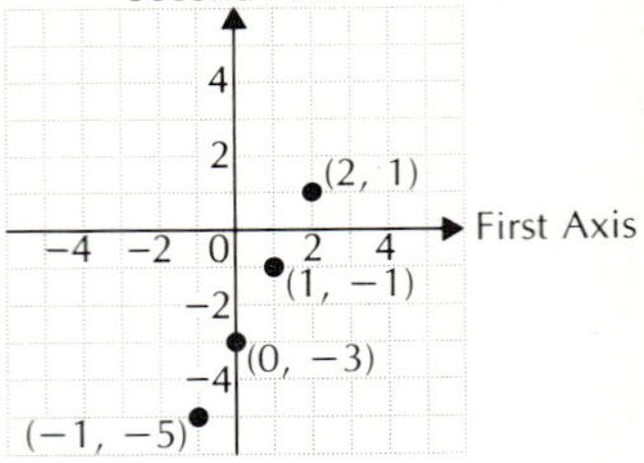

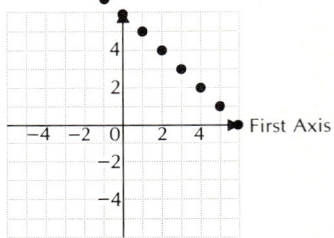

Exercise Set 6.2, p. 231

1. Yes **3.** No **5.** No
7. [Graph of $y = 4x$]
9. [Graph of $y = -2x$]
11. [Graph of $y = \frac{1}{3}x$]
13. [Graph of $y = -\frac{3}{2}x$]

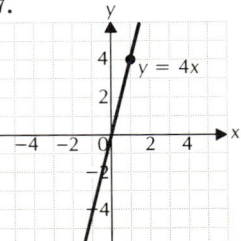

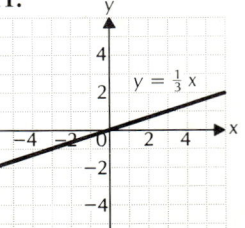

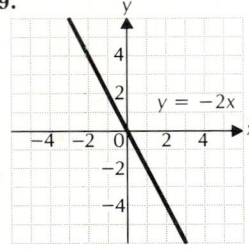

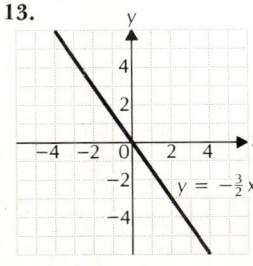

13. [graph: 4x + 5y = 20, through (0, 4) and (5, 0)]

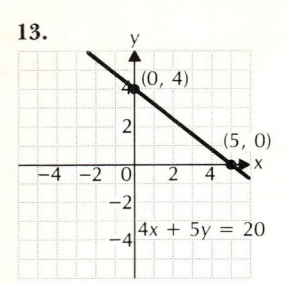

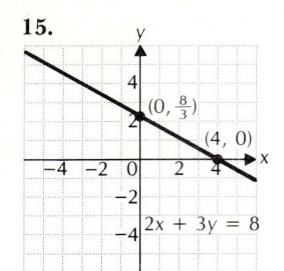

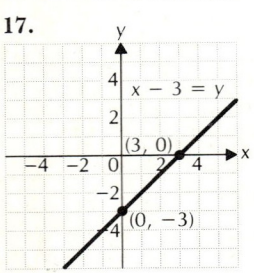

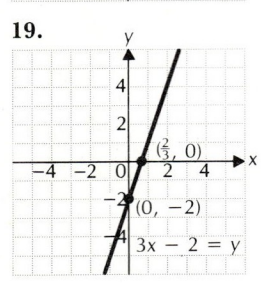

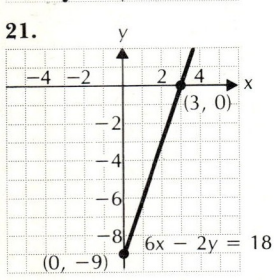

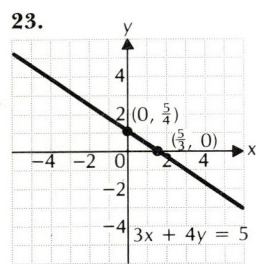

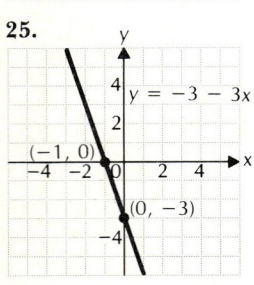

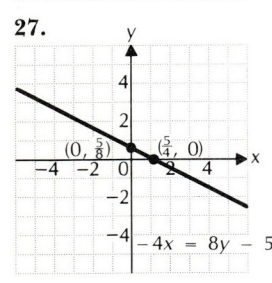

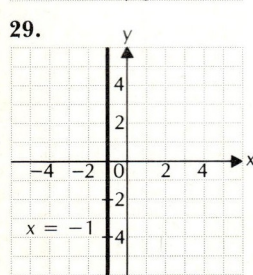

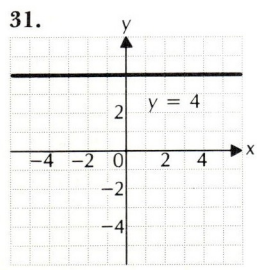

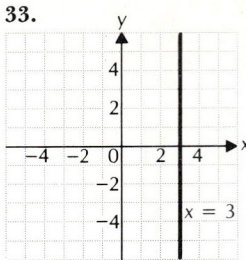

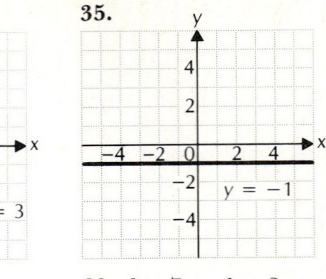

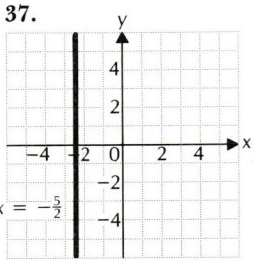

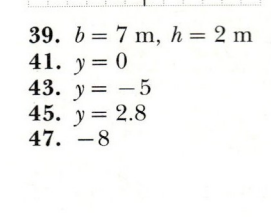

15. [graph: 2x + 3y = 8, through (0, 8/3) and (4, 0)]

17. [graph: x − 3 = y, through (3, 0) and (0, −3)]

19. [graph: 3x − 2 = y, through (2/3, 0) and (0, −2)]

21. [graph: 6x − 2y = 18, through (3, 0) and (0, −9)]

23. [graph: 3x + 4y = 5, through (0, 5/4) and (5/3, 0)]

25. [graph: y = −3 − 3x, through (−1, 0) and (0, −3)]

27. [graph: −4x = 8y − 5, through (0, 5/8) and (5/4, 0)]

29. [graph: $x = -1$]

31. [graph: $y = 4$]

33. [graph: $x = 3$]

35. [graph: $y = -1$]

37. [graph: $x = -\frac{5}{2}$]

39. $b = 7$ m, $h = 2$ m
41. $y = 0$
43. $y = -5$
45. $y = 2.8$
47. -8

Exercise Set 6.4, pp. 238–239

1. $x + y = 58$, $x - y = 16$, x is one number, y is the other number
3. $2l + 2w = 400$, $w = l - 40$, $l =$ length, $w =$ width
5. $53.95 + 0.30m = c$, $54.95 + 0.20m = c$, $m =$ mileage, $c =$ cost
7. $x - y = 16$, $3x = 7y$, $x =$ larger, $y =$ smaller
9. $x + y = 180$, $y = 3x + 8$, x and y are the angles
11. $x + y = 90$, $x - y = 34$, x and y are the angles
13. $x + y = 820$, $y = x + 140$, $x =$ hectares of Riesling, $y =$ hectares of Chardonnay
15. $x - y = 18$, $2y + 3x = 74$, $x =$ larger, $y =$ smaller
17. $2l + 2w = 76$, $l = w + 17$, $l =$ length, $w =$ width
19. $x = 0.2y$, $x + 20 = 0.52(y + 20)$, $x =$ Patrick's age, $y =$ his father's age
21. $\frac{1}{3}(b + 2) = h - 1$, $\frac{1}{2}(b + 2)(h - 1) = 24$, $b =$ base, $h =$ height

Exercise Set 6.5, pp. 242–243

1. Yes **3.** No **5.** Yes **7.** Yes **9.** Yes
11. (2, 1) **13.** (−12, 11) **15.** (4, 3)
17. (−3, −3) **19.** No solution **21.** (2, 2)
23. (5, 3) **25.** Infinitely many solutions
27. $A = 2$, $B = 2$ **29.** Yes
31. The solution is $(\frac{2}{3}, \frac{3}{7})$, but it would be very difficult to discover by graphing. The check would tell you whether the possible pair you found by graphing is a solution.

Exercise Set 6.6, pp. 247–248

1. $(1, 3)$ **3.** $(1, 2)$ **5.** $(4, 3)$ **7.** $(-2, 1)$
9. $(-1, -3)$ **11.** $(\frac{17}{3}, \frac{16}{3})$ **13.** $(\frac{25}{8}, -\frac{11}{4})$
15. $(-3, 0)$ **17.** $(6, 3)$ **19.** No solution
21. 15, 12 **23.** 37, 21 **25.** 28, 12
27. 120 m, 80 m **29.** 100 yd, $53\frac{1}{3}$ yd
31. $(7, -1)$ **33.** $(2, -1)$ **35.** $l = \frac{P + 10}{4}$
37. $(2, -1, 3)$
39. The x-terms drop out and leave $12 = 14$. The lines are parallel, so there is no solution.
41. Answers may vary. $x = 10 - z$, $y = 10 - z$, so $(10 - z) + (10 - z) + z = 10$ or $z = 10$. $x = 10 - z$ so $x = 0$, $y = 10 - z$ so $y = 0$, then $x + y = 0$. But we know $x + y = 10$, so there is no solution.

Exercise Set 6.7, pp. 252–254

1. $(9, 1)$ **3.** $(3, 5)$ **5.** $(3, 0)$ **7.** $(-\frac{1}{2}, 3)$
9. $(-1, \frac{1}{5})$ **11.** No solution **13.** $(-3, -5)$
15. $(4, 5)$ **17.** $(4, 1)$ **19.** $(4, 3)$ **21.** $(1, -1)$
23. $(-3, -1)$ **25.** $(2, -2)$ **27.** $(5, \frac{1}{2})$
29. $(2, -1)$ **31.** 10 miles **33.** 43° and 137°
35. 62° and 28°
37. 480 hectares Chardonnay, 340 hectares Riesling
39. 23 pheasants, 12 rabbits **41.** 45 yr; 10 yr
43. $(5, 2)$ **45.** $(1, -1)$ **47.** $(525, 1000)$
49. $\left(\frac{b - c}{1 - a}, \frac{b - ac}{1 - a}\right)$ **51.** $(4, 3)$
53. (a) $x = \frac{ce - bf}{ae - bd}$, $y = \frac{af - cd}{ae - bd}$; (b) $(2700, 0.5)$

Exercise Set 6.8, pp. 262–263

1. 350 cars, 160 trucks
3. Sammy: 44; daughter: 22
5. Marge: 18, Consuelo: 9
7. 70 dimes, 33 quarters
9. 300 nickels, 100 dimes
11. 203 adults, 226 children
13. 130 adults, 70 students
15. 40 g of A, 60 g of B
17. 43.75 L **19.** 80 L of 30%, 120 L of 50%
21. 6 kg of cashews, 4 kg of pecans
23. $12,500 at 12%, $14,500 at 13% **25.** 75
27. 10 at $20 per day, 5 at $25 per day **29.**
31. Glove: $79.95; bat: $14.50; ball: $4.55

Exercise Set 6.9, pp. 267–268

1. 2 hr **3.** 4.5 hr **5.** $7\frac{1}{2}$ hr after first train leaves
7. 14 km/h **9.** 384 km **11.** 330 km/h
13. 15 mi **15.** ≈ 317.03 km/h
17. 180 mi, 96 mi **19.** 144 mi **21.** 40 min

Exercise Set 6.10, pp. 276–277

1. 0 **3.** $-\frac{4}{5}$ **5.** 7 **7.** $-\frac{2}{3}$ **9.** -2 **11.** 0
13. No slope **15.** No slope **17.** 0 **19.** No slope
21. 0 **23.** $-\frac{3}{2}$ **25.** $-\frac{1}{4}$ **27.** 2 **29.** $\frac{4}{3}$ **31.** $\frac{1}{3}$
33. $\frac{1}{2}$ **35.** -4, $(0, -9)$ **37.** 1.8, $(0, 0)$
39. $-\frac{2}{3}$, $(0, 3)$ **41.** $-\frac{8}{7}$, $(0, -3)$ **43.** 3, $(0, -\frac{5}{3})$
45. $-\frac{3}{2}$, $(0, -\frac{1}{2})$ **47.** $y = 5x - 5$ **49.** $y = \frac{3}{4}x + \frac{5}{2}$
51. $y = x - 8$ **53.** $y = -3x - 9$ **55.** $y = \frac{2}{3}x + \frac{8}{3}$
57. $y = \frac{1}{4}x + \frac{5}{2}$ **59.** $y = -\frac{1}{2}x + 4$
61. $y = -\frac{3}{2}x + \frac{13}{2}$ **63.** $y = \frac{2}{5}x - 2$ **65.** $y = \frac{3}{4}x - \frac{5}{2}$
67. $y = 3x - 9$ **69.** $y = \frac{3}{2}x - 2$

Summary and Review: Chapter 6, pp. 277–279

1.–3. [6.1] **4.** [6.1] IV
5. [6.1] III **6.** [6.1] I
7. [6.1] $(-5, -1)$
8. [6.1] $(-2, 5)$
9. [6.1] $(3, 0)$
10. [6.2] No
11. [6.2] Yes

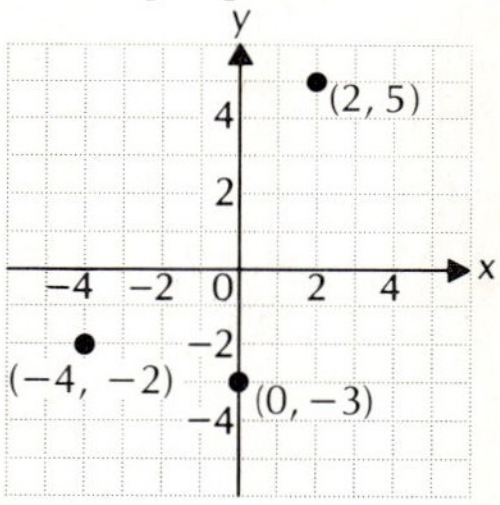

12. [6.2] $y = 2x - 5$

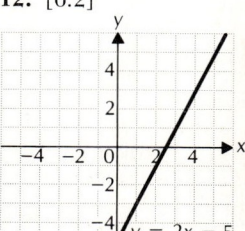

13. [6.2] $y = -\frac{3}{4}x$

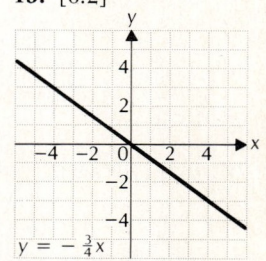

14. [6.2] $y = -x + 4$

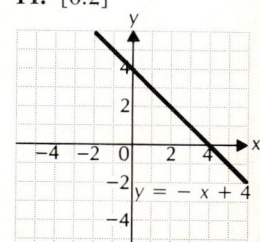

15. [6.2] $4x + y = 3$

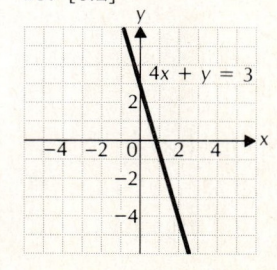

16. [6.3] **17.** [6.3]

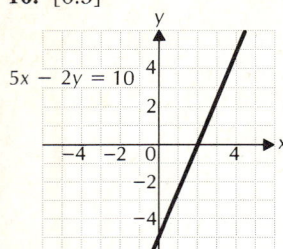

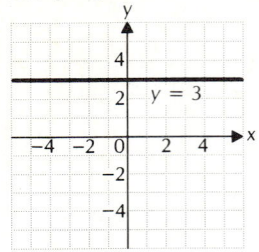

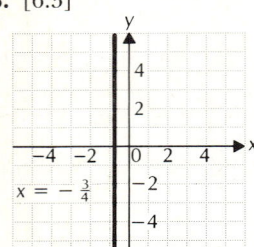

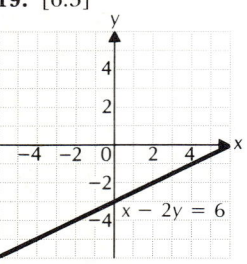

18. [6.3] **19.** [6.3]

20. [6.5] No **21.** [6.5] Yes **22.** [6.5] Yes
23. [6.5] No **24.** [6.5] $(6, -2)$ **25.** [6.5] $(6, 2)$
26. [6.5] $(2, -3)$
27. [6.5] No solution; lines are parallel.
28. [6.6] $(0, 5)$ **29.** [6.6] $(-2, 4)$
30. [6.6] $(1, -2)$ **31.** [6.6] $(-3, 9)$
32. [6.6] $(1, 4)$ **33.** [6.6] $(3, -1)$
34. [6.7] $(3, 1)$ **35.** [6.7] $(1, 4)$
36. [6.7] $(5, -3)$ **37.** [6.7] $(-4, 1)$
38. [6.7] $(-2, 4)$ **39.** [6.7] $(-2, -6)$
40. [6.7] $(3, 2)$ **41.** [6.7] $(2, -4)$
42. [6.6] $10, -2$ **43.** [6.7] $12, 15$
44. [6.6] $l = 37\frac{1}{2}$ cm, $w = 10\frac{1}{2}$ cm
45. [6.9] 135 km/h
46. [6.8] 297 orchestra seats, 211 balcony seats
47. [6.8] 40 L of each **48.** [6.8] Jeff: 27; son: 9
49. [6.5] $C = 1, D = 3$ **50.** [6.6] $(2, 1, -2)$
51. [6.7] $(2, 0)$ **52.** [6.7] 24 **53.** [6.8] $96

Cumulative Review: Chapters 1–6, pp. 281–282

1. [1.4] $\frac{4}{3c}$ **2.** [1.6] 49 **3.** [2.2] $>$
4. [2.3] $-\frac{43}{8}$ **5.** [2.5] $-\frac{4}{5}$ **6.** [2.6] -6.2
7. [2.6] 1 **8.** [2.8] $-8x + 28$ **9.** [3.4] 10
10. [3.3] -3 **11.** [3.2] $\frac{9}{2}$ **12.** [3.3] -2
13. [3.4] $\frac{8}{3}$ **14.** [3.4] $-\frac{1}{4}$ **15.** [3.3] -8
16. [3.1] $-\frac{1}{2}$ **17.** [1.7] $\frac{3}{2}x + 2y - 3z$
18. [4.3] $-4x^3 - \frac{1}{7}x^2 - 2$ **19.** [4.1] x^{10}
20. [4.1] $\frac{1}{z^3}$ **21.** [4.1] $-27x^6y^3$

22. [4.5] $-y^3 - 2y^2 - 2y + 7$
23. [1.7] $12x + 16y + 4z$ **24.** [4.7] $a^2 - 9$
25. [4.6] $2x^5 + x^3 - 6x^2 - x + 3$
26. [4.7] $2 - 10x^2 + 12x^4$
27. [4.7] $6x^7 - 12x^5 + 9x^2 - 18$ **28.** [4.7] $4x^6 - 1$
29. [4.8] $36x^2 - 60x + 25$ **30.** [4.6] $64 - \frac{1}{9}x^2$
31. [2.7] $9(4 - 9y)$
32. [2.7] $-2(3 + x + 6y)$ or $2(-3 - x - 6y)$
33. [5.4] $(x - 6)(x - 4)$
34. [5.5] $(2x + 1)(4x + 3)$
35. [5.1] $3x^2(2x^3 - 12x + 3)$
36. [5.2] $2(x + 3)(x - 3)$ **37.** [5.3] $(4x + 5)^2$
38. [5.5] $(3x - 2)(x + 4)$
39. [5.1] $(x^3 - 3)(x + 2)$
40. [5.1] $4x(-12x^3 - x + 3)$
41. [5.2] $(4x^2 + 9)(2x + 3)(2x - 3)$
42. [5.5] $2(3x - 2)(x - 4)$
43. [5.2] $3(1 + 2x^3)(1 - 2x^3)$ **44.** [5.7] $\frac{1}{2}, -4$
45. [5.9] $xy^3 - 2xy^2 - 4x^2y^2$ **46.** [5.9] $9x^4 - 16y^2$
47. [5.9] $4a^4b^2 - 20a^3b^3 + 25a^2b^4$
48. [5.9] $(2x^2 - 3y)^2$
49. [6.2] **50.** [6.3]

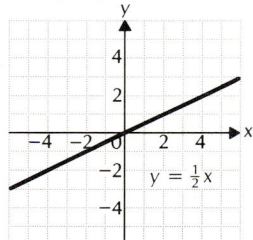

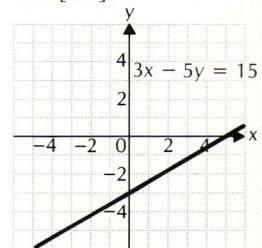

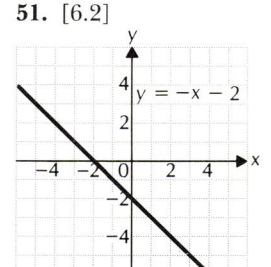

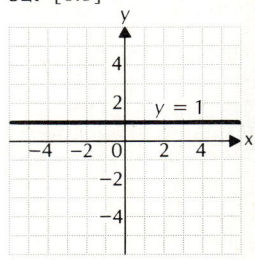

51. [6.2] **52.** [6.3]

53. [6.6] $(2, -1)$ **54.** [6.7] $(4, -5)$
55. [6.7] $(-1, 3)$ **56.** [6.7] $(0, 5)$
57. [1.9] 0.74 cm **58.** [5.8] 14, 16 or $-14, -16$
59. [3.5] $38°, 76°, 66°$
60. [3.5] Hamburger: $1.30; milkshake: $0.90
61. [6.8] Length: 80 ft, width: 30 ft **62.** [2.4] -7
63. [3.3] $-144, 144$ **64.** [4.7] $16y^6 - y^4 + 6y^2 - 9$
65. [5.6] $2(a^{16} + 81b^{20})(a^8 + 9b^{10})(a^4 + 3b^5) \times (a^4 - 3b^5)$

CHAPTER 7

Exercise Set 7.1, pp. 286–287

1. (a) No; (b) no; (c) no; (d) yes
3. (a) No; (b) no; (c) yes; (d) yes
5. (a) No; (b) no; (c) yes; (d) no
7. (a) Yes; (b) yes; (c) yes; (d) no 9. $\{x \mid x > -5\}$
11. $\{y \mid y > 3\}$ 13. $\{x \mid x \leq -18\}$ 15. $\{a \mid a < -6\}$
17. $\{x \mid x \leq 16\}$ 19. $\{x \mid x > 8\}$ 21. $\{y \mid y > -5\}$
23. $\{x \mid x > 2\}$ 25. $\{x \mid x \leq -3\}$ 27. $\{x \mid x \geq 13\}$
29. $\{x \mid x < 4\}$ 31. $\{y \mid y \geq -11\}$ 33. $\{c \mid c > 0\}$
35. $\{y \mid y \leq \frac{1}{4}\}$ 37. $\{x \mid x > \frac{7}{12}\}$ 39. $\{x \mid x > 0\}$
41. 0.6 km 43. $\{r \mid r < -2\}$ 45. $\{x \mid x \geq 1\}$
47. $\{x \mid x > -11.8\}$ 49. $\{x \mid x \leq -1.2\}$
51. $\{x \mid x \leq -18,058,999\}$

Exercise Set 7.2, pp. 289–291

1. $\{x \mid x < 7\}$ 3. $\{y \mid y \leq 9\}$ 5. $\{x \mid x < \frac{13}{7}\}$
7. $\{x \mid x > -3\}$ 9. $\{y \mid y \geq -\frac{2}{5}\}$ 11. $\{x \mid x \geq -6\}$
13. $\{y \mid y \leq 4\}$ 15. $\{x \mid x > \frac{17}{3}\}$ 17. $\{y \mid y < -\frac{1}{14}\}$
19. $\{x \mid \frac{3}{10} \geq x\}$ 21. $\{x \mid x < 8\}$ 23. $\{y \mid y \geq 6\}$
25. $\{x \mid x \leq 6\}$ 27. $\{y \mid y > 4\}$ 29. $\{x \mid x < -3\}$
31. $\{x \mid x \geq -2\}$ 33. $\{y \mid y < -3\}$
35. $\{x \mid x > -3\}$ 37. $\{y \mid y < -\frac{10}{3}\}$ 39. $\{x \mid x \leq 7\}$
41. $\{x \mid x > -10\}$ 43. $\{y \mid y < 2\}$ 45. $\{y \mid y \geq 3\}$
47. $\{y \mid y > -2\}$ 49. $\{y \mid y > -2\}$ 51. $\{y \mid y \leq \frac{33}{7}\}$
53. $\{x \mid x < \frac{9}{5}\}$ 55. 140°, 28°, 12° 57. $\{t \mid t \leq 0\}$
59. $\{y \mid y < \frac{2.2}{7}\}$ 61. $\{x \mid x \leq 9\}$ 63. $\{y \mid y \leq -3\}$
65. $\{x \mid x > \frac{8}{3}\}$ 67. $\{x \mid x \leq -4a\}$ 69. $\left\{x \mid x \geq \frac{3y}{2}\right\}$
71. All nonzero reals 73. Yes 75. No
77. $x \geq 6$ 79. $x \leq y$

Exercise Set 7.3, pp. 293–295

1. $\{s \mid s \geq 97\}$ 3. $\{t \mid t > 4\frac{2}{3}\}$ 5. $\{n \mid n \leq 0\}$
7. $\{m \mid m \leq 525.8 \text{ mi}\}$ 9. $\{l \mid l \geq 16.5 \text{ yd}\}$
11. $\{l \mid l \geq 92 \text{ ft}\}, \{l \mid l \leq 92 \text{ ft}\}$ 13. $\{c \mid c \geq 20\}$
15. $\{s \mid s \leq \$18.05\}$ The most the student can spend for the sweater is $18.05.
17. $\{b \mid b \leq 4\}$ 19. 47 and 49
21. $\{s \mid s < \$20,000\}$

Exercise Set 7.4, pp. 299–300

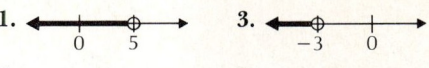

1. [number line: open circle at 5, shaded left, from 0]
3. [number line: open circle at −3, shaded right, to 0]
5. [number line: closed circle at 6, shaded left, from 0]
7. 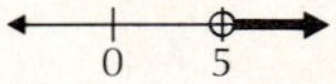 [number line: open circle at 5, shaded right, from 0]
9.  [number line: open circle at 7, shaded left, from 0]
11. 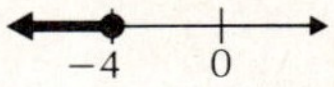 [number line: closed circle at −4, shaded right, to 0]
13.  [number line: closed circle at 8, shaded left, from 0]
15.  [number line: open circle at 7, shaded right, from 0]
17.  [number line: open circle at 3, shaded left, from 0]
19.  [number line: open circle at 3, shaded right, from 0]
21. 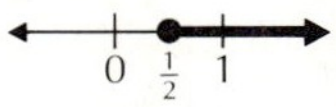 [number line: closed dot at $\frac{1}{2}$ to closed dot at 1]
23.  [number line: open circles at −6 and 6, shaded between]
25.  [number line: closed dots at −7 and 7, shaded between]
27. 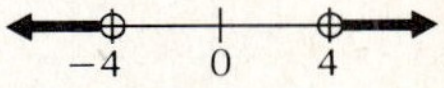 [number line: open circles at −4 and 4, shaded outside]
29. 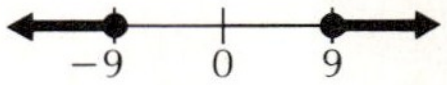 [number line: closed dots at −9 and 9, shaded between]
31. Yes 33. Yes

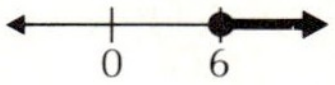

35. 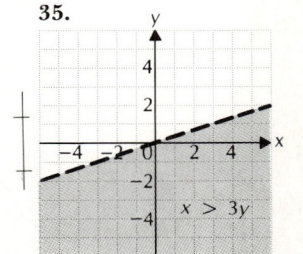 [graph: $x > 3y$]

37. 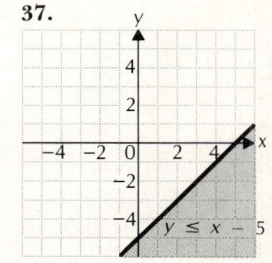 [graph: $y \leq x - 5$]

39. 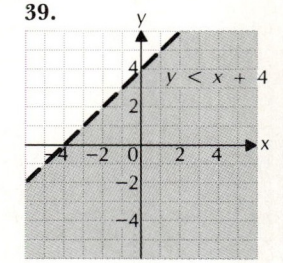 [graph: $y < x + 4$]

41. 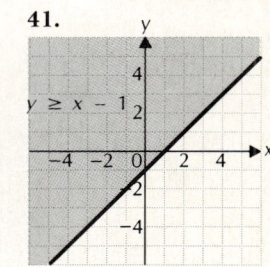 [graph: $y \geq x - 1$]

43. 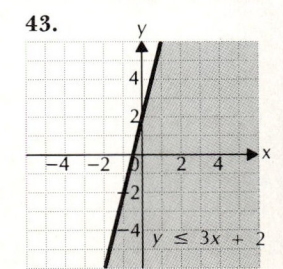 [graph: $y \leq 3x + 2$]

45. 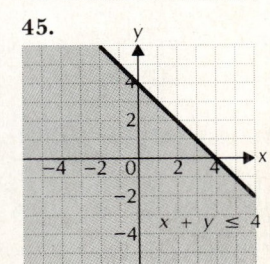 [graph: $x + y \leq 4$]

47. [graph: $x - y > -2$]

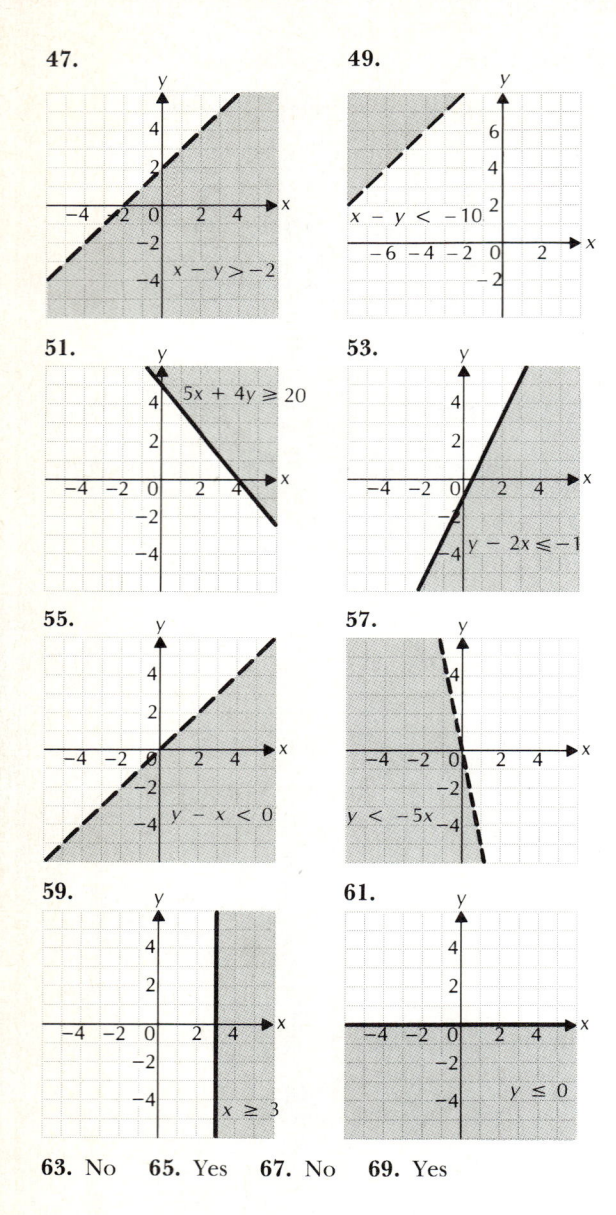

49. [graph: $x - y < -10$]

51. [graph: $5x + 4y \geq 20$]

53. [graph: $y - 2x \leq -1$]

55. [graph: $y - x < 0$]

57. [graph: $y < -5x$]

59. [graph: $x \geq 3$]

61. [graph: $y \leq 0$]

63. No   **65.** Yes   **67.** No   **69.** Yes

Exercise Set 7.5, p. 303

1. $\{3, 4, 5, 6, 7, 8\}$ **3.** $\{41, 43, 45, 47, 49\}$
5. $\{3, -3\}$ **7.** False   **9.** True   **11.** True
13. $\{c, d, e\}$ **15.** $\{1, 10\}$ **17.** $\varnothing$

19. $\{a, e, i, o, u, q, c, k\}$ **21.** $\{0, 1, 2, 5, 7, 10\}$
23. $\{a, e, i, o, u, m, n, f, g, h\}$
25. The set of integers   **27.** The set of real numbers
29. (a) A; (b) A; (c) A; (d) $\varnothing$

Summary and Review: Chapter 7, pp. 304–305

1. [7.1] Yes   **2.** [7.1] No   **3.** [7.1] Yes
4. [7.1] $\{y | y \geq -\frac{1}{2}\}$ **5.** [7.2] $\{x | x \geq 7\}$
6. [7.2] $\{y | y > 2\}$ **7.** [7.2] $\{y | y \leq -4\}$
8. [7.2] $\{x | x < -11\}$ **9.** [7.2] $\{y | y > -7\}$
10. [7.2] $\{x | x > -6\}$ **11.** [7.2] $\{x | x > -\frac{9}{11}\}$
12. [7.2] $\{y | y \leq 7\}$ **13.** [7.2] $\{x | x \geq -\frac{1}{12}\}$
14. [7.3] 86   **15.** [7.3] $\{w | w > 17 \text{ cm}\}$
16. [7.4] No   **17.** [7.4] No   **18.** [7.4] Yes
19. [7.4] [number line: 0 to 9, closed]
20. [7.4] [number line: 0 closed to 1 open]
21. [7.4] [number line: -2 to 2, closed]
22. [7.4] [graph: $x < y$]
23. [7.4] [graph: $x + 2y \geq y$]

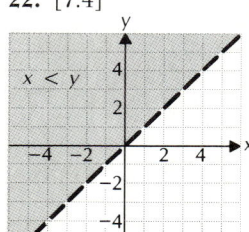

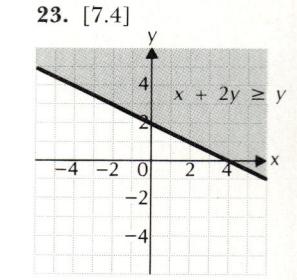

24. [7.5] $\{32, 36, 40, 44, 48, 52, 56\}$
25. [7.5] False   **26.** [7.5] False   **27.** [7.5] $\{2, 4\}$
28. [7.5] $\varnothing$ **29.** [7.5] $\{0, 1, 2, 3\}$
30. [7.5] $\{A, W, R, E, F, B\}$ **31.** [7.2] $\{y | y > \frac{8}{3}\}$
32. [7.2] $\{x | x \leq \frac{15}{2}\}$ **33.** [7.3] 63, 65, 67
34. [7.5] $\{7, 9, 11, 13\}$

CHAPTER 8

Exercise Set 8.1, pp. 312–314

1. Yes   **3.** Yes   **5.** No   **7.** 8, 12, −4
9. −6, 15, 72   **11.** 6, −10, 16   **13.** 2, 7, 4
15. 4, 5, 3   **17.** 1, 91, 98   **19.** −3, −2, 78
21. 1.606, 1.909, 4.03   **23.** 70°, 220°, 10,020°

25. $45.15 **27.** $54.24 **29.** 5, 8, 11, 14
31. 0, 2 **33.** 40 **35.** -186
37. $f(x) = \frac{15}{4}x - \frac{13}{4}$ **39.** No, one flip may correspond to 0 or 1 heads.

Exercise Set 8.2, pp. 317–318

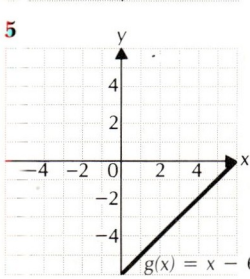

1. [graph of $f(x) = x + 4$]

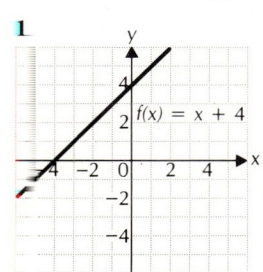

3. [graph of $h(x) = 2x - 3$]

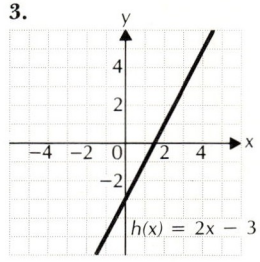

5. [graph of $g(x) = x - 6$]

7. [graph of $f(x) = 2x - 7$]

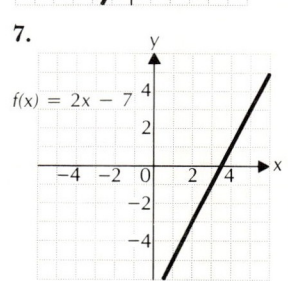

9. [graph of $f(x) = \frac{1}{2}x + 1$]

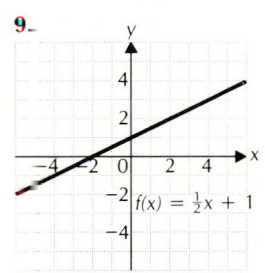

11. [graph of $g(x) = 2|x|$]

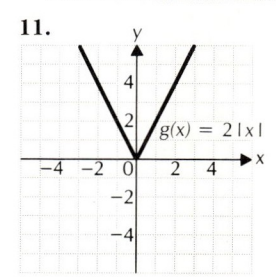

13. [graph of $g(x) = x^2$]

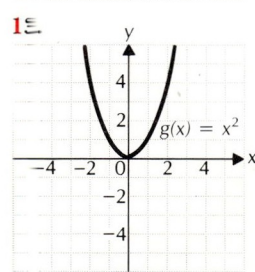

15. [graph of $f(x) = \frac{2}{x}$]

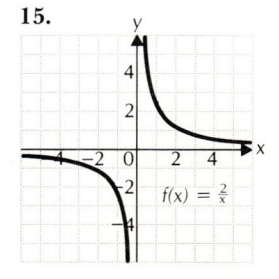

17. No **19.** Yes **21.** 390 km **23.** Answers may vary.

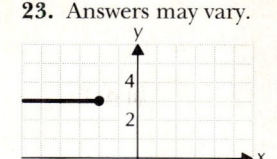

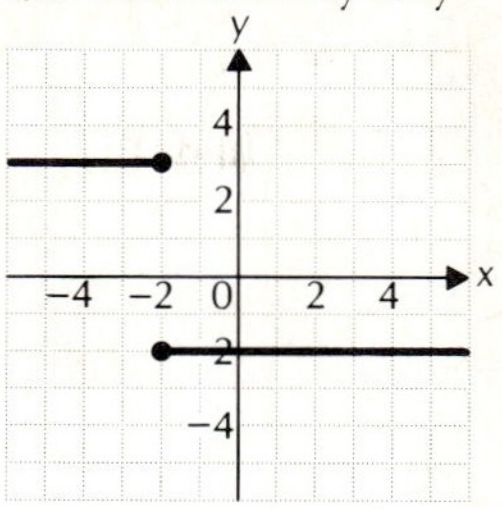

[graph]

25. No

[graph of $y^2 = x$]

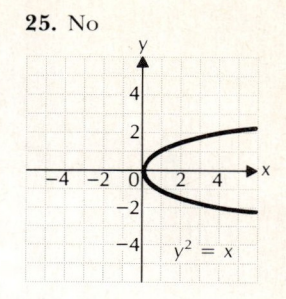

Exercise Set 8.3, pp. 320–322

1. $y = 4x$ **3.** $y = 1.75x$ **5.** $y = 3.2x$ **7.** $y = \frac{2}{3}x$
9. $183.75 **11.** $22\frac{6}{7}$ **13.** $16\frac{2}{3}$ kg **15.** 68.4 kg
17. $P = kS$ ($k =$ number of sides) **19.** $B = kN$
21. If $p = kq$, then $q = \frac{1}{k}p$. Since k is a constant, so is $\frac{1}{k}$, and q varies directly as p. **23.** $S = kV^6$
25. $P = kRI^2$

Exercise Set 8.4, pp. 324–325

1. $y = \frac{75}{x}$ **3.** $y = \frac{80}{x}$ **5.** $y = \frac{1}{x}$ **7.** $y = \frac{1050}{x}$
9. $y = \frac{0.06}{x}$ **11.** $5\frac{1}{3}$ hr **13.** 320 cm^3 **15.** 54 min
17. 2.4 ft **19.** $C = \frac{k}{N}$ **21.** $I = \frac{k}{R}$ **23.** $I = \frac{k}{d^2}$
25. Yes **27.** No **29.** $F = \frac{kS^2 m}{r}$

Summary and Review: Chapter 8, pp. 325–326

1. [8.1] No **2.** [8.1] Yes **3.** [8.1] 2, -4, -7
4. [8.1] 0, 2, -3 **5.** [8.1] -7, 1, 2
6. [8.1] 2700 calories
7. [8.2] [graph of $g(x) = x + 7$]
8. [8.2] [graph of $f(x) = x^2 - 3$]

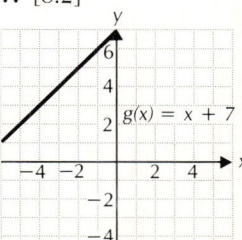

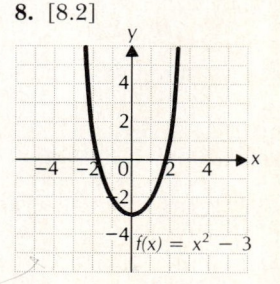

9. [8.2]

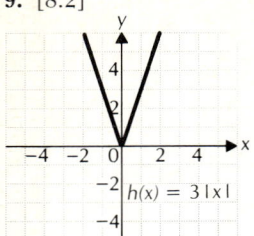

(graph of $h(x) = 3|x|$)

10. [8.2] No
11. [8.2] Yes
12. [8.3] $y = 3x$
13. [8.3] $y = \frac{1}{2}x$
14. [8.3] $y = \frac{4}{5}x$
15. [8.4] $y = \frac{30}{x}$
16. [8.4] $y = \frac{1}{x}$
17. [8.4] $y = \frac{0.65}{x}$
18. [8.3] $247.50   19. [8.4] 1 hour
20. [8.1] (a) 115 ft; (b) 179 ft   21. [8.1] -84
22. [8.2] No

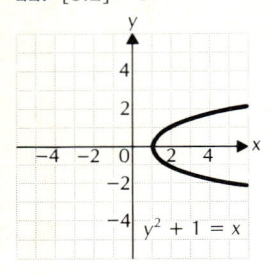

(graph of $y^2 + 1 = x$)

CHAPTER 9

Exercise Set 9.1, pp. 333–335

1. $\dfrac{(3x)(x+4)}{2(x-1)}$ 3. $\dfrac{(x-1)(x+1)}{(x+2)(x+2)}$
5. $\dfrac{(2x+3)(x+1)}{4(x-5)}$ 7. $\dfrac{(a-5)(a+2)}{(a^2+1)(a^2-1)}$
9. $\dfrac{(x+1)(x-1)}{(2+x)(x+1)}$ 11. $\dfrac{(3y-1)y}{(2y+1)y}$ 13. $\dfrac{a-3}{a+2}$
15. $\dfrac{t+2}{2(t-4)}$ 17. $\dfrac{x+5}{x-5}$ 19. $a+1$ 21. $\dfrac{x^2+1}{x+1}$
23. $\dfrac{3}{2}$ 25. $\dfrac{6}{t-3}$ 27. $\dfrac{a-3}{a-4}$ 29. $\dfrac{t-2}{t+2}$
31. $\dfrac{x+2}{x-2}$ 33. $\dfrac{a^2-9}{a^2+4a}$ 35. $\dfrac{2a}{a-2}$ 37. $\dfrac{x^2-4}{x^2-1}$
39. $\dfrac{t-2}{t-1}$ 41. $-20, -18$ and $18, 20$ 43. $x + 2y$
45. $\dfrac{(t-1)(t-9)^2}{(t^2+9)(t+1)}$ 47. $\dfrac{x-y}{x-5y}$ 49. -2
51. $0, 2, 7$ 53. The same number you selected.

Exercise Set 9.2, pp. 337–338

1. $\dfrac{x}{4}$ 3. $\dfrac{1}{x^2-y^2}$ 5. $\dfrac{x^2-4x+7}{x^2+2x-5}$ 7. $\dfrac{3}{10}$ 9. $\dfrac{1}{4}$
11. $\dfrac{y^2}{x}$ 13. $\dfrac{(a+2)(a+3)}{(a-3)(a-1)}$ 15. $\dfrac{(x-1)^2}{x}$ 17. $\dfrac{1}{2}$
19. $\dfrac{15}{8}$ 21. $\dfrac{15}{4}$ 23. $\dfrac{a-5}{3a-3}$ 25. $\dfrac{(x+2)^2}{x}$
27. $\dfrac{3}{2}$ 29. $\dfrac{c+1}{c-1}$ 31. $\dfrac{y-3}{2y-1}$ 33. $\dfrac{1}{(c-5)^2}$
35. $\dfrac{t+5}{t-5}$ 37. 4 39. $\dfrac{a}{(c-3d)(2a+5b)}$
41. $-\dfrac{1}{b^2}$ 43. x 45. $\dfrac{4}{x+7}$ 47. $\dfrac{3(y+2)^3}{y(y-1)}$
49. $\dfrac{1}{a/b} = 1 \cdot \dfrac{b}{a} = \dfrac{b}{a}$

Exercise Set 9.3, pp. 341–343

1. 1 3. $\dfrac{6}{3+x}$ 5. $\dfrac{2x+3}{x-5}$ 7. $\dfrac{1}{4}$ 9. $-\dfrac{1}{t}$
11. $\dfrac{-x+7}{x-6}$ 13. $y+3$ 15. $\dfrac{2b-14}{b^2-16}$
17. $-\dfrac{1}{y+z}$ 19. $\dfrac{5x+2}{x-5}$ 21. -1
23. $\dfrac{-x^2+9x-14}{(x-3)(x+3)}$ 25. $\dfrac{1}{2}$ 27. 1 29. $\dfrac{4}{x-1}$
31. $\dfrac{8}{3}$ 33. $\dfrac{13}{a}$ 35. $\dfrac{4x-5}{4}$ 37. $\dfrac{x-2}{x-7}$
39. $\dfrac{2x-16}{x^2-16}$ 41. $\dfrac{2x-4}{x-9}$ 43. $\dfrac{-9}{2x-3}$
45. $\dfrac{18x+5}{x-1}$ 47. 0 49. $\dfrac{20}{2y-1}$ 51. $140
53. 0 55. $\dfrac{2b-a-c}{a-b+c}$ 57. $\dfrac{x}{3x+1}$

Exercise Set 9.4, pp. 345–346

1. 108 3. 72 5. 126 7. 360 9. 420
11. $\dfrac{59}{300}$ 13. $\dfrac{71}{120}$ 15. $\dfrac{23}{180}$ 17. $8a^2b^2$ 19. c^3d^2
21. $8(x-1)$, or $8(1-x)$ 23. $x(x+3)(x-3)$
25. $(x+1)(x-2)(x+2)$ 27. $y^2(y+1)(y-1)$
29. $2(x+y)^2(x-y)$ 31. $(x+2)(x-1)(2x+1)$
33. $(2x-3)(2x+3)$ 35. $12a(a+2)(a+3)$
37. $x^3(x-2)(x+2)^2$ 39. $10x^3(x-1)(x+1)^2$
41. $120(x+1)(x-1)^2$
43. One expression is a multiple of the other.

Exercise Set 9.5, p. 349

1. $\dfrac{2x+5}{x^2}$ 3. $\dfrac{41}{24r}$ 5. $\dfrac{x^2+4xy+y^2}{x^2y^2}$

7. $\dfrac{6x}{(x-2)(x+2)}$ 9. $\dfrac{11x+2}{3x(x+1)}$

11. $\dfrac{x^2+6x}{(x-4)(x+4)}$ 13. $\dfrac{6}{z+4}$ 15. $\dfrac{3x-1}{(x-1)^2}$

17. $\dfrac{11a}{10(a-2)}$ 19. $\dfrac{2x^2+8x+16}{x(x+4)}$

21. $\dfrac{x^2+5x+1}{(x+1)^2(x+4)}$ 23. $\dfrac{2x^2-4x+34}{(x-5)(x+3)}$

25. $\dfrac{3a+2}{(a-1)(a+1)}$ 27. $\dfrac{2x+6y}{(x-y)(x+y)}$

29. $\dfrac{3x^2+19x-20}{(x+3)(x-2)^2}$ 31. $\dfrac{16y+28}{15}, \dfrac{y^2+2y-8}{15}$

33. $\dfrac{2z^2+9z-18}{(z+2)(z-2)}$ 35. $\dfrac{11z^4-22z^2+6}{(2z^2-3)(z^2+2)(z^2-2)}$

Exercise Set 9.6, pp. 351–352

1. $\dfrac{-(x+4)}{6}$ 3. $\dfrac{7z-12}{12z}$ 5. $\dfrac{4x^2-13xt+9t^2}{3x^2t^2}$

7. $\dfrac{2x-40}{(x+5)(x-5)}$ 9. $\dfrac{3-5t}{2t(t-1)}$ 11. $\dfrac{2s-st-s^2}{(t+s)(t-s)}$

13. $\dfrac{y-19}{4y}$ 15. $\dfrac{-2a^2}{x^2-a^2}$ 17. $\dfrac{3x+20}{(x-4)(x+4)}$

19. $\dfrac{1}{2}$ 21. $\dfrac{x-3}{(x+1)(x+3)}$ 23. $\dfrac{2}{y(y-1)}$

25. $\dfrac{z-3}{2z-1}$ 27. $\dfrac{-3x+1}{(2x-3)(x+1)}$ 29. $\dfrac{1}{2c-1}$

31. $\dfrac{2}{x+y}$ 33. $\dfrac{-3xy-3a+6x}{(a+2x)(a-2x)(y-3)^2}$

35. Answers may vary. $\dfrac{5x^2}{x^2+y^2} - \dfrac{2xy}{-x^2-y^2}$

Exercise Set 9.7, pp. 354–355

1. $\dfrac{25}{4}$ 3. $\dfrac{1}{3}$ 5. $\dfrac{1+3x}{1-5x}$ 7. -6 9. $\dfrac{5}{3y^2}$

11. 8 13. $-\dfrac{1}{a}$ 15. $\dfrac{x+y}{x}$ 17. $\dfrac{x-2}{x-3}$

19. 8.1 cm 21. $\dfrac{(x-1)(3x-2)}{5x-3}$ 23. $-\dfrac{ac}{bd}$

25. $\dfrac{5x+3}{3x+2}$

Exercise Set 9.8, pp. 357–358

1. $3x^4 - \tfrac{1}{2}x^3 + \tfrac{1}{8}x^2 - 2$ 3. $1 - 2u - u^4$
5. $5t^2 + 8t - 2$ 7. $-4x^4 + 4x^2 + 1$
9. $6x^2 - 10x + \tfrac{3}{2}$ 11. $-3rs - r + 2s$ 13. $x + 2$
15. $x - 5$, R (-50) 17. $x - 2$, R (-2)
19. $x - 3$ 21. $x^4 - x^3 + x^2 - x + 1$
23. $2x^2 - 7x + 4$ 25. $x^3 - 6$
27. $x^3 + 2x^2 + 4x + 8$ 29. $t^2 + 1$
31. $25{,}543.75$ ft² 33. $x^2 + 5$ 35. $a + 3$, R 5
37. $2x^2 + x - 3$
39. $a^5 + a^4b + a^3b^2 + a^2b^3 + ab^4 + b^5$
41. $3a^{2h} + 2a^h - 5$ 43. 2

Exercise Set 9.9, pp. 363–364

1. $\tfrac{47}{2}$ 3. -6 5. $\tfrac{24}{7}$ 7. $-4, -1$ 9. $4, -4$
11. 3 13. $\tfrac{14}{3}$ 15. 10 17. 5 19. $\tfrac{5}{2}$ 21. -1
23. $\tfrac{17}{2}$ 25. No solution 27. -5 29. $\tfrac{5}{3}$ 31. $\tfrac{1}{2}$
33. No solution 35. 7 37. $-\tfrac{1}{30}$ 39. $2, -2$
41. 4 43. $\tfrac{4}{3}, -\tfrac{4}{3}$

Exercise Set 9.10, pp. 372–373

1. $\tfrac{20}{9}$ 3. $20, 15$ 5. 30 km/h, 70 km/h
7. 20 mph
9. Passenger: 80 km/h; freight: 66 km/h
11. $2\tfrac{2}{9}$ hr 13. $5\tfrac{1}{7}$ hr 15. 9 17. 2.3 km/h
19. $581\tfrac{9}{11}$ 21. 702 km 23. 1.92 g 25. 287
27. (a) 1.92 tons; (b) 14.4 kg 29. $\tfrac{36}{68}$
31. $\tfrac{3}{4}$ 33. 2 mph
35. The product of the extremes is equal to the product of the means. 37. $9\tfrac{3}{13}$ days 39. 45 mph

Exercise Set 9.11, pp. 376–377

1. $r = \dfrac{S}{2\pi h}$ 3. $b = \dfrac{2A}{h}$ 5. $n = \dfrac{S+360}{180}$

7. $b = \dfrac{3V - kB - 4kM}{k}$ 9. $r = \dfrac{S-a}{S-l}$

11. $h = \dfrac{2A}{b_1+b_2}$ 13. $a = \dfrac{v^2pL}{r}$ 15. $b_1 = \dfrac{2A-hb_2}{h}$

17. $E = \dfrac{180A}{\pi r^2}$ 19. $M = \dfrac{-V-hB-hc}{4h}$

21. $L = \dfrac{ay}{v^2p}$ 23. $p = \dfrac{ar}{v^2L}$ 25. $n = \dfrac{a}{c(1+b)}$

27. $F = \dfrac{9C + 160}{5}$ **29.** $g = \dfrac{mf + t}{m}$ **31.** $p = \dfrac{qf}{q - f}$

33. $A = P(1 + r)$ **35.** $R = \dfrac{r_1 r_2}{r_1 + r_2}$ **37.** $D = \dfrac{BC}{A}$

39. $h_2 = \dfrac{p(h_1 - q)}{q}$ **41.** V is multiplied by 8

43. $0, -3$ **45.** $T = \dfrac{FP}{u + FE}$

Exercise Set 9.12, pp. 383–385

1. $\dfrac{1}{3^2}$, or $\dfrac{1}{3 \cdot 3}$, or $\dfrac{1}{9}$

3. $\dfrac{1}{10^4}$, or $\dfrac{1}{10 \cdot 10 \cdot 10 \cdot 10}$, or $\dfrac{1}{10{,}000}$ **5.** 4^{-3}

7. x^{-3} **9.** a^{-4} **11.** p^{-n} **13.** $\dfrac{1}{7^3}$ **15.** $\dfrac{1}{a^3}$

17. $\dfrac{1}{y^4}$ **19.** $\dfrac{1}{z^n}$ **21.** 2^7 **23.** 3^3 **25.** x^{-1}

27. x^7 **29.** x^{-13} **31.** 1 **33.** 7^3 **35.** x^2
37. x^9 **39.** z^{-4} **41.** x^3 **43.** 1 **45.** 2^6
47. 5^{-6} **49.** x^{12} **51.** $x^{-12}y^{-15}$ **53.** $x^{24}y^8$
55. $9x^6y^{-16}z^{-6}$ **57.** 7.8×10^{10} **59.** 9.07×10^{17}
61. 3.74×10^{-6} **63.** 1.8×10^{-8} **65.** 10^7
67. 10^{-9} **69.** $784{,}000{,}000$
71. 0.0000000008764 **73.** $100{,}000{,}000$
75. 0.0001 **77.** 6×10^9 **79.** 3.38×10^4
81. 8.1477×10^{-13} **83.** 2.5×10^{13}
85. 5.0×10^{-4} **87.** 3.0×10^{-21}
89. 2.478125×10^{-1} **91.** 3.5×10^{-10}

Summary and Review: Chapter 9, pp. 385–387

1. [9.1] $\dfrac{x - 2}{x + 1}$ **2.** [9.1] $\dfrac{7x + 3}{x - 3}$ **3.** [9.1] $\dfrac{y - 5}{y + 5}$

4. [9.1] $\dfrac{a - 6}{5}$ **5.** [9.1] $\dfrac{6}{2t - 1}$ **6.** [9.2] $-20t$

7. [9.2] $\dfrac{2x^2 - 2x}{x + 1}$ **8.** [9.4] $30x^2y^2$

9. [9.4] $4(a - 2)$, or $4(2 - a)$

10. [9.4] $(y^2 - 4)(y + 1)$ **11.** [9.3] $\dfrac{-3x + 18}{x + 7}$

12. [9.5] -1 **13.** [9.3] $\dfrac{4}{x - 4}$ **14.** [9.6] $\dfrac{x + 5}{2x}$

15. [9.3] $\dfrac{2x + 3}{x - 2}$ **16.** [9.5] $\dfrac{2a}{a - 1}$

17. [9.3] $d + c$ **18.** [9.6] $\dfrac{-x^2 + x + 26}{(x - 5)(x + 5)(x + 1)}$

19. [9.6] $\dfrac{2(x - 2)}{x + 2}$ **20.** [9.7] $\dfrac{z}{1 - z}$

21. [9.7] $c - d$ **22.** [9.8] $5x^2 - \tfrac{1}{2}x + 3$
23. [9.8] $3x^2 - 7x + 4$, R 1   **24.** [9.9] 8
25. [9.9] $3, -5$ **26.** [9.10] $5\tfrac{1}{7}$ hours
27. [9.10] 240 km/h, 280 km/h   **28.** [9.10] -2
29. [9.10] 160 **30.** [9.11] $s = \dfrac{rt}{r - t}$

31. [9.11] $C = \tfrac{5}{9}(F - 32)$, or $C = \tfrac{5}{9}F - \tfrac{160}{9}$

32. [9.12] y^{-4} **33.** [9.12] $\dfrac{1}{5^3}$, or $\dfrac{1}{125}$

34. [9.12] x^{-2} **35.** [9.12] t^9 **36.** [9.12] 7^{-10}
37. [9.12] 4^{-15} **38.** [9.12] 8^9
39. [9.12] $81a^{-24}$ **40.** [9.12] $x^{10}y^{-5}z^{-35}$
41. [9.12] 2.78×10^{-5} **42.** [9.12] 3.9×10^9
43. [9.12] 0.00000005 **44.** [9.12] $12{,}800$
45. [9.12] 2.09×10^4 **46.** [9.12] 5×10^{-5}

47. [9.1] $0, 5, 3$ **48.** [9.2] $\dfrac{5(a + 3)^2}{a}$

49. [9.3] $\dfrac{10a}{(a - b)(b - c)}$

50. [9.8] $y^5 - by^4 + b^2y^3 - b^3y^2 + b^4y - b^5$, R $2b^6$

Cumulative Review: Chapters 1–9, pp. 388–391

1. [4.2] 93 **2.** [2.6] -313.5 **3.** [2.3] 1
4. [2.8] $-5x + 7$ **5.** [2.8] $72 - 48x$
6. [2.8] $x - 1$ **7.** [3.4] -1 **8.** [3.3] $\tfrac{5}{21}$
9. [3.4] 0 **10.** [4.2] a **11.** [2.8] $9m - 28n$
12. [4.7] $2x^2 - 31x + 84$
13. [4.8] $1 - 10x + 25x^2$ **14.** [4.8] $25x^2 - 16$
15. [4.6] $2x^6 + x^4 - 19x^2 - 24$
16. [4.8] $n^2 - 8nk + 16k^2$
17. [4.7] $2x^2 - 16xy - 130y^2$
18. [5.9] $12(3x + 2y + z)$
19. [5.9] $-7(3x + 4w)$
20. [5.9] $xy(x^2y^2 + xy - 4)$
21. [5.1] $a^6(a^2 + a - 1)$
22. [5.2] $(2 - 3m^2)(2 + 3m^2)$
23. [5.2] $(10 - a)(10 + a)$
24. [5.9] $9(x - b)(x + b)$
25. [5.4] $(x - 6)(x + 2)$
26. [5.4] $(s - 15)(s - 1)$
27. [5.4] $t(t - 3)(t - 2)$
28. [5.5] $(7x + 1)(x - 1)$
29. [5.5] $(4y + 3)(5y + 1)$
30. [5.5] $3(2y + 5)(y - 1)$
31. [5.1] $a(a^3 - 5)(a + 1)$

32. [5.9] $(5a + 2m)(a - 6m)$
33. [5.9] $(x^4 + y^4)(x^2 + y^2)(x + y)(x - y)$
34. [6.2] 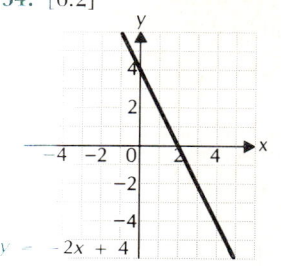
35. [6.3]

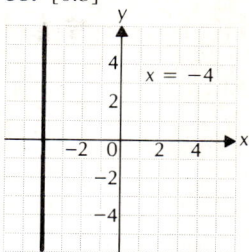

36. [6.3] 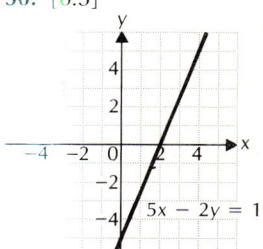
37. [6.7] $(-2, 2)$
38. [6.6] $(3, -3)$
39. [7.2] $\{x \mid x \geq 7\}$
40. [7.2] $\{x \mid x \leq -6\}$
41. [7.2] $\{x \mid x > -6\}$
42. [7.2] $\{x \mid x \leq 20\}$
43. [7.4] 
44. [7.4]

45. [7.4] 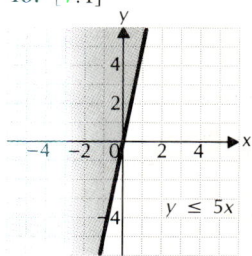
46. [7.4]

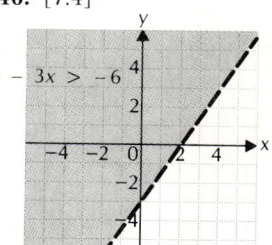

47. [7.5] $\{15, 27\}$ 48. [7.5] $\{-1, 0, 1, 3\}$
49. [8.1] 3   50. [8.1] 9   51. [8.1] 6
52. [8.3] $y = 11x$ 53. [8.3] $y = 500x$
54. [8.4] $y = \dfrac{15}{x}$ 55. [8.4] $y = \dfrac{0.05}{x}$
56. [9.1] $\dfrac{2x - 5}{x - 7}$ 57. [9.2] $\dfrac{3x^3(x - 5)}{x + 5}$
58. [9.3] $\dfrac{x^2 + 9}{x - 3}$ 59. [9.7] $\dfrac{2 + 12x}{9x - 6}$
60. [9.8] $3x^2 - 4$, $R(-15)$ 61. [9.9] 2
62. [9.9] 3   63. [9.11] $M = \dfrac{ER}{e} - R$
64. [9.11] $S = \dfrac{TR}{R - T}$ 65. [9.12] x^{-12}
66. [9.12] 6^{-13} 67. [9.12] $16a^{-12}$
68. [9.12] 3.28×10^{-5} 69. [9.12] 8.3×10^8

70. [3.5] 3   71. [1.9] 48   72. [3.5] $180
73. [9.10] 25 km/h
74. [5.8] Length: 12 cm; width: 9 cm
75. [5.9] 0.004392   76. [7.2] $\{x \mid x > \tfrac{105}{199}\}$
77. [9.1] $\dfrac{1}{x + 2}$ 78. [9.7] $\dfrac{(y - 4)(2y + 1)}{13 - y}$

CHAPTER 10

Exercise Set 10.1, pp. 398–399

1. 1, −1   3. 4, −4   5. 10, −10   7. 13, −13
9. 2   11. −3   13. −8   15. −15   17. 19
19. Irrational   21. Irrational   23. Rational
25. Irrational   27. Rational   29. Irrational
31. Rational   33. Rational   35. Rational
37. Rational   39. Rational   41. Irrational
43. Irrational   45. 2.449   47. 4.359   49. 6.557
51. 13, 24   53. 2   55. −5, −6   57. 14.071
59. 70.228   61. $\dfrac{x + y}{2}$ 63. $3\tfrac{3}{8}$ 65. $\dfrac{3z + w}{4}$

Exercise Set 10.2, p. 401

1. $a - 4$ 3. $t^2 + 1$ 5. $\dfrac{3}{x + 2}$ 7. Yes   9. No
11. $x \geq 0$ 13. $t \geq 5$ 15. $y \geq -8$ 17. $x \geq -20$
19. $y \geq \tfrac{7}{2}$ 21. Any value   23. $|t|$ 25. $3|x|$
27. 7   29. $4|d|$ 31. $|x + 3|$ 33. $|a - 5|$
35. $5330   37. 6, −6   39. 3, −3   41. $3|a|$
43. $2\dfrac{x^4}{|y^3|}$ 45. $\dfrac{13}{m^8}$ 47. $m \geq 0$ and $m \leq -3$
49. $x \geq 2$ and $x \leq -3$ 51. $x \geq 2$ and $x \leq -2$

Exercise Set 10.3, pp. 404–405

1. $\sqrt{6}$ 3. $\sqrt{12}$, or $2\sqrt{3}$ 5. $\sqrt{\tfrac{3}{10}}$ 7. 17
9. $\sqrt{75}$, or $5\sqrt{3}$ 11. $\sqrt{2x}$ 13. $\sqrt{0.72}$
15. $\sqrt{xt}$ 17. $\sqrt{x^2 - 3x}$ 19. $\sqrt{10x - 5}$
21. $\sqrt{x^2 + 3x + 2}$ 23. $\sqrt{2x^2 - 2x - 12}$
25. $\sqrt{x^2 - 16}$ 27. $\sqrt{x^2 - y^2}$ 29. No real value
31. $2\sqrt{3}$ 33. $5\sqrt{3}$ 35. $2\sqrt{5}$ 37. $10\sqrt{2}$
39. $\sqrt{3}\sqrt{x}$ 41. $3\sqrt{x}$ 43. $4\sqrt{a}$ 45. $8|y|$
47. $|x|\sqrt{13}$ 49. $2|t|\sqrt{2}$ 51. 11.180   53. 18.972
55. 17.320   57. 11.043   59. 20 mph, 54.8 mph
61. $\sqrt{3}\sqrt{x - 1}$ 63. $\sqrt{x + 2}\sqrt{x - 2}$
65. $|x|\sqrt{x - 2}$ 67. 0.1   69. x^2
71. 7, $7\sqrt{10}$, 70, $70\sqrt{10}$, 700; each is $\sqrt{10}$ times the last.   73. =   75. >   77. >   79. =
81. $z^4|w^5|$

Exercise Set 10.4, pp. 407–408

1. $6\sqrt{5}$ **3.** $4\sqrt{3x}$ **5.** $12\sqrt{2}y$ **7.** $2x\sqrt{5}$
9. $\sqrt{2}(2x+1)$ **11.** $\sqrt{y}(6+y)$ **13.** x^3 **15.** x^6
17. $x^2\sqrt{x}$ **19.** $t^9\sqrt{t}$ **21.** $(y-2)^4$
23. $2(x+5)^5$ **25.** $6m\sqrt{m}$ **27.** $2a^2\sqrt{2a}$
29. $8x^3y\sqrt{7y}$ **31.** $3\sqrt{6}$ **33.** $3\sqrt{10}$ **35.** $6\sqrt{7x}$
37. $6\sqrt{xy}$ **39.** 10 **41.** $5b\sqrt{3}$ **43.** $2t$
45. $a\sqrt{bc}$ **47.** $2xy\sqrt{2xy}$ **49.** $6xy^3\sqrt{3xy}$
51. $10ab^2\sqrt{5ab}$ **53.** $a^2 - 5\sqrt{a}$
55. $6(x-2)^2\sqrt{10}$ **57.** $2^{54}x^{158}\sqrt{2x}$ **59.** $0.2x^{2n}$
61. Only if $B=0$ **63.** $y^{[(n-1)/2]}\sqrt{y}$ **65.** 2
67. -3 **69.** x

Exercise Set 10.5, pp. 411–412

1. $\dfrac{3}{7}$ **3.** $\dfrac{1}{6}$ **5.** $-\dfrac{4}{9}$ **7.** $\dfrac{8}{17}$ **9.** $\dfrac{13}{14}$ **11.** $\dfrac{6}{a}$
13. $\dfrac{3a}{25}$ **15.** $\dfrac{\sqrt{10}}{5}$ **17.** $\dfrac{\sqrt{6}}{4}$ **19.** $\dfrac{\sqrt{21}}{6}$
21. $\dfrac{\sqrt{2}}{6}$ **23.** $\dfrac{\sqrt{2}}{2}$ **25.** $\dfrac{2\sqrt{6}}{3}$ **27.** $\dfrac{\sqrt{3x}}{x}$
29. $\dfrac{\sqrt{xy}}{y}$ **31.** $\dfrac{x\sqrt{2}}{6}$ **33.** 0.577 **35.** 0.935
37. 0.289 **39.** 0.707 **41.** 0.592 **43.** 0.850
45. 1.57, 3.14, 8.88, 11.10 **47.** 1 sec **49.** $\dfrac{\sqrt{30}}{100}$
51. $\dfrac{\sqrt{3yx}}{ax^2}$ **53.** $\dfrac{\sqrt{5z}}{5zw}$ **55.** $\dfrac{\sqrt{30}}{50}$ **57.** $\dfrac{(z^2-1)\sqrt{2}}{z^2}$
59. $\dfrac{\sqrt{6x}}{4x}$ **61.** $\dfrac{\sqrt{3y(2+y^2)}}{2+y^2}$

Exercise Set 10.6, pp. 414–415

1. 3 **3.** 2 **5.** $\sqrt{5}$ **7.** $\dfrac{1}{5}$ **9.** $\dfrac{2}{5}$ **11.** 2
13. $3y$ **15.** $x^2\sqrt{5}$ **17.** 2 **19.** $\dfrac{\sqrt{5}}{2}$ **21.** $\dfrac{\sqrt{21}}{3}$
23. $\dfrac{3\sqrt{2}}{4}$ **25.** $\dfrac{\sqrt{10}}{5}$ **27.** $\sqrt{2}$ **29.** $\dfrac{\sqrt{55}}{11}$
31. $\dfrac{\sqrt{21}}{6}$ **33.** $\dfrac{\sqrt{6}}{2}$ **35.** 5 **37.** $\dfrac{\sqrt{3x}}{x}$ **39.** $\dfrac{4y\sqrt{3}}{3}$
41. $\dfrac{a\sqrt{2a}}{4}$ **43.** $\dfrac{\sqrt{42x}}{3x}$ **45.** $\dfrac{3\sqrt{6}}{8c}$ **47.** $\dfrac{y\sqrt{xy}}{x}$
49. $\dfrac{\sqrt{2}}{4a}$ **51.** 303.6 kg **53.** $\dfrac{\sqrt{6}}{9}$ **55.** $\dfrac{\sqrt{10}}{3}$

57. $6\sqrt{2}$ **59.** $\dfrac{\sqrt{30}}{2x}$

Exercise Set 10.7, pp. 417–419

1. $7\sqrt{2}$ **3.** $4\sqrt{5}$ **5.** $13\sqrt{x}$ **7.** $-2\sqrt{x}$
9. $25\sqrt{2}$ **11.** $\sqrt{3}$ **13.** $\sqrt{5}$ **15.** $13\sqrt{2}$
17. $3\sqrt{3}$ **19.** $-24\sqrt{2}$ **21.** $6\sqrt{2} - 6\sqrt{3}$
23. $(2+9x)\sqrt{x}$ **25.** $(3-2x)\sqrt{3}$ **27.** $3\sqrt{2x+2}$
29. $(x+3)\sqrt{x^3-1}$ **31.** $(-x^2+3xy+y^2)\sqrt{xy}$
33. $-2(a+b)\sqrt{2(a+b)}$ **35.** $\dfrac{2\sqrt{3}}{3}$ **37.** $\dfrac{13\sqrt{2}}{2}$
39. $\dfrac{\sqrt{6}}{6}$ **41.** $\dfrac{\sqrt{3}}{18}$ **43.** (a) None; (b) $\sqrt{10}+5\sqrt{2}$
45. $16\sqrt{2}$ **47.** $11\sqrt{3} - 10\sqrt{2}$ **49.** 0
51. $\dfrac{x+1}{x}\sqrt{x}$ **53.** $\left(\dfrac{2}{b} - \dfrac{2}{a^2} + \dfrac{5a}{4}\right)\sqrt{2ab}$
55. Any pairs of numbers a, b such that $a=0$, $b \geq 0$ or $a \geq 0$, $b=0$ **57.** -4 **59.** -4
61. $7\sqrt{2} - 5\sqrt{5}$ **63.** $b^3 + z\sqrt{z}$
65. x or y is 0; the other can be any number.
67. $5 - 2\sqrt{7}$ **69.** $\dfrac{6+x+y-5\sqrt{x+y}}{9-x-y}$

Exercise Set 10.8, pp. 423–426

1. $c=17$ **3.** $c=\sqrt{32} \approx 5.657$ **5.** $b=12$
7. $b=4$ **9.** $c=26$ **11.** $b=12$ **13.** $a=2$
15. $b=\sqrt{2} \approx 1.414$ **17.** $a=5$
19. $\sqrt{75} \approx 8.660$ m **21.** $\sqrt{208} \approx 14.422$ ft
23. $\sqrt{7200} \approx 84.853$ ft **25.** $h=\dfrac{a}{2}\sqrt{3}$ **27.** $\sqrt{181}$
29. $\dfrac{2}{3}$ **31.** 8 ft **33.** 3, 4, 5 **35.** $12 - 2\sqrt{6}$
37. 6 **39.** 50 ft^2

Exercise Set 10.9, pp. 429–430

1. 25 **3.** 38.44 **5.** 397 **7.** $\dfrac{621}{2}$ **9.** 5 **11.** 3
13. $\dfrac{17}{4}$ **15.** No solution **17.** No solution
19. ≈ 346.48 km **21.** 11,236 m
23. 125 ft, 245 ft **25.** 49 **27.** 12 **29.** ≈ 7.30 ft
31. 0 **33.** $\dfrac{1}{2}$ **35.** 5 **37.** 0, 4 **39.** $s = \dfrac{t^2 g}{2}$
41. 1610 ft

Summary and Review: Chapter 10, pp. 430–432

1. [10.1] $-8, 8$ **2.** [10.1] $-5, 5$
3. [10.1] $-14, 14$ **4.** [10.1] $-20, 20$

5. [10.1] 6 **6.** [10.1] −9 **7.** [10.1] 7
8. [10.1] −13 **9.** [10.1] Irrational
10. [10.1] Rational **11.** [10.1] Irrational
12. [10.1] Rational **13.** [10.1] Rational
14. [10.1] Rational **15.** [10.1] Rational
16. [10.1] Irrational **17.** [10.1] 1.732
18. [10.1] 9.950 **19.** [10.1] 10.392
20. [10.3] 17.889 **21.** [10.5] 0.354
22. [10.5] 0.742 **23.** [10.2] $x^2 + 4$
24. [10.2] $5ab^3$ **25.** [10.2] Yes **26.** [10.2] No
27. [10.2] Yes **28.** [10.2] Yes
29. [10.2] $x \geq -7$ **30.** [10.2] $y \geq 10$
31. [10.2] $|m|$ **32.** [10.2] $|7t|$ **33.** [10.2] $|p|$
34. [10.2] $|x - 4|$ **35.** [10.3] $\sqrt{21}$
36. [10.3] $\sqrt{at}$ **37.** [10.3] $\sqrt{x^2 - 9}$
38. [10.3] $\sqrt{6xy}$ **39.** [10.4] $-4\sqrt{3}$
40. [10.4] $4t\sqrt{2}$ **41.** [10.4] $x + 8$
42. [10.4] $\sqrt{t - 7}\sqrt{t + 7}$ **43.** [10.4] x^4
44. [10.4] $m^7\sqrt{m}$ **45.** [10.4] $2\sqrt{15}$
46. [10.4] $2x\sqrt{10}$ **47.** [10.4] $5xy\sqrt{2}$
48. [10.4] $10a^2b\sqrt{ab}$ **49.** [10.5] $\frac{5}{8}$ **50.** [10.5] $\frac{2}{3}$
51. [10.5] $\frac{7}{t}$ **52.** [10.5] $\frac{\sqrt{2}}{2}$ **53.** [10.5] $\frac{\sqrt{2}}{4}$
54. [10.6] $\frac{\sqrt{5y}}{y}$ **55.** [10.6] $\frac{2\sqrt{3}}{3}$ **56.** [10.6] $\frac{\sqrt{15}}{5}$
57. [10.6] $\frac{x\sqrt{30}}{6}$ **58.** [10.7] $13\sqrt{5}$
59. [10.7] $\sqrt{5}$ **60.** [10.7] $\frac{1}{2}\sqrt{2}$
61. [10.8] $b = 20$ **62.** [10.8] $c = \sqrt{3}$
63. [10.8] $7\sqrt{2}$ **64.** [10.9] 52
65. [10.9] No solution **66.** [10.9] 405 ft
67. [10.1] 2 **68.** [10.9] No solution
69. [10.7] $(x - \sqrt{5})(x + \sqrt{5})$
70. [10.9] $b = \sqrt{A^2 - a^2}$

CHAPTER 11

Exercise Set 11.1, pp. 439–440

1. 1, −3, 2 **3.** 2, 0, −3 **5.** 7, −4, 3
7. 2, −3, 5 **9.** 3, −2, 8 **11.** $\sqrt{10}, -\sqrt{10}$
13. $\sqrt{10}, -\sqrt{10}$ **15.** $\frac{\sqrt{10}}{2}, -\frac{\sqrt{10}}{2}$ **17.** $\frac{2}{3}, -\frac{2}{3}$
19. $\frac{4\sqrt{5}}{5}, -\frac{4\sqrt{5}}{5}$ **21.** $\frac{4}{7}, -\frac{4}{7}$ **23.** $2\sqrt{5}, -2\sqrt{5}$
25. 7.9 sec **27.** 3.3 sec **29.** 49.9, −49.9
31. $\sqrt{3}, -\sqrt{3}$ **33.** $\sqrt{11}, -\sqrt{11}$ **35.** 3, −3
37. $3\sqrt{1 + a}, -3\sqrt{1 + a}$

Exercise Set 11.2, pp. 443–444

1. 0, −7 **3.** 0, −2 **5.** 0, $\frac{2}{5}$ **7.** 0, −1 **9.** 0, 3
11. 0, $\frac{1}{5}$ **13.** 0, $\frac{3}{14}$ **15.** 0, $\frac{81}{2}$ **17.** −12, 4
19. −5, −1 **21.** −9, 2 **23.** 5, 3 **25.** −3
27. 4 **29.** $-\frac{2}{3}, \frac{1}{2}$ **31.** $\frac{1}{3}, \frac{4}{3}$ **33.** $\frac{5}{3}, -1$
35. $-\frac{1}{4}, \frac{2}{3}$ **37.** 7, −2 **39.** −5, 4 **41.** 4
43. 3, −2 **45.** 9 **47.** 7 **49.** $-\frac{1}{3}, 1$
51. 0, $\frac{\sqrt{5}}{5}$ **53.** $-\frac{5}{2}, 1$ **55.** 0, $-\frac{b}{a}$
57. $\sqrt{2}, -\sqrt{2}$

Exercise Set 11.3, pp. 447–448

1. 9, −5 **3.** $-3 \pm \sqrt{21}$ **5.** $-13 \pm 2\sqrt{2}$
7. $7 \pm 2\sqrt{3}$ **9.** $-9 \pm \sqrt{34}$ **11.** $\frac{-3 \pm \sqrt{14}}{2}$
13. $x^2 - 2x + 1$ **15.** $x^2 + 18x + 81$
17. $x^2 - x + \frac{1}{4}$ **19.** $t^2 + 5t + \frac{25}{4}$
21. $x^2 - \frac{3}{2}x + \frac{9}{16}$ **23.** $m^2 + \frac{9}{2}m + \frac{81}{16}$ **25.** 10%
27. 18.75% **29.** 8% **31.** 20% **33.** −10, 8
35. $-2 \pm \sqrt{29}$ **37.** 2, −5 **39.** $\frac{4 \pm \sqrt{2}}{3}$
41. $\frac{7}{8}, -\frac{13}{8}$ **43.** 12.6% **45.** $2239.13
47. $3b, -b$ **49.** 0, $\frac{4}{5}$
51. $x^2 + (2b - 4)x + (b - 2)^2$

Exercise Set 11.4, p. 451

1. −2, 8 **3.** −21, −1 **5.** $1 \pm \sqrt{6}$
7. $11 \pm \sqrt{19}$ **9.** $-5 \pm \sqrt{29}$ **11.** $\frac{7 \pm \sqrt{57}}{2}$
13. −7, 4 **15.** $\frac{-3 \pm \sqrt{17}}{4}$ **17.** $\frac{-3 \pm \sqrt{145}}{4}$
19. $\frac{-2 \pm \sqrt{7}}{3}$ **21.** $-\frac{1}{2}, 5$ **23.** $-\frac{7}{2}, \frac{1}{2}$ **25.** 5670
27. 12, −12 **29.** $\pm 16\sqrt{2}$ **31.** $\pm 2\sqrt{c}$
33. $3a, -2a$ **35.** $c + 1, -c$ **37.** $\frac{-2 \pm \sqrt{4 - 3a}}{a}$

39. $\dfrac{-m \pm \sqrt{m^2 - 4nk}}{2k}$

Exercise Set 11.5, pp. 455–456

1. $-3, 7$ **3.** 3 **5.** $-\frac{4}{3}, 2$ **7.** $-\frac{7}{2}, \frac{1}{2}$
9. $-3, 3$ **11.** $1 \pm \sqrt{3}$ **13.** $5 \pm \sqrt{3}$
15. $-2 \pm \sqrt{7}$ **17.** $\dfrac{-4 \pm \sqrt{10}}{3}$ **19.** $\dfrac{5 \pm \sqrt{33}}{4}$
21. $\dfrac{1 \pm \sqrt{2}}{2}$ **23.** $-\frac{5}{3}, 0$
25. No real-number solutions **27.** $-5, 5$
29. $\dfrac{5 \pm \sqrt{73}}{6}$ **31.** $\dfrac{3 \pm \sqrt{29}}{2}$ **33.** $\dfrac{-1 \pm \sqrt{10}}{3}$
35. $-2.7, 0.7$ **37.** $-6.7, -3.3$ **39.** $-0.2, 1.2$
41. $0.3, 2.4$ **43.** $0, 2$ **45.** $\dfrac{3 \pm \sqrt{5}}{2}$
47. $\dfrac{-7 \pm \sqrt{61}}{2}$ **49.** $\dfrac{-2 \pm \sqrt{10}}{2}$
51. No real-number solutions **53.** $\dfrac{-1 \pm \sqrt{3a+1}}{a}$
55. $\dfrac{c \pm \sqrt{3}d}{2}$ **57.** $\dfrac{a}{b}, -\dfrac{c}{d}$
59. (a) yes, no; (b) yes, no; (c) when $b^2 \geq 4ac$
61. 0.3

Exercise Set 11.6, pp. 458–459

1. $6, -\frac{2}{3}$ **3.** $6, -4$ **5.** 1 **7.** $5, 2$
9. No solution **11.** $\sqrt{7}, -\sqrt{7}$ **13.** $-2 \pm \sqrt{3}$
15. $2 \pm \sqrt{34}$ **17.** 9 **19.** 7 **21.** $1, 5$ **23.** 3
25. $25, 13$ **27.** No solution **29.** 3 **31.** 7 mi
33. $\dfrac{-3 \pm \sqrt{201}}{8}$ **35.** $\dfrac{-1 \pm 3\sqrt{5}}{6}$ **37.** 13
39. 10 **41.** 0

Exercise Set 11.7, pp. 461–462

1. $A = \dfrac{N^2}{6.25}$ **3.** $T = \dfrac{cQ^2}{a}$ **5.** $c = \sqrt{\dfrac{E}{m}}$
7. $d = \dfrac{c \pm \sqrt{c^2 + 4aQ}}{2a}$ **9.** $a = \sqrt{c^2 - b^2}$
11. $t = \dfrac{\sqrt{s}}{4}$ **13.** $r = \dfrac{-\pi h \pm \sqrt{\pi^2 h^2 + \pi A}}{\pi}$
15. $r = 6\sqrt{\dfrac{10A}{\pi S}}$ **17.** $a = \sqrt{c^2 - b^2}$ **19.** $a = \dfrac{2h}{\sqrt{3}}$

21. (a) $r = \dfrac{C}{2\pi}$; (b) $A = \dfrac{C^2}{4\pi}$
23. $\dfrac{2 \pm \sqrt{4 - am + an}}{a}$ **25.** $\dfrac{1}{3a}, 1$ **27.** $r = \sqrt[3]{\dfrac{3v}{4\pi}}$

Exercise Set 11.8, pp. 466–468

1. 3 cm **3.** 7 ft, 24 ft
5. Width: 8 cm; length: 10 cm
7. Width: 16 cm; length: 20 cm
9. Width: 5 m; length: 10 m
11. 4.6 m, 6.6 m
13. Width: 3.6 in.; length: 5.6 in.
15. Width: 2.2 m; length: 4.4 m **17.** 7 km/h
19. 8 mph **21.** 4 km/h **23.** 36 mph
25. 1 km/h **27.** $3 + 2\sqrt{2} \approx 5.828$ **29.** 2.41 cm
31. 14.14 in.; two 10-in. pizzas **33.** 2 ft
35. 7.2 cm

Exercise Set 11.9, pp. 472–473

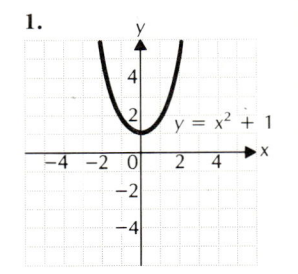

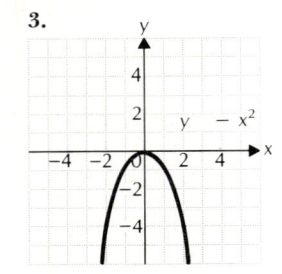

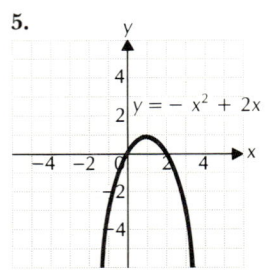

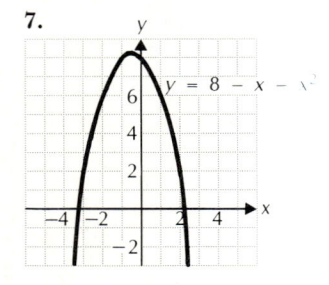

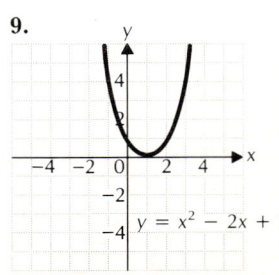

1. $y = x^2 + 1$

3. $y = -x^2$

5. $y = -x^2 + 2x$

7. $y = 8 - x - x^2$

9. $y = x^2 - 2x + 1$

19. [11.5] $\dfrac{-3 \pm \sqrt{29}}{2}$ 20. [10.9] [11.6] 2
21. [7.2] $\{x|x \geq -1\}$ 22. [3.3] 8 23. [3.3] -3
24. [7.2] $\{x|x < -8\}$ 25. [7.2] $\{y|y \leq \tfrac{35}{22}\}$
26. [11.1] $-\sqrt{10}, \sqrt{10}$ 27. [11.3] $3 \pm \sqrt{6}$
28. [9.9][11.6] $\tfrac{2}{9}$ 29. [9.9][11.6] -5
30. [9.9][11.6] No solution 31. [10.9] 12
32. [9.11] $b = \dfrac{At}{4}$ 33. [9.11] $m = \dfrac{tn}{t+n}$
34. [11.7] $A = \pi r^2$ 35. [11.7] $x = \dfrac{b \pm \sqrt{b^2 + 4ay}}{2a}$
36. [4.1][9.12] x^{-4} 37. [4.1][9.12] y^7
38. [4.1][9.12] $4y^{12}$ 39. [4.3] $10x^3 + 3x - 3$
40. [4.4] $7x^3 - 2x^2 + 4x - 17$
41. [4.5] $8x^2 - 4x - 6$ 42. [4.6] $-8y^4 + 6y^3 - 2y^2$
43. [4.6] $6t^3 - 17t^2 + 16t - 6$ 44. [4.8] $t^2 - \tfrac{1}{16}$
45. [4.8] $9m^2 - 12m + 4$
46. [5.9] $15x^2y^3 + x^2y^2 + 5xy^2 + 7$
47. [5.9] $x^4 - 0.04y^2$
48. [5.9] $9p^2 + 24pq^2 + 16q^4$ 49. [9.1] $\dfrac{2}{x+3}$
50. [9.2] $\dfrac{3a(a-1)}{2(a+1)}$ 51. [9.5] $\dfrac{27x - 4}{5x(3x-1)}$
52. [9.6] $\dfrac{-x^2 + x + 2}{(x+4)(x-4)(x-5)}$
53. [5.1] $4x(2x - 1)$ 54. [5.2] $(5x - 2)(5x + 2)$
55. [5.5] $(3y + 2)(2y - 3)$ 56. [5.3] $(m - 4)^2$
57. [5.1] $(x^2 - 5)(x - 8)$
58. [5.6] $3(a^2 + 6)(a + 2)(a - 2)$
59. [5.2] $(2x + 1)(2x - 1)(4x^2 + 1)$
60. [5.9] $(7ab - 2)(7ab + 2)$
61. [5.9] $(3x + 5y)^2$ 62. [5.9] $(2a + d)(c - 3b)$
63. [5.9] $(5x - 2y)(3x + 4y)$ 64. [9.7] $-\tfrac{42}{5}$
65. [10.1] 7 66. [10.1] -25 67. [10.2] $8|x|$
68. [10.3] $\sqrt{a^2 - b^2}$ 69. [10.4] $8a^2b\sqrt{3ab}$
70. [10.4] $5\sqrt{6}$ 71. [10.4] $9xy\sqrt{3x}$
72. [10.5] $\tfrac{10}{9}$ 73. [10.5] $\dfrac{8}{x}$ 74. [10.7] $16\sqrt{3}$
75. [10.6] $\dfrac{2\sqrt{10}}{5}$ 76. [10.8] 40

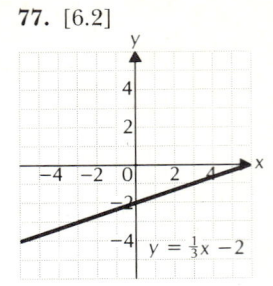

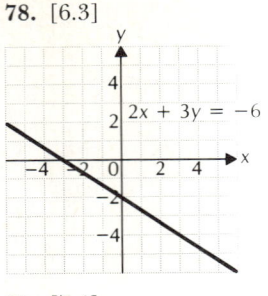

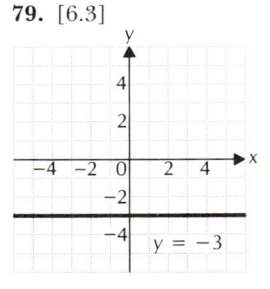

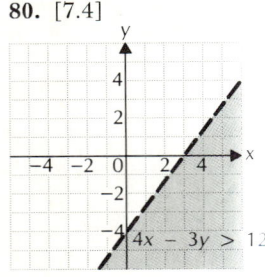

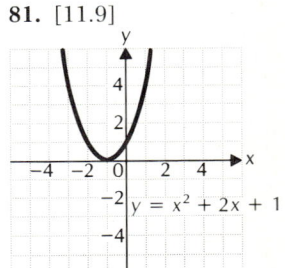

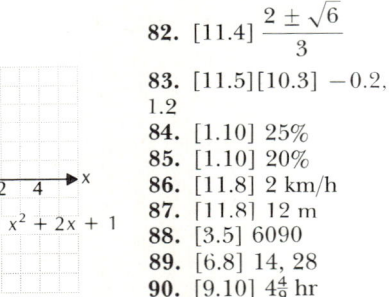

77. [6.2] $y = \tfrac{1}{3}x - 2$

78. [6.3] $2x + 3y = -6$

79. [6.3] $y = -3$

80. [7.4] $4x - 3y > 12$

81. [11.9] $y = x^2 + 2x + 1$

82. [11.4] $\dfrac{2 \pm \sqrt{6}}{3}$
83. [11.5][10.3] $-0.2, 1.2$
84. [1.10] 25%
85. [1.10] 20%
86. [11.8] 2 km/h
87. [11.8] 12 m
88. [3.5] 6090
89. [6.8] 14, 28
90. [9.10] $4\tfrac{4}{9}$ hr
91. [8.3] $451.20, variation contant is the amount earned per hour 92. [8.1] $-4, 0, 0$
93. [7.5] $\{c, d\}$ 94. [7.5] $\{a, b, c, d, e, f, g, h\}$
95. [2.1] $-12, 12$ 96. [10.4] 2 97. [11.4] 30
98. [10.8] $\dfrac{\sqrt{6}}{3}$ 99. [4.8] Yes 100. [9.6] No
101. [4.8] No 102. [10.4] No 103. [10.2] Yes

INDEX

Miscellaneous formulas of general interest

Accidents (average number of terms, of age), 145
Account balance with simple interest, 36
Batting average, 6
Compound interest, 139, 446–447, 460
Converting Celsius temperature to Fahrenheit, 48
Distance light travels, 381
Distance to horizon, 427–428, 459
Distance of lightning (meteorology), 309–310
Distance of projectile, 207
Falling object, 146, 438–439
Force (physics) in terms of mass and acceleration, 128
Height (anthropology) from humerus, 311–312
Horsepower (mechanics), 128

Intelligence quotient, 126
Karat and percentage of gold, 320
Magic number (sports), 208
Number of games in league play (sports), 143, 205
Planet orbits, 15
Possible handshakes, 207
Pressure in atmosphere, 313
Relativity (physics), 128
Safest position of ladder, 45
Simple interest, 53, 127, 375
Skid marks and speed of car, 404
Temperature and cricket chirps, 122
Tire tread and replacement costs, 314
Torricelli's theorem, 460
World track record, 122, 129, 294

Abscissa, 251
Absolute value, 62–63, 66
    and addition, 69
    and distance, 62–63
    equations with, 66–67
    inequalities with, 296
    and multiplications, 80
    of numbers, 62–63
    and radicals, 400

Addition, 17–18, 21
    associative law of, 28–30, 70. *See also* Solving problems; Fractional expressions; Radical expressions.
    commutative law of, 17, 70
    and the distributive law, 32–34, 88
    of exponents, 136
    of fractional expressions, 21, 338–339, 344, 346–348

INDEX

Addition (cont.)
  integers, 69
  on number line, 67–68
  numbers of arithmetic, 21
  polynomials, 151, 210
    in columns, 151
  of radical expressions, 415–417
  rational numbers, 69
  signed numbers, 69
  using LCM, 344, 346–348
Addition method for solving equations, 248–249, 254. See also Systems of equations.
Addition principle
  for equations, 102–103. See also Systems of equations.
  for inequalities, 284–285
  using with multiplication principle, 108–112
Additive inverses, 70–71, 91–92
  of denominators, 339–340
  of least common multiples, 345
  of polynomials, 155–156
  of sums, 91–92
Additive property of zero, 18, 69
  and subtraction, 76, 156
Algebraic expression, 5. See also Simplifying; Fractional expressions; and specific names.
Angle measures of triangles, 121
Angles, sum of measures
  in polygon, 129
  in triangle, 121
Applications, see Applied problems; Problems; Solving problems.
Applied problems, 45, 61, 71, 77, 115–121, 124–126, 139, 140, 142, 152, 157–158, 246, 252, 254–262, 264–267, 275–276, 291–293, 309–312, 318–320, 322–323, 364–371, 374–376, 381, 397–398, 412, 420–423, 427–429, 438–439, 442–443, 446–447, 459–466. See also Problems; Solving problems.
Approximating irrational numbers, 395–396. See also Irrational numbers.
Approximating square roots, 396, 403, 454. See also Square roots.
  of fractions, 408–409
  of products, 403
Areas
  and polynomials, 152–153, 157, 163, 166, 171, 189
  of circles, 128
    of sector, 128
  of parallelograms, 128
  of rectangles, 3, 127, 152
  of right circular cylinders, 214
  of trapezoids, 374
  of triangles, 127, 204

Arrows, 308
Ascending order, 148
Associative law, 28–30, 81
  of addition, 28–30
  of multiplication, 28–30, 81
Axes of graphs, 222

Base, 24, 136
BASIC (computer language), 94–96
Binomials, 149
  differences of squares, 179–180
  products of, 161, 164, 167–168, 169
    FOIL method, 164, 169, 186–187, 192
  squares of, 168, 182
    factoring, 183
Braces, 94. See also Parentheses.
Brackets, 94. See also Parentheses.

Carry out, 38
Changing the sign, see Additive inverses.
Checking. See also Solving problems.
  solutions of equations, 362, 427
Circles. See also Problems.
  area of, 128
    sector, 128
  circumference of, 125
  diameter of, 125
  radius of, 125
Circumference, 125
Coefficients, 148, 209
  factoring, 176
Collecting like terms, 90, 93, 109–111, 144, 148, 209–210, 415–417. See also Specific operations.
Combining like terms, see Collecting like terms.
Common denominators, 21, 338
Commutative law
  of addition, 17
  of multiplication, 17, 81
Completing the square, 444–445, 448–452. See also Quadratic equations.
  in deriving the quadratic formula, 451–452
  factoring by, 444–445, 448–450
Complex fractional expressions, 352–353
Composite, 10
Compound interest, 139
Computer language, 94–96
Consecutive integers, 118
  even, 122
  odd, 122
Constant, 311
Constant of variation, 318–320
Constants, 35
  as like terms, 35
Coordinates, 223. See also Ordered pairs.
Correspondence, 308. See also Function.
Cube, volume of, 142

INDEX

Decimal notation
  converting to percent notation, 50
  converting to scientific notation, 381
  for irrational numbers, 396
  for rational numbers, 396
  repeating, 66
Degrees
  of polynomials, 148, 209
  of terms, 148, 209
Denominators, 18–22
  additive, inverse of, 339–340
  least common multiple of, 21, 111
  rationalizing, 410–411, 413–414
Descending order, 147
Diagonals, number of, 442, 461. *See also* Square; Rectangle; etc.
Differences of squares, 179–180. *See also* Factoring.
Direct variation, 318. *See also* Problems; Variation.
Discriminant, 453
Distance, 128, 146, 264–267, 318–322, 365–366, 438–439, 464–466. *See also* Problems.
Distributive law, 32–34, 88. *See also* Parentheses; Multiplication.
  as basis for multiplication, 32–33, 88, 160
  and collecting like terms, 34–35, 90, 144, 210
  and factoring, 33–34, 89, 177
Division
  of decimals, 85
  definition of, 82
  using exponents, 137, 379–380
  of fractional expressions, 22–23, 85, 335–336
  integers, 85. *See also* Rational numbers.
  as multiplication by reciprocal, 22, 105
  of numbers of arithmetic, 85
  of polynomials, 355–357
  of radical expressions, 409, 413
  rational numbers, 82–85
  and reciprocals, 22–23, 83–85
  using scientific notation, 382–383
  by zero, 18, 83
Divisor, 22. *See also* Division.
Domain, 308–309. *See also* Function.

Eliminating variables, *see* Systems of equations.
Empty set, 302. *See also* Sets.
Equations, 37, 221, 359–363, 435–436, 456. *See also* Formulas; Solving equations, systems of.
  containing parentheses, 113–114. *See also* Parentheses.
  first-degree, 221, 232
  fractional, 359–363, 456
  graphs of, *see* Graphing
  linear, 221, 232. *See also* Graphing; Slope.
  of lines
    point–slope, 274–275
    slope–intercept, 273
  with missing variables, 233–234. *See also* Graphing.
  quadratic, 435–436. *See also* Quadratic equations.
  with radical expressions, 426–429, 457. *See also* Radical expressions.
  second-degree, *see* Quadratic equations.
  solving, *see* Solving equations.
  systems of, 237, 239. *See also* Systems of equations.
    addition principle, 102–103, 248–249, 254
    by factoring, 200, 441–443
    multiplication principle, 104–106
  translating to, 37, 238. *See also* Translating.
  of variation, 318, 322
Equilateral triangle, 425
Equivalent expressions, 17–18, 89, 156, 331, 415–417. *See also* Simplifying; Factoring.
Equivalent inequalities, 285
Equivalents, fractional, decimal, and percent, 479
Evaluating expressions, 5, 26, 142, 208
Evaluating polynomials, 142, 208
Exponential notation, 24–26. *See also* Exponents.
Exponents, 24–26, 136, 377–380. *See also* Scientific notation.
  dividing using, 137, 379–380
  evaluating, 26
  integers as, 136
  multiplying using, 136, 378–379
  negative, 377–378
  notation, 24–26, 136
  one as, 25
  raising a power to a power, 138, 380
  raising a product to a power, 138, 380
  zero as, 25, 137
Expressions
  algebraic, 5
  equivalent, 17–18, 89, 156, 331, 415–417
  fractional, *see* Fractional expressions
  simplifying, 19. *See also* Simplifying.

Factoring, 9–11, 175
  completely, 89, 176, 181
  finding LCM by, 13–14
  general strategy, 196
  hints, 181, 190
  numbers, 10

Factoring (*cont.*)
  polynomials, 176–177, 180
    by completing the square, 444–445, 448–450
    differences of squares, 179–180
    by grouping, 177–178, 190–192
    monomials, 176
    squares of binomials, 183
    terms with a common factor, 176
    trinomial squares, 183
    trinomials, 185–188, 189–194
    with several variables, 212–214
  radical expressions, 403. *See also* Radical expressions.
  removing common factor, 19–21, 89, 176, 193–194
  as reverse of multiplying, 89, 176
  solving equations by, 200, 441–443. *See also* Quadratic equations.
Factorizations, 10–11. *See also* Prime factorization; Factoring.
Factors, 7. *See also* Exponents; Factoring.
Falling object, 146, 438–439
Familiarize, 1–4
First-degree equations, 221, 273–275. *See also* Line.
Five-step process for problem solving, *see* Solving problems.
FOIL method, 164, 169. *See also* Factoring, trinomials.
Formulas, 123–127, 374–375, 459–460. *See also* Problems.
  distance, 128
  falling object, 146, 438–439
  geometric, 120–121, 125, 142, 152, 157, 162, 204, 214, 234, 246, 374, 442–443, 461–462
  interest, 53, 127, 139, 375, 446–447, 460
  pendulum, 412, 459
  products of certain kinds of polynomials, 160–161, 164, 167–169
  quadratic, 435, 451–452
  solving for given letter, 126
  speed, 264–267, 318, 322, 365–366, 369, 464–466
Fractional equations, 359–363, 456
  applied problems, *see* Problems.
Fractional expressions, 329
  addition of, 21, 338–339, 344, 346–348
  complex, 352–353
  division of, 22–23, 335–336
  equivalence of, 331
  factoring, 20, 331–333
  least common multiple of, 344–345
  multiplication of, 85, 330, 332
  multiplying by 1, 330–331
  reciprocals of, 22, 83–84, 335
  simplifying, 19–20, 331–333, 350

  subtraction of, 21–22, 339–341, 350
Fractional notation, 85, 394. *See also* Fractional expressions.
  and addition, 21
  converting to decimal notation, 66
  converting to percent notation, 50
  and division, 22–23, 84–85
  and multiplication, 85–86
  and reciprocals, 22–23, 84–85
  sign changes in, 86
  simplifying, 19–20, 23, 330
  and subtraction, 21
Fractional radicand, 409, 413. *See also* Radical expressions.
Fractions, *see also* Fractional expressions.
  adding of, 21
  division of, 22
  multiplication of, 21
Function, 307
  applications, 309–314. *See also* Problems.
  constant, 311
  correspondence, 308
  domain, 308–309
  formulas, 309–310
  graphing, 315–317
  identifying, 308–309, 316
  input, 308–309
  "machine," 310
  notation, 310
  outputs, 308–309
  range, 308
  values, 311
  variation, 318, 322–323
  vertical-line test, 316

Geometric formulas, 481. *See also* Formulas, geometric.
Graphing
  equations, 221, 226, 234. *See also* Systems of equations; Inequalities; Quadratic equations.
    with missing variable, 233–234
    in two variables, 222, 226–231
  inequalities, 295–299
    with absolute value, 296
    in one variable, 284
    in two variables, 296
  using intercepts, 232–233
  with missing variable, 233–234
  on a plane, 222
  quadratic equations, 468–471
  slope, 228
  systems of equations, 240–242
Gravitational force, 374
Greater than ($>$), 61–62, 66. *See also* Inequalities.
Greater than or equal to ($\geq$), 61–62, 66. *See also* Inequalities.

INDEX

Grouping
    in addition, *see* Addition, associative law of.
    factoring by, 177–178, 190–192
    in multiplication, *see* Multiplication, associative law of.

Half-plane, 298
Halfway between rational numbers, 395
Handshakes, 207
Horizontal lines, 233–234. *See also* Graphing, with missing variable; Slope.
Hypotenuse, 419. *See also* Triangles.

Improper symbols, 21
Inequalities, 62, 283
    addition principle for, 284–285
    equivalent, 285
    graphs of, 295–299
      absolute value, 296
      in one variable, 295–296
      in two variables, 297–298
    linear, 298
    multiplication principle for, 288
    problem solving using, 291–293
    solutions of, 284, 296–297
      in one variable, 284
      in two variables, 296
Inputs, 308
    using principles together, 288–289
    problem solving using, 291–293
Integer problems, *see* Problems, number.
Integers, 60
    absolute value of, 62–63, 66
    addition of, 17, 28–30
    additive inverses of, 60, 70–71, 91–92
    consecutive, 118
    consecutive even, 122
    consecutive odd, 122
    division of, 82. *See also* Rational numbers.
    as exponents, 136
    multiplication of, 17–18, 28–30
    negative, 60
    order of, 61–62
    positive, 60
    quotient of, *see* Rational numbers.
    subtraction of, 74–76
Intercepts, 229, 232–233
    graphs using, 232. *See also* Graphing.
Interest, *see also* Problems.
    compound, 139, 446–447, 460
    simple, 53, 127, 375
      and account balance, 36
Intersection, 301–302
Inverse variation, *see* Variation.
Inverses
    additive, 70–71, 91–92, 155–156, 339–340, 345
      and subtraction, 76, 155–156

multiplicative, 22–23, 83–85. *See also* Rational numbers; Division.
Irrational numbers, 395
    approximating, 396–397, 403, 410–411, 454
    decimal notation for, 396
Irreducible, 182

Karat and percentage of gold, 320. *See also* Problems.

LCM, *see* Least common multiple
Least common denominator, 21
Least common multiple, 12–15, 21. *See also* Solving equations.
    additive inverse of, 345
    of algebraic expressions, 344–345
    of planet orbits, 15
    of polynomials, 343
Legs of right triangles, 419. *See also* Triangles.
Less than ($<$), 61–62, 66. *See also* Inequalities.
Less than or equal to ($\leq$), 61–62, 66. *See also* Inequalities.
Like radicals, 415–417
Like terms, 144, 415–417
Line, *see also* Graphing; Slope; Equations, linear.
    horizontal, 271
    point–slope equation, 274–275
    slope, 269–270
    slope–intercept equation, 273
    vertical, 272
Line of symmetry, 468
Linear equations, 221. *See also* Slope.
    graphing, *see* Graphing, equations.
Linear inequalities, 298
Lines, parallel, 241–242, 251, 302. *See also* Systems of equations, having no solution; Slope.

Magic number, 208
Mean, 126, 291–292
Meaningless radical expression, 399
Member of, 301
Missing terms, 149
Missing variables in equations, 233–234
Mixture problems, 259–262. *See also* Problems.
Monomials, 136, 149
    factoring, 176
    products of, 136, 160, 164
    quotients of, 137
Motion problems, *see* Problems, distance; Problems, speed.
Multiples, least common, 12–15, 21, 343–345. *See also* Prime factorization.

Multiplication
  associative law of, 28–30, 81
  commutative law of, 17, 81
  using the distributive law, 32–34, 88, 160
  using exponents, 136, 378–379
  formulas and methods, 169
  of fractional expressions, 21, 85–86, 330–333
  integers, 17–18, 28–30
  numbers of arithmetic, 18, 21, 86. *See also* Rational numbers.
  polynomials, 160–161, 164, 167–168, 211
  property of zero, 80
  of radical expressions, 402
  rational numbers, 79–81, 85–86
  using scientific notation, 382–383
Multiplication principle
  for equations, 104–106. *See also* Systems of equations, solving by addition method.
  using with addition principle, 108–112
  for inequalities, 288
Multiplicative inverse, 22–23, 83–85. *See also* Reciprocals.
Multiplying by 1, 18–19, 145
Multiplying by −1, 86, 91–93. *See also* Additive inverses.

Natural numbers, 9, 16
Negative numbers, 60, 84
Negative square root, 394
Nonsensible replacements, 18, 83, 399
Notation
  computer language (BASIC), 95–96
  decimal, 50, 66, 381, 396
  exponential, 24–26, 136
  fractional, 85, 394
  percent, 49–50
  scientific, 380–383
Number of games played, 143, 205
Number line, 60, 222
  and addition, 67–68
  and graphing rational numbers, 65
  order on, 61–62
  and subtraction, 74–75
Number pair, 222–223
Number patterns, 79–80
Numbers
  of arithmetic, 5, 16, 22, 59–60, 85
    addition of, 21
    division of, 22–23
    multiplication of, 21, 86
    reciprocals of, 22
    subtraction of, 21–22
  integers, 60
  irrational, 395
  natural, 9, 16
  negative, 60

  order of, 61–62, 65–66
  positive, 60
  prime, 10
  rational, 64–66, 394. *See also* Rational numbers.
  real, 66
  relationships between, 66
  whole, 16, 60
Numerals, fractional, *see* Fractional expressions.
Numerator, 18. *See also* Fractions.

One, as exponent, 25
Opposite of a number, *see* Additive inverses.
Order
  in addition, 94–95
  in multiplication, 94–95
  on number line, 61–62, 65–66
  of numbers, 61–62, 65–66
  of operations, 28, 94–95
Ordered pairs, 222–223
Origin, 222. *See also* Graphing.
Outputs, 308. *See also* Function.

Pairs, ordered, 222–223
Parabolas, 469–471
Parallel lines, *see* Systems of equations.
Parallelograms, areas of, 128
Parentheses, 28–30, 95
  in equations, 113–114
  within parentheses, 94
  product within, 138
  removing, 91–93
Parking, 397–398
Pendulums, periods of, 412, 459
Percent, problems involving, 50–53, 117
Percent notation, 48
  converting to decimal, 48
  converting to fractional, 49–50
Perimeter, 120
  of rectangle, 3, 120, 127, 238, 246
  of square, 146
Pi (π), 396
Plotting points, 222–223
Points, coordinates of, 222–223
Polygon
  measure of interior angles, 129
  number of diagonals, 442–443, 461
  perimeter, 120
Polynomials, 135, 141–145. *See also* Integers; Applied problems; Quadratic equations.
  addition of, 151, 210
  additive inverses of, 155–156
  in ascending order, 148
  binomials, 149
  coefficients in, 148, 209

INDEX

degrees of, 148, 209. *See also* Quadratic equations.
in descending order, 147
division of, 355–356
equivalent, 143, 156
evaluation, 142, 208
factoring, *see* Factoring, polynomials.
irreducible, 182
least common multiple of, 343–345
missing terms in, 149
monomials, 136, 149
multiplication of, 160–161, 164, 167, 169, 211
prime, 182
problem solving with, 142, 157–159
in several variables, 208
special products of, 164, 167–168
subtraction of, 156, 210
terms of, 143
trinomials, 149, 182
Positive numbers, 60, 84. *See also* Rational numbers.
Powers, 24–26. *See also* Exponents.
raising to a power, 138
square roots of, 406
Prime factorization, 10–11. *See also* Factoring.
LCM, 12–15
Prime numbers, 10. *See also* Numbers.
Prime polynomial, 182
Principal, *see* Interest
Principal square root, 394
Principal of squaring, 426, 457
Principle of zero products, 198–199, 441
Problems, *see also* Applied problems; Solving problems.
age, 256
altitude, 77
anthropology, 311–312
average, 126, 291–292
ball diamonds, size of, 422, 424
board cutting, 115–116
conversions, 292
costs, 6
diagonals of polygon, number of, 442, 461
distance, 128, 146, 264–267, 318, 322, 365–366, 464–466
of free falling object, 146, 438–439
geometric
angles, 120, 121
area, 152, 157, 204, 214, 374, 462
circumference, 125
diagonals of polygon, 442–443, 461
perimeter, 120, 234, 246
triangle, 121
grade of road, 275
gravitational force, 374
horizon, distance to, 427–428, 459

integer, 7, 43, 116, 118, 202–203, 205, 236, 364–365
interest
compound, 139, 446–447, 460
simple, 53, 127, 375
I.Q. (intelligence quotient), 126
karat rating of gold, 319–320
ladder
height of, 421–422
safe position of, 45, 124
meteorology, 309–310
mixture, 259–262
motion, *see* Problems, distance; Problems, speed.
number, 7, 43, 116, 118, 202–203, 205, 236, 364–365
parking-lot arrival spaces, 397–398
pendulum, period of, 412, 459
percent, 50–53, 117
profit–loss, 71
proportions, 369–371
ratios, 369–371
rent-a-car, 119, 237–238, 252
right triangles, 420–423, 463–464
sight, distance from given height, 427–428, 459
skidding car, speed of, 404
speed, 264–267, 318, 322, 365–366, 369, 464–466. *See also* Problems, distance.
sports, 6, 44, 143, 205, 208, 254–255, 369–371, 422, 424
ticket, 257–258
Torricelli's theorem, 460
wildlife population, estimating, 371–372
work, 323, 367–369
Products, *see also* Multiplication.
formulas and methods, 169
raising to a power, 138
of sums and differences, 167
of two binomials, 161, 164, 167–168
Properties, *see also* Simplifying; specific property.
of 0, 18, 80
of 1, 18
of −1, 92
Proportion, 369–371
Pythagorean property, 419. *See also* Triangles.

Quadrants in graphs, 223
Quadratic equations, 435–436
in applied problems, *see* Problems
approximating solutions of by graphing, 471–472
graphs of, 468–471
solving, 436–439, 441–443, 444–454
by completing the square, 444–445, 448–450
by factoring, 441–443

INDEX

Quadratic equations, solving *(cont.)*
    by quadratic formula, 451–454
    in standard form, 436
Quadratic formula, 435, 451–452
    discriminant, 453
Quotient, of integers, 85. *See also* Rational numbers; Square roots.

Radical equations, 426–429, 457
Radical expressions, 393, 399. *See also* Square roots.
    adding, 415–417
    approximating, 403
    division of, 409, 413
    in equations, 426–429
    factoring, 403
    meaningless, 399
    multiplying, 402
    simplifying, 400, 403, 405–407, 415–417
Radical symbol, 394
Radicals, like, 415–417. *See also* Radical expressions.
Radicand, 399
    meaningless, 399
    perfect square, 400. *See also* Radical expressions.
Radius, 125
Range, 308. *See also* Function.
Rate, *see* Speed; Distance.
Ratio, *see* Proportion.
Rational numbers, 64–66, 394
    absolute value of, 66
    addition, *see* Addition.
    additive inverses of, 70–71, 91
    decimal notation for, 66, 396
    division of, 82–85
    graphing, 65
    fractional expressions, 335
    halfway between two, 395
    multiplication of, 79–81
    order, 61–62, 65–66
    as quotient of integers, 85
    reciprocal of, 83–84. *See also* Multiplicative inverse.
    subtraction, 74–76
Rationalizing denominators, 410–411, 413–414
Real numbers, 66, 396. *See also* Rational numbers; Irrational numbers.
Reciprocals, 22–23, 83–85
    and division, 22–23, 83–85
    of fractional expressions, 335
Rectangles
    areas of, 127, 152, 462–463
    perimeters of, 120, 127, 238, 246
Rectangular solids
    diagonal of, 425
    volumes of, 142, 163

Relatively prime, 179
Repeating decimals, 66
Right circular cylinder, area of, 214
Right triangles, 419–423. *See also* Triangles.
Roots, square, *see* Square roots; Radical expressions.
Roster method, 301

Same denominators, 21, 338
Scientific notation, 380–383
    converting from decimal notation, 381
    converting to decimal notation, 381
    dividing using, 383–384
    multiplying using, 382–383
Second-degree equations, 435–436. *See also* Quadratic equations.
Second step in problem solving, *see* Translating.
Set notation, 300–303
Sets, 300, 307
    belongs to, 301
    empty, 302
    intersection of, 301
    member of, 301
    membership, 301
    naming, 300
        roster method, 300–301
    union of, 302–303
Signs, 71
Similar terms, *see* Like terms.
Simple interest, 53, 127, 375
Simplifying, *see also* Collecting like terms; Solving equations.
    complex fractional expressions, 352–353
    expressions with exponents, 138–139. *See also* Exponents.
    fractional expressions, 19–23, 331–333
    fractional notation, 21
    radical expressions, 400, 403, 405–407, 415–417
    removing factor of, 331–332
    removing parentheses, 91–93
Slant, *see* Slope
Slope, 228, 269. *See also* Graphing.
    applications, 275
    from equation, 272
    of horizontal line, 271
    of vertical line, 272
    of any line, 269–270
    point–slope equation of line, 274
    slope–intercept equation, 273
Solution set, *see* Solutions
Solutions
    of equations, 37–38, 102, 225–226
    of inequalities, 284. *See also* Inequalities.
    of systems of equations, 239–240. *See also* Systems of equations.

INDEX

Solving equations, *see also* Solving problems; Problems.
  using the addition principle, 102–103
    and checking, 102–103
    containing parentheses, 113–114. *See also* Parentheses.
  by factoring, 200
  fractional, 359–363
  using the multiplication principle, 104–106
  with parentheses, 113–114
  using principle together, 108–112
  quadratic, 436–439, 441
    by completing the square, 444–445, 448–450
    by factoring, 441–443
    by quadratic formula, 451–454
  with radicals, 426–429, 457
  systems of, 237
    by addition method, 248–249, 254
    by graphing, 241–242
    by substitution method, 243–245, 254
  of type $x + a = b$, 38–39, 102–103
  of type $ax = b$, 40, 104–106
Solving formulas, *see* Formulas.
Solving inequalities, *see* Inequalities.
Solving problems, 42–46, 115–121, 202–206. *See also* Applied problems; Problems; Systems of equations.
  five-step process, 42, 115, 202–206
    first step, 1–4
    second step, 6–8, 44, 236. *See also* Translating.
    third step, 38
    with integers, 60–61
    with rational numbers, 71, 77
Special products of polynomials, 164, 167–168
  squaring binomials, 168
  sum and difference of two expressions, 167
  two binomials, 164
Speed, 365–366. *See also* Problems.
  of a skidding car, 404
Sphere, volume of, 462
Square
  of binomial, 168
  perimeter of, 146
Square roots, 394–397. *See also* Radical expressions.
  approximating, 396–397, 403, 410–411, 454
  finding on a calculator, 397
  of fractions, 410–411, 413
  negative, 394
  of powers, 406–407
  principal, 394
  products, 402
  of quotients, 408–409
Squares, differences of, *see* Factoring.

Squaring, principle of, 426–427
Standard form of quadratic equations, 436
Substituting, *see* Evaluating expressions.
Substitution method for solving equations, 243–245, 254. *See also* Systems of equations.
  solving for the variable first, 245
Subtraction
  by adding inverses, 76, 155–156
  definition, 74
  and the distributive law, 88, 90
  of exponents, 137, 379–380
  of fractional expressions, 21–22
  integers, 74–76
  numbers of arithmetic, 21–22
  on number line, 74
  polynomials, 156, 210
  polynomials in columns, 157
  of radicals, 415–417
  rational numbers, 74–76
Subtrahend, 76
Symmetry, line of, 468
Systems of equations, 237
  graph of, 240–242
  having no solution, 241–242, 251, 302
  solution of, 239, 250
  solving
    by addition method, 248–249, 254
    by graphing, 241–242
    by substitution method, 243–245, 254

Table, *see* Solving problems, first step.
  of prime, 10
Terms
  coefficients of, 148, 209
  collecting like, 90, 93, 109–111, 144, 148, 209–210
  degrees of, 148, 209
  like, 144, 209, 415–417
  missing, 149
  of polynomials, 143, 209
  similar, 144, 209, 415–417
Torricelli's theorem, 460
Translating, *see also* Applied problems; Problems; Solving problems, second step.
  to algebraic expression, 6–8
  to equations, 37–38, 236–238
  to polynomials, 152
  with integers, 60–62
Trapezoids, areas of, 374
Triangles
  angles, measures of, 121
  areas of, 127
  equilateral, 425
  Pythagorean property of, 419
  right, 419–423
    hypotenuse of, 419
    legs of, 419

Triangles, right (*cont.*)
   and problem solving, 421–423, 463–464
Trinomial squares, 182
  factoring, 183
Trinomials, 149
  factoring, 185–188, 189–194. *See also* Factoring.
  multiplying, 161

Union, 302–303

Value of expression, 5
Variables, 5, 7. *See also* Polynomials; Translating.
  eliminating, 249
Variation
  constant of, 318–320
  direct, 318
  equations of, 318, 322
  inverse, 322–323
  joint, 322

Vertical lines, *see* Graphing, with missing variable; Slope.
Vertical-line test, 316
Volumes of rectangular solids
  cube, 142

Whole numbers, 16
Wildlife populations, 371–372
Work problems, 323, 367–369, 375–376

x-intercept, 232–233

y-intercept, 229, 232–233, 273. *See also* Graphing.

Zero
  additive property of, 18, 69
  degree of, 148
  division by, 18, 83
  as exponent, 25
  property of, 18, 80
Zero products, principle of, 198–199